STRUCTURAL ANALYSIS

The Verrazano-Narrows Bridge across the entrance to New York Harbor, completed 1965. (*American Institute of Steel Construction.*)

STRUCTURAL ANALYSIS

Ronald L. Sack

Professor of Civil Engineering
University of Idaho

McGraw-Hill Book Company

New York   St. Louis   San Francisco   Auckland   Bogotá   Hamburg
Johannesburg   London   Madrid   Mexico   Montreal   New Delhi
Panama   Paris   São Paulo   Singapore   Sydney   Tokyo   Toronto

This book was set in Times Roman by Beacon Graphics Corporation.
The editors were Kiran Verma and Madelaine Eichberg;
the production supervisor was Leroy A. Young.
The drawings were done by J & R Services, Inc.
The cover was designed by Myrna Sharpe.
Halliday Lithograph Corporation was printer and binder.

STRUCTURAL ANALYSIS

1 2 3 4 5 6 7 8 9 0 H A L H A L 8 9 8 7 6 5 4 3

ISBN 0-07-054392-5

Library of Congress Cataloging in Publication Data

Sack, Ronald L.
    Structural analysis.

    Includes bibliographies and index.
    1. Structures, Theory of.   I. Title.
TA645.S23   1984   624.1'71   83-701
ISBN 0-07-054392-5

CONTENTS

Part 4 Analysis of Statically Indeterminate Structures

Part 5 Matrix Structural Analysis

Appendixes

PREFACE

This book is intended to appeal to a broad spectrum of students, ranging from those beginning a formal education in structural engineering to seasoned practitioners who wish to broaden or deepen their knowledge of the subject. I have attempted to write primarily for the student by progressing from the specific to the general case and using an ample number of clarifying examples. The precedent for this approach was set long ago by such great teachers of structural engineering as August Föppl and Stephen P. Timoshenko.

The goal of most structural engineering work is to produce an efficiently constructed facility. This requires (1) an accurate description of loads, (2) an analysis for stresses and deflections, and (3) the design of members, subsystems, and their connections. The behavior of a structure frequently depends upon the sizes and materials of individual members; response influences design and vice versa. Thus, analysis and design should be considered as companion disciplines of structural engineering.

Over the past two decades marked progress has been made in structural engineering, and many complex contemporary structures owe their existence to these changes. The discoveries in structural analysis have been greatly stimulated by the development of the electronic digital computer, which in turn has had a profound effect on how structural engineering is practiced and taught.

For purposes of discussion we can subdivide the various methods of structural analysis into three basic categories: classical, approximate, and numerical, i.e., computer-oriented. All three stem from the same fundamental precepts of structural analysis and are interrelated. In this book each distinct method is described, the relationship of approaches is explained, and no attempt is made to cast all methods in a form amenable to computer implementation. Thus the distinct role of each method can be clarified. Many classical and approximate methods serve as extremely useful tools for verifying computer results, and some problems can be solved elegantly with a hand-calculation method. Furthermore, many of the classical methods should be studied for pedagogical reasons.

The book is divided into five main parts. Part One contains a description of the many factors that influence the design, analysis, and response of various structural systems. The static considerations of structures is presented in Part Two in order to extend the readers' knowledge of statics to the field of structural analysis. Part Three presents the study of structural deflections by both the direct and energy methods. Part Four, covering classical compatibility and equilibrium methods for statically indeterminate structures, uses the information from Parts Two and Three. In Part Five the general equilibrium method is derived and cast in matrix form.

The overview of structural engineering in Part One is frequently omitted in many courses because of time constraints, but the student can gain a broad understanding of structural engineering by reading and rereading this discussion. The survey of structural engineering in Appendix B gives a historical, social, and functional perspective. This information can also best be understood completely by periodic rereading.

The book is designed to be used in various types of structural analysis courses. A logical progression of topics with a uniform and continuous flow of information can be obtained by exercising modest discretion in selecting individual chapters or sections. For example, a one-semester introductory course on the theory of structures can be constructed using material from Part Two (Chaps. 2 to 5), Part Three (Chaps. 9 and 10), Part Four (Chaps. 11 and 12), and Part Five (Chaps. 15 to 17). This skeletal outline can be expanded with additional topics for a multicourse sequence, and the book is also arranged so that a cover-to-cover study is possible.

Optional text paragraphs in smaller type can be ignored during a first reading with no loss of continuity since they contain material intended to enhance the readers' knowledge once the fundamentals have been mastered. A broad collection of examples and a generous number of problems (both in SI and USCS units) are offered for demonstrating the principles and assisting the reader in developing an active understanding of the concepts. Answers for selected problems (Appendix C) are intended as a guide, not a mental crutch.

Many people have played a direct or indirect role in the writing of this book. I express my sincere thanks to Professor Robert G. Oakberg, of Montana State University, for his careful reading of the manuscript and for his valuable suggestions. I express my gratitude to my former colleague, George Dewey, for his many comments. The encouragement and support of Professor James H. Milligan, Chairman of the Department of Civil Engineering at the University of Idaho, is also greatly appreciated. Finally, I extend special thanks to the classes of third-year civil engineering students at the University of Idaho for enduring the inconvenience associated with studying from the manuscript form of this book.

Ronald L. Sack

LIST OF SYMBOLS

Symbols are generally defined where they first appear. Some symbols have been used in different contexts to define several quantities.

A = cross-sectional area of a member

$\mathbf{B}$ = coefficient matrix for expressing joint equilibrium of a truss ($\mathbf{P} = \mathbf{BF}$)

b = member width

C = carryover factor used in the moment-distribution method (prismatic members)

C_{ab}, C_{ba} = carryover factors used in the moment-distribution method (nonprismatic members)

C_e = exposure factor for snow loads

C_p = external pressure coefficient for wind loads

C_t = thermal factor for snow loads

C_1, C_2, C_3 = angular coefficients for nonprismatic members

DF_{ab} = distribution factor for member ab used in the moment-distribution method

E = modulus of elasticity (Young's modulus)

$\mathbf{F}$ = structural flexibility matrix with elements F_{ij} expressed in structural coordinates; column matrix of member forces and reactions

F = axial force or axial resisting force acting on a cross section of a member

F_P = axial resisting force of a member due to the actual P forces acting on the structure

F_Q = axial resisting force of a member due to the virtual Q forces acting on the structure

F_{ij} = axial resisting force of member ij

F_x, F_y, F_z = axial resisting forces of a member in the x, y, and z directions, respectively

FEM_{ab} = fixed-end moment for member ab at end a used in the slope-deflection and moment-distribution methods

$\mathbf{f}$ = member flexibility matrix with elements f_{ij} expressed in structural coordinates

G = modulus of elasticity in shear

G_h = gust response factor for wind loads

GC_{pi} = product of internal pressure coefficient and gust response factor for wind loads

H = horizontal component of cable tension

h = cable sag

$\bar{h}$ = cable sag at midspan

I = moment of inertia of cross-sectional area

$\mathrm{IL}(x)$ = ordinate of an influence line

$\mathbf{K}$ = structural stiffness matrix with elements K_{ij} expressed in structural coordinates

K_{ab} = stiffness factor (I_{ab}/L_{ab}) of flexural member ab used for prismatic members in the slope-deflection and moment-distribution methods

K_{ab}^M = modified stiffness factor of flexural member ab, for prismatic members in the moment-distribution method

K_z = velocity pressure coefficient for wind loads

k_{aa}, k_{ab}, k_{bb} = stiffness factors for nonprismatic members

$\mathbf{k}$ = member stiffness matrix with elements k_{ij} expressed in structural coordinates

$\bar{\mathbf{k}}$ = member stiffness matrix with elements $\bar{k}_{ij}$ expressed in member coordinates

L = length of a member; span length

ΔL = change in length of a member

M = moment of force (couple); bending moment; resisting moment; end moment of member

M_P = resisting moment of a member due to the actual P forces acting on the structure

M_Q = resisting moment of a member due to the virtual Q forces acting on the structure

m = bending moment acting on a member used in conjunction with matrix structural analysis

$\mathbf{P}$ = matrix of forces applied on a truss

P = forces applied on a structure externally

P_x, P_y, P_z = externally applied forces on a structure in the x, y, and z directions, respectively

p = wind design pressure

p_f = flat-roof design snow load

p_s = sloped-roof design snow load

Q = virtual externally applied force used in conjunction with the method of virtual work

q = distributed applied loading on a beam expressed in force per length

q_z, q_h = velocity pressure for wind loads

R = reaction force; radius of curvature

R_i = redundant reaction force

S = section modulus of cross-sectional area

s = variable of integration along the length of a member

$\mathbf{s}_{ij}$ = column matrix of generalized forces in members used in conjunction with matrix structural analysis

$\mathbf{s}$ = member stress matrix used in conjunction with matrix structural analysis

$\bar{\mathbf{s}}^0$ = matrix of initial member forces

$\mathbf{T}$ = coordinate transformation matrix used in conjunction with matrix structural analysis

T = cable tension; thrust in an arch; temperature

T_P = resisting torsional moment due to the actual P forces acting on the structure

T_Q = resisting torsional moment due to the virtual Q forces acting on the structure

ΔT = change in temperature from ambient conditions

t_{ij} = tangent offset distance of a beam at point i relative to a tangent at point j

U_i, V_i, W_i = displacements at node i in the x, y, and z directions, respectively, expressed in structural coordinates used in conjunction with matrix structural analysis

u_i, v_i, w_i = displacements at node i in the x, y, and z directions, respectively, expressed in structural coordinates

u, v, w = continuous functions of displacement in the three coordinate directions

$\bar{u}_i$, $\bar{v}_i$, $\bar{w}_i$ = displacements at node i in the x, y, and z directions, respectively, expressed in coordinates used in conjunction with matrix structural analysis

$\mathbf{U}$ = column matrix of nodal displacements for the entire structure expressed in structural coordinates used in conjunction with matrix structural analysis

$\mathbf{U}_f$ = column matrix of unknown displacements in structural coordinates

$\mathbf{U}_s$ = column matrix of known displacements in structural coordinates

$\mathbf{u}$ = column matrix of member displacements expressed in structural coordinates

$\mathbf{u}$ = column matrix of member displacements expressed in member coordinates

V = shear force in a member; resisting shear force; basic wind speed in miles per hour used in calculating velocity pressure

V_P = resisting shear force due to the actual P forces acting on the structure

V_Q = resisting shear force due to the virtual Q forces acting on the structure

W_e = external work done by the structural displacements

W_i = internal work done by the deformation of members

W_e^* = external complementary work done by the external force system

W_i^* = internal complementary work done by the resisting forces acting in members

$\mathbf{X}$ = column matrix of nodal forces for the entire structure expressed in structural coordinates used in conjunction with matrix structural analysis

$\mathbf{X}_f$ = column matrix of known forces in structural coordinates

$\mathbf{X}_s$ = column matrix of unknown forces in structural coordinates

$\mathbf{X}^0$ = column matrix of initial forces in structural coordinates

X_i = redundant force or moment used in analysis of statically indeterminate structures

X, Y = scalar multipliers used in the moment-distribution method for sidesway solutions

X_i, Y_i, Z_i = forces acting on a structure at node i in the x, y, and z directions, respectively, expressed in structural coordinates used in conjunction with matrix structural analysis

$\mathbf{x}$ = column matrix of member forces expressed in structural coordinates

$\mathbf{\bar{x}}$ = column matrix of member forces expressed in member coordinates

$\mathbf{x}^0$ = column matrix of initial member forces expressed in structural coordinates

$\mathbf{\bar{x}}^0$ = column matrix of initial member forces expressed in member coordinates

x, y, z = orthogonal cartesian coordinates; structural coordinates

x_i, y_i, z_i = forces acting on a member at node i in the x, y, and z directions, respectively, expressed in structural coordinates used in conjunction with matrix structural analysis

$\bar{x}_i, \bar{y}_i, \bar{z}_i$ = forces acting on a member at node i in the $\bar{x}$, $\bar{y}$, and $\bar{z}$ directions, respectively, expressed in member coordinates used in conjunction with matrix structural analysis

$\bar{x}, \bar{y}, \bar{z}$ = orthogonal cartesian coordinates; member coordinates

Subscripts

ap = applied

E = quantity associated with an error in fabrication used in analysis of statically indeterminate structures

i = the point associated with the quantity

ij = the interval (member) associated with the quantity

P = force (or moment) resulting from the actual forces acting on the structure

Q = force (or moment) resulting from the virtual forces acting on the structure

S = quantity associated with support movement used in analysis of statically indeterminate structures

T = quantity associated with temperature change used in analysis of statically indeterminate structures

0 = quantity associated with the primary structure used in analysis of statically indeterminate structures

Superscripts

E = force (or moment) at a point that is equivalent in an energy sense to a distributed loading used in conjunction with matrix structural analysis

F = force (or moment) required to give zero displacement at the point, i.e., a fixed-end force used in conjunction with matrix structural analysis

L = left

M = modified stiffness factor used in moment distribution

R = right

0 = quantity initially introduced by temperature, fabrication error, precambering, etc.

Greek Symbols

α (alpha) = coefficient of linear thermal expansion

α, β, γ = angles measured to a vector from the positive x, y, and z axes, respectively

Γ (gamma) = matrix of direction cosines, an orthogonal transformation ($\Gamma^T = \Gamma^{-1}$)

γ (gamma) = shearing strain

Δ (delta) = deflection; total change in a quantity

δ (delta) = deflection; increment of a quantity; first variation of a quantity (a virtual quantity)

ε (epsilon) = lineal strain

Θ (theta) = rotation of tangent elastic curve with respect to original direction of axis of member; denotes reference to structural coordinates in matrix structural analysis

θ (theta) = angle; rotation of tangent elastic curve with respect to original direction of axis of member; denotes reference to member coordinates in matrix structural analysis

κ_T (kappa) = St. Venant's torsion constant, which depends upon member cross section

κ_V (kappa) = shear constant, which depends upon member cross section

λ, μ, ν = direction cosines (cos α, cos β, cos γ, respectively)

ν (nu) = Poisson's ratio

Σ = summation of quantities

σ (sigma) = normal stress

τ (tau) = shear stress

ϕ (phi) = angle; angle between axis of a member and the structural coordinates used in conjunction with matrix structural analysis

ψ (psi) = angle; rotation of chord of an elastic curve with respect to original direction of axis of a member

Miscellaneous

{ } = Column matrix written as a row matrix to conserve space in text

PART
ONE

STRUCTURES

ONE

AN OVERVIEW

1.1 GENERAL DISCUSSION

A building is made up of individual structural elements such as trusses, beams, girders, columns, shear walls, and the foundation. The interaction between components establishes the structural integrity and response of the building to forces imposed by earthquake, wind, snow, ice, rain, and thermal effects, in addition to those represented by the building's contents and its own weight. The characteristics of the system must be such that it performs its intended design function. Thus, strength requirements are accompanied by stiffness constraints to prevent excessive deflections, bouncy floors, outward-tilting walls, uncomfortable structural oscillations, and the like.

Buildings, bridges, dams, airframes, ships, offshore oil platforms, cars, buses, and cranes are all examples of engineered structures that must be analyzed, designed, and constructed to yield an economical, safe, and serviceable system. In addition, many less imposing but common manufactured products require structural analysis and design to ensure reliability. For example, a paraboloidal microwave dish antenna must be stiff enough to maintain a constant shape when subjected to environmental loads; office equipment must be designed to function without excessive vibration; and snow skis must be sufficiently strong and have the proper bending and torsional characteristics to perform properly. This book is oriented toward the analysis of conventional structures (buildings, bridges, etc.), but the concepts are equally applicable to more general structural types.

This chapter presents a survey of structures with emphasis on structural analysis; it provides an opportunity to observe the broad outlines of the subject before concentrating on details. These qualitative descriptions are intended to impart an understanding of many of the specific considerations involved in planning, investigating, and constructing structures. The topics include use of materials, structural loads, design methods, common structural systems, and criteria for selecting an efficient structural system.

1.2 STRUCTURAL LOADING AND BEHAVIOR

Loads are imposed on a structure by people and/or stored goods, vehicles, environmental effects (wind, snow, ice, rain, and earthquake), and the self weight of the materials. The structure must transmit the forces of these loads in such a way that each individual component, as well as the total system, remains near the equilibrium configuration. During this process, the structure must remain intact, i.e., neither local nor general material failure can occur, and the stiffness and mass distribution must be such that displacements are acceptable and vibration frequencies and amplitudes tolerable. Thus, the structural *behavior* or *response* is measured by considering several parameters.

The transfer of loads throughout a simple structure can be envisaged by examining the flexible cable structure in Fig. 1.1. Since the cable changes shape to accommodate the applied loads, the configuration varies as the two gondolas move across the span. The forces from these two suspended loads are transferred to the supports at each end by the internal tensile forces in the cable; i.e., the *load path* for this structure coincides with the deflected cable shape. A more complex example of load transfer occurs in an arch bridge (Fig. 1.2). A schematic diagram of this type of structure is shown in Fig. 1.3. Although the primary structural element is the curved arch, the vertical struts and the horizontal roadway are both vital to the functioning of the bridge. Vehicles impose both static and dynamic forces on the bridge deck, which are transferred to the main structural arch through the vertical struts. Subsequently, these forces must be transmitted to the reactions or foundations by a combination of internal thrust (force component parallel to the arch axis), shear (force component perpendicular to the arch axis), and flexure (bending moment).

1.3 CRITERIA FOR ACCEPTABILITY

Structural integrity can be partially detected by the occupants of a building, airframe, etc. Uncomfortable oscillations, floors with excessive bounce, and outward-tilting walls are examples of structural behavior perceived to be unacceptable. However, the catastrophic collapse of entire floors and bridges while fully occupied indicates that occupants cannot always evaluate all aspects of structural integrity intuitively. It is the obligation of the structural engineer thoroughly to understand, investigate, and assess the predicted response of each structure using the appropriate criteria for acceptability.

Figure 1.1 Transfer of forces within a flexible cable.

Figure 1.2 An arch bridge.

The structure must have sufficient *strength* to avoid collapse when full design loading is imposed. This implies that the unit material stresses remain at an acceptable level as the loads are transferred throughout the structure. The properties of various common structural materials are discussed in Sec. 1.6; the stress levels in the structure that are judged acceptable depend upon the design approach followed (see Sec. 1.7). Failure of the material in one local region does not necessarily imply that the strength of the structure is inadequate. How the loads will be transmitted can be predicted using the equations of static equilibrium unless the structure is statically indeterminate, i.e., there are structural redundancies. In such cases, various load paths exist for transmitting the forces, and local failure requires that an alternate load path exist in order to preclude overall structural failure. If sufficiently few additional local regions of failure occur, the structure may still be able to function adequately. The *strength criterion* is one of the most obvious for judging structural integrity, but it must be accompanied by other considerations.

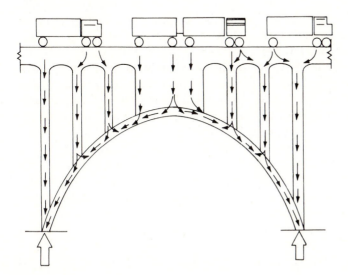

Figure 1.3 Transfer of forces within an arch bridge.

Figure 1.4 Deflection of a cantilever beam.

Even though a structure has sufficient strength, it may exhibit excessive deflections under full loading, which in turn can result in significant problems. A second criterion for structural acceptability is the limitation of deflections, accomplished by providing the structure with sufficient *stiffness*. The analytical concepts of stiffness will be developed in Parts Three to Five, but here we can visualize this property by examining the beam in Fig. 1.4. If a unit displacement is imposed at the tip of this cantilever beam, the required force in the direction of the displacement is a measure of stiffness; i.e., in this case stiffness is expressed as the force per unit imposed displacement. Inadequate stiffness can lead to various deleterious effects. For example, excessive deflections of a supporting beam may render a delicate piece of machinery inoperative if it is required to remain in a perfectly horizontal position. A high-rise structure that experiences large lateral deflections due to wind loading may not suffer primary structural damage, but the interior and exterior cladding and such appurtenances as masonry walls, may be totally destroyed.

Typically, deflections are limited by the applicable design code. For example, the maximum deflection of a roof member (or floor member) supporting plaster is limited by the *Uniform Building Code*[1][†] (UBC) to the span length divided by 360 when loaded with live load only and by 240 when loaded with live load plus dead load. A structure that is subjected to dynamic loading such as that imposed by earthquake must also have sufficient stiffness to limit the displacements; however, in this case the displacements are related to the *distribution of mass* throughout the structure. Although structural vibrations per se will not be discussed in this book, they are an important consideration for structures in regions of high seismic activity or subjected to dynamic forces such as those imparted by wind and waves.

The *stability* of a structure is another quality that must be carefully studied in judging acceptability. For example, if the roof on a house is not anchored against upward loading, it can be blown away during a windstorm with sufficient uplift (see Sec. 2.6). A type of geometric instability is exhibited by a truss with improperly arranged members; because it is incapable of transmitting the loads throughout the structure, such a truss will collapse under the effect of small loads (see Sec. 3.5). Another type of unstable behavior is exemplified by a slender compression member that *buckles* into a curved configuration, significantly degrading the ability of the member to support compressive loading. In some cases a horizontal beam with loading applied only in the vertical plane can deflect in a direction normal to this plane and display *lateral torsional buckling behavior*.

Thus, the general criteria of *strength, stiffness, stability,* and sometimes *mass distribution* must all be used to determine structural acceptability. These attributes are

---

[†]Numbered references appear at the end of the chapter.

influenced by basic material behavior, the cross-sectional properties of structural members, and general structural geometric configuration. Although some of the methods for studying these effects are described in the ensuing chapters and others are more properly classed as design criteria, the structural engineer must be aware of them all during analysis and design.

1.4 STRUCTURES AND THEIR COMPONENTS

Many types of structures consist of three-dimensional assemblages of individual members or subassemblies with applied forces that generally have components in all three orthogonal cartesian coordinate directions. One of the primary tasks associated with analyzing them is to determine how the forces are transferred throughout the structure; however, an investigation of a general three-dimensional assemblage can be extremely cumbersome and time-consuming. With some insight into structural behavior it is usually possible to identify the various individual elements and the associated loads that constitute the total structure, reducing the analysis to an investigation of a set of individual one- and two-dimensional systems. The following descriptions for several typical structures illustrate how subassemblies should be identified before initiating any formal analysis.

The pedestrian truss bridge in Fig. 1.5 must resist vertical, lateral, and longitudinal loads that arise from various independent sources. The occupants walking between the trusses on the bridge floor transmit primarily vertical loading to the floor beams, which

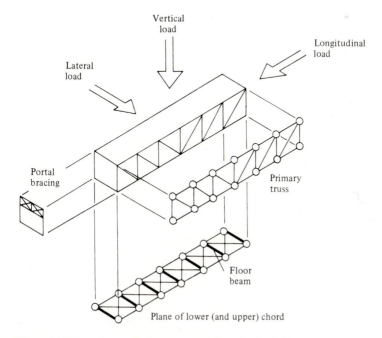

Figure 1.5 The main structural components of a pedestrian bridge.

span between the two vertically oriented longitudinal trusses. These two primary trusses, in turn, via truss action transfer the forces applied by the beam reactions to the structural foundations at either end of the bridge. These primary trusses are also capable of resisting longitudinal loads such as those imposed by earthquakes. Since the forces are transferred from the floor to the floor beams, to the longitudinal trusses, and finally to the foundation, we can analyze each of these structural components independently with the proper transfer of forces between each subassembly. Lateral forces, e.g., those induced by wind, are applied normal to the primary trusses, but since these trusses are incapable of resisting loads in this direction, the planes of the upper and lower chords (top and bottom of the structure) contain horizontal trusses with double diagonals to transfer these loads to the foundations. These two horizontal trusses can be analyzed independently. Note that the horizontal reactions of the top truss must be transferred down to the foundation at the ends of the structure; this is typically accomplished by including a vertical subassembly at this position that can resist lateral forces. Here it consists of a portal frame fabricated from a shallow truss and columns (capable of transmitting bending moments). Thus, this three-dimensional structure can be visualized as having been synthesized from four trusses, two portal frames, and other secondary members such as floor beams, etc., each of which can be analyzed as a two-dimensional component.

The one-story building shown in Fig. 1.6 resists applied loads in a different fashion. In this case, the forces are generally applied to the external walls and roof (cladding) of the structure, and small bending members spanning between the main rigid frames transfer the loads to the frames. Longitudinal bending members located in the plane of the roof are called *purlins,* and those in the vertical side walls are called *girts.* Each rigid frame consists of two columns and two roof beams, which are rigidly connected to form a subassembly capable of transferring forces and bending moments.

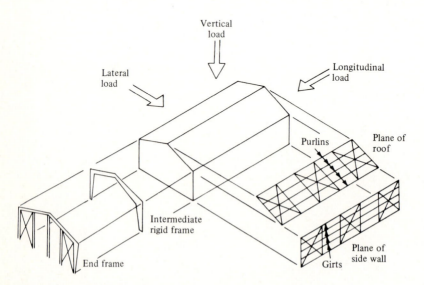

Figure 1.6 The main structural components of a single-story building.

The lateral and vertical loads are transferred to these frames, which in turn deliver them to the foundations. Since the rigid frames are individually incapable of resisting forces applied in the longitudinal direction of the building, wind bracing, consisting of light diagonal tension rods, is incorporated into several bays (the space between rigid frames). In conjunction with the frames, these rods form trusses which resist the longitudinal forces. (In some cases, the combined resistance of the cladding, fastened to the frames and girts and/or purlins, is considered as an assemblage which resists longitudinal forces like a diaphragm.) Although end frames are constructed using columns and roof members that are lighter than the corresponding parts in the intermediate frames, they incorporate intermediate columns and diagonal bracing to resist both vertical and lateral forces. This three-dimensional building can be investigated fully by analyzing the cladding, purlins, girts, and rigid frames as two-dimensional subassemblages, with an understanding of load transfer from element to element.

Unlike the pedestrian bridge and the one-story building, the structures illustrated in Figs. 1.7 and 1.8 do not lend themselves to the analysis of individual two-dimensional components. The space truss in Fig. 1.7 is a simplified version of an offshore drilling tower; the three sides form equilateral triangular cross sections that increase in size from top to bottom. Since the three side planes are not orthogonal, the forces in the members of any one plane are not independent of those in the other two planes; i.e., a single force applied parallel to one plane at the top of the tower will be propagated into the other two planes as it is transferred to the foundations of the structure. This structure must be investigated by considering the entire three-dimensional assemblage. The saddle-shaped structure in Fig. 1.8 is a hyperbolic paraboloid, so named since sections passed through the structure perpendicular to a line connecting the two high points or the two low points reveal that the surface is a parabola and planes through axes oriented 45° to these directions indicate hyperbolas. This structure has edge beams to anchor the cables, which are stretched in two orthogonal directions, and the transfer of loads occurs in a three-dimensional manner that cannot be resolved into sets of independent two-dimensional subassemblies.

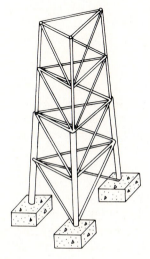

Figure 1.7 A space truss.

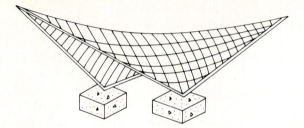

Figure 1.8 A hyperbolic paraboloid cable structure.

Thus, we see that in some cases structural analysis can be simplified considerably by identifying the individual two-dimensional component systems that form the complete three-dimensional structure. In other cases, e.g., a concrete cooling tower or an inflatable dome, the structure functions by transferring the forces in a manner too complex to be analyzed by subdividing the system into a number of two-dimensional subassemblies. Part of this process of simplification requires an ability to understand and recognize the basic structural components of a building assemblage, a skill that can be sharpened by studying existing structures with an eye toward identifying their components.

1.5 LOADS

The primary purpose of a structure is to resist applied loads while performing its intended function in an acceptable fashion. Most structures are required to support the weight of their own materials; the type and origin of additional loads depend upon the intended use and environmental exposure. For example, a building subjected to the loads imposed by stored goods and occupants may also be required to resist the environmental effects of temperature, snow, ice, rain, wind, and earthquake. A highway bridge must support the traffic and environmental loads; an airframe must withstand aerodynamic loading; and an offshore oil-drilling tower must sustain the usual operational forces plus potentially severe wave action.

The loads on structures can be broadly classified as *static* (fixed with respect to time and position) or *dynamic*. For example, the transfer of forces within a structure induced by the weight of stored goods and snow can be investigated using the equations of static equilibrium, whereas moving vehicles on a bridge and rotating machinery in a building both impose dynamic forces and a thorough investigation should consider the mass distribution of the structure. In some cases it is possible to calculate the effects of dynamic loads by considering them as quasi-static (their frequency is much less than the critical frequencies of the structure) and using an equivalent static design force.

A distributed load acts over a prescribed area of the building. External vertical, transverse, and longitudinal distributed loads, along with their respective projected areas, are illustrated in Fig. 1.9. The forces applied to specific structural components must be computed using the applied loads and an understanding of how loads are transferred through the structure. Since the building in Fig. 1.10 has rigid frames

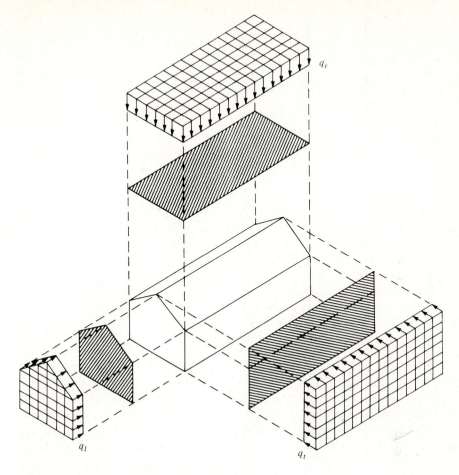

q_r

q_l

q_t

Figure 1.9 Uniform loads and associated projected areas.

spaced a distance L apart, it is reasonable to assume that each intermediate frame will resist the load applied over an area with a width of $L/2$ on either side of the frame of interest; this area is referred to as the *tributary area*.

A *deterministic load* is one that can be prescribed with a high degree of precision; i.e., the direction and magnitude (or frequency, amplitude, and function of time) are known exactly. If there is uncertainty because of a natural variation or incomplete knowledge, the load should be considered as a probabilistic quantity, which must be investigated using probability theory, and the magnitude, direction, frequency, etc., must be treated as *random variables*. For example, if we attempt to establish the weight of a carefully controlled product such as structural steel, the weight of individual samples of a set will not all be precisely the same. The resulting distribution of the sample weights can be described with the proper graphic display (e.g., the probability density function) and the corresponding statistics (e.g., the mean, variance, and standard deviation). Since most loads cannot be prescribed exactly, it is proper to regard them all as random.

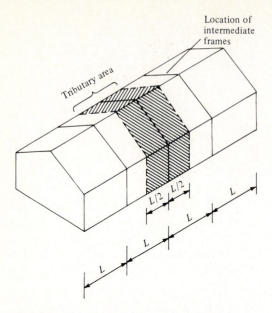

Figure 1.10 Tributary area for an intermediate frame.

Usually a structure must be designed in accordance with regulations provided by a jurisdiction through some type of building code. Typical codes or specifications also define the design loads. One of the most widely adopted building codes in the United States is the *Uniform Building Code*.[1] The American National Standards Institute, Inc., publishes a standard[2] that is used by many code-writing bodies in satisfying their own unique requirements, and the Applied Technology Council has carefully defined[3] the earthquake forces for the design of buildings in the United States. The American Association of State Highway and Transportation Officials (AASHTO) prescribes[4] the loads for the design of highway bridges in the United States; the loads and performance standards for manufactured products such as automobiles, aircraft, ships, etc., are defined by various agencies. The latest editions of codes and specifications are cited throughout this book, but since they are revised periodically, the current version should always be used. While codes are necessary for any design, they should be used primarily as a guide.

Thus, the design loads must be carefully prescribed and calculated for each structure, and the final responsibility rests with the structural engineer. The various codes help in this process but should be augmented with any available information germane to the structure under consideration. There are many categories of design loads, each with its own special properties; some typical ones are discussed in the ensuing subsections.

Dead Loads

Dead load (also called *gravity load*) acts vertically downward and consists of the weight of all structural and nonstructural components permanently fixed in position. In a building this includes the self weight of all beams, columns, trusses, and other structural members plus the loads of partitions, exterior and interior walls, etc. Dead load also includes the mechanical, electrical, and other service equipment supported by the structure. Densities of some common building materials are shown in Table 1.1 and typical loads imposed by a few building components in Table 1.2. A more complete listing of dead loads appears in various codes and standards, and additional data can usually be obtained from product manufacturers.

Even though gravity loads can be established to a relatively high degree of accuracy, values still have a certain amount of dispersion; i.e., in general the coefficient of variation of dead load is approximately 0.10. Although these loads display statistical regularity, the reader is reminded that dead loads are nondeterministic.

Since the analysis or design process requires the dead load to be included in the total load to be supported by a structure, these values must be determined a priori. That is, before starting the investigation, the weights (vis-a-vis the member sizes) must be known. They can be estimated using past experience or various available guidelines. Approximate formulations exist for estimating the dead load of trusses, which are

Table 1.1 Typical mean value for some material densities†

Material	lb/ft^3	kg/m^3
Aluminum, cast	165	2643
Concrete, reinforced, stone aggregate	150	2402
Soil, dry	75–95	1201–1762
Steel, rolled	490	7849
Wood, 15% moisture content	40	641

†SI units are described in Appendix A.

Table 1.2 Typical mean-value loads for some building components

Component	lb/ft^2	Pa
Floors:		
Concrete slab, stone aggregate, per inch (25.4 mm)	12.0	575
Hardwood, per inch (25.4 mm)	4.0	192
Hollow-core concrete planks, per 6 in (152.4 mm)	45–50	2155–2394
Plywood, per inch (25.4 mm)	3.0	144
Steel decking	2–10	96–479
Roofs:		
Asphalt shingles	3	144
Corrugated steel	1–5	47–239
5-ply felt and gravel	6	287
Insulation, rigid, per inch (25.4 mm)	1.5	72
Walls:†		
Brick, 4 in (102 mm)	40	1915
Hollow concrete block (heavy aggregate), 8 in (203 mm)	55	2633

†Load per wall width; i.e., total load is obtained by multiplying the quantity given by the width of the wall.

typically derived as loads to be applied to the joints of the structure. Although dead loads can be estimated for buildings using the known values of materials and component weights, it is frequently necessary to make preliminary estimates of the dead loads to investigate overall structural response such as propensity for overturning, etc. For these purposes the averaged dead load from existing structures can be used; some typical values for various types of construction are timber framing 50 lb/ft^2, steel framing 70 lb/ft^2, and reinforced concrete framing 120 lb/ft^2. The value for prestressed concrete is approximately 75 percent that for reinforced concrete.

Typically, the dead load is a small percentage of the total design load for simple structures such as an isolated beam or a single rigid frame, but in some cases it is necessary to calculate dead loads carefully. For example, the lower columns in a high-rise structure must support the aggregate of the loads imposed by all the upper-level stories; these values can be considerably in error if all the dead loads have not been calculated accurately. It is generally advisable to use the actual value of the dead loads in carrying out the final structural design calculations.

Building Live Loads

Live loads are induced during the construction, use, and occupancy of a building and do not include environmental loads produced by wind, snow, ice, rain, temperature, or earthquake. Live loads can be either uniformly distributed or concentrated and are imposed by people, files, furniture, and computers in a building; stored goods in a warehouse; books in a library; children in a schoolroom; vehicles in a garage; etc. If a load survey of a building's contents is conducted, a *histogram* (Fig. 1.11) can typically be constructed. In this sample survey, 236 objects were investigated for load and corresponding floor area occupied; 133 of them produced loading between 20 and 50 lb/ft^2. This set of sustained loadings can be described using standard statistical measures such as the mean, variance, standard deviation, coefficient of variation, etc.;

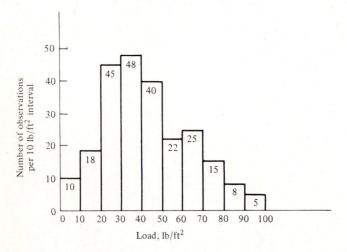

Figure 1.11 Histogram of typical floor-load data.

furthermore, it is possible to represent the histogram as a continuous mathematical function termed the *probability density function*.

In addition to the sustained loading, a building is generally subjected to several extreme loadings, produced by emergencies, crowding, etc. Statistically combining the sustained loads with the extreme transient loadings gives the maximum total load predicted to occur during a specified reference period, and the mean of these maxima can be used to obtain a set of minimum design loads. Some typical values for minimum uniformly distributed live loads are presented in Table 1.3, but more specific design values can be obtained from the various applicable building codes. Most building codes allow a reduction in the design live loads based upon the floor area supported by a member; i.e., the probability that a large surface will be uniformly loaded to capacity decreases in proportion to the area.

Table 1.3 Typical minimum live loads

Occupancy or use	lb/ft^2	Pa
Assembly areas and theaters:		
Fixed seats	60	2,873
Lobbies	100	4,788
Movable seats	100	4,788
Dance halls and ballrooms	100	4,788
Hospitals:		
Operating rooms and laboratories	60	2,873
Private rooms and wards	40	1,915
Corridors above first floor	80	3,830
Libraries:		
Reading rooms	60	2,873
Stacks	150	7,182
Manufacturing:		
Light	125	5,985
Heavy	250	11,970
Office buildings:		
Lobbies	100	4,788
Offices	50	2,394
Residential:		
Habitable attics and sleeping areas	30	1,436
All other areas	40	1,915
Hotels and multifamily houses:		
Private rooms	40	1,915
Public rooms	100	4,788
Schools:		
Classrooms	40	1,915
Corridors above first floor	80	3,830
Stores, retail:		
First floor	100	4,788
Upper floors	75	3,591

Snow Loads

The static forces induced by *snow loads* can impose the most severe test of a building constructed in deep snow country. Ground snow loads are influenced by climatic and geographical features, and the building geometry and siting strongly affect snow deposition on the roof. The United States has a vast store of ground snow data systematically accumulated over a period of years for many stations located throughout the country. Typically, the extreme ground snow loads for each station are predicted by fitting the annual maximum values with a standard extreme-value statistical distribution. Many building codes use the extreme ground snow loads with a 50-year mean recurrence interval; i.e., there is a 2 percent annual probability of these values being exceeded. These site-specific ground snow loads are usually spatially extrapolated in a consistent manner to predict the snow loads for an entire region. Typically, this process yields a *ground-snow-load zoning map* which is used as the basis for structural design. The range of values for the United States with a 50-year mean recurrence interval is shown in Table 1.4. Since extreme local variations in the ground snow loads usually exist for mountainous areas, high country, etc., it is incumbent on the engineer to use all available information for a rational prediction of the maximum design snow load in those areas.

The deposition and distribution of snow on a roof is a complex phenomenon influenced by various factors such as siting and several critical building characteristics. No single code is used for snow-load design in the United States, local jurisdictions typically adopting the one judged applicable for their particular locale. We can understand the snow-load roof-to-ground conversion factors by studying the standard presented as ANSI A58.1-1982. The roof snow load for the contiguous United States is calculated as follows:

$$p_f = 0.7 C_e C_t I p_g \qquad (1.1)$$

where p_f is the flat-roof design snow load and p_g the site-specific ground snow load, both in pounds per square foot. The dimensionless exposure factor C_e is 0.8 for a very windy and exposed site and increases in increments of 0.1 through five different siting categories to a maximum value of 1.2 for a highly sheltered location. The dimensionless thermal C_t factor is 1.0 for a heated structure, 1.1 for a structure heated just above freezing, and 1.2 for an unheated structure. The building importance factor I is 0.8 for agriculture buildings, 1.2 for essential facilities, 1.1 for buildings with more than 300 people in one area, and 1.0 for all other structures. The fact that snow is not

Table 1.4 Typical ground snow loads†

Region of United States	lb/ft²	Pa
Eastern, northern tier	10–100	479–4788
Central, northern tier	10–80	479–3830
Western, northern tier	10–510	479–24,419
Alaska	20–400	958–19,152

†Data from Ref. 2.

totally retained on a sloped roof is accounted for by computing the sloped-roof snow load p_s as follows:

$$p_s = C_s p_f \qquad (1.2)$$

where the roof-slope factor C_s depends upon whether the roof is warm ($C_t = 1.0$) or cold ($C_t > 1.0$). For example, for a warm roof

$$C_s = \begin{cases} 1.0 & \text{for slippery surfaces with a roof angle} \leq 15° \\ & \text{and all other surfaces with a roof angle} \leq 30° \\ 0.0 & \text{for all surfaces with a roof angle} \geq 70° \end{cases}$$

C_s for each type of roof surface varies linearly with roof angle between 1.0 and 0.0. In addition, values of C_s for other roof configurations (curved, vaulted, etc.) are also prescribed. The effects of unloaded roof portions, unbalanced snow load, drifting, sliding snow, and the additional load from rain on snow must also be considered in defining roof snow loads.

Wind Loads

Wind loads are imposed on both structural and nonstructural system components. In addition to large member forces, they can introduce severe structural vibrations and sometimes structural instability. The effects of wind-induced structural loads depend upon the wind speed, the mass density of the air, and the location, geometry, and vibrational characteristics of the structure.

Accurate wind speed information is routinely gathered at about 150 stations throughout the United States, and additional data are recorded by various local, regional, state, and federal agencies. Unfortunately, since there is no universal uniform standard for gathering these data, anemometer heights are not constant at the various stations, wind speed averaging times differ, and many agencies record the wind speed in terms of the fastest mile. (The *fastest-mile wind speed* corresponds to the shortest time required for an imaginary 1-mi length of air to pass the anemometer. Since the computed speed is the average for this imaginary 1-mi stream length, the averaging time varies with the speed.)

Typically, the data from all reliable sources are corrected to a standard height of 10 m (32.8 ft), a common method is used for time averaging, and the maximum annual values at each station are obtained. They are analyzed using a standard extreme-value statistical distribution, and the extrema for a specified mean recurrence interval are spatially extrapolated over an area to give a *wind-speed zoning map*, which shows the geographical variation of wind speed for a specified mean return period. Wind zoning done for much smaller geographical regions, e.g., a city, is referred to as *micro-zonation*. Statistical simulation procedures are used to include wind speeds for special conditions such as hurricanes. Typical fastest-mile speeds recorded at 10 m (32.8 ft) above the ground for a mean recurrence interval of 50 years (i.e., an annual probability of 0.02) are designated in ANSI A58.1-1982 (for the contiguous United States and Alaska they are 70 to 110 mi/h and for Hawaii 80 mi/h). Of course, extreme local variations of these values can occur, and it is the responsibility of the structural engineer to augment any standard code information with additional local data when wind is a significant factor.

The wind speed, mass density of the air, and certain structural characteristics all influence the wind pressure acting on a structure. There is no single codification of structural wind loads in the United States, and, as for snow loads, local jurisdictions adopt the code considered applicable for their specific locale. An understanding of wind pressure can be gained by considering the ANSI A58.1-1982 standard. In general, the *wind pressure* at a point on the exterior of a structure is calculated as the product of the velocity pressure, a pressure coefficient, and a gust factor.

The *velocity pressure* q_h at a given height above ground is

$$q_h = 0.00256K_z(IV)^2 \qquad \text{lb/ft}^2 \tag{1.3}$$

where K_z = velocity-pressure exposure coefficient, dimensionless
V = basic wind speed, mi/h
I = structure-importance factor with values slightly different from those described under snow loads

The constant 0.00256 is the mass density of air at standard conditions. K_z, used to adjust the wind speed to that for a specific structure, depends upon both the height and siting of the structure. Values of this coefficient range from 0.12 for a low building in a sheltered location to 2.41 for a high-rise structure in an unobstructed location.

The *pressure coefficient* incorporates the effects of the wind direction and the siting and geometry of the building. Such coefficients have been established for many typical buildings, and an example is shown in Fig. 1.12; however, for the design of a new high-rise structure in a metropolitan area, it is not uncommon to obtain these coefficients from wind tunnel tests which investigate the modeled structure and its surroundings.

The *gust factor,* which incorporates the effect of both the dynamic variations in the wind speed and the structural vibrational characteristics, depends upon the height and siting of the structure. It is possible to obtain these factors analytically using the concepts of structural dynamics and considering detailed wind-speed data. Typical values of these factors for the main wind-force-resisting structural system range from 2.36 for a low building surrounded by many other structures to 1.00 for a high-rise structure in a location with no obstructions for the wind.

The equation adopted by ANSI A58.1-1982 for calculating the wind pressure p is

$$p = q_h G_h C_p - q_h(GC_{pi}) \tag{1.4}$$

where G_h = gust response factor
C_p = external pressure coefficient
GC_{pi} = product of internal pressure coefficient and gust response factor for building

Note that the total pressure on a building component depends upon the external and internal pressure. Values of GC_{pi} range from +0.75 to ±0.25, depending upon the number of openings in the structure. The plus or minus sign indicates that the pressure may be directed toward the component or away from it, in which case it represents a suction.

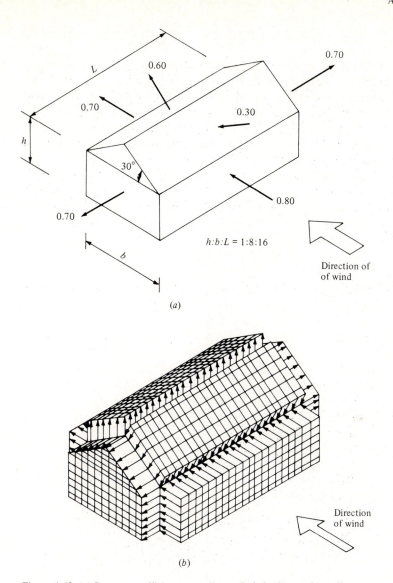

L

0.60

0.70

0.70

0.30

h

30°

0.80

0.70

$h{:}b{:}L = 1{:}8{:}16$

b

Direction of
of wind

(a)

Direction
of wind

(b)

Figure 1.12 (a) Pressure coefficients according to Ref. 2; (b) exterior wall pressures.

Earthquake Loads

Earthquake loads depend upon time-varying ground motions and the stiffness and mass distribution of the structure. During an earthquake, the transmitted ground motions induce accelerations of the various system components. Earthquake loads are defined by (1) the type and severity of the ground movement and (2) the inertial response of the structure.

It is generally accepted that the movement of large surface *plates* of the earth's crust move along discontinuities called *faults,* resulting in ground motions or earthquake waves which are potentially destructive. Earthquakes are frequently characterized using the *Richter magnitude,* obtained from an open-ended logarithmic scale and related to the energy released. A *strong-motion accelerograph* is often used to obtain *accelerograms* of the three orthogonal components of ground acceleration, providing a wealth of critical design information for the structural engineer. A set of simple one-degree-of-freedom oscillators, with various natural periods and a common degree of damping, are analytically excited with an accelerogram to give *spectral response plots,* which display the peak displacement, velocity, and/or acceleration as functions of the natural period of the oscillators. These spectral response parameters can be related to the traditional earthquake measures such as the Richter magnitude and distance from the earthquake using *attenuation relationships.*

The statistical nature of earthquake ground-movement information is quite different from that of snow and wind data, for which measurements are taken at predetermined periodic intervals. The seismic record for a region can be statistically analyzed to yield the return periods (the inverse of the annual probability of occurrence) for either a given Richter magnitude or peak ground acceleration. From these results, a *probabilistic seismic zoning map* can be constructed, showing, for example, zones of peak ground accelerations. The common approach in the United States consists of combining the historical seismic record with a knowledge of the fault locations and soil characteristics of a region and evaluating the seismic risk. Each region is assigned a single number reflecting the severity of earthquake activity, and the numbers are plotted to give a *seismic risk map* portraying the zonal distribution of assigned risks. The UBC and ANSI A58.1-1982 both use a risk scale ranging from 0 to 4, with a 4 indicating the highest seismic risk.

Ideally, a structure subjected to dynamic earthquake ground movement should be analyzed using the theory of structural dynamics. Spectral response plots can be used as base excitation for the structure and the corresponding dynamic response calculated using the *response-spectrum earthquake design approach.* Common design practice, however, consists of imposing a set of equivalent static loads on the structure and calculating the resulting member forces. This approach is embodied in most major codes used in the United States. For example, ANSI A58.1-1982 specifies that the minimum total lateral earthquake design force V for a building is

$$V = ZIKCSW \tag{1.5}$$

where Z = dimensionless seismic-zone coefficient (0.125 to 1.000)
   I = dimensionless occupancy importance factor (1.0 to 1.5)
   K = dimensionless horizontal force factor associated with lateral structural load-resisting system (0.67 to 1.33 for buildings)
   S = dimensionless soil-profile coefficient (1.0 to 1.5)
   W = total dead load of structure (units of V correspond to those of W)
and where

$$C = \frac{1}{15\sqrt{T}} \le 0.12$$

T is the natural period of the structure. These coefficients have been established by structural engineers, using the best information available and continue to be updated. The response-spectrum earthquake design approach can be used to demonstrate that Eq. (1.5) characterizes the dynamic behavior of simple vibrational models.

The behavior of structures subjected to earthquakes is a complex phenomenon that has traditionally been investigated using a quasi-static approach. As our knowledge of seismicity and structural dynamics grows, the structural engineer becomes equipped to produce more rationally designed structures and mitigate the damage of earthquakes.

Soil and Hydrostatic Pressure

The structural loads, which are ultimately transmitted to the foundation or substructure, must be reacted by *soil pressure*. In addition, lateral soil pressures are applied to foundation walls and various retaining structures. Similarly, *hydrostatic pressure* causes uplift on slabs, applies lateral forces to such structures as dams, storage tanks, and retaining structures, and constitutes a significant load that must be resisted by ship structures.

The soil pressures exerted on the cantilever retaining wall in Fig. 1.13*a* represent major design loads for this structure. The mass of soil above the horizontal base exerts vertical loads. *Active lateral pressure* is imposed by the soil behind the wall; the *passive lateral pressure* in front of the wall results from the impending lateral motion of the structure. Generally, the passive pressure is significantly larger than the active pressure, and the direction of the action exerted upon the wall by the soil can differ from the normal by the angle of cohesion. The overturning tendency induced by the lateral forces exerted on the vertical stem of the retaining wall results in nonuniform bearing soil pressures under the base. The anchored sheet-pile bulkhead in Fig. 1.13*b* is a type of retaining structure widely used in waterfront construction. The interlocking steel sheet piling is driven into the soil and anchored near the top with steel tie rods, which are connected to other piles or deadmen in the firm soil. In this case, the backfill, which typically consists of sand, exerts active pressure on the bulkhead and tends to push it forward. The wall is restrained by the anchor and the passive pressure of the ground in front. Hydrostatic pressure acts on the sheet piling below the water table behind the wall and below the water surface on the other side.

Soil and hydrostatic pressure constitute significant structural loads which depend upon soil characteristics. Since soil is not uniform, these loads are probabilistic.

Thermal Loads

A building fabricated during the heat of the summer may experience significant member deformations due to temperature changes in a climate with severe winter weather. Similarly, a supersonic aircraft typically experiences intense thermal changes in flight which can have a significant effect on the structure. If potential thermal deformations are resisted or prevented by the inherent structural behavior, member forces will develop throughout the structure. These *thermal loads* can be significant. Even if thermal loads do not develop, it is necessary to consider thermal effects in providing

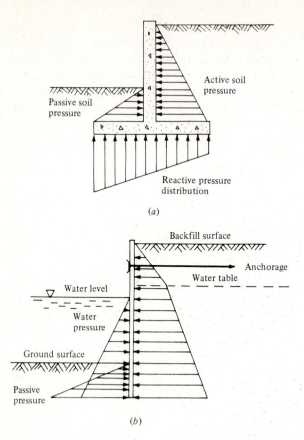

Figure 1.13 (*a*) Soil pressure on a cantilever retaining wall; (*b*) soil and hydrostatic pressure on an anchored bulkhead.

construction details at the supports and other critical points in the structure, to prevent damage from thermal expansion and contraction.

Highway Loads

Many of the loads imposed on bridges depend on use, structural configuration, and siting. In addition to the dead load, bridge structures must resist forces introduced by the loads of the vehicles, impact, and such environmental effects as wind, earthquake, and temperature. Bridges on the state or federal highway system in the United States must be designed in accordance with the specifications of the American Association of State Highway and Transportation Officials (AASHTO).

The live loads imposed by the vehicles are considered to consist of either a single standard truck load or a corresponding lane load. The truck load represents a single heavy vehicle on the span, which may be a two-axle truck or a two-axle truck plus a one-axle semitrailer. A lane load corresponding to each truck load is made up of a line of moderate-weight vehicles with a single heavy load in their midst. Both the truck- and lane-loading cases must be investigated to establish the critical design load, and the maximum is determined by the type and span of the structure. An example of truck loads and the corresponding lane loading is shown in Fig. 5.16. AASHTO acknowledges the presence of dynamic loads imposed by moving vehicles and replaces them by equivalent static loads. The impact I, expressed as a fraction of the live load, is

$$I = \frac{50}{L + 125} \le 0.30 \tag{1.6}$$

where L is the length of the structure (in feet) that must be loaded to produce the maximum live-load force. Longitudinal loads imposed by traffic accelerations are also specified. If a bridge is designed to carry more than two lanes of traffic, the AASHTO standard allows a reduction in the design live load, since the probability of simultaneously loading all lanes to capacity decreases as the number of lanes increases beyond 2. The AASHTO standard also specifies maximum design values for loads on the superstructure and for ice loading on bridge piers. The effects of earthquakes can be investigated using either equivalent static loads or the response-spectrum method.

It is almost certain that the original design loads for a bridge will be exceeded during the lifetime of the structure. For example, cranes and earthmoving equipment must invariably be transported across a span that is designed for smaller vehicular loads; in such cases the structure must be reviewed by a responsible structural engineer before an overload permit is issued. With improved technology, vehicle size and load capacity increase, and as traffic patterns change, the use of a bridge may be changed; thus, bridge overloads are almost inevitable. Considered over the useful life of the structure, the magnitude and the distribution of bridge loads are probabilistic.

Load Combinations

There is a distinct possibility that several of the loads discussed previously will be imposed on a structure simultaneously, but the likelihood that the maximums of all loads will occur at the same time is remote. Individual load cases must be combined on some rational basis, typically established by an applicable standard or code. The reasoning behind these analysis and design guides is dictated largely by the structural material used and the basic design approach (structural design methods are discussed in Sec. 1.7).

In *working-stress design* (also termed *service-load design*), the computed elastic stress must not exceed a specified limiting value. Some typical load combinations are

1. Dead load
2. Dead, live, and wind load
3. Dead and wind (or earthquake) load
4. Dead, live, snow, and wind (or earthquake)

The critical effects from both wind and earthquake loads are considered, where appropriate, but these two loads are usually assumed not to act simultaneously. When several loads are considered to act together, the total combined load effect is usually multiplied by a *load-combination factor,* which is less than unity, for example, 0.66 for dead plus wind load.

The objective of *limit-states design,* or *load-factor design,* is to ensure that a structure or a component will not be unfit for use with respect to serviceability (i.e., deflection constraints) or safety (i.e., a strength constraint). In this approach, individual load cases are multiplied by distinct load factors to account for (1) unavoidable deviations of the actual loads from their nominal values and (2) uncertainties in the analysis. The modified loads are combined to yield design loads. Examples of load combinations and their corresponding load factors are

1. $1.4 \times$ dead
2. $1.2 \times$ dead $+ 1.6 \times$ live $+ 0.5 \times$ snow
3. $1.2 \times$ dead $+ 1.3 \times$ wind $+ 0.5 \times$ live $+ 0.5 \times$ snow
4. $1.2 \times$ dead $+ 1.5 \times$ earthquake $+ 0.5 \times$ live

1.6 STRUCTURAL MATERIALS

Steel, reinforced concrete, prestressed concrete, and *wood* are the most commonly used structural materials in the building industry, whereas *aluminum, titanium,* and *fiber-reinforced composites* find structural applications in the aircraft and automotive industries. Each material responds to applied loads in a unique fashion, and the various properties and characteristics must be carefully considered in the design process. Individual material properties, e.g., compressive strength, display statistical variation, but they can be accurately described using an appropriate probability density function. Since complete understanding of material properties and behavior is not a prerequisite for the study of structural analysis, a brief qualitative survey of material behavior follows.

Steel

Steel is commonly used for structural applications ranging from high-rise buildings to offshore oil platforms. *Hot-rolled shapes* that are formed from hot billet steel are typically manufactured from *mild carbon steel* (0.15 to 0.29 percent). A sample stress-strain curve for A36 structural steel is shown in Fig. 1.14. The complete curve typically extends to a maximum strain of about 0.035 and exhibits an ultimate strength in the vicinity of 60 kips/in^2 (414 MPa). In the *elastic region* the stress is below the *yield point.* A decrease in stress is accompanied by unloading along the curve followed during loading; no permanent deformation is evident when the stress is reduced to zero. In the *linear range* the stress is below the *proportional limit.* The initial slope of the stress-strain diagram is the *modulus of elasticity E.* The subsequent response, wherein strain increases with no increase in stress, represents the *plastic region,* and in the *strain-hardening region,* the curve again has a constant positive slope.

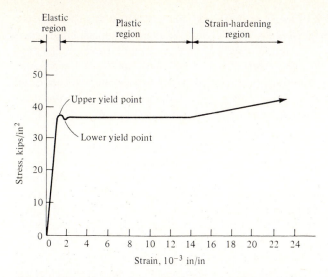

Figure 1.14 Stress-strain diagram for A36 steel at room temperature.

Hot-rolled steel structures in the United States are usually designed using the specifications[5] of the American Institute of Steel Construction (AISC), which gives rules for the design of structural members, connections, etc., along with the material design stresses. Material properties for hot-rolled steel have a coefficient of variation between 0.03 and 0.11, depending upon the type of test, and typical mean values for A36 steel are given in Table 1.5. The design of cold-formed steel structures in the United States is specified by the American Iron and Steel Institute (AISI).[6]

The horizontal plastic region of the curve in Fig. 1.14, along with its length, is a measure of the *ductility* of the material. *Fatigue behavior* occurs when the material is

Table 1.5 Typical mean-value properties for hot-rolled steel

Modulus of elasticity	29×10^3 kips/in² (200 GPa)
Minimum yield stress	36 kips/in² (248 MPa)
Minimum ultimate tensile stress	60 kips/in² (414 MPa)
Coefficient of linear expansion	6.5×10^{-6} in/in · °F (1.2×10^{-5} mm/mm · °C)
Allowable stresses:	
Tension	22 kips/in² (151 MPa)
Compression	21.6 to 3.7 kips/in² (149 to 26 MPa), decreasing with increasing slenderness
Flexure	24 kips/in² (165 MPa), decreasing for members not adequately braced laterally and with thin outstanding flanges
Shear	14.5 kips/in² (100 MPa)

subjected to repeated cycles of stress reversal, and the behavior in this case depends upon both the mean stress level and the range of stress. Under certain conditions, steel can display *brittle fracture,* wherein a relatively small amount of energy is absorbed in propagating the failure. *Residual stresses* can be introduced into steel structures by welding and can cause distortions and brittle failure. Steel is susceptible to *corrosion,* which can be minimized using coatings of paint, asphalt, or plastic; *weathering steels* develop an oxidized layer of protection.

Concrete

Concrete offers more flexibility in construction than other structural materials since it can easily be shaped to resist the applied loads in an optimal manner. Concrete is efficient in compression but inherently weak in tension. Reinforced concrete is produced by incorporating round steel bars with surface deformations in the concrete to resist the tensile stresses.

Concrete is a mixture of fine and coarse aggregate bound together by a paste made from portland cement and water that hardens as a result of exothermic chemical reactions in the paste. The compressive strength depends upon the properties of the aggregate, the proportions of water and cement, and the time and quality of the curing. The compressive strength commonly quoted is based on the strength of a standard cylinder 6 in in diameter by 12 in long (150 by 300 mm) cured under standard laboratory conditions for 28 days and tested at a specified rate of loading. It is possible to attain strengths of 14 kips/in^2 (96 MPa) and greater, but commercial concrete is commonly in the range of 3000 to 5000 lb/in^2 (21 to 34 MPa). The stress-strain curve for concrete in Fig. 1.15 indicates that this material has neither a definite yield-point stress nor a distinct straight-line elastic response. Typically, the slope of the initial tangent is interpreted as the modulus of elasticity. The coefficient of variation of typical concrete compressive strengths tested at a low rate of loading varies from 0.15 to 0.18, and typical values for stone concrete properties are given in Table 1.6.

The initial behavior of concrete under design loads is nearly elastic, but the strain increases with time even with no increase in the load. This time-dependent deformation under a constant stress is called *creep.* Under sustained load the added deformations can become greater than the elastic response. Creep behavior decreases (approximately exponentially) with time and can be minimized by using some steel in compression. Time-dependent deformation that reduces the internal stresses is called *relaxation;*

Table 1.6 Typical properties of stone concrete[7]

Compressive strength	4000 lb/in^2 (27.6 MPa)
Modulus of elasticity	3.6×10^3 kips/in^2 (24.8 GPa)
Tensile strength	380 lb/in^2 (2.6 MPa), for prestressed members only
Coefficient of linear expansion	6.5×10^{-6} in/in · °F (1.2×10^{-5} mm/mm · °C)

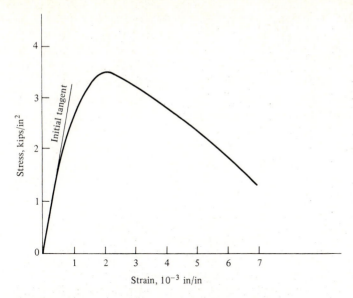

Figure 1.15 Stress-strain diagram for concrete in compression.

e.g., the stresses induced in a relatively new structure due to differential support settlement are significantly reduced over a period of time.

As concrete cures, it loses moisture by evaporation and *shrinks,* but shrinkage can be minimized if moist curing conditions are maintained. Uniform shrinkage causes compressive stresses in the concrete and tension in the reinforcing steel. Usually the moisture loss is not uniform; differential shrinkage is induced, which imposes significant internal shrinkage stresses. The deflections induced by shrinkage generally increase with time.

The steels used for conventional reinforced concrete have yield stresses from 40 to 60 kips/in^2 (276 to 414 MPa), with corresponding ultimate tensile stresses of 70 to 90 kips/in^2 (483 to 621 MPa) and a modulus of elasticity of 29×10^3 kips/in^2 (200 GPa).

Another method for combining the respective strengths of steel and concrete is to induce large initial compressive stresses in the concrete using high-strength steel cable (*tendons*), resulting in *prestressed concrete.* The method of producing *pretensioned* prestressed members consists of stretching the tendons between external anchorages before the concrete is placed. On curing, the concrete bonds to the steel and the initial prestressed force is transmitted to the member by releasing the jacking force. *Posttensioned* prestressed concrete members, are produced by stretching the tendons, jacking against the ends of the member, after the concrete has been properly cured and hardened. The concrete used in prestressed construction generally has compressive strengths ranging from 4000 to 6000 lb/in^2 (27.6 to 41.4 MPa), and the steel consists of seven-wire stranded cable with diameters from 0.25 to 0.60 in (6.4 to 15.2 mm) manufactured from steel with an ultimate strength of 250 or 270 kips/in^2 (1724 or

1862 MPa). The apparent modulus of elasticity for stranded cable is 27×10^3 kips/in^2 (186 GPa).

Wood

The housing industry commonly uses wood as a structural material, and it is also used in large structures such as industrial buildings and multifamily units. As the demand for wood increases and the supply dwindles, solid-sawn timber and plywood are being supplanted by innovative products such as glued-laminated members, particle board, fabricated I beams, and wood-steel trusses. Only about 20 of the many species of hardwoods and softwoods that grow in the United States are used for structural purposes, the most common being Douglas fir (*Pseudotsuga menziesii*), hemlock (*Tsuga heterophylla*), larch (*Larix occidentalis*), and loblolly pine (*Pinus taeda*).

The mechanical properties of wood depend upon the density, the slope of the grain, the presence of imperfections, e.g., knots, moisture content, the rate at which the load is applied, the duration of the load, and other factors. In general, wood displays linear elastic behavior with a well-defined proportional limit. Statistical information for wood properties is limited, and, for example, data for tension of glued-laminated beams show a coefficient of variation from 0.09 to 0.23. Typical coefficients of linear expansion in the longitudinal direction are 1.7×10^{-6} to 2.5×10^{-6} in/in · °F (3.0×10^{-6} to 4.5×10^{-6} mm/mm · °C). Typical properties for construction grade Douglas fir are given in Table 1.7.

Aluminum

Aluminum is used as a structural material for aircraft, heavy-duty structures, and applications where a combination of light weight, good corrosion resistance, and high strength is essential. Pure aluminum is a soft, ductile material with a relatively low tensile strength. It is commonly alloyed with other elements and heat-treated to enhance its structural properties. Aluminum alloys are designated by a four-digit number, in which the first digit indicates the major alloying group. For example, alloys in the 6000 series contain silicon and magnesium to form magnesium silicide, which makes them heat-treatable; 6061 is a major alloy in this series. Zinc is the major alloying element in the 7000 group, and 7075 is a high-strength alloy used in airframe struc-

Table 1.7 Typical properties for construction grade Douglas fir[8]

Modulus of elasticity	1.5×10^3 kips/in^2 (10.3 GPa)
Tension parallel to grain	625 lb/in^2 (4.3 MPa)
Horizontal shear	95 lb/in^2 (0.66 MPa)
Extreme fiber in bending	1200 lb/in^2 (8.3 MPa)
Compression parallel to grain	385 lb/in^2 (2.7 MPa)
Compression perpendicular to grain	1150 lb/in^2 (7.9 MPa)

Table 1.8 Properties of 6061-T6 aluminum alloy[9]

Modulus of elasticity	10×10^3 kips/in² (69 GPa)
Ultimate tensile strength	42 kips/in² (290 MPa)
Tensile yield strength	37 kips/in² (255 MPa)
Allowable stresses:	
Tension	19 kips/in² (131 MPa)
Compression	19 kips/in² (131 MPa), decreasing for increasing slenderness
Shear	12 kips/in² (83 MPa), decreasing for slender webs

tures. The alloy designation is followed by a hyphen and the temper designation; thus, 7075-T6 indicates a solution heat treatment followed by artificial aging.

In general, aluminum can be riveted, welded, brazed, or soldered, but welding is not recommended for some alloys (e.g., 7075). Because of its superior corrosion resistance, coating is usually unnecessary, but in some situations chemical, electrochemical, or paint finishes are used to give the desired appearance or added protection. When aluminum is used in contact with other metals, damage may occur by galvanic action and the aluminum must be insulated from the other material.

Aluminum has a specific gravity of 2.7 and a coefficient of linear expansion of 1.3×10^{-5} in/in · °F (2.4×10^{-5} mm/mm · °C); the Aluminum Association specifies the properties shown in Table 1.8 for the commonly used 6061-T6 alloy.

Masonry

Masonry construction, consisting of masonry units bonded together with mortar, are used to create exterior and interior load-bearing walls, partitions, panel walls, etc., both reinforced and unreinforced. Examples of masonry units are brick, tile, stone, and block; the mortar is a mixture of sand, masonry cement, portland cement, and hydrated lime or lime putty. The UBC specifies four distinct mortar mixtures that can be used for plain masonry; they have 28-day strengths ranging from 350 to 2500 lb/in² (2.4 to 17.2 MPa). Both masonry and mortar are compressive materials. For solid brick masonry with a minimum compressive strength of 4500 lb/in² (31.0 MPa) the combined allowable working stress on the net area of the wall is 250 lb/in² (1.7 MPa).

1.7 DESIGN METHOD

The principles of structural analysis and design are combined with experience and judgment in the design of a building. In a broad sense, the structural integrity of a system is established by (1) calculating the generalized stresses (*loads*) sustained by the components, (2) determining the strength or stiffness (*resistance*) of the total system and each component, and (3) systematically comparing and modifying the corresponding resistances and loads to ensure acceptable structural behavior. Although the methods for doing this depend to a certain extent upon the code in use, there are several distinct methods of design.

The *working-stress design method* requires that the computed stresses in the members be limited to a specified value when the structure is subjected to basic load combinations referred to as the *service loads*. The various applicable codes for steel, reinforced concrete, timber, etc., have each established the limiting-stress values as a percentage of the strength of the material. In steel design, for example, the limiting stress is typically based on the yield-point stress.

Acceptability for the *strength design method* is based on a comparison of factored loads and member strengths to provide safety against structural failure. The ultimate strengths of critical sections form the basis for the *ultimate-strength method* for reinforced concrete, the load-carrying capacity of members and/or entire frames forms the basis for the *plastic design method* used with steel. Individual *load factors* are used as multipliers on the loads, and the results are compared with factored resistances, the products of *resistance factors* and the corresponding strengths. The load factors are typically greater than 1, while the resistance factors are usually less than 1.

The broad objective of *limit-states design* is to ensure that a structure or individual component will not become incapable of functioning acceptably. *Serviceability limit state* is exemplified by excessive deflections and local damage, whereas *ultimate limit state* results from full or partial structural collapse due to inadequate strength, overall instability, etc. Nominal loads are multiplied by *load factors* to account for unavoidable deviations of the actual load from the nominal value and uncertainties in the analysis for the member forces. Multipliers on the maximum load-carrying capacity of a member, called *resistance factors,* incorporate the effects of unavoidable deviations of the actual member strengths from the nominal values and the manner of failure and its consequences. Similar factors, motivated somewhat differently, are also used in the strength design method.

In general, loads and resistances are probabilistic variables which can be described using appropriate probability density functions and the corresponding statistical measures of dispersion. In Fig. 1.16 the load and resistance for a typical structure are represented by the probability density functions $f(l)$ and $g(r)$, respectively. The ratio of the mean values $\bar{r}/\bar{l}$ is the conventional *factor of safety* (also called the *central factor of safety*). A high central factor of safety does not imply a low probability of failure

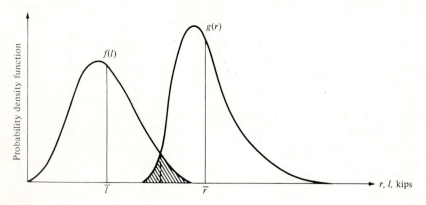

Figure 1.16 A load-resistance interference diagram.

because the latter depends upon the dispersion of the two distributions. We also observe from Fig. 1.16 that regardless of the value of the central factor of safety there is still a finite probability of failure. That is, the degree of overlap of the curves in the load-resistance interference diagram of Fig. 1.16 is a measure of the probability of structural failure. If R and L are random variables representing respectively the resistance and the load of the structure or component, then the *safety ratio* can be described as R/L and the *margin of safety S* is another random variable which is defined as the difference $R - L$.

In general, the conventional factor of safety is obtained by dividing the resistance by the corresponding load. Even though this load is regarded by some codes as a deterministic quantity, we can observe that it has a probabilistic interpretation, a viewpoint that is helpful in making transitions between various design methods.

1.8 ACTUAL AND IDEALIZED STRUCTURES

Structural analysis predicts the response of an idealized model of an actual structure, and the success of an investigation is critically dependent upon the relationship between the analytical predictions and true physical behavior. The process of identifying individual structural components, as described in Sec. 1.4, must be augmented with an appropriate simplified sketch before an analysis can be started. Usually, these drawings portray individual members, joints, reactions, and other components using the appropriate symbols that closely approximate physical behavior and permit tractable analytical descriptions.

For example, if the simple foundation detail in Fig. 1.17 is to be modeled, we note that the pin will transmit forces from the superstructure and that these forces are transferred to the concrete by the steel-support interface. If the support has suffi-

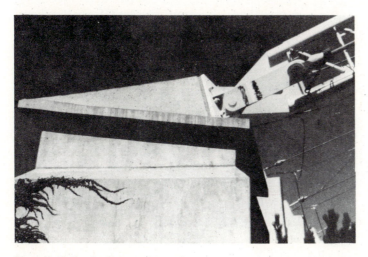

Figure 1.17 Detail of a support bracket.

cient stiffness, a concentrated compressive force applied to the pin by the superstructure is transmitted to the concrete, as represented by the model in Fig. 1.18. Treating the actual support in this manner will result in an excellent prediction of the true force distribution.

An analysis of the early truss design in Fig. 1.19 must begin with identifying and idealizing the independent subsystems. The detail of a typical truss joint in Fig. 1.20 reveals that if it functions as portrayed, free rotation is permitted about the pin and the forces will be transmitted axially along the centroids of each of the members. The truss can be idealized as a set of simple two-force members connected at the joints by smooth frictionless pins, as shown in Fig. 1.21. An analysis of this idealized model of the actual structure will portray true physical behavior accurately. Frequently, however, the idealization process is more complex. For example, in the truss joint of Fig. 1.20 the pin may not function as a smooth frictionless pin because of lack of lubrication, the presence of rust, or some other complicating factor. If a structure involves a complex interaction or combination of discrete members and continuous subsystems, it may defy an explicit analytical idealization. In such cases it is still necessary to represent the structure using approximations, the quality of which determines the accuracy of the prediction of the true physical behavior.

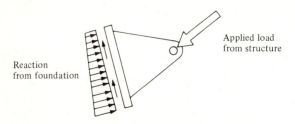

Figure 1.18 Structural idealization of a reaction.

Figure 1.19 A through truss bridge.

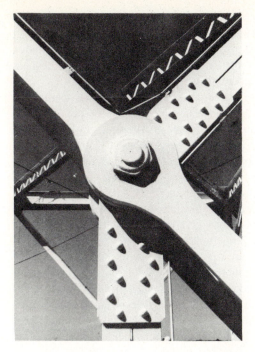

Figure 1.20 Construction detail of a truss joint.

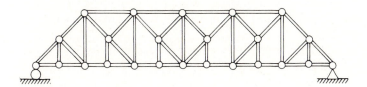

Figure 1.21 Structural idealization of the through truss bridge.

Throughout this book attempts will be made to relate the sketch of the idealized structure to the actual details. When no explicit discussion is given, the reader should attempt to envision the constructed structural configuration. The ability to make accurate structural idealizations which correlate analytical behavior and reality is acquired and improved through constant practice.

1.9 STRUCTURAL ELEMENTS

The structural engineer must function in concert with professionals from other disciplines to produce a logical and efficient system that satisfies the basic performance objectives. For a building this process involves blending architectural thinking with the disciplines of electrical, mechanical, and structural engineering. It is necessary to consider various structural systems and the associated elements (subsystems) in order to identify the one most appropriate for a given situation. A preliminary configuration should be selected carefully before the structural concept is complete and the analysis of a specific structural arrangement is accomplished.

An enlightened selection of an overall structural configuration which is crucial to the success or failure of any building design, can be accomplished only with a complete understanding of the tenets of structural analysis and a thorough grasp of design concepts in steel, reinforced concrete, prestressed concrete, masonry, and wood. It is difficult to know precisely where in the continuum of structural engineering thought this understanding should begin, but obviously the beginning student must have some idea of how the various structural elements function and interrelate in a total system. The following qualitative discussion is offered as a brief introduction intended to give an impression of the availability and capabilities of some of the common structural elements for buildings and bridges. It is important to note, however, that analysis of a number of the elements in this survey are beyond the scope of this book.

Trusses consist of an array of members typically made from steel, wood, or concrete, which are connected only at their ends and so arranged that the internal forces developed in response to the applied loads are either tensile or compressive. This behavior generally occurs if all the applied loads are concentrated at the joints, where the axes of the members are carefully aligned. These structural elements often function as planar trusses, and it is frequently possible to identify and analyze the two-dimensional trusses which are components of a total structural system, as shown in Fig. 1.5. The truss of Fig. 1.22a has been idealized as a set of smooth pins connecting the members; if the loads are applied as shown, the members must sustain either tension or compression (Fig. 1.22b).

Planar truss elements are used extensively for bridges, roofs, building floors, and other purposes, e.g., resisting lateral loads applied to high-rise structures. The possibilities for planar truss elements are broad. One common example is the type used in light floor or roof systems, with parallel top and bottom chord members and a relatively shallow depth. Special trusses of various configurations and member arrangements are capable of supporting large loads and spanning great distances. Some characteristic parameters are shown in Table 1.9.

Beam elements are long slender components with a length-to-depth ratio ranging typically from 4 to 25. The example in Fig. 1.23a transmits vertical forces into the two end supports, and a transverse section at an arbitrary point along the beam gives the free-body diagram in Fig. 1.23b. Only internal shear V and bending moment M occur in this case, but generally three independent forces and three moments can exist as internal resisting effects. They consist of two mutually perpendicular shear forces, an axial force, two perpendicular bending moments and torsion (moment along the longitudinal axis).

Typical beams used in most buildings and bridges are constructed from steel, reinforced concrete, prestressed concrete, or wood. The names of various beam elements are determined by their position and/or function. In a building, a *girder* is a main supporting beam framing into the columns; the smaller beams connected and oriented

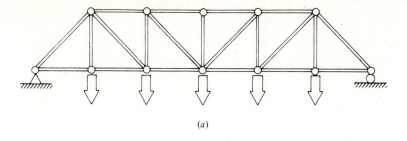

(a)

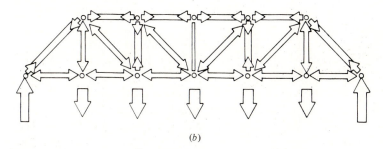

(b)

Figure 1.22 (*a*) A planar truss with applied loads; (*b*) the member forces.

at right angles to the girder are *floor beams;* lighter *joists* or *stringers* frame into the floor beams and deliver their reactions to them. In a bridge, *floor beams* span between the two main trusses, and the *stringers,* which are connected to the floor beams, are positioned in the longitudinal direction (see Fig. 5.13). The *spandrel beams* in a building are positioned on the extreme outside edges of the building; *rafters* are parallel beam elements that support the roof; *girts* and *purlins* are secondary beam elements positioned in the wall and roof, respectively, of a building, transmitting loads from the exterior cladding to the main structure (see Fig. 1.6). *Headers, lintels,* and *trimmers*

Table 1.9 Typical approximate truss parameters

| Type of truss | Depths | Spans (*L*) | |
		ft	m
Parallel chord:†			
All steel	*L*/18–*L*/22	10–120	3–37
Steel and wood	*L*/18–*L*/20	30–100	9–30
Wood	*L*/10–*L*/15	40–110	12–34
Special:			
Steel	*L*/4–*L*/10	30–600	9–183
Wood-trussed rafters	*L*/5–*L*/7	20–70	6–21
Wood	*L*/7–*L*/10	60–150	18–46

†The top and bottom members are referred to as the *chords*.

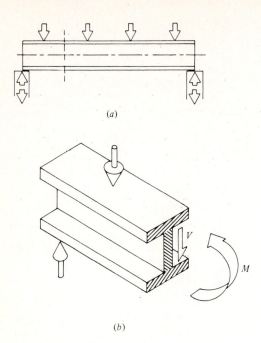

(a)

(b)

Figure 1.23 (*a*) Beam element with applied loads; (*b*) free-body diagram of a portion of the beam.

are special secondary beam elements used primarily for wood construction. Beam elements vary significantly in size and load capacity, since they are fabricated to meet specific requirements. The typical parameters in Table 1.10 are offered simply to give an idea of the range of capabilities and sizes.

The *rigid frame* in Fig. 1.24*a* responds as a continuous assemblage, the angles between the members at the points of connectivity remaining unchanged under load. Rigid-frame behavior implies that the forces and moments are transmitted continuously throughout the structural element, and this response is exhibited in a wide variety of configurations, ranging from the simple planar example illustrated in Fig. 1.24*a* to

Table 1.10 Typical beam parameters

Type of beam	Depths	Spans (L)	
		ft	m
Steel:			
Wide flange	$L/18-L/28$	10–70	3–21
Plate girders	$L/18-L/20$	25–400	8–122
Reinforced concrete, rectangular	$L/20-L/26$	15–100	5–30
Wood:			
Joists	$L/18-L/20$	8–24	2–7
Glued-laminated	$L/18-L/20$	10–80	3–24

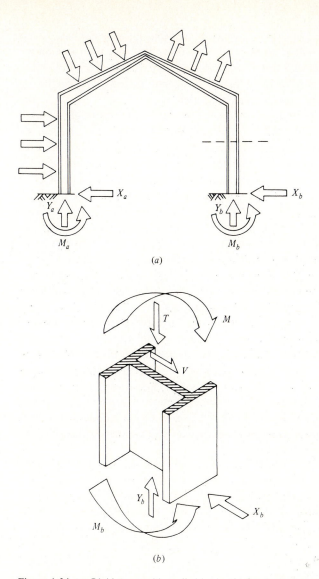

(a)

(b)

Figure 1.24 (a) Rigid frame with applied loads; (b) free-body diagram of a portion of the right column.

three-dimensional multiple-story and multiple-span arrangements. An example of the internal forces transmitted by the members of a plane rigid frame in response to applied loading is shown in Fig. 1.24b. Note that the shear force V, axial load T, and bending moment M are similar to the components found in beam elements. Since the complex behavior displayed by these structural elements usually requires analysis procedures beyond the simple equations of static equilibrium, only simple rigid frames are discussed before Parts Four and Five of this book.

The rigid frames used in buildings are commonly constructed from steel or reinforced concrete, but prestressed concrete and wood are also used. Since this type of structural element can be fabricated in various configurations for resisting both lateral and vertical loads, it is difficult to list typical parameters for rigid frames. Column heights range from 12 to 20 ft (3.7 to 6.1 m), and span lengths, which are influenced by roof slope, are characteristic of those listed for beam elements. The depths are usually somewhat shallow, due to the rigid connections.

A *cable* suspended between two points and loaded only by its own weight assumes the form of a *catenary*. If concentrated forces are applied to the cable (Fig. 1.25a), the loads are transmitted to the supports in pure tension (Fig. 1.25b). Steel cables are used to guy various types of towers; they are the primary structural component for suspension and cable-stayed bridges and are used for spans ranging from 80 to 4000 ft (24 to 1220 m). A set of suspended cables defines a surface like the one in Fig. 1.26a; such configurations can be used for roof structures. The applied loads, which are transmitted to the supports by tension in the main cables, are reacted by the compression struts and the guy cables, which are in tension (Fig. 1.26b). *Tensile-surface structures* can also be formed by a network of freely suspended or prestressed cables. Their shapes depend upon the direction of the cables and the configuration of the edge supporting members. Such structures are commonly used for relatively long spans [beyond 100 ft (30 m)] supporting light distributed loads.

Lightweight *membranes* are also used to construct tension structures for spans up to approximately 150 ft (46 m). For example, a simple saddle surface can be formed by prestressing a square light membrane between edge cables, with two supports at opposite corners at a higher elevation than the other two. The suspending system for *pneumatic structures* consists of internal air pressure. The *air-supported dome* in Fig. 1.27 is constructed from a lightweight membrane maintained in a state of pure tension by the presence of relatively low internal pressure. An *air-inflated structure* is formed by dual membranes which are fastened together with internal baffles, and the

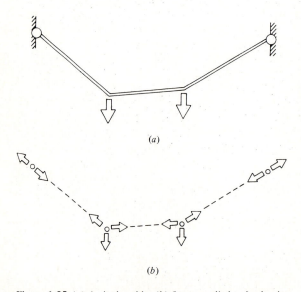

(a)

(b)

Figure 1.25 (a) A single cable; (b) forces applied to load points.

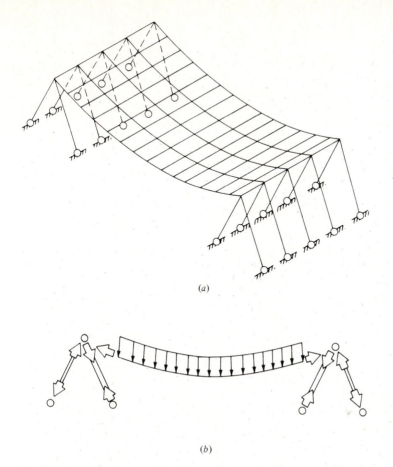

(a)

(b)

Figure 1.26 (*a*) Freely suspended cable structure; (*b*) internal forces in the members.

structural integrity is attained by pressurizing the void between them. Although these structures are still somewhat experimental, they provide economical coverings over swimming pools and sports arenas.

An *arch* attains much of its structural integrity through its shape. When a uniformly distributed load is placed on a parabolic arch, it will sustain only internal compressive forces. Since the continuous arch of Fig. 1.28*a* is loaded with a series of concentrated forces, the structure is subjected to internal shear V, axial force T, and bending moment M, as revealed by the free-body diagram of Fig. 1.28*b*. Note that these internal forces are similar to those for a rigid frame. Arch and truss behavior are sometimes combined into a *trussed arch* to achieve long spans. Although arches are generally most efficient for spans in excess of 100 ft (30 m), arches made from glued-laminated wood are commonly used for spans ranging from 40 to 140 ft (12 to 43 m). Arches are also frequently used for long-span bridge structures. Spans of 1000 ft (300 m) have been attained for arches constructed from steel and reinforced

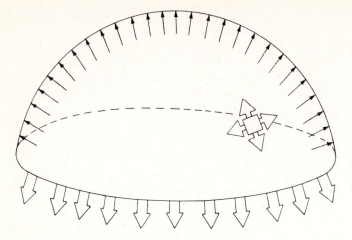

Figure 1.27 An air-supported dome, a pneumatic structure.

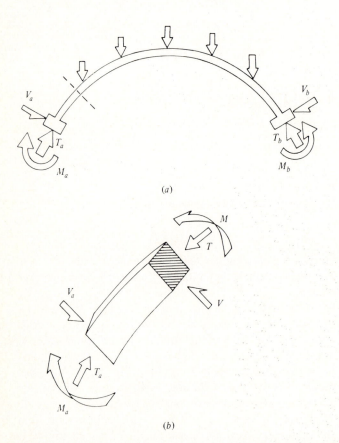

(a)

(b)

Figure 1.28 (a) An arch with applied loads; (b) free-body diagram near the reaction.

concrete, and a steel trussed arch has been used for a main span of approximately 1600 ft (488 m).

The overall geometry of a *folded-plate structure* considerably enhances the capabilities of the relatively thin cross section (see Fig. 1.29*a*). The structural integrity of these forms can be demonstrated by folding a piece of paper longitudinally into a series of vees (or into the illustrated geometry) and contrasting the stiffness of this configuration with that of the unfolded paper. A folded plate transfers loads in both the longitudinal and transverse directions as a series of interconnected beams. The *longitudinal beam action* (Fig. 1.29*b*) indicates that each individual section functions as a beam. For gravity loads the top and bottom horizontal plates will be in compression and tension, respectively, as shown in Fig. 1.29*a*. The *transverse beam action* (Fig. 1.29*c*) involves an imaginary transverse strip of the structure, and since the deflections at points of intersection of the plates are minimal, these are regarded as support points. Typically, folded plates are stiffened with end diaphragms (transverse stiffener plates filling the cross-sectional voids at the ends), and edge beams (not shown in Fig. 1.29)

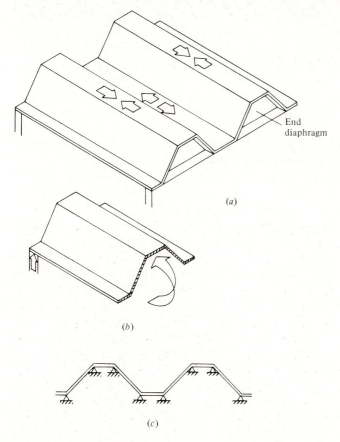

End diaphragm

(*a*)

(*b*)

(*c*)

Figure 1.29 (*a*) A folded-plate structure; (*b*) longitudinal beam action; (*c*) transverse beam action.

are used longitudinally to ensure the structural integrity. Concrete folded plates are typically 3 to 4 in (75 to 100 mm) thick with spans ranging from 50 to 100 ft (15 to 30 m). They may include prestressing cables in order to minimize deflections. Wooden folded plates are used for spans from 30 to 100 ft (9 to 30 m). The analysis of folded plates is not discussed in this book.

Various combinations of trusses are used for constructing flat roof and floor systems subjected to relatively light loads. For example, a simple *truss-on-truss system* is formed by two parallel heavy main trusses with smaller secondary trusses spanning between them. Since a system like this tends to transmit the floor loads in one direction along the secondary trusses, it is a one-way system which can be analyzed as a number of individual planar trusses. The somewhat more complex *truss-grid system,* formed by a series of similar intersecting trusses arranged in an orthogonal grid, is a two-way system since a load applied at a point of intersection of two trusses will be transmitted via truss action in two orthogonal directions.

A *space truss* (Fig. 1.30*a*) is composed of two-force members arranged as a group of interconnecting tetrahedra, and applied loads are transferred in all three principal directions. If a number of these trusses are combined and interconnected, a two-way system called a *space frame* results. In the typical space frame shown in Fig. 1.30*b* a vertical force applied in the middle of the floor will be transmitted to all four supports by the three-dimensional interaction of the individual members. The space-frame geometry typically offers efficient use of material and a somewhat stiffer structure than either the truss-on-truss or truss-grid systems. Space frames are fabricated from steel, aluminum, concrete, and wood. They are used when a flat roof or floor with clear spans in excess of 100 ft (30 m) is required.

The analysis of truss grids and space frames usually requires computer-oriented methods. An introductory analytical treatment for space trusses is presented in Chap. 8, and computer methods of matrix structural analysis are described in Part Five.

One- and two-way behavior exists in slabs, which are constructed from reinforced concrete and used for floor and roof systems. A *one-way slab* has a length-to-width ratio of 1.5 or greater and is supported on stiff beams or masonry walls along the longitudinal edge. This system functions as a series of strip beams, which transfer the loads in bending across the width of the slab. A *flat slab* consists of a uniform slab supported on a grid of equally spaced columns. In Fig. 1.31*a* the slab will deflect, creating compression on top near the center and tension on top along the column grid lines (Fig. 1.31*b*). Relatively complex behavior occurs in the vicinity of the columns, and additional strength must be introduced in this area to prevent the columns from punching through the slab. The two-way response of this system involves both bending and shear behavior. A *two-way slab* results if stiffening beams are introduced into the structure in Fig. 1.31*a* along the column grid lines. *Waffle slabs,* supported by an orthogonal grid of closely spaced shallow beams, are used for floor systems supporting heavy loads, e.g., garages and warehouses. The range of typical spans for slabs is 20 to 60 ft (6 to 18 m), and the depths vary from $L/24$ to $L/54$. Generally, one-way slabs have the greatest depths, and two-way slabs are the thinnest.

The capabilities of flat space frames can be enhanced by forming them into surfaces with single or double curvature. The structures are capable of covering large areas

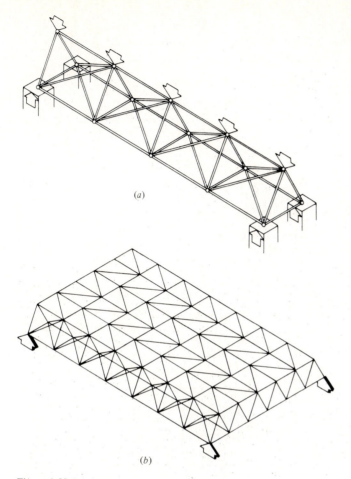

(a)

(b)

Figure 1.30 (a) A space truss; (b) a space frame.

without intermediate supports and represent an economical alternative to cable-net and thin-shell structures. The short individual members of these *curved space frames* can be arranged in one or in two layers. *Ribbed* and *lamella cylinders* are examples of reticulated surfaces with single curvature. The latter is a roof structure composed of a skew grid of intersecting arch forms made up of a single layer of individual beam members. The *Schwedler dome* in Fig. 1.32 is an example of a single-layer reticulated space frame with double curvature made up of individual two-force members. The equal-length two-force members of a *geodesic dome* are arranged in a series of isosceles triangles. The analysis of some of these structures is discussed in Chap. 8, and they can be investigated using the matrix structural analysis described in Part Five.

Continuous *shell structures* represent another alternative for long-span systems subjected to distributed loads. They are typically made from reinforced concrete, and span-to-thickness ratios of 500 are not uncommon. The forms can be generated by ro-

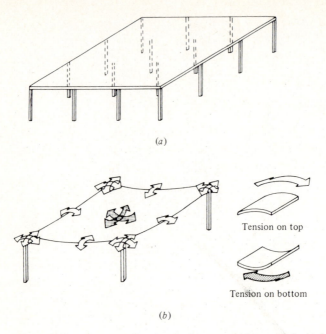

(a)

(b)

Tension on top

Tension on bottom

Figure 1.31 (a) Flat slab with columns; (b) response due to gravity loads.

tating a plane curve about a line or by translating a plane curve along a line or another curve. The *spherical shell* in Fig. 1.33 is an example of a shell of revolution in which a circular curve has been rotated about a vertical axis. *Cylindrical shells* can be produced by translating a circular shape along a straight line, and a *hyperbolic paraboloid* (saddle shape) is produced by translating a convex parabolic shape along an orthogonal concave parabola. A shell structure transmits normal stresses parallel to its surface and behaves like a membrane, but these membrane stresses are accompanied by tangential shear stresses. The spherical shell under uniform vertical load transmits the loads to the foundation as a series of compression arches, primarily by meridional forces like those shown in Fig. 1.33. Under partial loading the shape of a shell tends

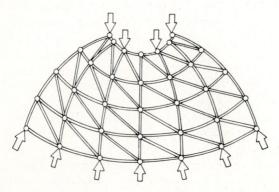

Figure 1.32 The Schwedler dome.

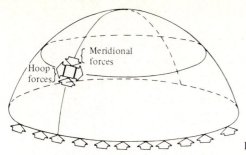

Figure 1.33 A hemispherical thin-shell structure.

to change. This is prevented by hoop forces acting perpendicular to the meridional forces (Fig. 1.33). Suitable edge-support conditions ensure that minimal bending stresses will be introduced into the shell. The analysis of shells, which display complex behavior, is beyond the scope of this book.

The function of the building *foundation* is to transmit the loads from the superstructure to the ground in such a way that settlement is minimized and the long-term bearing capacity of the soil is not exceeded. Many configurations exist for accomplishing these objectives. For example, a wall is typically supported on a continuous *wall footing,* which has a width greater than the wall. Columns are placed on single square *spread footings,* or the reactions from several columns may be combined on one rectangular footing. Sometimes individual footings are connected with footing ties to enhance their resistance to earthquake loading. If the soil has a low bearing capacity or the induced loads are high, a *mat foundation* is used under the entire building. The reactions for high-rise structures can be resisted with *piles,* which support the load primarily by friction, or *caissons,* which transfer the forces to subsurface strata with high bearing capacity.

Other types of substructures are also of interest to the structural engineer. *Retaining walls* are an integral part of many building designs. *Cofferdams* are used during construction to prevent the earth from collapsing into an excavation. The analysis and design of substructures is beyond the scope of this book, but structural engineers must understand these subsystems to develop a total functional building system.

The basic elements described in this section must individually meet the structural-integrity requirements of serviceability, e.g., deflection constraints, and strength. Furthermore, they must function in combination as a structural system. For example, the system of a high-rise structure is composed of rigid frames, the foundation, external cladding, internal stiffening core, and various secondary subsystems which must be joined together with appropriate structural connections to produce a functional whole. The interactions between individual elements within the total system play an extremely important role in the overall structural behavior.

It is not sufficient to view the structure in vacuo. The overall requirements of the entire design concept must be considered so that the structure functions in concert with the mechanical and electrical systems, producing a safe, functional system that satisfies both interior and exterior spatial requirements. Finally, all these considerations must be overlaid with the economic facts, since the cost of a system is an essential factor in determining its feasibility.

REFERENCES

1. International Conference of Building Officials: *Uniform Building Code,* Whittier, Calif., 1982.
2. American National Standards Institute, Inc.: *American National Standard Building Code Requirements for Minimum Design Loads for Buildings and Other Structures,* ANSI A58.1-1982, New York, 1982.
3. Applied Technology Council: *Tentative Provisions for the Development of Seismic Regulations for Buildings,* ATC 3-06, National Bureau of Standards Special Publication 510, Washington, 1978.
4. American Association of State Highway and Transportation Officials: *Standard Specifications for Highway Bridges,* 12th ed., Washington, 1977.
5. American Institute of Steel Construction, Inc.: *Manual of Steel Construction,* 8th ed., Chicago, 1980.
6. American Iron and Steel Institute: *Specifications for the Design of Cold-Formed Steel Structural Members,* New York, 1980.
7. American Concrete Institute: *Building Code Requirements for Reinforced Concrete* (ACI 318-77), Detroit, 1979.
8. National Forest Products Association: *National Design Specifications for Wood Construction,* Washington, 1982.
9. The Aluminum Association, Inc.: *Specifications for Aluminum Structures,* 3d ed., Washington, April, 1976.

ADDITIONAL READING

Benjamin, J. R., and C. A. Cornell: *Probability, Statistics and Decision for Civil Engineers,* McGraw-Hill, New York, 1970.

Bury, K. V.: *Statistical Models in Applied Science,* Wiley, New York, 1975.

Ellingwood, B., T. V. Galambos, J. C. MacGregor, and C. A. Cornell: *Development of a Probability Based Load Criterion for American National Standard A58,* National Bureau of Standards Special Publication 577, Washington, 1980.

Ghiocel, D., and D. Lungu: *Wind, Snow and Temperature Effects on Structures Based on Probability,* Abacus, Tunbridge Wells, Kent, England, 1975.

Hart, G. C.: *Uncertainty Analysis, Loads, and Safety in Structural Engineering,* Prentice-Hall, Englewood Cliffs, N. J., 1982.

Haugen, E. B.: *Probabilistic Approaches to Design,* Wiley, New York, 1968.

Lin, T. Y., and S. D. Stotesbury: *Structural Concepts and Systems for Architects and Engineers,* Wiley, New York, 1981.

Salvadore, M., and M. Levy: *Structural Design in Architecture,* 2d ed., Prentice-Hall, Englewood Cliffs, N. J., 1981.

Schodek, D. L.: *Structures,* Prentice-Hall, Englewood Cliffs, N. J., 1980.

TWO

THE STATICS OF STRUCTURES

2.1 BASIC CONCEPTS

Forces are applied to a structure either by direct mechanical contact or by remote action. These forces have an *internal* and *external effect*. For the body of Fig. 2.1 the external effects of the *applied forces* P_1 to P_5 (denoted by →) are the *reactive forces* R_1, R_2, and R_3 (↔). These reactive forces, which are exerted on the body by the foundation, maintain *external equilibrium* of the body. The action of the applied forces internal to the body is influenced by the distribution and properties of the materials.

The emphasis throughout this book is upon two-dimensional, or *planar,* structures; the associated force systems are termed *coplanar*. Figure 2.2a shows a general coplanar force system and the *resultant P* of these four forces. Since force is a vector quantity, it can be uniquely specified by its magnitude and line of action. A set of forces intersecting at a point, as in Fig. 2.2b, forms a *concurrent* force system. If the forces of the *parallel* force system shown in Fig. 2.2c were all aligned on a single line of action it would be a *collinear* force system. The *couple* shown in Fig. 2.2d consists of two parallel forces which are equal in magnitude, opposite in direction, and not collinear. It can be demonstrated that any system of forces and couples can be replaced by a single force with a specific line of action plus an associated couple.

The reactive forces must *stabilize* the structure externally. A planar structure is capable of exhibiting three independent components of motion; i.e., it has three degrees of freedom. With orthogonal cartesian coordinates the structure in Fig. 2.3 can potentially translate in both the x and y directions and rotate about an axis parallel to the z axis. The points where the reactive forces (*reactions*) are applied to the structure are called the *supports*. From Fig. 2.3 it can be observed that R_{ax} (the reaction at point a in the x direction) is designed to prevent translation in the x direction, and R_{ay} and R_{by} are intended to prevent y translation. Both y reactions are required: if R_{by} were omitted, the structure could rotate about point a as the instantaneous center, and if R_{ay} were omitted, rotation could occur about point b.

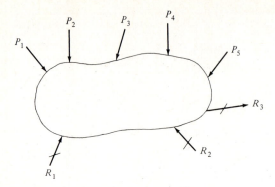

Figure 2.1 Applied and reactive forces on a structure.

2.2 SUPPORT DETAILS

The constructed configuration of the supports depends upon (1) the type of structure being restrained, (2) the material used, and (3) the number of kinematic degrees of freedom to be constrained at the reaction point. Examples of support details for beams, trusses, frames, etc., are given later as each structural type is discussed in detail. There are a number of generic supports, however, whose characteristics will be discussed in this section.

If only one component of displacement is to be constrained, use can be made of the *roller support,* shown in its symbolic form in Fig. 2.4a. Its idealized behavior can be envisioned as consisting of a simple cylindrical roller placed between the structure and an unyielding flat surface in such a way that normal to this surface motion is constrained and reactive forces are transmitted. An example of a construction detail functioning this way is shown in Fig. 2.5. If the material of the bearing pad has a low coefficient of static friction and is stiff enough to not deform significantly under the reactive force, the small horizontal force component can be neglected. A support constructed in this manner will not resist uplift forces; thus it functions like the idealized roller. When the symbol for a roller support is used, it generally implies that

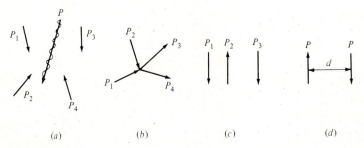

(a)   (b)   (c)   (d)

Figure 2.2 Coplanar force systems: (a) general, showing the resultant, (b) concurrent, (c) parallel, (d) couple.

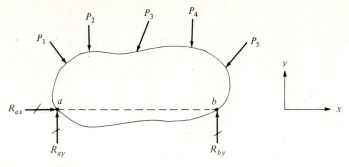

Figure 2.3 Reactive forces necessary to stabilize a structure.

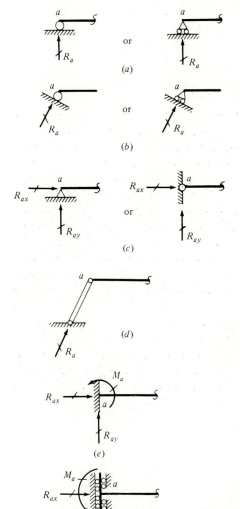

Figure 2.4 Symbols for various support types: (*a*) roller support for horizontal surface, (*b*) roller support for inclined surface, (*c*) hinge support, (*d*) link support, (*e*) fixed support, (*f*) guide support.

the reactive forces can act normal to the surface in either direction (toward or away from the surface). If the support in Fig. 2.5 were modified to behave in this way, it would require an anchor bolt that would not offer any resistance to horizontal motion. In steel construction, this is accomplished in a number of ways, using a vertical anchor bolt embedded in the pier and fitted into a slotted hole in the beam (see Chap. 4 for a typical detail).

Figure 2.4c shows the symbols for a *hinge support;* it totally constrains all translation, but the structure is free to rotate about the point of attachment. A simplified sketch of this type of support is presented in Fig. 2.6. The portion of the device anchored to the concrete is attached to the mating part on the superstructure with a smooth bearing pin, which for purposes of analysis is assumed to be frictionless. Thus, the reactive force R and the force transmitted by the structure F will be collinear and will act through the center of the bearing pin. This support will not transmit any bending moment.

The *link support* shown in Fig. 2.4d consists of a member which is capable of transmitting only tension and compression and which is connected to the structure at a with a smooth frictionless pin. If the link has negligible weight, we know the direction of the force exerted upon the structure by the link. Only the magnitude of the force is unknown. If this is a rigid link, there will be no displacement of the beam in the direction of the link axis, but for an elastic link the deformation must be computed using the principles of mechanics of solids.

The *fixed support* in Fig. 2.4e constrains the point from translation and rotation. With this type of support, there are three unknowns, which can be considered to be the two components of force in the x and y directions and a couple. The point of application is the point of connectivity between the support and the superstructure. The fixed support is commonly used with beams and frames.

The support condition shown in Fig. 2.4f is not common but should be understood since the behavioral concept is used frequently in carrying out analyses. The *guide support* is constrained from translation normal to the slot, and rotation is prevented. Thus, while there is no reactive force parallel to the slot, there is an unknown force normal to the slot and an unknown bending moment at the support.

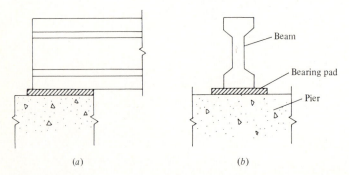

Figure 2.5 A constructed support that functions as a roller: (*a*) side view; (*b*) end view.

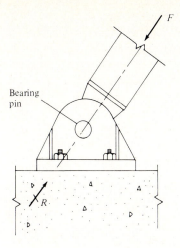

Figure 2.6 Detail view of a hinge support.

2.3 THE FREE-BODY DIAGRAM

Typically, the first step in an analysis is to envision the structural behavior, including a visualization of how the loads are transmitted throughout the structural system. These load paths can be established by isolating each of the major systems and determining the forces that act on it. The engineer can carry out the analysis after constructing an accurate sketch of the isolated system showing all the external forces; such a figure is called a *free-body diagram*. The ability to construct accurate free-body diagrams is the key to understanding structural behavior.

Assume that the reactions for the truss in Fig. 2.7a are to be calculated. Since it is of no interest at this point how the loads are transmitted to the supports, the truss as a whole is isolated. After the support symbols have been interpreted from Fig. 2.4, the reaction forces are sketched on the truss and the free-body diagram in Fig. 2.7b is obtained. Note that reactions have been shown acting in the positive coordinate direc-

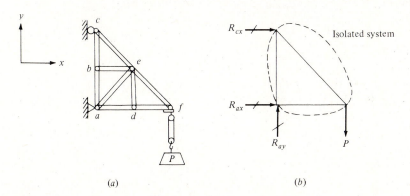

Figure 2.7 (a) A truss with applied loading, (b) free-body diagram.

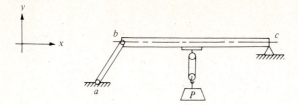

Figure 2.8 A beam with applied loading.

tions. If the calculations yield a positive sign for the numerical result, the direction of the force has been assumed correctly. If the result is negative, this implies that the force is acting in the negative coordinate direction. This convention is arbitrary, but regardless of the procedure used, it is essential to show the assumed direction of the reactions on the free-body diagram when it is first sketched and to interpret the algebraic sign of the calculated results.

It is often necessary to determine the forces acting on more than one component of a structure. For example, for the structure in Fig. 2.8 it may be required to establish the reactions at the ends of beam bc and the force in the support link ab. The free-body diagrams for each component of the structure are shown in Fig. 2.9. The beam is isolated, with the reaction forces acting in their assumed directions, in Fig. 2.9a. Using this free-body diagram and the appropriate equations, we can calculate R_b, R_{cx}, and R_{cy}. We know from Newton's third law that forces occur in equal and opposite pairs, i.e., *for every action there is an equal and opposite reaction,* where an action and a reaction constitute a force pair. Since there is no external force upon the pin that connects the beam to its support, forces R_{cx} and R_{cy} exerted on the support form force pairs with those on the beam. The two forces R'_{cx} and R'_{cy} represent the reaction of the foundation on the hinge support (Fig. 2.9b). Note that if the hinge height is small, R_{cy} and R'_{cy} will be

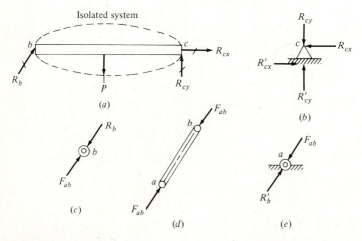

Figure 2.9 Free-body diagrams: (a) the beam, (b) the support at point c, (c) the pin at point b, (d) the link ab, (e) the pin at point a.

collinear; otherwise R'_{cy} will be offset slightly to the left. The pin connecting the beam and the strut at b are shown in the free-body diagram in Fig. 2.9c, where R_b is the reaction of the beam on the pin; thus, by Newton's third law, the direction of this force is opposite that shown on the beam. Newton's third law can also be applied to obtain the directions of F_{ab} at each end of link ab, as shown in the free-body diagram in Fig. 2.9d. The free-body diagram of the pin at the support point a is shown in Fig. 2.9e. R'_b is the reaction of the foundation on the pin, and by Newton's third law it must be equal in magnitude to F_{ab}.

2.4 THE EQUATIONS OF STATIC EQUILIBRIUM

Equilibrium involves the balance of forces. If a body is in a state of *static equilibrium,* the resultant of all forces acting on the body is zero. *Dynamic equilibrium* incorporates the relationships between force and acceleration from Newton's second law; this concept will not be discussed in this book although the possibility of motion with a constant linear velocity of the center of gravity must be admitted when using the static-equilibrium equations.

The equations of equilibrium for a force system in the xy plane can be written in scalar form as

$$\sum F_x = 0 \qquad \sum F_y = 0 \qquad \sum M_o = 0 \qquad (2.1)$$

The third equation is the algebraic sum of the moments of all the forces about an axis which is parallel to the z axis and passes through some arbitrary point o. For complete equilibrium in two dimensions, all three of the independent conditions in Eq. (2.1) must be satisfied. There are four categories of equilibrium for two-dimensional systems:

Case 1: Collinear force system Only one equation of equilibrium in the direction of the forces is required since the other two equations are identically satisfied.

Case 2: Concurrent force system Two force equations are required since the moment equation about the point of concurrency is identically satisfied.

Case 3: Parallel force system If all forces are aligned with the x axis, force equilibrium in the x axis and the moment equation are the only two equations.

Case 4: General force system All three equations are required.

The equilibrium equations (2.1) can also be expressed in two alternate forms for coplanar force systems, namely,

$$\sum F_x = 0 \qquad \sum M_a = 0 \qquad \sum M_b = 0 \qquad (2.1a)$$

where points a and b do not lie on a line that is perpendicular to the x axis, and

$$\sum M_a = 0 \qquad \sum M_b = 0 \qquad \sum M_c = 0 \qquad (2.1b)$$

where points a, b, and c are three arbitrary points not on the same line. Example 2.1 demonstrates the use of Eqs. (2.1), (2.1a), and (2.1b) for calculating the reactions for a structure.

Example 2.1 Calculate the reactions for the beam shown.

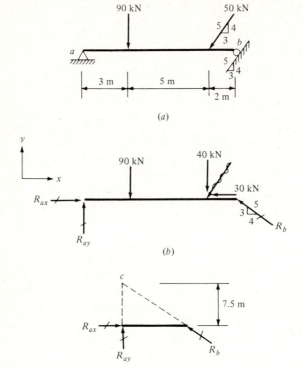

Figure E2.1 (a) Loaded beam, (b) free-body diagram, (c) intersection of R_b and R_{ay}.

SOLUTION From Fig. E2.1b reactions using Eq. (2.1) are

$$\sum M_b = 0 \curvearrowleft : \qquad 10R_{ay} - 40(2) - 90(7) = 0$$
$$R_{ay} = 71 \text{ kN } (\uparrow)$$

$$\sum F_y = 0 + \uparrow : \qquad \tfrac{3}{5}R_b + 71 - 90 - 40 = 0$$
$$R_b = 98.33 \text{ kN } (\nwarrow)$$

$$\sum F_x = 0 \xrightarrow{+} : \qquad R_{ax} - 30 - \tfrac{4}{5}R_b = 0$$
$$R_{ax} = 108.67 \text{ kN } (\leftrightarrow)$$

Reactions using Eq. (2.1a) are

$$\sum M_b = 0 \curvearrowright : \qquad\qquad R_{ay} = 71 \text{ kN } (\uparrow)$$

$$\sum M_a = 0 \curvearrowright : \qquad\qquad 90(3) + 40(8) - 10R_{by} = 0$$

$$R_{by} = 59 \text{ kN}$$

hence $\qquad\qquad R_b = \tfrac{5}{3}(59) = 98.33 \text{ kN } (\nwarrow)$

$$\sum F_x = 0 \xrightarrow{+} : \qquad\qquad R_{ax} = 108.67 \text{ kN } (\leftrightarrow)$$

From Fig. E2.1c using Eq. (2.1b)

$$\sum M_c = 0 \curvearrowright : \qquad 90(3) + 40(8) + 30(7.5) - R_{ax}(7.5) = 0$$

$$R_{ax} = 108.67 \text{ kN } (\leftrightarrow)$$

DISCUSSION In writing an equilibrium equation it is imperative to select a positive direction for the forces and be consistent throughout the calculation. The positive directions are indicated for each equation. The directions of the reactions assumed are indicated on the free-body diagram. The 50-kN force is resolved into its rectangular cartesian components to simplify the calculations. The reactions are calculated in three different ways to illustrate the various forms of the equations of static equilibrium.

In the first solution, using Eq. (2.1), point b is selected for the moment center. Since both R_{ax} and R_b pass through this point, the moment equation yields R_{ay} directly. It is important to attempt to obtain each of the reactive forces independently so that previous numerical errors will not be introduced into the reaction being computed. Unfortunately, in this solution the other two reactive forces depend upon this first result. Summation of forces in the x and y directions gives R_{ax} and R_b, respectively.

In the second solution, using Eq. (2.1a), the equation $\sum M_b = 0$ gives the same value for R_{ay} as that obtained in the first solution. Point a is selected as the second moment center so that R_b can be calculated independently of R_{ay}. In writing this moment equation the reaction R_b is resolved into its x and y components, and since the x component passes through the moment center, only R_{by} occurs in this equation. After R_{by} has been obtained, the total reaction is calculated using $R_b = \tfrac{5}{3}R_{by}$. The equation $\sum F_x = 0$ gives the same result for R_{ax} as that obtained in the first solution.

In using Eq. (2.1b) three moment centers must be selected. Points a and b give R_{ay} and R_b, respectively, as shown in the second solution. Point c, which is the intersection of R_{ay} and R_b, was selected for the third moment center. Using this point, we calculate R_{ax} independently from the results obtained for R_{ay} and R_b.

2.5 EQUATIONS OF CONDITION

Some structures incorporate unique internal features of construction that significantly affect the behavior of the system and require special attention from the structural engineer. For example, the beam in Fig. 2.10a has an internal *hinge* built in at point b. A hinge can be designed in various ways, depending upon the material of

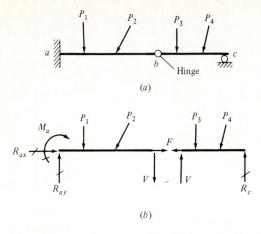

(a)

(b)

Figure 2.10 A beam with an internal hinge: (a) the beam with loading, (b) free-body diagram of the two beam segments.

construction, but it functions ideally as a smooth frictionless pin. The result is that no bending moment can be transmitted through the beam at point b. The free-body diagrams for the two segments of the beam are shown in Fig. 2.10b. Note that there are two internal components of force at point b, one parallel to the axis of the member (F) and another perpendicular to the axis (V). Since no moment is transmitted through the hinge, the equation $\Sigma M_b = 0$ can be imposed for the two individual free-body diagrams. The one independent equation introduced by the condition of construction is referred to as an *equation of condition*.

The three common construction details that introduce equations of condition are illustrated in Fig. 2.11. The hinge in Fig. 2.11a implies that the moment at this point

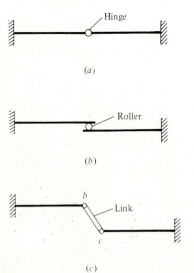

Figure 2.11 Construction details that introduce equations of condition: (a) the internal hinge, (b) the roller, (c) the link.

is zero and yields one equation of condition. For the roller of Fig. 2.11*b*, only the force normal to the axes of the beams will be transmitted. This gives two equations of condition since the force parallel to the longitudinal axes of the beams and the moment are both zero at the roller. The link (Fig. 2.11*c*) can transmit force only along its axis, and it introduces two conditional equations. If the link were positioned vertically, only vertical forces could be transmitted to the two beam segments at points *b* and *c* and it would function like a roller. Examples 2.2 and 2.3 illustrate the calculation of reactions for structures with internal hinges.

Example 2.2 Calculate the reactions for the beam illustrated.

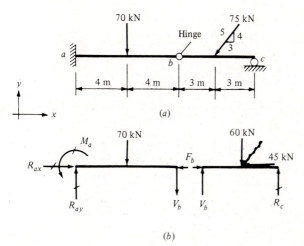

(a)

(b)   **Figure E2.2**

SOLUTION From Fig. E2.2*b* using the free-body diagram *bc* gives

$$\sum M_b = 0 \curvearrowright :$$

$$60(3) - 6R_c = 0$$
$$R_c = 30 \text{ kN } (\uparrow)$$

$$\sum M_c = 0 \curvearrowright :$$

$$6V_b - 60(3) = 0$$
$$V_b = 30 \text{ kN } (\uparrow)$$

$$\sum F_x = 0 \xrightarrow{+} :$$

$$F_b - 45 = 0$$
$$F_b = 45 \text{ kN } (\rightarrow)$$

and using the free-body diagram for *ab* gives

$$\sum F_x = 0 \xrightarrow{+} :$$

$$R_{ax} - 45 = 0$$
$$R_{ax} = 45 \text{ kN } (\leftrightarrow)$$

$$\sum M_a = 0 \curvearrowright :$$

$$70(4) + 30(8) - M_a = 0$$
$$M_a = 520 \text{ kN} \cdot \text{m } (\curvearrowright)$$

$$\sum F_y = 0 + \uparrow : \qquad\qquad R_{ay} - 70 - 30 = 0$$
$$R_{ay} = 100 \text{ kN} (\uparrow)$$

DISCUSSION Since the hinge at point b cannot transmit bending moment, the structure can be cut at that point to give the two free-body diagrams. The reactive forces have been shown acting in the positive coordinate directions (the positive direction of M_a is selected using the right-hand rule). The equations of static equilibrium in the form of Eqs. (2.1a) and (2.1) are used for the right and left free-body diagrams, respectively. This selection gives the computed results with the minimum amount of influence from previously obtained forces. As a check, the beam as a whole must also satisfy the equations of equilibrium. Using the calculated reactions gives

$$\sum F_x = R_{ax} - 45 = 45 - 45 = 0$$
$$\sum F_y = R_{ay} - 70 - 60 + R_c = 100 - 70 - 60 + 30 = 0$$
$$\sum M_c = 14R_{ay} - M_a - 70(10) - 60(3)$$
$$= 14(100) - 520 - 700 - 180 = 0$$

Example 2.3 Calculate the reactions at points a and c for this three-hinge arch.

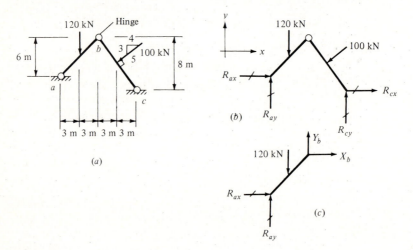

Figure E2.3

SOLUTION From Fig. E2.3b using the whole structure as a free-body diagram gives

$$\sum M_c = 0 \curvearrowleft : \qquad -100(5) - 120(9) + 2R_{ax} + 12R_{ay} = 0$$
$$R_{ax} + 6R_{ay} = 790 \qquad\qquad (1)$$
$$\sum F_x = 0 \xrightarrow{+} : \qquad\qquad R_{ax} + R_{cx} = 80 \qquad\qquad (2)$$
$$\sum F_y = 0 + \uparrow : \qquad\qquad R_{ay} + R_{cy} = 180 \qquad\qquad (3)$$

From Fig. E2.3c using the free-body diagram of member ab gives

$$\sum M_b = 0 \ \curvearrowright : \qquad\qquad -120(3) + 6R_{ay} - 6R_{ax} = 0$$

$$R_{ax} - R_{ay} = -60 \qquad\qquad (4)$$

Solving Eqs. (1) and (4) gives

$$R_{ax} = 61.4 \text{ kN } (\leftrightarrow) \qquad R_{ay} = 121.4 \text{ kN } (\updownarrow)$$

Substituting these results into Eqs. (2) and (3) gives

$$R_{cx} = 18.6 \text{ kN } (\leftrightarrow) \qquad R_{cy} = 58.6 \text{ kN } (\updownarrow)$$

As a check, the free-body diagram of member bc can be used and moments summed about point b to give

$$100(5) - 8R_{cx} - 6R_{cy} = 500 - 148.8 - 351.6 \approx 0$$

DISCUSSION Since the hinge at point b cannot transmit bending moment, the free-body diagram in Fig. E2.3c will have only two components of force at that point. The three equations of static equilibrium for the whole structure are augmented with the equation obtained by taking moments about point b for the free-body diagram for the left portion of the three-hinge arch. The four reactive forces have been assumed to act in the positive coordinate directions, and all results are positive, indicating that the assumed and actual directions coincide. Any direction could have been assumed for these reactions at the beginning of the calculations, and the signs of the results would have revealed the actual directions. The values of neither Y_b nor X_b were computed, but they could have been obtained using Fig. E2.3c. The calculation check performed by using the right free-body diagram (not shown) provides an independent method for writing the equilibrium equations. The result is only approximately zero since only one decimal place was used in calculating the answers.

2.6 CLASSIFICATION OF STRUCTURAL BEHAVIOR

Before performing the calculations for a structure, it is prudent to visualize how the loads are transmitted to the supports. This can be done using the free-body diagrams for the structural system. The engineer must determine whether the reactions can be uniquely established using only the equations of static equilibrium and whether the structure can transmit the loads to the supports without experiencing a catastrophic collapse or a significant change in geometry.

Overall structural behavior can be assessed by examining the reactive forces plus the effect of any hinges, links, and/or rollers. For example, the planar structure in Fig. 2.12a has three independent unknown reactive forces, as shown in the free-body diagram. Since three equations of static equilibrium are available, the reactions can be defined uniquely. Consequently, this is termed a *statically determinate* structure. If the beam is supported as shown in Fig. 2.12b, there is one additional reactive force. The three equations of static equilibrium must be augmented with an equation of deformation in order to establish the reactive forces. This structure is *statically indeterminate with one redundant reaction;* i.e., the redundancy of the structure is the number of independent reaction forces minus the number of equations of static equilibrium. For

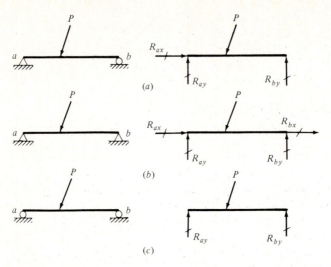

Figure 2.12 Loaded beam and free-body diagram: (*a*) statically determinate, (*b*) statically indeterminate, (*c*) mechanism.

the beam supported as shown in Fig. 2.12*c* there is no reactive force to equilibrate the horizontal component of the applied force; consequently the beam is free to translate horizontally and is a simple *mechanism;* i.e., the structure is statically *unstable,* having insufficient supports to ensure static equilibrium.

This assessment procedure can be expanded to include cases where equations of condition are introduced by construction details. The structure in Fig. 2.13*a* has a total of five unknown reactive forces. Note that the term "force" has been expanded to imply *generalized force,* which, in this case includes the bending moment at point *a* (this designation will be used in all subsequent chapters). There are three equations of static equilibrium for the structure, and the roller introduces two additional equations of condition; i.e., the moment and horizontal force are both zero at point *b*. Since there are five unknowns and five independent equations, the reactions can be uniquely calculated and the structure can be classified as statically determinate.

The structure in Fig. 2.13*b* has six unknown reactive forces, as shown in the free-body diagram. Three equations of static equilibrium can be written for the structure as a whole, and the roller introduces two additional equations of condition. Therefore, the number of unknowns is 1 greater than the number of equations, and the beam is classified as statically indeterminate with one redundancy.

If the structure is supported as shown in Fig. 2.13*c,* there is no reactive force to prevent rotation of the two segments about their respective supports; i.e., the structure is a mechanism. This can also be verified by noting that there are four unknown reactive forces and five equations.

Thus, the behavior of structures can be classified according to their reactions by comparing the number of unknown reactive forces and the number of equations. The criteria alluded to previously can be summarized as follows. If

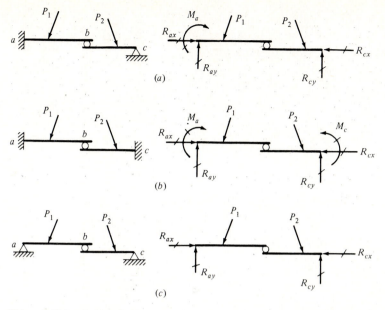

Figure 2.13 Loaded beam with a roller and free-body diagram: (*a*) statically determinate, (*b*) statically indeterminate, (*c*) mechanism.

$$r \begin{cases} = 3 + c & \text{structure is statically determinate} \\ > 3 + c & \text{structure is statically indeterminate} \\ < 3 + c & \text{structure is a mechanism (unstable)} \end{cases} \tag{2.2}$$

where r is the number of independent unknown reactive forces and c is the number of equations of condition introduced by the details of construction as discussed in Sec. 2.5.

Equations (2.2) are *necessary* conditions for satisfying the equations of static equilibrium, but even though these conditions are met, it is not *sufficient* to state that the structure is in a state of equilibrium. The beams in Fig. 2.14 both have three reactive forces, and by the test of Eq. (2.2) they are statically determinate. Further examination reveals that neither beam can support its loading. Since the beam of Fig. 2.14*a* has no reaction to equilibrate the horizontal component of the applied loading, the structure will translate horizontally if the illustrated load is transmitted to the beam. The beam in Fig. 2.14*b* has the supports arranged so that the lines of action of all three reactions intersect at point *o*, as shown on the free-body diagram. The structure is free to rotate about this point, as can be verified by summing moments about point *o* and noting that the force P produces a moment equal to Pd. This moment tends to rotate the entire structure counterclockwise about point *o*. These are examples of structures that are *geometrically unstable;* a condition resulting from the arrangement of the supports. Note that these geometrically unstable structures can be stable under certain loading conditions; i.e., the beam in Fig. 2.14*a* can support vertical loads and

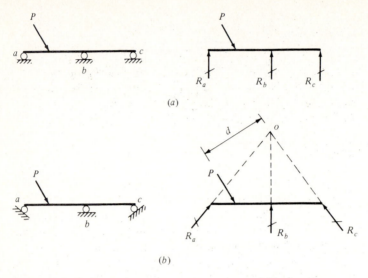

Figure 2.14 Statically determinate beams that are geometrically unstable.

the beam of Fig. 2.14b will support loads that pass through point o. Example 2.4 demonstrates the classification of six structures, and Example 2.5 illustrates the calculations for a statically determinate frame.

Example 2.4 Classify each of the structures illustrated.

SOLUTION Each structure must first be investigated with respect to Eq. (2.2) and then inspected for geometric instability.

In Fig. E2.4a since $r = 4$ and $c = 1$, Eq. (2.2) gives $r = 3 + c$ and the structure is statically determinate.

In Fig. E2.4b since $r = 4$; and $c = 1$, Eq. (2.2) gives $r = 3 + c$ and the structure should be classified as statically determinate. Further inspection reveals that there is nothing to resist a horizontally applied force on member ab, and the structure is geometrically unstable.

In Fig. E2.4c since $r = 3$ and $c = 0$, Eq. (2.2) gives $r = 3 + c$ and the structure is statically determinate.

In Fig. E2.4d since $r = 3$ and $c = 0$, Eq. (2.2) gives $r = 3 + c$ and the structure should be classified as statically determinate. Note that the reactive forces all meet at point c, indicating that the structure is geometrically unstable, since the application of any force that does not pass through point c will cause the structure to rotate about that point.

In Fig. E2.4e since $r = 4$ and $c = 0$, Eq. (2.2) gives $r > 3 + c$ and the structure is statically indeterminate. Even though the reactive forces all meet at point c, there is a moment at point c to resist rotation about that point.

In Fig. E2.4f since $r = 4$ and $c = 1$, Eq. (2.2) gives $r = 3 + c$ and the structure is statically determinate. Again, as in the previous structure, there is a moment at point c to resist rotation about that point even though the three reactive forces intersect at c. Since the support at a has both a horizontal and vertical component, the instability demonstrated by member ab of the structure in Fig. E2.4b does not exist here.

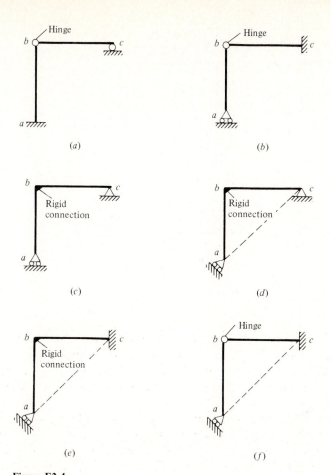

Figure E2.4

Example 2.5 Calculate the reactions at points a and c for this structure.

SOLUTION From Fig. E2.5b

$$\sum F_x = 0 \xrightarrow{+} : \qquad\qquad R_{ax} + 30 = 0$$
$$R_{ax} = -30 \text{ kips } (\leftrightarrow)$$

$$\sum F_y = 0 +\uparrow : \qquad\qquad R_{ay} + R_c = 40 \qquad\qquad (1)$$

$$\sum M_a = 0 \curvearrowright : \qquad 40(6) + 30(12) - 12R_c - M_a = 0$$
$$12R_c + M_a = 600 \qquad\qquad (2)$$

From Fig. E2.5d

$$\sum M_b = 0 \curvearrowright : \qquad\qquad 40(6) - 12R_c = 0$$
$$R_c = 20 \text{ kips } (\updownarrow)$$

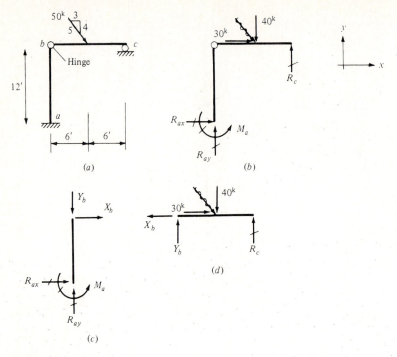

Figure E2.5

Substituting into Eqs. (1) and (2) gives

$$R_{ay} = 20 \text{ kips } (\updownarrow) \qquad M_a = 360 \text{ kip} \cdot \text{ft } (\curvearrowright)$$

The other two equations of static equilibrium applied to Fig. E2.5*d* give

$$Y_b = 20 \text{ kips } (\uparrow) \qquad X_b = 30 \text{ kips } (\leftarrow)$$

but directions reversed in Fig. E2.5*c*. As a check on the results, using Fig. E2.5*c* gives

$$\sum F_x = R_{ax} + X_b = -30 + 30 = 0$$

$$\sum F_y = R_{ay} - Y_b = 20 - 20 = 0$$

$$\sum M_b = 12R_{ax} + M_a = 12(-30) + 360 = 0$$

DISCUSSION This structure was classified in the previous example as statically determinate. First the three equations of static equilibrium are written for the whole structure, and they give the value for R_{ax} and the relationships between R_{ay}, R_c, and M_a. Summing moments about point *b* using the free-body diagram for member *bc* establishes the value for R_c, and the other two reactive forces can be calculated. The check on the numerical accuracy of the results is carried out using the free-body diagram for member *ab*. The free-body diagrams for members *ab* and *bc* could also have been used to calculate the reactions.

PROBLEMS

2.1 Calculate the reactions for the illustrated beams.

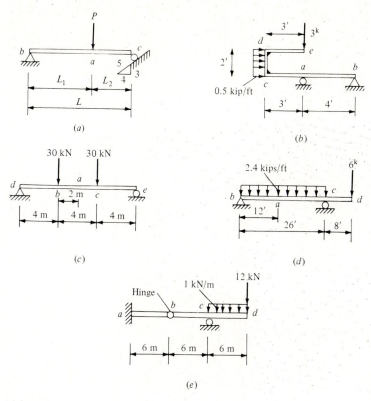

(a)

(b)

(c)

(d)

Hinge

(e)

Figure P2.1

2.2 Calculate the reactions at *a* and *d* for the girder. Note that the top beams are not connected midway between points *bc* and *ef* or at point *d*. The 20-kip load is applied to the beam to the right.

2.3 Calculate the location of the right support such that the reactions at points *a* and *b* will be equal.

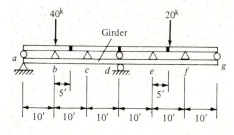

Figure P2.2

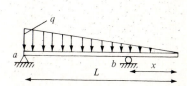

Figure P2.3

2.4 Classify the behavior of each of these structures.

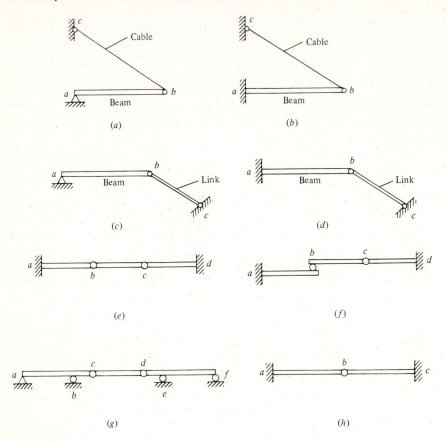

Figure P2.4

ADDITIONAL READING

Beer, F. P., and E. R. Johnston, Jr.: *Vector Mechanics for Engineers: Statics,* 2d ed., chap. 4, McGraw-Hill, New York, 1972.

Higdon, A., W. B. Stiles, A. W. Davis, and C. R. Evces: *Engineering Mechanics, Second Vector Edition,* vol. I, *Statics,* chap. 4, Prentice-Hall, Englewood Cliffs, N. J., 1976.

Malvern, L. E.: *Engineering Mechanics,* vol. I, *Statics,* chap. 4, Prentice-Hall, Englewood Cliffs, N. J., 1976.

Meriam, J. L.: *Engineering Mechanics,* vol. 1, *Statics, SI Version,* chap. 3, Wiley, New York, 1980.

THREE

ANALYSIS OF TRUSSES USING STATICS

3.1 DESCRIPTION OF BEHAVIOR

In Chap. 2 the external equilibrium of structural systems was investigated on the assumption that the body transmitting the forces was capable of doing so without collapsing. In this chapter we shall investigate the state of *internal equilibrium* for truss structures. Trusses are composed of members (typically steel, timber, or concrete) joined only at their ends and arranged to give efficient transfer of forces to the supports. This type of structural system is used extensively for bridges, roofs, building floors, and lateral-load resistance for high-rise structures. Typical applications are shown in Fig. 3.1.

Steel offshore oil-drilling towers, domes, and space trusses are examples of structures that can be designed to function as three-dimensional truss structures. In these cases the structure consists of a composite of trusses in many planes, and the structural integrity depends on the three-dimensional interaction of the members (for a discussion of three-dimensional trusses see Chap. 8). Since the truss bridge in Fig. 3.2 does not depend upon any three-dimensional truss interaction for the transfer of forces, the path of the loads can be synthesized using two-dimensional considerations. For clarity the deck of the bridge has not been shown, but the forces induced by the vehicles will be transmitted from the deck to the stringers. The reactions of the stringers are reversed and applied as loads on the floor beams; in turn, the reactions of the floor beams are reversed and applied to the trusses at the joints. Thus each of the two trusses can be investigated as a planar structure.

Although there is disparity in a truss between actual and ideal behavior, the design and fabrication are usually carried out to minimize such differences. Generally, it is assumed that

1. All members function as *two-force members;* i.e., they are axially loaded and transmit the forces in either tension or compression.
2. The members are connected with *smooth frictionless pins.*
3. The *lines of action* of all members at the pin connections *meet at a point.*
4. All *loads* are *applied at* the pin *connections.*

Figure 3.1 (*a*) Peyton Bridge, Jackson County, Oregon. (*American Institute of Steel Construction.*) (*b*) Staggered truss system. (*United States Steel.*)

If all these conditions are met, the *primary stresses* will be introduced by either axial tension or compression. Any spurious stresses resulting from bending moments and/or shear (forces normal to the member axis) are termed *secondary stresses*. Assumptions 1 and 4 imply that the members are weightless, but if the engineer must include this effect, half the weight is lumped to each end of the member. A structure is usually analyzed for dead loads and live loads separately. The member weights are not included

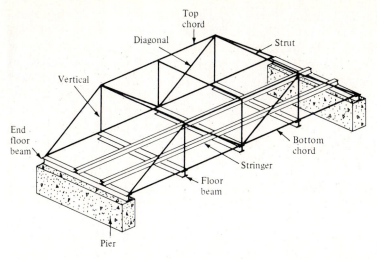

Figure 3.2 Structural arrangement for a parallel-chord through a bridge truss; all secondary bracing in the planes of the upper and lower chords is not shown; deck and portal bracing not shown for clarity.

in a live-load analysis, but they can be lumped into the joint loads in a dead-load analysis if the member weights are small compared with the other dead loads.

The pin connections are termed the *panel points* or *joints* of the truss. A typical construction detail for the joint of the upper chord of a steel truss is shown in Fig. 3.3. The members are designed so that the centroidal axes intersect at a common point. They are welded, bolted, or riveted so that the connections will transmit primarily axial forces with negligible bending moments, resulting in a concurrent force system. The illustrated connection, which does not conform strictly to assumption 2 for ideal truss behavior, introduces secondary stresses. If the connections are properly designed and fabricated, the secondary stresses are usually small compared with the primary stresses and can be ignored. Secondary stresses are discussed in Sec. 12.10, but in this book it will generally be assumed that they are nonexistent.

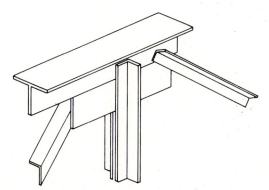

Figure 3.3 Typical joint for the upper chord of a truss for welded or bolted construction in steel.

The arrangement of truss members is critical to formation of a stable configuration. For example, the arrangement in Fig. 3.4a will collapse inward if a downward force is placed at point d, but the triangle in Fig. 3.4b forms a stable structure. For this triangle, member ab fixes point a relative to point b, and point c is fixed relative to points a and b by members ac and bc, respectively. Thus, point c is fixed relative to both points a and b to yield a configuration that will not collapse when loaded. The triangular arrangement is commonly called a *rigid truss*, and the unstable four-member structure in Fig. 3.4a is *nonrigid*. Actually, if the members are all deformable, the points will move relative to each other by small amounts because of the material deformations under the effect of the loading. These deformations will be ignored in this chapter.

A new rigid truss is formed by starting with a basic rigid triangle and adding two members and an additional joint (the two added members must not be parallel). The truss in Fig. 3.4c is formed by first adding members ad and cd to the triangle abc. Since points a and c are initially fixed relative to each other, the addition of members ad and cd fixes point d relative to points a and c, respectively. Point d is fixed relative to points a and c, and by extending the reasoning used for the original triangle it can be established that point d is also fixed relative to point b. Similarly, a larger rigid truss extending to point e is obtained by adding members de and ce. Configurations that can be formed by consecutively adding two members and an additional joint to a rigid truss are termed *simple trusses*. They are not always composed of triangular shapes. The simple truss in Fig. 3.4d originated with triangle abc, and joints d, e, and f were added in consecutive order.

A *compound truss* consists of two or more simple trusses joined so that relative motion of the components is constrained. Three conditions of connectivity are required to preclude relative translation and rotation. A pin prohibits relative translation in two mutually perpendicular directions, and a truss member prevents displacement along its

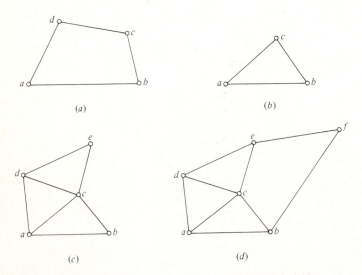

(a)

(b)

(c)

(d)

Figure 3.4 Examples of simple trusses: (a) nonrigid. (b) to (d) rigid.

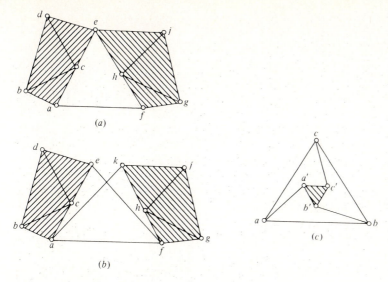

Figure 3.5 Examples of compound trusses.

axis. For example, the pin at *e* of Fig. 3.5*a* prevents relative horizontal and vertical translation of the two simple trusses (shown shaded), and member *af* prohibits relative rotation about the pin. Note that it is impossible to envision the formation of this truss with the same reasoning used for a simple truss.

Alternatively, simple trusses can be connected using three truss members, provided the members are not all parallel. This is demonstrated in Fig. 3.5*b*, where simple trusses *abcde* and *fghjk* are connected with the three members *ak, ef,* and *af.* Members *af* and *ef* fix point *f* relative to points *a* and *e*, respectively. In the absence of member *ak* the simple truss could rotate about joint *f*, and joint *k* would be free to move relative to joints *a* and *e*. However, member *ak* fixes joint *k* relative to joint *a*. Since joint *k* is fixed relative to joint *f*, and *f* in turn is fixed relative to joint *e*, it follows that joint *k* is also fixed relative to joint *e*. This three-member connection yields a rigid compound truss. The two simple trusses *abc* and *a'b'c'* in Fig. 3.5*c* are similarly connected together by the three nonparallel members *aa'*, *bb'*, and *cc'* to form a compound truss.

3.2 MEMBER FORCES

One of the primary tasks associated with the analysis of trusses is to calculate the forces sustained by the individual members. Squire Whipple wrote the first book describing how to do this analytically and graphically in 1847.[†] Typically, the entire structure is first considered, and the reaction forces are calculated. For example, consider the loaded truss in Fig. 3.6*a*. The equations of static equilibrium for a planar structure, as

[†]A historical summary appears in Appendix B.

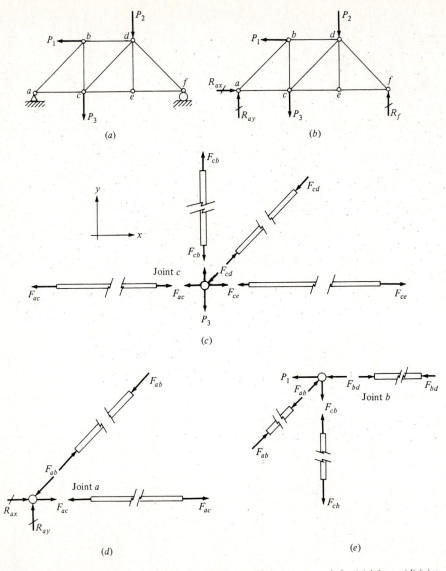

Figure 3.6 (*a*) Truss with applied loads; free-body diagram of (*b*) truss as a whole, (*c*) joint *c*, (*d*) joint *a*, and (*e*) joint *b*.

described in Chap. 2, are applied to the free-body diagram for the structure as a whole (Fig. 3.6*b*) to obtain the reaction forces R_{ax}, R_{ay}, and R_f.

If every part of the structure is in equilibrium, the entire truss must be in equilibrium. For example, if member *ac* of Fig. 3.6*c* is removed from the structure, it will be maintained in a state of equilibrium by the collinear forces F_{ac} applied at each end of the member, as shown (member forces have been sketched in the directions in which they actually act). The forces in the members connected at joint *c*, together with the applied force P_3, form a concurrent force system that must also be in equilibrium. By Newton's third law, the member force exerted upon the joint by the member acts in the direction opposite from the force exerted upon the member by the joint.

Two sign conventions are used to indicate the effect of the member forces. One designation, used to specify how a member is deformed, is the *strength-of-materials sign convention*. In Fig. 3.6c F_{ac} is tending to elongate the member, placing it in tension. This is designated as a positive axial force. We also note in Fig. 3.6c that members cb and ce are in tension while cd is in compression. A second, coordinate-based, sign convention generally used in writing the equations of static equilibrium is referred to as the *statical sign convention*. Thus, equilibrium of joint c in Fig. 3.6c would be expressed in the x and y coordinate directions. The relationship between these two sign conventions can be observed from Fig. 3.6c. That is, a force associated with a tensile member, when shown on the joint, is directed away from the joint; the force from a compressive member acts toward the joint.

Typically, the equations of static equilibrium are used to investigate the joints, and neither the direction nor magnitude of the member forces is known a priori. It is therefore necessary at the outset to indicate a direction for each of the member forces by making one of the following assumptions:

1. All members are in tension.
2. All member forces act in the direction perceived to be correct based upon one's knowledge of truss behavior, e.g., the bottom chord of a simply supported truss with downward vertical loading is in tension.
3. All forces applied to the joint act in the positive coordinate directions.
4. Forces act in various directions.

The assumed directions for the forces are initially indicated on the free-body diagrams. After the calculations have been carried out, the assumed direction of a force is correct if the algebraic sign of the result is positive and incorrect if the sign is negative. Throughout this book, all member forces will initially be assumed to be in tension, and the solutions will be interpreted accordingly. The reader is free to experiment with the various options available, but it is imperative to adopt a consistent procedure in order to avoid confusion and mistakes in the calculations.

The equations of static equilibrium are used to calculate the member forces. If each joint is individually isolated and analyzed, this procedure is the *method of joints*. The *method of sections* involves analyzing the free-body diagram of a portion of the truss to obtain the member forces. These two methods will be discussed individually in the next two sections.

3.3 THE METHOD OF JOINTS

The member forces in a statically determinate truss can be calculated using the equations of static equilibrium for each joint. Both the analytical and graphical forms of this method were described by Whipple. Since the member forces acting on a joint form a concurrent planar force system, two nontrivial independent equilibrium equations can be formulated at each joint. For simple trusses it is possible to calculate the member forces by investigating the joints in a sequence that gives only two unknowns at the joint under consideration. Alternatively, the equilibrium equations can be formulated for all the joints in the truss, yielding a set of equations to be solved for the member forces, but this approach is relatively inefficient.

For the truss in Fig. 3.6a, after the reaction forces have been calculated, the analysis for member forces can be started at joint a, since there are only two unknown member forces there. The free-body diagrams for both the members and the joint have been sketched in Fig. 3.6d, but generally it is necessary to work only with the joint. The calculations reveal that members ab and ac are in compression and tension, respectively; the forces have been drawn accordingly on both the members and the joint. The order in which the equilibrium equations are invoked determines how the member forces are to be numerically calculated. For example, at joint a, if we first apply the equation $\Sigma F_y = 0$, F_{ab} can be calculated independently of the unknown member force F_{ac}. However, if $\Sigma F_x = 0$ is used first, this equation will involve both F_{ab} and F_{ac}; therefore, $\Sigma F_y = 0$ must also be written and these equations solved simultaneously to obtain the two member forces. Thus, several complications are avoided by using the first approach.

In selecting the next joint to be investigated we observe that joint c has three unknowns; therefore, it is logical to proceed to joint b. Since F_{ab} has been calculated using joint a, there are only two unknowns at joint b and they can be obtained using the two equations of equilibrium. The analysis is completed by proceeding to joints c, d, e, and f in order. Forces F_{ef} and F_{df} are calculated before joint f is investigated; therefore, the analysis of this joint serves to check the numerical accuracy of these two member forces.

A number of frequently occurring *special conditions,* if recognized, can be used to shorten the calculations. For example, it can be concluded for the truss in Fig. 3.6a that $F_{ed} = 0$. This fact can be verified by constructing the free-body diagram for joint e and summing forces in the vertical direction. Since there are no forces applied to this joint, ed must be a zero-force member because it is the only force acting in the vertical direction. This same logic can be applied to the joint in Fig. 3.7a to conclude that $F_{ik} = 0$. By summing forces in the y direction for the joint in Fig. 3.7b we observe that $F_{ij} = 0$; furthermore, $F_{ik} = 0$ since there are no forces in the x direction. For the joint in Fig. 3.7c the forces in opposite members must be equal; that is, $F_{ij} = F_{il}$ and $F_{im} = F_{ik}$. This can be demonstrated by summing forces first in

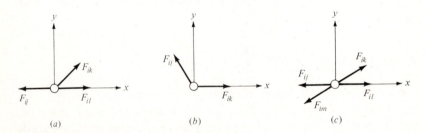

$$(a) \qquad\qquad (b) \qquad\qquad (c)$$

Figure 3.7 Special conditions.

the y direction and then in the x direction. A truss analysis using the method of joints for a horizontal Pratt truss is carried out in Example 3.1.

Example 3.1 Calculate the member forces F_{ab}, F_{ac}, F_{bd}, F_{bc}, F_{cd}, F_{ce}, F_{de}, and F_{df} using the method of joints.

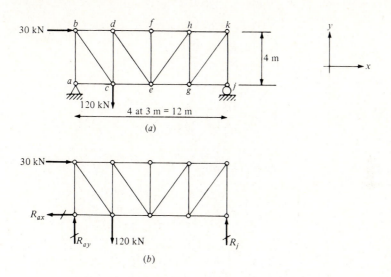

(a)

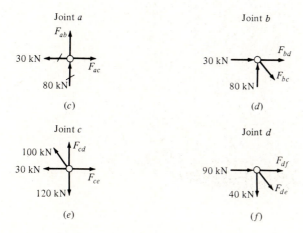

(b)

Joint a

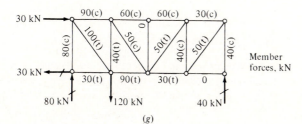

(c)

Joint b

(d)

Joint c

(e)

Joint d

(f)

(g)

Member forces, kN

Figure E3.1

SOLUTION From Fig. E3.1*b*

$$\sum M_a = 0 \; \curvearrowright : \qquad\qquad 120(3) + 30(4) - 12R_j = 0$$
$$R_j = 40 \text{ kN } (\uparrow)$$

$$\sum M_j = 0 \; \curvearrowright : \qquad\qquad 12R_{ay} - 120(9) + 30(4) = 0$$
$$R_{ay} = 80 \text{ kN } (\uparrow)$$

$$\sum F_x = 0 \; \xrightarrow{+} : \qquad\qquad R_{ax} = 30 \text{ kN } (\leftrightarrow)$$

From Fig. E3.1*c*

$$\sum F_x = 0 \; \xrightarrow{+} : \qquad\qquad F_{ac} = 30 \text{ kN (t)}$$
$$\sum F_y = 0 \; {+}\uparrow : \qquad\qquad F_{ab} = -80 \text{ kN (c)}$$

where (t) stands for tension and (c) for compression.
   From Fig. E3.1*d*

$$\sum F_y = 0 \; {+}\uparrow : \qquad\qquad -\tfrac{4}{5}F_{bc} + 80 = 0$$

$$F_{bc} = 100 \text{ kN (t)}$$

$$\sum F_x = 0 \; \xrightarrow{+} : \qquad F_{bd} = -\tfrac{3}{5}(100) - 30 = -90 \text{ kN (c)}$$

From Fig. E3.1*e*

$$\sum F_y = 0 \; {+}\uparrow : \qquad\qquad F_{cd} - 120 + \tfrac{4}{5}(100) = 0$$

$$F_{cd} = 40 \text{ kN (t)}$$

$$\sum F_x = 0 \; \xrightarrow{+} : \qquad\qquad F_{ce} - 30 - \tfrac{3}{5}(100) = 0$$

$$F_{ce} = 90 \text{ kN (t)}$$

From Fig. E3.1*f*

$$\sum F_y = 0 \; {+}\uparrow : \qquad\qquad -\tfrac{4}{5}F_{de} - 40 = 0$$

$$F_{de} = -50 \text{ kN (c)}$$

$$\sum F_x = 0 \; \xrightarrow{+} : \qquad\qquad F_{df} + 90 - \tfrac{3}{5}(50) = 0$$

$$F_{df} = -60 \text{ kN (c)}$$

DISCUSSION This truss is statically determinate, and the reactions are obtained by using the three equations of static equilibrium as shown. The member forces can be calculated next using the method of joints by starting at either joint *a* or *j* because each has only two unknowns.

Joint a is isolated by making an imaginary cut through members ab and ac, and the resulting free-body diagram (Fig. E3.1c) is shown acted upon by the calculated reactions R_{ax} and R_{ay} plus the two unknown member forces. Note that F_{ab} and F_{ac} have been assumed to act in such a direction that the members are in tension. This concurrent force system is analyzed using the only two available equations of statics, $\sum F_x = 0$ and $\sum F_y = 0$, to give the values shown for the member forces. F_{ac} comes out plus, which implies that the member is in tension, whereas F_{ab} is in compression since the algebraic sign of the force is negative; i.e., the assumed direction of the force is opposite the calculated direction.

It can be observed that there are two unknown member forces at joint b, as contrasted with joint c with its three unknowns; thus, joint b is investigated next. Cutting members ab, bc, and bd and applying the 30-kN force plus the member forces gives the illustrated free-body diagram of the joint (Fig. E3.1d). The member force for ab has been shown acting toward the joint since this is known to be the case from the calculations performed using joint a. In applying the equations of static equilibrium $\sum F_y = 0$ was used first since it yields F_{bc} independently from F_{bd}. Joints c and d are subsequently investigated to obtain the four other required member forces.

All the member forces are shown in Fig. E3.1g for readers who may want to continue with the investigation. Such a summary sketch with the results of an analysis showing the magnitudes and type (compressive or tensile) of the member forces is a convenient format for use in the design of the structure.

It is worth noting that some member forces can be obtained by the method of joints without proceeding through the truss joint by joint. For example, from observation of joint f it can be deduced that fe is a zero-force member by summing forces in the y direction.

Zero-force members, such as de in the truss of Fig. 3.6a, seem completely unnecessary, but there are good reasons why such members must exist:

1. Most structures are designed after analyzing a number of possible load combinations; hence, a zero-force member in one load case may be heavily loaded in another critical loading configuration.
2. In most truss analyses it is assumed that all members are weightless; thus, zero-force members can serve to maintain the structure in its intended shape under the effect of gravity loading.
3. Sometimes a long member must be stabilized laterally, and this can be accomplished by introducing an additional joint and a member at this point. In the truss in Fig. 3.6a the length of the tensile member between c and f has been reduced by the presence of panel point e, and member de is required in this case.
4. The structure could be rendered unstable if a zero-force member were omitted (stability of trusses will be discussed in Sec. 3.5).

Although the method of joints can be used to find all the member forces for any statically determinate simple truss by selectively proceeding from joint to joint, compound trusses are not amenable to solution by this method. For example, the analysis of the Fink roof truss in Fig. 3.8 can begin at joint a, and then joints b and c can subsequently be investigated. Since both joints d and e have three unknowns, the investigation must be stopped at this point unless the joint equilibrium equations are to be formulated and solved as a set of simultaneous linear algebraic equations.

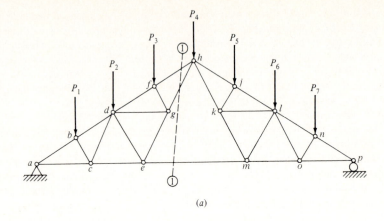

(a)

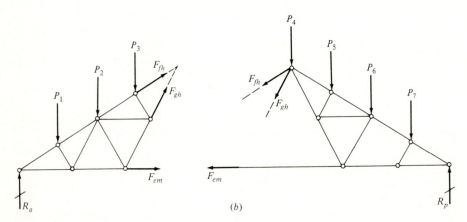

(b)

Figure 3.8 A Fink roof truss; (a) the loaded truss, (b) free-body diagrams obtained by cutting three members.

3.4 THE METHOD OF SECTIONS

An alternative analysis procedure for trusses is to consider an entire portion of the structure as a free-body diagram. Generally, for a given section through the truss, three (or fewer) members are cut; if they do not have a point of concurrency, three independent equations of static equilibrium are available to calculate these member forces. This approach, known as the *method of sections,* is attributed to both J. W. Schwedler and A. Ritter, German engineers. The method of sections differs from the method of joints in that it is not necessary to start at one of the supports and proceed through the truss until all members of interest have been investigated. Also, the method of sections is extremely useful for investigating compound trusses, where connecting forces must be found between the simple trusses that make up the compound structure.

The Fink truss in Fig. 3.8a is a compound truss composed of the two simple trusses aeh and hmp, connected by the point at joint h and by member em. After the reactions have been computed by considering the truss as a whole, the member forces in the vicinity of the connections can be obtained by making an imaginary cut through the structure at section 1-1 giving the two free-body diagrams shown in Fig. 3.8b. All the member forces have been assumed initially to be in tension and are shown accordingly in Fig. 3.8b. In this case, since the three members that have been cut do not have a point of concurrency, these forces can be calculated by using the three equations of static equilibrium.

It is generally most convenient to use the free-body diagram with the fewest forces (in this case the one on the left). It is most advantageous to calculate the member forces by choosing the equations of statics in the form of Eq. (2.1b), i.e., by summing moments about three different points. In this way, each unknown member force can be calculated independently of the other two unknown forces. For example, if we take moments about point h to find F_{em}, the calculations will involve neither F_{fh} nor F_{gh} since the line of action for both of these forces passes through the moment center. Similarly, F_{fh} and F_{gh} can be obtained by summing moments about joints e and a, respectively. Since any point can be used as the moment center, it is immaterial whether the point selected lies on or off the free-body diagram being investigated, e.g., point h. Also, by using only the moment form of the equilibrium equations it is possible to anticipate the sign of some of the member forces. Hence, by summing moments about point a to fine F_{gh} we observe that since the forces P_1, P_2, and P_3 all tend to rotate the free-body diagram clockwise about the point, member F_{gh} must be in tension in order to balance these moments. Without knowing the magnitudes of the reaction and the applied forces it is difficult to apply this simplistic approach to the other two members since the reactions and the applied forces have the tendency to rotate the free-body diagram in opposite directions. Calculations for the method of sections are demonstrated for a simple truss in Example 3.2 and for a compound truss in Example 3.3.

Example 3.2 Calculate the member forces F_{df}, F_{de} and F_{ce} for the truss of Example 3.1 using the method of sections.

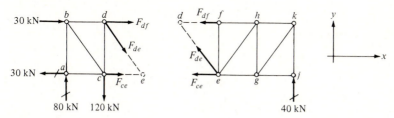

Figure E3.2

SOLUTION From the right-hand free-body diagram

$$\sum M_e = 0 \curvearrowright:$$

$$-4F_{df} - 6(40) = 0$$

$$F_{df} = -60 \text{ kN (c)}$$

$$\sum M_d = 0 \curvearrowright : \qquad\qquad 4F_{ce} - 9(40) = 0$$

$$F_{ce} = 90 \text{ kN (t)}$$

$$\sum F_y = 0 + \uparrow : \qquad\qquad \tfrac{4}{5}F_{de} + 40 = 0$$

$$F_{de} = -50 \text{ kN (c)}$$

DISCUSSION: After making an imaginary cut through members df, de, and ce, the right-hand free-body diagram was selected for use since it has the fewest forces to consider. In the method of sections it is usually advisable to consider the form in which the three equations of static equilibrium are to be expressed because a judicious choice will usually minimize the calculations. For example, if we choose joint e as the moment center, neither F_{de} nor F_{ce} will be involved in the calculations for finding F_{df}. Similarly, using point d as the moment center for calculating F_{ce} will result in a single equation with one unknown. Point d can be used even though this joint is not physically on the free-body diagram being considered since it is the intersection of the forces F_{df} and F_{de}. Finally, the force in member de can be found independently of the other two member forces by summing forces in the y direction.

Example 3.3 Calculate the member forces F_{bd}, F_{ac}, F_{bc}, F_{ce}, and F_{cd} for this Fink truss using the method of sections. Note that members bc and ef are perpendicular to the upper chord.

SOLUTION From Fig. E3.3b

$$\sum M_c = 0 \curvearrowright : \qquad\qquad 5(6.25) - 4(1.25) + 6.25(F_{bd})_y = 0$$

$$(F_{bd})_y = \frac{1}{\sqrt{5}}F_{bd} = -\frac{21}{5}$$

$$F_{bd} = -9.39 \text{ kips (c)}$$

$$\sum M_b = 0 \curvearrowright : \qquad\qquad 5(5) - 2.5F_{ac} = 0$$

$$F_{ac} = 10.00 \text{ kips (t)}$$

$$\sum M_a = 0 \curvearrowright : \qquad\qquad 4(5) + 2.5\sqrt{5}\, F_{bc} = 0$$

$$F_{bc} = -3.58 \text{ kips (c)}$$

From Fig. E3.3c

$$\sum M_d = 0 \curvearrowright : \qquad\qquad 5(10) - 4(5) - 5F_{ce} = 0$$

$$F_{ce} = 6.00 \text{ kips (t)}$$

$$\sum M_a = 0 \curvearrowright : \qquad\qquad 4(5) - 6.25(F_{cd})_y = 0$$

$$(F_{cd})_y = \tfrac{4}{5}F_{cd} = 3.20$$

$$F_{cd} = 4.00 \text{ kips (t)}$$

DISCUSSION To conserve space the reactions are indicated on the sketch of the truss. The free-body diagram (Fig. E3.3b) is obtained by passing an imaginary cut through members bd, bc, and ac and using the portion of the truss to the left of this cut. If ingenuity is used

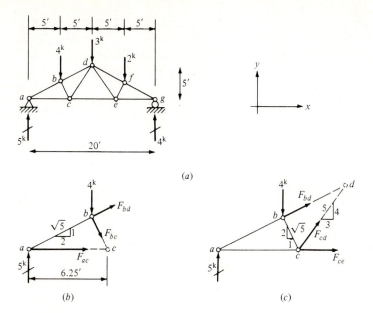

Figure E3.3

in writing the equations of statics, each of the required member forces can be found independently of all the others. Since forces F_{bc} and F_{ac} both pass through c, this point can be used to write the moment equation for calculating F_{bd}. Instead of computing the length of member bc, which is the lever arm for F_{bd}, the force F_{bd} is resolved into its x and y components at joint a. Since $(F_{bd})_x$ passes through the moment center, only $(F_{bd})_y$ is involved in the equation. Similarly, joints b and a were chosen to obtain F_{ac} and F_{bc}, respectively. Note that in computing F_{bc} the lever arm ab is known since members bc and ab are perpendicular to each other.

The free-body diagram in Fig. E3.3c results from cutting the truss through members bd, cd, and ce and using the part of the truss to the left of the cut. Point d is used to calculate F_{ce}, since both F_{bd} and F_{cd} pass through it. Note that in calculating F_{cd}, rather than computing the perpendicular distance from the moment center a to F_{cd}, the force is resolved into its x and y components at point c. Thus, the resulting equation includes only $(F_{cd})_y$ since $(F_{cd})_x$ passes through the moment center.

3.5 STATICAL DETERMINACY AND STABILITY

Since the emphasis in the two previous sections was on using the equations of static equilibrium to calculate the member forces in trusses, only statically determinate and stable trusses were presented, but before attempting to investigate a truss it is necessary to classify the behavior of the structure. Sometimes it is difficult to detect whether the structure is statically determinate, indeterminate, or a mechanism. In Sec. 3.1 it was pointed out that a rigid simple truss is formed by starting with a triangle consisting of three bars (members) and three joints and subsequently adding two bars and a joint to form another triangle contiguous with the first (with the provision that these two added bars not be collinear). The arrangement of members for the truss is continued by adding two members and one joint at a time until the final configuration is attained. For a truss with n joints it requires $n - 3$ additional joints to expand the structure beyond the initial three-joint triangle. Since each of these joints is formed by the intersection of two members, the total number of bars b required to form this simple rigid truss is

$$b = 2(n - 3) + 3 = 2n - 3 \qquad (3.1)$$

This is the minimum number of bars needed to form a rigid simple truss. More bars are unnecessary, and fewer bars will yield a mechanism.

This same logic can be extended to the kinematic behavior of a compound truss. For example, if two rigid simple trusses forming a compound truss have n_1 and n_2 joints, they must have b_1 and b_2 bars, respectively, where

$$b_1 = 2n_1 - 3 \qquad \text{and} \qquad b_2 = 2n_2 - 3$$

If the two simple trusses are joined by three additional members that are neither parallel nor concurrent to form a rigid compound truss, the resulting truss must have b members, such that

$$b = b_1 + b_2 + 3 = (2n_1 - 3) + (2n_2 - 3) + 3$$

$$= 2(n_1 + n_2) - 3 = 2n - 3$$

where the total number of joints in the compound truss n is equal to $n_1 + n_2$. Therefore, we observe that the criterion for the minimum number of bars required to form a rigid compound truss is identical to that for a rigid simple truss. This requirement for a rigid system in a plane is one of the fundamental theorems advanced by A. F. Möbius, mathematician and professor of astronomy at the University of Leipzig.

This criterion can be developed and extended by considering the equilibrium, instead of the topology, of the truss. At each joint of a truss there is a concurrent force system that is composed of the member forces plus any applied loads and reactions at the joint. Since the equation of moment equilibrium is trivially satisfied for a concurrent force system, the equations $\Sigma F_x = 0$ and $\Sigma F_y = 0$ ensure equilibrium at each joint of the truss. For a truss with n joints, each joint can be individually isolated and $2n$ independent equations of equilibrium can be written.† The total number of unknowns, which should simultaneously satisfy the $2n$ equations, is equal to the sum of the number of bars b and the number of independent reaction forces r. By comparing the number of unknowns with the number of equations using the ideas of linear algebra it is possible

---

†It may appear that more than $2n$ independent equations exist, since a typical truss analysis starts by considering the structure as a whole and writing the appropriate three equations of static equilibrium to calculate the reaction forces. This initial step does not yield additional independent equations beyond the $2n$ equations; however, the reactions are generally calculated first for convenience so that the member forces can be obtained one at a time by proceeding from joint to joint or using the appropriate free-body diagrams of various sections of the truss. If only the $2n$ joint equilibrium equations are formulated, they must be used to calculate the reaction forces, as well as the member forces; i.e., the set of linear joint equilibrium equations must be solved simultaneously. Furthermore, if the reactions satisfy the joint equilibrium equations, they must also satisfy equilibrium for the truss as a whole; therefore, we conclude that there are only $2n$ independent equations of static equilibrium that can be formulated for a truss.

to come to some conclusions about the structural behavior. If the number of unknowns is equal to the number of equations, the bar forces can be uniquely calculated and the structure is *statically determinate*. If the number of unknowns exceeds the number of equations, the bar forces cannot be determined independently unless additional equations of deformation are introduced and the structure is *statically indeterminate* (the analysis of statically indeterminate trusses is presented in Chap. 11). If the number of unknowns is less than the number of equations, the system is a *mechanism* and will collapse when subjected to a general loading condition. Thus, trusses can be classified according to their behavior by comparing the number of unknowns (reactive and member forces) and the number of equations of static equilibrium. The criteria used for classification can be summarized as follows; if

$$b + r - 2n \begin{cases} = 0 & \text{truss is statically determinate} \\ > 0 & \text{truss is statically indeterminate} \\ < 0 & \text{truss is a mechanism} \end{cases} \tag{3.2}$$

Since three independent support forces are required to prevent rigid-body motions in a plane, that is $r = 3$, the first criterion is identical to that described in Eq. (3.1).

These criteria include the degree of indeterminacy of both the internal (the number of members) and the external (the number of reactions) forces, but they do not ensure that the structure will be stable. That is, the first two criteria in Eq. (3.2), in which $b + r - 2n$ is greater than or equal to zero, are *necessary* conditions for establishing that a structure is statically indeterminate or determinate, respectively; however, even when these conditions are met, they are *not sufficient* to state that the structure is stable. For example, using Eq. (3.2), each of the trusses shown in Fig. 3.9 should be classified as either statically determinate or indeterminate. The truss in Fig. 3.9a

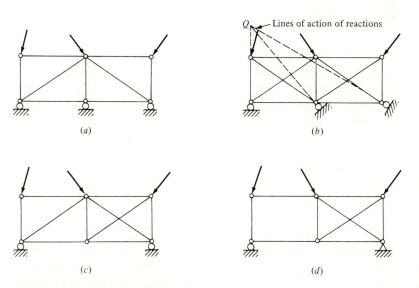

Figure 3.9 Examples of geometrically unstable trusses.

has $b + r - 2n = 0$, but there is no reaction to equilibrate the horizontal component of the applied loading; hence, the structure will translate horizontally if the illustrated load is transmitted to the truss. From the criteria of Eq. (3.2) the truss in Fig. 3.9b is statically indeterminate to the second degree (i.e., it has two redundant members), but the supports have a point of concurrency at point Q, which implies that the structure is free to rotate about this point (this can be verified by summing moments about this point of rotation). Also, the number of unknown reactions and member forces is equal to the number of equilibrium equations for the truss in Fig. 3.9c, but the structure will translate horizontally because there is no reaction component to resist the forces in this direction. The truss in Fig. 3.9d should also be classified as statically determinate according to the criteria of Eq. (3.2), but the truss would collapse under the loading because of the internal unstable arrangement of members in the left panel, i.e., the two parallel members cannot resist vertical forces. These are all examples of trusses that are *geometrically unstable,* a condition resulting from the arrangement of the supports and/or the members. From these counterexamples to the above criteria, we can state that if a structure is judged to be either statically determinate or indeterminate using Eq. (3.2), it is *geometrically stable* if

> There are three or more reactions that are neither parallel nor concurrent and the bars in the truss are arranged in an adequate manner, i.e., such that internal instability will not occur.

The trusses in Fig. 3.10 are presented to illustrate the use of the criteria for classifying trusses. The fan truss in Fig. 3.10a and the modified scissors truss in Fig. 3.10b are both examples of statically determinate roof trusses, and Eq. (3.2) confirms this fact. Note that the former is a simple truss, while the latter is a compound truss consisting of two simple trusses joined together with a pin at the peak and a horizontal tie member. The rhomboid truss in Fig. 3.10c is classified as indeterminate with one internal redundancy. Examples of this truss situated in a vertical orientation can be observed as the external bracing on high-rise structures to resist the lateral forces induced by earthquake and wind loads. It can be established that this is a simple truss by starting with the triangle abc and sequentially locating joints d, f, e, etc., by adding two members at a time. If one of the two end vertical members were omitted, the structure would be statically determinate and stable. The truss shown in Fig. 3.10d was patented by Albert Fink in 1851 in the United States. This compound truss is statically determinate and stable in spite of the fact that it has no bottom chord, a characteristic of this truss. The simple truss of Fig. 3.10e is statically indeterminate to the second degree (externally) because of the two reactions in excess of the three required to prevent rigid-body translation and rotation.

Equations of condition can be introduced into trusses just as they are in beams and frames by virtue of conditions of design and construction. For example, the truss in Fig. 3.11a is composed of two rigid simple trusses (shaded) connected together with a hinge, and this condition, along with the proper arrangement of the supports, results in a stable structure. If an imaginary cut is made through the structure at the hinge, the additional equation introduced is that the sum of the moments about the hinge equals zero. For the structure as a whole the three equations of static equilibrium relating the reactions can be augmented with this additional equation of condition. Consequently,

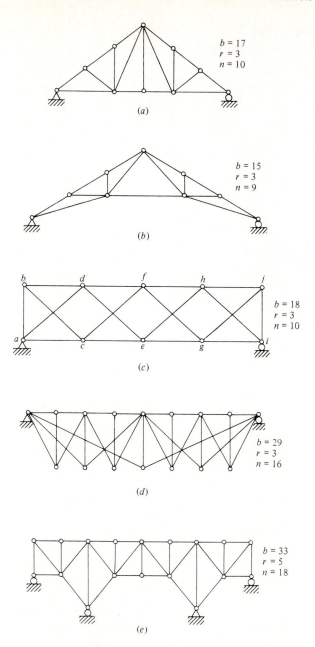

$b = 17$
$r = 3$
$n = 10$

(a)

$b = 15$
$r = 3$
$n = 9$

(b)

$b = 18$
$r = 3$
$n = 10$

(c)

$b = 29$
$r = 3$
$n = 16$

(d)

$b = 33$
$r = 5$
$n = 18$

(e)

Figure 3.10 Examples for classification.

the structure is statically determinate in spite of the fact that there are four reactions. Note that if the number of unknowns were compared with the number of equations using Eq. (3.2), the structure would also be classified as statically determinate. Similarly, the structure of Fig. 3.11b is statically determinate by the criteria of

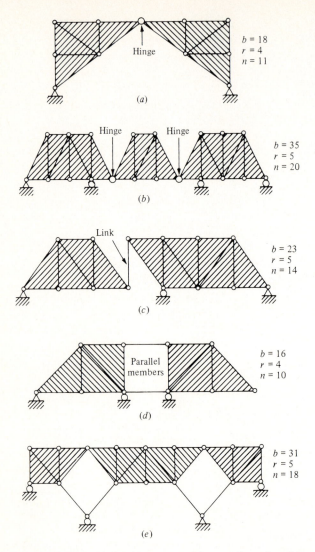

Figure 3.11 Classification of trusses involving equations of condition.

Eq. (3.2) even though there are five reactions. In this case the two hinges joining the three simple trusses allow two equations of condition to be formulated; they can be used in conjunction with the three equations of equilibrium for the structure as a whole to calculate the reactions. The link connecting the two simple trusses in Fig. 3.11c introduces two equations of condition; i.e., the direction and point of application of the connecting force between the two component simple trusses is known. Alternatively, the number of equations of condition can be deduced by summing moments about both ends of the link (two equations of condition can similarly be introduced into a structure by connecting two trusses with a roller). Thus, for the truss of Fig. 3.11c the three equations of static equilibrium for the structure as a whole are augmented by the two equations of condition, and the reactions can be calculated at the onset of the analysis. The presence of two parallel members with no diagonal members in the panel (Fig. 3.11d) introduces an equation of condition since forces perpendicular to these members cannot be transmitted through the panel. In general, this type of arrangement of members yields a truss that is internally unstable, but when used in conjunction with the proper support conditions, the resulting structure is stable. Thus, for the truss in Fig. 3.11d, this one equation of condition can be used with the three equations of equilibrium obtained for the truss as a whole to calculate the four reactions. The truss in Fig. 3.11e is a Wichert truss; its members are arranged in the same configuration as those in the truss of Fig. 3.10e with the vertical members over the supports omitted. The two intermediate reactions must be directed along the axes of the two truss members framing into the support points; therefore, by considering these four members as support reactions, this truss can be observed to be composed of three simple trusses joined together with two hinges and supported with the four two-force members plus the two end reactions. Either using Eq. (3.2) or noting that the five reactions can be calculated from the three equations of equilibrium plus the two equations of condition introduced by the two hinges, we can deduce that the structure is statically determinate. There is no apparent unstable arrangement of either the members or the reactions for the Wichert truss, but when computing the member forces, the values may be determinate, inconsistent, infinite, or indeterminate. In general, the Wichert truss is unstable by virtue of the geometry of the structure; therefore, the structure is categorized as being geometrically unstable. This instability, which is introduced by the arrangement of the members, can sometimes be visually detected, but it will become evident when computing the reactions.†

The configuration of some trusses can be categorized as neither simple nor compound. For example, the truss in Fig. 3.12a is a simple truss, as can be demonstrated by starting with the triangle abc and subsequently expanding the structure by adding two members and one joint at a time to locate joints e, d, f, and g in that order. If member eb is removed and replaced by member ag (see Fig. 3.12b), a different rigid truss is created that is neither simple nor compound but *complex*. The truss in Fig. 3.12b can be analyzed using the method of joints with the method of sections, but the sections and sequence of joints analyzed must be chosen discretely to minimize calculations. Consider the truss in Fig. 3.13, where $b = 9$, $r = 3$, and $n = 6$; thus, the structure is statically determinate by Eq. (3.2) and contains no apparent geometrically unstable features. Therefore, it should be

†Text in small type, like that which follows, is optional and may be omitted without loss of continuity.

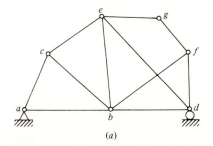

(a)

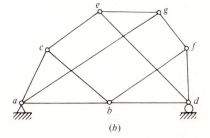

(b)

Figure 3.12 (a) Simple truss, (b) complex truss.

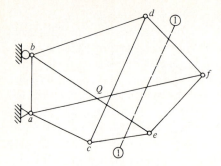

Figure 3.13 A complex truss.

possible to write 12 equations of equilibrium and solve them simultaneously to obtain the member forces and the reactions. Note that if the three reactions are first obtained, the number of simultaneous equations to be solved is reduced to 9.

An alternative to this approach can be posed by using a combination of the method of joints and the method of sections. For example, if the member force F_{ce} is denoted as X, by using the method of joints at joint c both F_{ac} and F_{cd} can be calculated in terms of X. Also, by applying the method of joints to joint d, F_{bd} and F_{df} can be found as a function of X. Now by cutting through the truss at section 1-1 and summing moments about point Q (the intersection of members af and be), using the right free-body diagram, one equation is obtained. This equation will be in terms of the unknown applied forces and X; therefore, X can be explicitly calculated. Also, in turn, F_{bd}, F_{df}, F_{cd}, and F_{ac} can be obtained, and the remaining member forces can be obtained by using the method of joints for joints a, b, f, and e.

Several special methods have been developed for solving complex trusses. Otto Mohr suggested using the method of virtual displacements for solving this type of structure, and further refinements were described by N. E. Joukowski and H. Müller-Breslau (the method of virtual work is described in Chap. 10 for calculating displacements, but it will not be discussed for the analysis of member forces in statically determinate trusses). A method of replacement members was developed by L. Henneberg.

Since the solution of 12 equations using the digital computer is a relatively simple task, the structural engineer must make the choice of using a special method of analysis for investigating complex trusses or treating them like simple and compound trusses. It is interesting to note that the truss in Fig. 3.13 is geometrically unstable for certain arrangement of the members, e.g., when the outside members form an equilateral hexagon, and this instability can be detected by examining the coefficient matrix obtained from the equations of equilibrium for the joints. In this case, it appears logical to use the digital computer to solve the structure since this singularity can easily be detected, thus avoiding the problem of finding a set of member forces to be inconsistent after carrying out long calculations by hand.

PROBLEMS

3.1 Determine whether each of these trusses is simple, compound, or complex.

3.2 Determine whether each of these trusses is statically determinate, indeterminate, a mechanism, or stable. Indicate the number of redundants for statically indeterminate trusses; if the structure is unstable, designate the reason, e.g., geometrically unstable because of support arrangement.

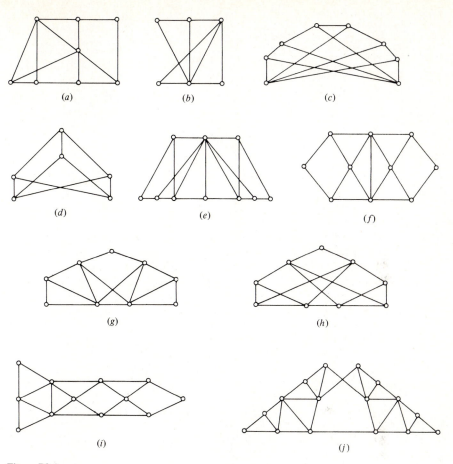

Figure P3.1

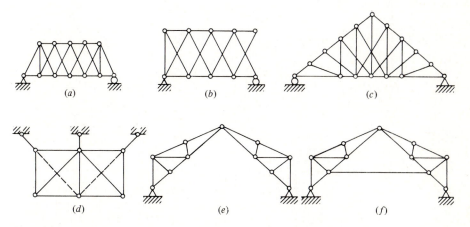

Figure P3.2

92 THE STATICS OF STRUCTURES

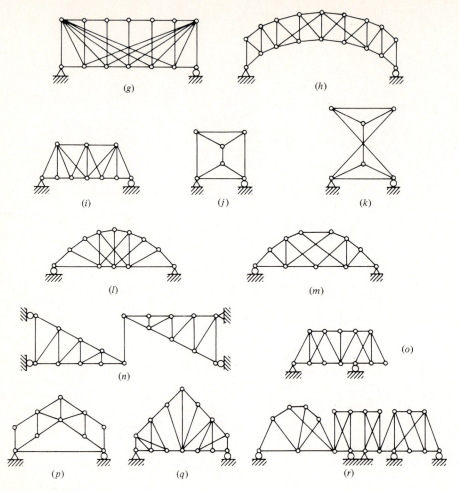

Figure P3.2 Continued

3.3 Calculate the forces in the marked members using the method of joints.

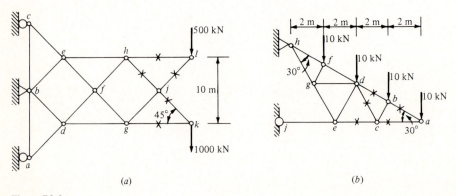

Figure P3.3

3.4 Calculate the ratio P_1/P_2 given that the force in member gj is zero for $P_2 \neq 0$.

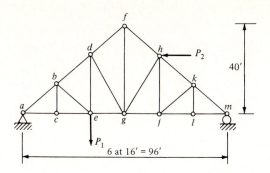

Figure P3.4

3.5–3.20 Calculate the forces in all members of these trusses.

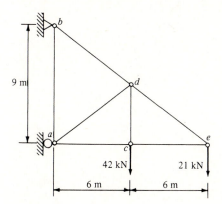

Figure P3.5

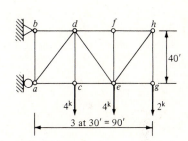

Figure P3.6

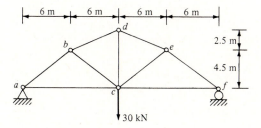

Figure P3.7

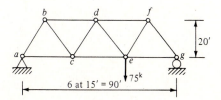

Figure P3.8

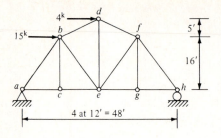

Figure P3.9

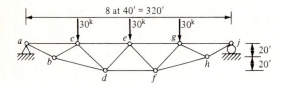

Figure P3.10

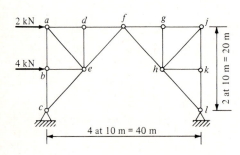

Figure P3.11

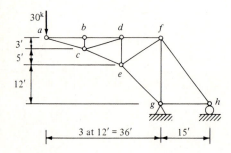

Figure P3.12

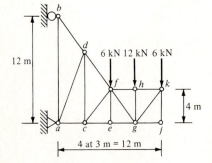

Figure P3.13

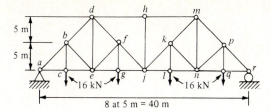

Figure P3.14

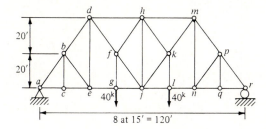

Figure P3.15

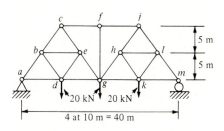

Figure P3.16

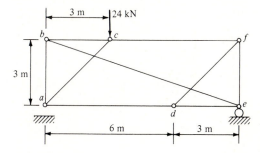

Figure P3.17

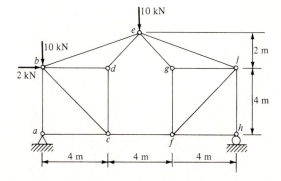

Figure P3.18

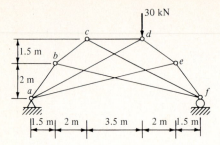

Figure P3.19

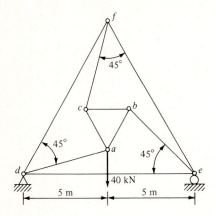

Figure P3.20 Triangles *abc* and *def* are equilateral.

3.21–3.24 Calculate the reactions for each of these truss structures.

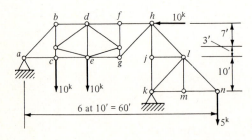

Figure P3.21

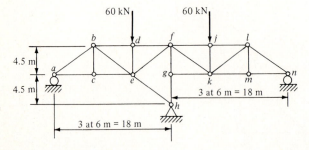

Figure P3.22

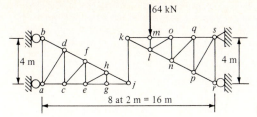

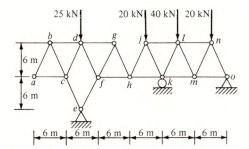

Figure P3.23

Figure P3.24 Triangles are isosceles.

3.25 Calculate the forces in all members of these complex trusses.

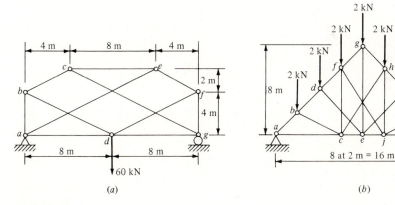

(a)   (b)

Figure P3.25

ADDITIONAL READING

Beer, F. P., and E. R. Johnston, Jr.: *Vector Mechanics for Engineers: Statics,* 2d ed., chap. 6, McGraw-Hill, New York, 1972.

Higdon, A., W. B. Stiles, A. W. Davis, and C. R. Evces: *Engineering Mechanics, Second Vector Edition,* vol. I, *Statics,* chap. 4, Prentice-Hall, Englewood Cliffs, N.J., 1976.

Malvern, L. E.: *Engineering Mechanics,* vol. I, *Statics,* chap. 5, Prentice-Hall, Englewood Cliffs, N.J., 1976.

Meriam, J. L.: *Engineering Mechanics,* vol. 1, *Statics, SI Version,* chap. 4, Wiley, New York, 1980.
Norris, C. H., J. B. Wilbur, and S. Utku: *Elementary Structural Analysis,* 3d ed., chap. 4, McGraw-Hill, New York, 1976.
Timoshenko, S. P., and D. H. Young: *Theory of Structures,* 2d ed., chap. 2, McGraw-Hill, New York, 1965.

FOUR

ANALYSIS OF STATICALLY DETERMINATE BEAMS

4.1 DESCRIPTION OF BEHAVIOR

A slender member more than 4 to 5 times as long as either of its cross-sectional dimensions and subjected to transverse loading acts as a beam. Unlike the truss members discussed in Chap. 3, beams are capable of withstanding axial forces, bending moments, and/or torsional moments. Each of these responses is independent if the member behavior is linear, i.e., if deflections and rotations are small. The equilibrium requirements for statically determinate beams will be described in this chapter.

The horizontal supporting members commonly used in buildings and bridges are examples of beams, and they are typically designated according to their position and/or function. Components of the primary structure are referred to as floor beams, girders, joists, spandrel beams, and stringers, whereas girts, headers, lintels, purlins, and rafters are considered to be secondary structural beams. Descriptions of various flexural members are given throughout the text where appropriate (e.g., Figs. 1.5 and 1.6). An example of beams incorporated into a high-rise structure is shown in Fig. 4.1.

Beams used in most buildings and bridges are constructed from steel, wood, or a combination of concrete and steel. Beams are produced using a number of grades of structural steel, together with an appropriate process (cold forming, hot forming, cold rolling, or hot rolling). The W and *tee* sections, as designated by the American Institute of Steel Construction,[1] shown in Fig. 4.2 are two standard cross sections. A structure constructed using a unique type of steel beam is illustrated in Fig. 4.3.

Wood beams are made primarily from Douglas fir (*Pseudotsuga menziesii*), western hemlock (*Tsuga heterophylla*), western larch (*Larix occidentalis*), and various pines (*Pinus*), but other species of structural grade lumber are also permitted by the *National Design Specification*.[2] Solid-sawn beams are commonly used for light construction, rectangular glued-laminated sections are fabricated for longer spans, and various innovative manufactured cross sections such as an I configuration are becoming increasingly popular as the supply of high-quality solid timber diminishes (see Fig. 4.4). An example of a structure with wood beam construction is shown in Fig. 4.5.

Figure 4.1 Seattle First Bank, Seattle, Washington. (*American Institute of Steel Construction.*)

The standard reinforced concrete section in Fig. 4.6*a* uses carbon steel to resist the tensile stresses that typically occur in the extreme bottom fibers of a singly reinforced beam. The U-shaped steel is incorporated to withstand the shear stresses and prevent diagonal cracking in the concrete. Design criteria for reinforced concrete beams are specified by the American Concrete Institute.[3] The prestressed concrete section illus-

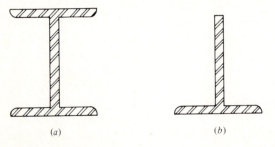

Figure 4.2 Typical hot-rolled steel beam cross sections: (*a*) W shape (AISC designation), (*b*) structural tee.

Figure 4.3 Castellated steel beams, Grange Mutual Parking Structure. (*Bethlehem Steel Corporation*.)

trated in Fig. 4.6*b* is typical of the shapes used for highway bridges. The high-strength steel strands are given an initial tensile load to prevent the concrete from being stressed in tension, and the result is an effective beam. The Prestressed Concrete Institute designates design methods for these beams.[4] Generally, concrete sections are formed to the optimal shape for resisting the applied loads, and the steel is placed to sustain the tensile stresses. Thus, the cross section may be rectangular, circular, tee-shaped, or almost any other configuration. An example of a prestressed concrete structure is shown in Fig. 4.7. Although materials such as aluminum, titanium, and fiber-reinforced composites are also used to construct beams, their use is generally limited to specialized applications such as those in the airframe and automobile industries. An example of a composite structural component is shown in Fig. 4.8.

(*a*)

(*b*)

Figure 4.4 Typical wood beam cross sections: (*a*) solid sawn, (*b*) glued laminated.

Figure 4.5 Wood beams. (*Western Wood Products Association.*)

4.2 IDEALIZED BEAM-SUPPORT CONDITIONS

Various support details are constructed for beams, but for purposes of analysis they can usually be idealized as a *roller,* a *hinge,* or a *clamped condition.* Some specific details are described in this section to illustrate how they are constructed and to show the appropriate symbols designating idealized behavior.

The detail shown in Fig. 4.9 for a steel beam resting on a concrete foundation will transmit forces in a fashion similar to that of the idealized roller support shown. The anchor bolt resists forces in the vertical direction; in some cases, the anchor bolt is omitted, and the reaction is incapable of resisting upward forces. Unless specified to the contrary, the roller symbol will imply that forces can be resisted in both directions normal to the surface upon which it acts. Used in conjunction with the bearing pad, the slotted hole in the beam minimizes forces transmitted parallel to the concrete

(*a*)

(*b*)

Figure 4.6 Typical concrete beam cross sections: (*a*) reinforced, (*b*) prestressed.

Figure 4.7 The Cowlitz River prestressed concrete bridge. (*ABAM Engineers, Inc., Tacoma, Washington.*)

surface. Low-friction pads made from tetrafluoroethylene (coefficient of friction from 0.03 to 0.07) ensure low sliding resistance. Bearing pads can also be made from elastomers, laminated fabric, laminated fabric-rubber, asbestos-cement, tempered hardboard, heavy felt, lead, and plastics, depending upon the magnitude of the bearing

Figure 4.8 Fabricating a graphite-epoxy composite panel. (*Boeing Commercial Airplane Co.*)

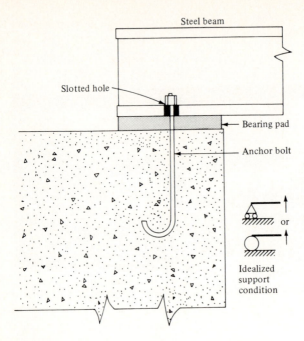

Figure 4.9 Typical construction details for a roller support.

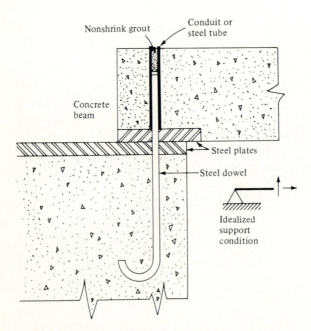

Figure 4.10 Doweled connection for precast concrete beam.

stress. For some applications the bearing pad is omitted altogether and the beam placed directly on the concrete. Bearing details that approximate an idealized roller also exist for beams made from reinforced concrete, prestressed concrete, and wood.

The hinge support can resist forces in all directions, but moments cannot be transmitted. The structure is free to rotate about the pin, and the reaction force passes through the pin at an angle that depends upon the direction of the applied forces. Figures 4.10 and 4.11 illustrate construction details for two different types of hinge support. The doweled connection used with precast concrete construction transmits forces to the concrete foundation: horizontally through shear in the dowel and vertically through the grouting. We can observe, however, a certain amount of resistance to rotation, which should not be present in idealized behavior. Since the moment that can be transmitted by this support is minimal, it is usually ignored. The detail shown in Fig. 4.11 for a wood arch also acts like a hinge support and appears to support no moment. Inevitably the bearing surface and the bridge pin will rust, impairing the rotational capacity of the connection; the slight moment resistance this represents is usually ignored by the structural analyst. Various configurations of the hinge support are constructed for steel, concrete, and wood beams; thus, the details can differ greatly.

An idealized clamped or fully fixed support condition is capable of resisting both forces and moments with neither translational nor rotational deformations. Figure 4.12

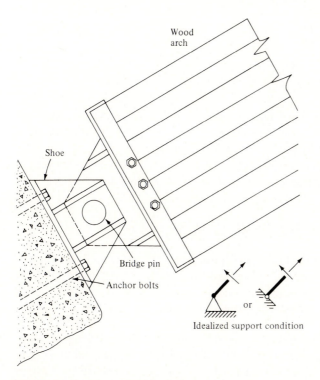

Figure 4.11 True hinge anchorage for large wood arches.

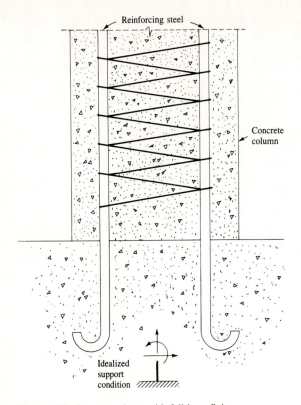

Figure 4.12 Concrete column with full base fixity.

shows a typical construction detail that approximates this behavior. The vertical reinforcing steel transmits both horizontal and vertical forces, as well as bending moments
to the support. The clamped-support condition is also attained in a variety of other ways
using common construction materials. Figure 4.13 shows two constructed support
details, one of which can be idealized as a roller and the other as a hinge support.

4.3 EQUILIBRIUM OF INTERNAL FORCES

A loaded beam in equilibrium transfers the applied forces and moments to the supports
through several independent modes of behavior. An imaginary cut through the beam
perpendicular to the longitudinal axis reveals an internal force and moment which resist
the applied loads and keep the beam segment in equilibrium. In Fig. 4.14 the three
orthogonal components of the force and the moment are each shown acting in the
positive coordinate directions. Note that the x coordinate is directed along the longitudinal axis and the yz plane coincides with the plane of the imaginary cut. The
moments can be designated using either a double-headed vector (as shown) or a curved
arrow with the understanding that the right-hand rule must be used to relate these two
symbols. The axial force is denoted as F, the shear forces consist of V_y and V_z, and the

(b)

(a)

Figure 4.13 (a) Bridge support detail approximating a roller; (b) hinge support detail (*ABAM Engineers, Inc., Tacoma, Washington*).

three moment components are made up of the torsional moment M_x plus the two bending moments M_y and M_z.

Two sign conventions, similar to those in truss analysis, must be used to specify the internal forces and moments in a beam. That is, a coordinate-based statical sign convention is observed in writing the equations of static equilibrium. The forces and moments are shown acting in their positive directions in Fig. 4.14. In addition, a strength-of-materials sign convention is used to indicate how the member is deformed, and to do so for a beam it is necessary to examine the internal forces. The two-force member in Fig. 4.15a will be used to illustrate. If an imaginary cut is made through

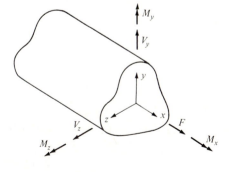

Figure 4.14 Internal forces and moments on a section of a beam.

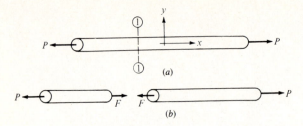

Figure 4.15 (a) A two-force member, (b) free-body diagrams created by an imaginary cut at section 1-1.

the member at section 1-1, two segments are created and each must be in equilibrium. According to Newton's third law, the force F on the right free-body diagram is equal in magnitude but opposite in direction to F on the left free-body diagram, and the former acts in the negative coordinate direction. Note, however, that both internal forces F tend to elongate the member and place it in tension. This is a positive axial force, defined intuitively in Chap. 3. The deformations induced by the other internal forces and moments in a beam acting on a cut segment can be analyzed and described in a similar manner. A positive shear force is defined to be one that deforms the beam segment as shown in Fig. 4.16b. That is, the shear force on the cut face forms a clockwise couple with the resultant of all the applied forces. A positive bending moment is one that compresses the top fibers, places the bottom fibers in tension (see Fig. 4.16c), and distorts the beam into a concave-upward curve; i.e., the beam assumes the shape associated with a positive second derivative, or a smiling mouth.

The deformation (or strength-of-materials) sign convention described, though not the only one that exists for the resisting forces and moments, is compatible with the various methods presented throughout this book (see, for example, Secs. 9.6 and 9.7). The convention is probably the most widely adopted by structural engineers and is popular in introductory texts on the mechanics of solids. The reader is cautioned, however, that it is not used universally by all authors or by engineers in all countries.

All beams discussed in this chapter bend only in the plane of the applied forces. For this to occur, two conditions must be met: (1) the shear center and the centroid of

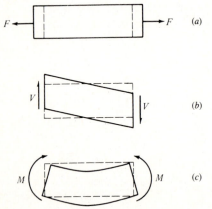

Figure 4.16 (a) Positive axial force, (b) positive shear force, (c) positive bending moment; deformation, or strength-of-materials, convention.

the cross section must coincide. The shear center is that point through which transverse forces must act in order to avoid twisting the beam. (2) The loads and reactions must all lie in the plane which contains the centroidal axis of the member and one of its principal axes for every cross section (the principal axes coincide with an orientation such that the product of inertia is zero). For a discussion of these concepts see, for example, Ref. 5. The calculations for obtaining the internal resisting forces and bending moment at a point in a beam are presented in Example 4.1

Example 4.1 Calculate the internal resisting axial force, shear force and bending moment at section 1-1 for this simply supported beam.

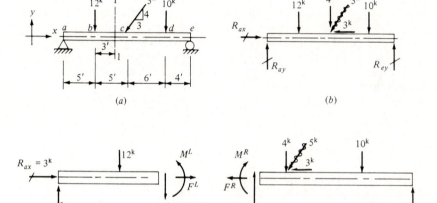

Figure E4.1

SOLUTION From Fig. E4.1*b*

$$\sum M_a = 0 \curvearrowleft: \qquad 20R_{ey} - 10(16) - 4(10) - 12(5) = 0 \qquad R_{ey} = 13 \text{ kips} (\uparrow)$$

$$\sum F_y = 0 +\uparrow: \qquad R_{ay} + R_{ey} - 12 - 4 - 10 = 0 \qquad R_{ay} = 13 \text{ kips} (\uparrow)$$

$$\sum F_x = 0 \xrightarrow{+}: \qquad R_{ax} - 3 = 0 \qquad R_{ax} = 3 \text{ kips} (\rightarrow)$$

From Fig. E4.1*c*

$$\sum F_y = 0 +\uparrow: \qquad 13 - 12 - V^L = 0 \qquad V_L = 1 \text{ kip}$$

$$\sum F_x = 0 \xrightarrow{+}: \qquad 3 + F^L = 0 \qquad F^L = -3 \text{ kips (c)}$$

$$\sum M_1 = 0 \curvearrowleft: \qquad M^L + 12(3) - 13(8) = 0 \qquad M^L = 68 \text{ kip} \cdot \text{ft}$$

From Fig. E4.1*d*

$$\sum F_y = 0 +\uparrow: \qquad V^R - 4 - 10 + 13 = 0 \qquad V^R = 1 \text{ kip}$$

$$\sum F_x = 0 \xrightarrow{+} : \qquad\qquad -F^R - 3 = 0 \qquad F^R = -3 \text{ kips (c)}$$

$$\sum M_1 = 0 \curvearrowleft : \qquad -M^R - 4(2) - 10(8) + 13(12) = 0 \qquad M^R = 68 \text{ kip} \cdot \text{ft}$$

DISCUSSION The reaction forces at points a and e are calculated using the entire beam. The equation of equilibrium involving the sum of the moments about a different point, such as e, could have been used but would have yielded an additional equation that is a linear combination of the other two. The resisting axial force, shear force, and bending moment at section 1-1 can be calculated using either the free-body diagram to the left or to the right of the imaginary cut. The segment yielding the simplest form of the equilibrium equations is generally used, but for purposes of illustration, the values of F, V, and M are calculated in this example using both free-body diagrams. Note that the superscripts L and R are used to denote that the quantity is associated with the left or right free-body diagram, respectively. All quantities are shown in the free-body diagrams acting in the positive direction, using the deformation sign convention (Fig. 4.16). Thus the algebraic sign of the various calculated quantities indicates how they should be plotted. As a general practice the reader is urged to make a judgment on the directions of the axial force, shear, and bending moment before any calculations and check the anticipated directions with the sign of the computed results. If these two do not coincide, the discrepancy should be reconciled. This approach is useful for finding calculation errors and also for developing an intuitive under-standing of beam behavior.

4.4 AXIAL-FORCE, SHEAR-FORCE, AND BENDING-MOMENT DIAGRAMS

The values of the axial force, shear force and bending moment are frequently plotted at all points along the beam for a full understanding of the behavior. Typically, the abscissa is drawn along the beam length, and the values of axial force, shear force, and bending moment are plotted as the ordinates on three separate diagrams, using the deformation sign convention (Fig. 4.16). It is important to identify all points where extrema and/or zero values occur. These diagrams are used to locate critical design values and to visualize how the beam resists the applied forces and moments.

Examples 4.2 and 4.3 illustrate construction of the diagrams. The axial-force diagram is presented in Example 4.2 but not in Example 4.3 since this quantity is zero throughout that beam. Because the axial force in many beams is zero, commonly only the shear and moment diagrams are constructed.

Example 4.2 Draw the axial-force, shear-force, and bending-moment diagrams for the beam of Example 4.1.

SOLUTION From Fig. E4.2b

$$F = -3 \text{ kips} \qquad V = +13 \text{ kips} \qquad M = +13x \text{ kip} \cdot \text{ft}$$

From Fig. E4.2c

$$F = -3 \text{ kips} \qquad V = -12 + 13 = +1 \text{ kip}$$

$$M = 13x - 12(x - 5) = (x + 60) \text{ kip} \cdot \text{ft}$$

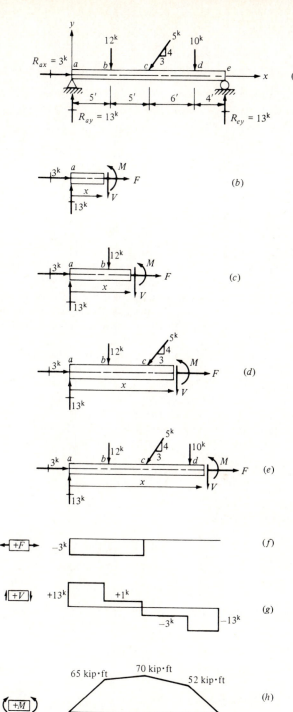

Figure E4.2 (f) Axial force, (g) shear force, (h) bending moment.

From Fig. E4.2d

$$F = -3 + 3 = 0 \text{ kip} \qquad V = -4 - 12 + 13 = -3 \text{ kips}$$

$$M = 13x - 12(x - 5) - 4(x - 10) = (-3x + 100) \text{ kip} \cdot \text{ft}$$

From Fig. E4.2e

$$F = -3 + 3 = 0 \text{ kip} \qquad V = -10 - 4 - 12 + 13 = -13 \text{ kips}$$

$$M = 13x - 12(x - 5) - 4(x - 10) - 10(x - 16) = (-13x + 260) \text{ kip} \cdot \text{ft}$$

Fig. E4.2h

$x = 5$ ft:	$M = 13(5) = 65 \text{ kip} \cdot \text{ft}$
$x = 10$ ft:	$M = 10 + 60 = 70 \text{ kip} \cdot \text{ft}$
$x = 16$ ft:	$M = -3(16) + 100 = 52 \text{ kip} \cdot \text{ft}$

DISCUSSION The reactions obtained in Example 4.1 and shown in Fig. E4.2a are used to calculate the axial force, shear force, and bending moment at all points of the beam. For expedience, the reaction forces are shown on the drawing of the beam instead of using a separate free-body diagram. All four free-body diagrams make use of the left segment of the beam after the cut has been made. Although this may not always be the most efficient method for performing the calculations, it should help to clarify how the equations change from point to point on the beam and to plot the diagrams.

The resisting forces and moment on the cut face of the free-body diagram are shown acting in the positive direction as defined by the deformation sign convention (Fig. 4.16). For convenience this convention is sketched on the left side of Fig. E4.2f to h. The three diagrams are plotted using the equations developed. Note that at the points where concentrated forces are applied there is a jump in the shear diagram equal to the magnitude of the force. The moment diagram is continuous at these locations, but the slope experiences a jump.

Example 4.3 Draw the shear and moment diagrams for the simply supported beam illustrated.

SOLUTION From Fig. E4.3a

$$\sum M_a = 0 \curvearrowright : \qquad -20(6)(6) + 120 + 15R_e = 0$$

$$R_e = +40 \text{ kN} (\uparrow)$$

$$\sum M_e = 0 \curvearrowright : \qquad 120 + 20(6)(9) - 15R_a = 0$$

$$R_a = +80 \text{ kN} (\uparrow)$$

From Fig. E4.3b

$$V = +80 \text{ kN} \qquad M = 80x \text{ kN} \cdot \text{m}$$

From Fig. E4.3c

$$V = +80 - 20(x - 3) = (+140 - 20x) \text{ kN}$$

$$M = 80x - \frac{20(x - 3)^2}{2} = (-10x^2 + 140x - 90) \text{ kN} \cdot \text{m}$$

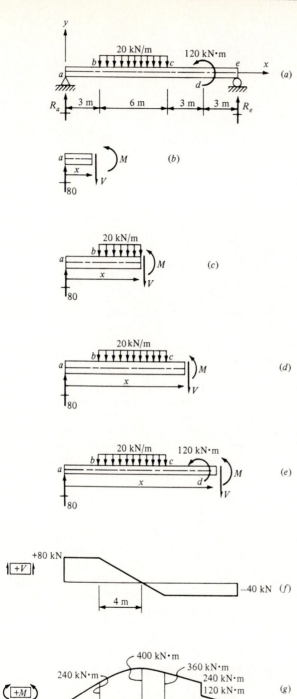

Figure E4.3 (*f*) Shear and (*g*) bending moments.

From Fig. E4.3*d*

$$V = +80 - 20(6) = -40 \text{ kN}$$

$$M = 80x - 20(6)(x - 6) = (720 - 40x) \text{ kN} \cdot \text{m}$$

From Fig. E4.3*e*

$$V = +80 - 20(6) = -40 \text{ kN}$$

$$M = 80x - 20(6)(x - 6) - 120 = (600 - 40x) \text{ kN} \cdot \text{m}$$

Fig. E4.3*f*

$x = 9 \text{ m}:$ $\qquad\qquad\qquad\qquad\qquad\qquad$ $V = -40 \text{ kN}$

Fig. E4.3*g*

$x = 3 \text{ m}:$ $\qquad\qquad\qquad\qquad$ $M = +240 \text{ kN} \cdot \text{m}$

$x = 9 \text{ m}:$ $\qquad$ $M = -10(9)^2 + 140(9) - 90 = +360 \text{ kN} \cdot \text{m}$

$x = 12^- \text{ m}:$ $\qquad$ $M = 720 - 40(12) = 240 \text{ kN} \cdot \text{m}$

$x = 12^+ \text{ m}:$ $\qquad$ $M = 600 - 40(12) = 120 \text{ kN} \cdot \text{m}$

DISCUSSION The reactions for this example are computed using the total beam, and for convenience these results are illustrated acting on the beam drawing. In this case moment equations were used, but invoking equilibrium in the y direction would simply give a linear combination of the two moment equations and would not impart any new information. The left segment of the cut beam has been used in all cases, and the resisting forces are shown in the positive direction according to the deformation sign convention (Fig. 4.16). Since both shear and moment are functions of the variable x, it is necessary to evaluate the appropriate expressions at critical points. The moment diagram is quadratic between points b and c, and we obtain the slope by differentiating the expression for moment with respect to x, which gives

$$\frac{dM}{dx} = -20x + 140$$

An extremal point is obtained by equating the derivative to zero; hence the moment diagram has a zero slope at the point $x = 7$. Evaluating the moment expression at this point yields the maximum moment of 400 kN $\cdot$ m.

Note that in the vicinity of the uniformly distributed load the slope of the shear diagram equals the value of the loading and the concentrated moment introduces a jump in the moment diagram that equals the magnitude of the couple; these facts are discussed in detail in Sec. 4.5.

4.5 DIFFERENTIAL EQUILIBRIUM RELATIONS

As an alternative to the method presented in Sec. 4.4 we can consider a small element of the beam as a free-body diagram. The equilibrium conditions, combined with a limiting process, give the differential equations relating the load, shear force, and bending moment. Integrating these differential equations of equilibrium gives a convenient method for calculating the shear forces and bending moments.

A distributed load $q(x)$ is applied over the length of the beam and directed in the positive y direction (Fig. 4.17a). The shear forces and bending moments are shown in Fig. 4.17b acting on the small beam element in their respective positive directions according to the deformation sign convention (Fig. 4.16). The applied load on the small element is $q(x^*) \Delta x$, where $x \le x^* \le x + \Delta x$; this force acts at a point $\beta \Delta x$ from the right face, $0 \le \beta \le 1$ (Fig. 4.17b). If there are neither concentrated forces nor concentrated couples in the interval $[x, x + \Delta x]$, the equations of equilibrium give

$$\sum F_y = V - (V + \Delta V) + q(x^*) \Delta x = 0$$

$$\sum M_O = -M + (M + \Delta M) - q(x^*) \Delta x \beta \Delta x - V \Delta x = 0 \qquad (4.1)$$

Rearranging terms gives

$$\frac{\Delta V}{\Delta x} = q(x^*) \qquad \frac{\Delta M}{\Delta x} = V + q(x^*) \beta \Delta x \qquad (4.2)$$

ΔV and ΔM decrease as Δx decreases, and in the limit as $\Delta x \to 0$, $q(x^*) \to q(x)$, $\beta \to 0$; therefore,

$$\lim_{\Delta x \to 0} \frac{\Delta V}{\Delta x} = \frac{dV}{dx} = q(x) \qquad (4.3)$$

and

$$\lim_{\Delta x \to 0} \frac{\Delta M}{\Delta x} = \frac{dM}{dx} = V \qquad (4.4)$$

Thus, these equations of differential equilibrium can be stated as follows:

If there are no concentrated forces or moment couples at a point on a beam, the slope of the shear diagram is equal to the applied load, and the slope of the bending moment diagram is equal to the shear force.

The use of these equations is illustrated for the uniformly loaded beam in Example 4.4.

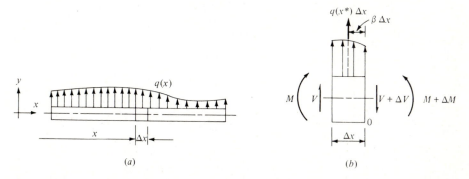

Figure 4.17 A beam with distributed load: (a) the loaded beam, (b) free-body diagram of a beam segment.

Example 4.4 Draw the shear and moment diagrams using the differential equations of equilibrium (4.3) and (4.4).

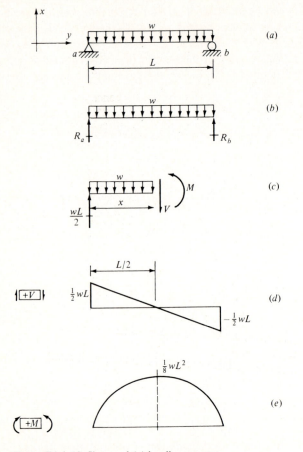

Figure E4.4 (*d*) Shear and (*e*) bending moment.

SOLUTION From Fig. E4.4*b*

$$\sum M_a = 0 \curvearrowright: \qquad R_b L - wL\frac{L}{2} = 0 \qquad R_b = \tfrac{1}{2}wL \;(\uparrow)$$

$$\sum F_y = 0 + \uparrow: \qquad R_a + R_b - wL = 0$$

$$R_a = wL - R_b = wL - \tfrac{1}{2}wL = \tfrac{1}{2}wL \;(\uparrow)$$

From Fig. E4.4*c*

$$\sum F_y = 0 + \uparrow: \qquad V = -wx + \tfrac{1}{2}wL$$

$$\sum M_x = 0 \curvearrowright: \qquad M = \tfrac{1}{2}wLx - \tfrac{1}{2}wx^2$$

DISCUSSION In this case, $q = -w$ since the uniformly distributed load is directed in the negative coordinate direction. From the reactions R_a and R_b and the free-body diagram in Fig. E4.4b, $V_a = +wL/2$. From Eq. (4.3) the shear diagram must have a constant slope equal to $-w$, and it must start with a value of V_a at point a, as shown in Fig. E4.4d. Since $R_b = |V_b|$, this is the correct value necessary to complete the diagram after decreasing the shear from point a to point b at the rate of $-w$. From Eq. (4.4) we conclude that the maximum value of the moment diagram will be attained at $x = L/2$ because the shear is zero at this point and the moments at the ends of the beam are zero.

Figure 4.18a illustrates a beam segment subjected to a concentrated load P_{ap} and a concentrated couple M_{ap} at two points a and b on the beam. The free-body diagram of a small element of the beam in the vicinity of P_{ap} is shown in Fig. 4.18b. The shears immediately to the left and right of $x = x_a$ are denoted V_{a-} and V_{a+}, respectively, and defined as

$$V_{a-} = \lim_{\varepsilon \to 0} V(x_a - \varepsilon) \qquad V_{a+} = \lim_{\varepsilon \to 0} V(x_a + \varepsilon)$$

Equilibrium of forces in the y direction gives the discontinuity or jump condition for the shear

$$V_{a+} = V_{a-} + P_{ap} \tag{4.3a}$$

We note that $M_{a-} = M_{a+}$.

The free-body diagram of a small element of the beam in the vicinity of M_{ap} is shown in Fig. 4.18c, where the respective moments immediately to the left and right of $x = x_b$ are

$$M_{b-} = \lim_{\varepsilon \to 0} M(x_b - \varepsilon) \qquad \text{and} \qquad M_{b+} = \lim_{\varepsilon \to 0} M(x_b + \varepsilon)$$

Moment equilibrium for this element gives the jump condition for the moment at this point

$$M_{b+} = M_{b-} + M_{ap} \tag{4.4a}$$

Equilibrium of forces in the y direction dictates that $V_{b-} = V_{b+}$. From Eqs. (4.3a) and (4.4a) these conditions can be stated as follows:

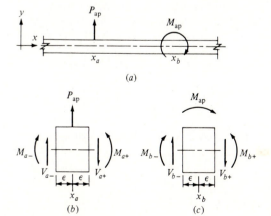

(a)

(b)

(c)

Figure 4.18 (a) Beam segment with concentrated force and couple; (b) free-body diagram at P_{ap}, (c) free-body diagram at M_{ap}.

In the presence of concentrated forces and moment couples, the jump (discontinuity) in the shear force is equal to the magnitude of the concentrated force or reaction, and the shear force is undefined at this point. The jump in the moment is equal to the magnitude of the concentrated couple or reaction, and the bending moment is undefined at this point.

From Eq. (4.4) we note that the extrema of the bending-moment diagram can occur (1) at the ends of the beam; (2) where the shear force is zero; or (3) where the shear force is undefined, i.e., at a point where the shear is discontinuous. Example 4.5 illustrates the use of the jump conditions of Eqs. (4.3a) and (4.4a).

Example 4.5 Draw the shear and moment diagrams for this simply supported beam.

SOLUTION From Fig. E4.5b

$$V = 9 \text{ kips} \qquad M = 9x \text{ kip} \cdot \text{ft}$$

From Fig. E4.5c

$$V = 9 - 21 = -12 \text{ kips}$$
$$M = 9x - 21(x - 10) = (-12x + 210) \text{ kip} \cdot \text{ft}$$

From Fig. E4.5d

$$V = 9 - 21 = -12 \text{ kips}$$
$$M = 9x - 21(x - 10) + 150 = (-12x + 360) \text{ kip} \cdot \text{ft}$$

DISCUSSION The reactions for this beam have been calculated using the entire structure as a free-body diagram and applying the equilibrium equations. The free-body diagrams in Fig. E4.5b to d show the resisting shear and moment acting in a positive direction according to the sign convention in Fig. 4.16. Regardless of whether the free-body diagram to the left or right of the imaginary cut is used, a positive resisting shear forms a clockwise couple with the resultant of the applied forces, and a positive moment induces compression in the top of the beam. The jump condition for shear at a concentrated load in Eq. (4.3a) is illustrated in the shear diagram at point b, and the jump condition for moment at a couple expressed in Eq. (4.4a) is shown in the moment diagram at point c. The slope of the shear diagram is zero, as dictated by Eq. (4.3), and the slope of the moment diagram is equal to the shear where there are no concentrated forces [see Eq. (4.4)].

Equations (4.3) and (4.4) can also be expressed in differential form as

$$dV = q \, dx \qquad (4.5)$$
$$dM = V \, dx \qquad (4.6)$$

If q and V are continuous functions on the interval $[x_i, x_j]$, where x_i and x_j are two points on the beam, then by the fundamental theorem of calculus

$$V_j - V_i = \int_{x_i}^{x_j} q \, dx \qquad (4.7)$$

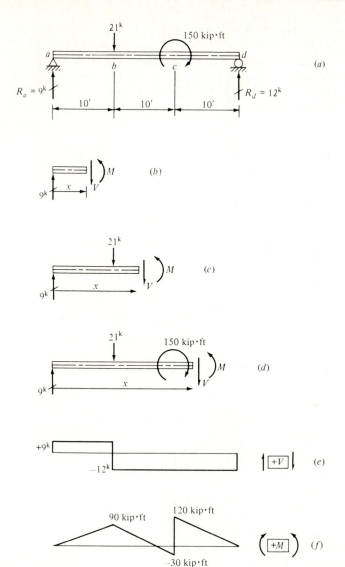

Figure E4.5 (e) Shear and (f) bending moment.

and

$$M_j - M_i = \int_{x_i}^{x_j} M \, dx \qquad (4.8)$$

where the integrals in Eqs. (4.7) and (4.8) represent the area under the loading diagram and the shear diagram, respectively, between points x_j and x_i. These two equations can be expressed as follows:

On the intervals between points of application of forces and concentrated couples, the change in shear equals the area under the loading diagram between the same two points, and the change in the bending moment equals the area under the shear diagram between the same two points.

Example 4.6 illustrates the application of these equations in constructing the shear and moment diagrams.

Example 4.6 Construct the shear and moment diagrams for this beam with a suspended span.

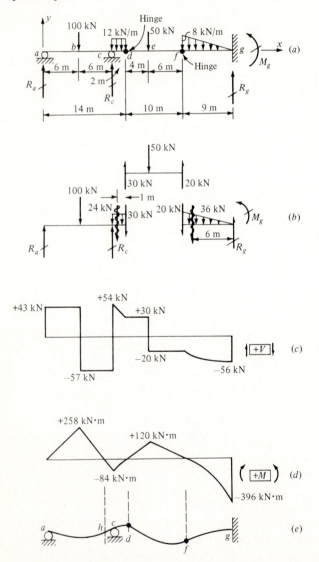

Figure E4.6 (c) Shear, (d) bending moment, and (e) deflected shape.

SOLUTION From Fig. E4.6a

$$\sum F_y = 0 + \uparrow : \qquad R_a + R_c + R_g = 100 + 24 + 50 + 36 = 210$$

From Fig. 4.6b, for segment fg

$$\sum F_y = 0 + \uparrow : \qquad -20 - 36 + R_g = 0 \qquad R_g = 56 \text{ kN} (\uparrow)$$

$$\sum M_g = 0 \curvearrowright : \qquad M_g = -20(9) - 36(6) = -396 \text{ kN} \cdot \text{m} \curvearrowright$$

and for segment ad

$$\sum M_c = 0 \curvearrowright : \qquad -12R_a + 100(6) - 24(1) - 30(2) = 0 \qquad R_a = 43 \text{ kN} (\uparrow)$$

$$\sum F_y = 0 + \uparrow : \qquad R_a + R_c - 100 - 24 - 30 = 0 \qquad R_c = 111 \text{ kN} (\uparrow)$$

Fig. E4.6d

$$M_a = 0$$

$$M_b = M_a + 6V_a = 0 + 6(43) = 258 \text{ kN} \cdot \text{m}$$

$$M_c = M_b + 6V_{b+} = 258 + 6(-57) = -84 \text{ kN} \cdot \text{m}$$

$$M_d = M_c + \tfrac{1}{2}(2)(V_{c+} + V_d) = -84 + \tfrac{1}{2}(2)(84) = 0 \text{ kN} \cdot \text{m}$$

$$M_e = M_d + 4V_d = 0 + 4(30) = 120 \text{ kN} \cdot \text{m}$$

$$M_f = M_e + 6V_{e+} = 120 + 6(-20) = 0 \text{ kN} \cdot \text{m}$$

$$M_g = M_f + 9(-20) + \tfrac{2}{3}(9)(-36) = -396 \text{ kN} \cdot \text{m}$$

DISCUSSION This structure has four unknown reaction components, but only two independent equations of statics are available for analyzing the total structure. If we isolate the span between points d and f, two conditional equations can be written and can be used to augment the equations of statics, to give a total of four equations for obtaining the reactions at points a, c, and g. The shear and moment diagrams are constructed using the integrated form of the differential equations of equilibrium shown in Eqs. (4.7) and (4.8) and the jump conditions in Eqs. (4.3a) and (4.4a). Thus, the shear at point a is equal to the value of R_a, and the values of the shear are obtained by combining the magnitudes of the applied forces with the value of the shear ordinates. For example, the shear just to the left of point b is equal to 43 kN, but just to the right of this point the diagram must decrease by the magnitude of the applied load. This gives an ordinate of -57 kN between points b and c. Also, the ordinate at point d is obtained by decreasing the ordinate at point c by the area under the loading diagram between these two points, i.e.,

$$V_d = 54 \text{ kN} - (12 \text{ kN/m})(2 \text{ m}) = 30 \text{ kN}$$

Similarly, the change in the moment diagram between two points is obtained by taking the area under the shear diagram between those points. For example,

$$M_c = M_b + (6 \text{ m})(-57 \text{ kN}) = 258 \text{ kN} \cdot \text{m} - 342 \text{ kN} \cdot \text{m} = -84 \text{ kN} \cdot \text{m}$$

The value of M_g was obtained during the initial step of calculating the reactions, but it could also have been determined using Eq. (4.8) as follows:

$$M_g - M_f = \int_0^9 [-20 - \tfrac{1}{2}(16 - \tfrac{8}{9}x)x]\,dx$$

$$= \left[-20x - 4x^2 + \frac{4}{9}\left(\frac{x^3}{3}\right) \right]_0^9$$

$$= -180 - 324 + 108 = -396 \text{ kN} \cdot \text{m}$$

But since M_f is zero, $M_g = -396$ kN · m.

The deflected shape for this beam can be sketched approximately by noting from the bending-moment diagram that the beam will have a positive curvature between points a and h and also between d and f. The beam must have no displacement at the supports and it cannot rotate at g. The hinges at points d and f imply that there is no slope continuity; thus, the slope of the beam will be discontinuous across them. In the absence of the deflection calculations it is impossible to ascertain whether point d moves up or down from the undeflected position (see Sec. 4.6).

4.6 APPROXIMATE DEFLECTED SHAPES

It is useful to attempt to visualize the general deflected shape of a loaded structure. The detailed calculation of deflections will be discussed in Part Three, but it is possible at this point to begin developing a qualitative understanding of deflection behavior. This can be done using the information from the bending-moment diagram. In Fig. 4.16 we observe that a positive bending moment deforms the beam concave upward and a negative moment convex upward. We must also observe the deformation conditions imposed by the various support conditions. A fixed end prevents both rotation and translation, whereas a roller or a hinge constrains only translations. The deflected shape of the beam in Example 4.6 is sketched in Fig. E4.6e.

PROBLEMS

4.1 In Fig. P2.1a to e compute the magnitude of the axial force (where applicable), shear, and bending moment at point a and indicate whether the quantity should be plotted as positive or negative on a diagram (point a is left of the load and reaction for Fig. P2.1a and b, respectively).

4.2 The differential relations shown as Eqs. (4.3) and (4.4) were derived for a slender member extending in the x direction with loading in the y direction. To emphasize this fact we can write these equations as

$$\frac{dV_y}{dx} - q_y = 0 \quad \text{and} \quad \frac{dM_z}{dx} - V_y = 0 \tag{a}$$

Show that for a slender member extending in the y direction with loading in the x direction the corresponding equations are

$$\frac{dV_x}{dy} - q_x = 0 \quad \text{and} \quad \frac{dM_z}{dy} + V_x = 0 \tag{b}$$

There are six different combinations of slender members extending in one coordinate direction with transverse loading in another coordinate direction. Verify that for three of them the differential equations corresponding to Eqs. (4.3) and (4.4) have the sign pattern of Eq. (*a*) and that for the other three the sign pattern is that of Eq. (*b*).

4.3 For Fig. P2.1*a* to *e* construct the shear and bending-moment diagrams and sketch the approximate deflected shape using the moment diagram.

4.4 Draw the shear and bending-moment diagrams for the girder in Fig. P2.2. Sketch the approximate deflected shape.

4.5 Draw the thrust (axial-force), shear, and bending-moment diagrams for this rigid frame. Plot moments on the compressive side of the members. Sketch the approximate deflected shape.

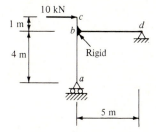

Figure P4.5

4.6 Locate the supports (calculate *x*) such that the magnitudes of the largest positive and negative bending moments are equal.

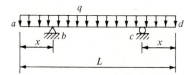

Figure P4.6

4.7 Draw the deflected shape of the beam, the shear diagram, and the loaded beam corresponding to the given bending-moment diagram.

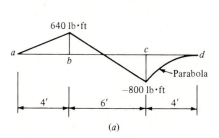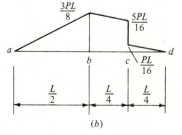

(*a*)   (*b*)

Figure P4.7

REFERENCES

1. American Institute of Steel Construction: *Steel Construction Manual*, 8th ed., Chicago, 1980.
2. National Forest Products Association: *National Design Specification for Wood Construction*, Washington, 1982.
3. American Concrete Institute: *Building Code Requirements for Reinforced Concrete* (ACI 318-77), Detroit, 1979.
4. Prestressed Concrete Institute: *Design Handbook, Precast and Prestressed Concrete*, 2d ed., Chicago, 1978.

5. Crandall, S.H., N.C. Dahl, and R.J. Lardner: *An Introduction to the Mechanics of Solids,* 2d ed., McGraw-Hill, New York, 1978.

ADDITIONAL READING

Beer, F.P., and E.R. Johnston, Jr.: *Mechanics of Materials,* McGraw-Hill, New York, 1981.
Higdon, A., E.H. Ohlsen, W.B. Stiles, J.A. Weese, and W.F. Riley: *Mechanics of Materials: SI Version,* 3d ed., Wiley, New York, 1978.
Muvdi, B.B., and J.W. McNabb: *Engineering Mechanics of Materials,* Macmillan, New York, 1980.
Popov, E.P.: *Mechanics of Materials (SI Version),* 2d ed., Prentice-Hall, Englewood Cliffs, N.J., 1978.
Timoshenko, S.P., and J.M. Gere: *Mechanics of Materials,* Van Nostrand Reinhold, New York, 1972.
Willems, N., J.T. Easley, and S.T. Rolfe: *Strength of Materials,* McGraw-Hill, New York, 1981.

FIVE

MOVING LOADS AND INFLUENCE LINES

5.1 EFFECTS OF MOVING LOADS

Vehicles moving across a bridge constitute a unique type of live loading on a structure. While crossing a highway bridge, a truck applies forces to all possible positions of the bridge. For the structure shown in Fig. 5.1 the left support reaction will (1) be zero before a truck enters the span, (2) have a very large value when a truck is first driven onto the structure, and (3) decrease to zero as a truck continues to the right across the bridge. The internal forces and moments will also vary as a function of the position of the loading. In addition, if the roadway surface is bumpy or the bridge has vertical or horizontal curves, the vehicle will introduce dynamic forces into the structure, but the treatment of these effects is beyond the scope of this book. The problem of moving loads is indigenous to highway and railway bridges, but the flexural members supporting a movable crane and many other structures must also be similarly investigated. Examples of structures subjected to the effects of moving loads are shown in Fig. 5.2.

The effects on the structure of dead loading and fixed live loading can be assessed by the methods of analysis described in Chaps. 2 to 4. When the position of the loading varies, special considerations must be introduced to analyze the reactions and internal-member responses. In this chapter we investigate the problem of calculating the reactions, internal axial forces, shear forces, and bending moments as loads move across statically determinate structures.

5.2 DESCRIPTION OF AN INFLUENCE LINE

As the applied loading on a structure changes location, the reactions, member axial loads, shear forces, and bending moments change value correspondingly. Since the problem is more complex than if the load were fixed in position, a systematic method of investigation is required. Typically, a single concentrated load is positioned at every possible load-application point on the structure, and the various forces (reactions, member axial forces, shear forces and/or bending moments) are calculated. Hence, a

Figure 5.1 A timber bridge. (*BKBM Consulting Engineers, Minneapolis.*)

Figure 5.2 (*a*) The George Washington Bridge (*The Port Authority of New York and New Jersey*).

Figure 5.2 (*b*) Priest Point Park prestressed concrete bridge (*Arvid Grant & Assoc., Olympia, Washington*).

Figure 5.2 (*c*) New River Gorge Bridge, Fayette County, West Virginia (*United States Steel*).

Figure 5.2 (*d*) Disney World monorail (*ABAM Engineers, Inc., Tacoma, Washington*).

description is obtained of a particular force response of the structure expressed in terms of the location of the applied loading. The single concentrated force applied to the structure in this procedure usually has the value of unity. That is, it may be 1 N, 1 lb, 1 kN, 1 kip, or whatever unit of force is convenient. The resulting description of the reaction, member axial force, etc., can be displayed as a graph with the abscissa representing distance along the structure and the ordinate representing the value of the force. This diagram is referred to as an *influence line*.

For example, if the influence line for the reaction at point *a* of the beam shown in Fig. 5.3*a* is required, the single concentrated load is moved across the span from point *a* to *b* and the reaction at point *a* is calculated. For this case, a general expression for the reaction can be obtained and the corresponding diagram constructed. Placing the unit load in Fig. 5.3*a* at a typical position located a distance x from point *a* and summing moments about point *b* gives

$$R_a = \frac{(L - x)(1)}{L} \tag{5.1}$$

From the expression in Eq. (5.1) the influence line for R_a, shown in Fig. 5.3*b* can be constructed and used to study the effect of the moving load. Thus, for example, if the load is positioned at the left reaction ($x = 0$), the value of R_a is unity. As the load moves across the span and reaches the midspan ($x = L/2$), the influence ordinate in Fig. 5.3*b* shows that R_a equals 0.5. Furthermore, when the unit load is at the right support ($x = L$) the influence ordinate has the value of zero. From Eq. (5.1) it can also be observed that the influence line for R_a is a linear function of x. In constructing an influence line it is imperative to note that the load moves across the structure and the

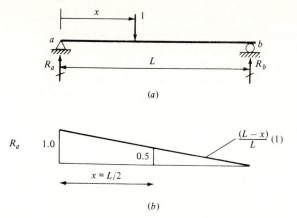

(a)

(b)

Figure 5.3 A simply supported beam with the influence line for left reaction.

point being investigated remains fixed at the location where the function is to be determined. The above discussion leads to the following description:

> An influence line is a diagram showing the change in the value of a particular function (reaction, member axial force, internal shear, or bending moment) as a unit concentrated load moves across the structure.

The dimensional units of the ordinates of the influence line shown in Fig. 5.3 can be interpreted in the following ways:

1. If the unit applied load is considered to be dimensionless, the ordinates representing values of the reaction force are also dimensionless.
2. If the unit applied load is considered to have a value of 1 N (lb, etc.), the ordinates of the influence line would have dimensions equal to those of the applied force.
3. If the loading has dimensions of force (newton, pound, etc.), the reaction will have identical units of force. The ordinates of the influence line can be interpreted to have dimensions of force per unit applied force, e.g., newtons per newton; therefore, they are dimensionless. In this case, the ordinates of the influence line are sometimes referred to as *influence coefficients*.

The first interpretation will be used in this book, and the point will be discussed further in Sec. 5.4, where influence lines are used with the actual applied loading to obtain the values of the reaction, member forces, etc.

5.3 INFLUENCE LINES FOR BEAMS

For most typical beams, the structural analyst is interested in the influence lines for the reactions, as well as the change in the internal quantities in the beam as the loading moves across the structure. Therefore, influence lines for the shear and moment at a specific cross section must also be constructed for beam structures. In order to do so it is necessary to make an imaginary cut through the beam at the point of interest and then compute the value of the shear and moment at this cross section as the unit concentrated load traverses the beam. Obtaining the influence lines for these internal quantities is similar to obtaining the reaction influence line, but the shear and bending moment are plotted as the ordinates of the influence lines. The sign convention used to plot these quantities is the same as that observed previously for the shear and moment diagrams (see Sec. 4.3), but it is important to note the differences between the shear and moment influence lines and the shear and moment diagrams. The former involve a fixed point of reference (the function is plotted at a single point) and a moving load, whereas the latter involve a loading that is fixed in position while the point of reference moves across the beam, i.e., an ordinate represents the function at one particular location on the beam.

The reaction influence line discussed in the previous section for the simply supported beam in Fig. 5.3 is shown again in Fig. 5.4. The influence line for the reaction at point b can be obtained in a manner similar to that used for the other reaction. Thus placing the unit concentrated load at a typical point on the beam and summing moments about the point a gives

$$R_b = \frac{x(1)}{L} \tag{5.2}$$

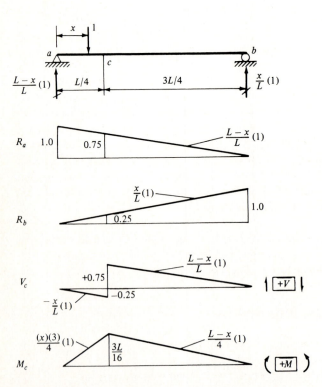

Figure 5.4 A simply supported beam with various influence lines.

The influence line for R_b is sketched in Fig. 5.4 using Eq. (5.2). A review of this diagram, keeping the meaning of the ordinates in mind, shows that it is intuitively correct. That is, when the load is at point a, the influence ordinate is zero, and when the load is at point b, the influence ordinate is unity; i.e., there should be no reaction at b when the load is at point a, and one would anticipate that the load would be totally reacted by R_b when the load is positioned at point b. It is of interest to note that the sum of the influence ordinates for R_a and R_b is 1 for a given abscissa of their respective influence lines. Using the equation of statics and summing forces in the vertical direction gives

$$R_a + R_b - 1 = 0 \quad \text{or} \quad R_a + R_b = 1$$

This observation is simply a verification of one of the equations of static equilibrium.

The free-body diagrams in Fig. 5.5 can be used to investigate the shear and moment at point c as the unit loading moves across the beam. Figure 5.5a is correct if the unit load is located between points a and c, and the free-body diagrams in Fig. 5.5b are appropriate for the unit load situated between points c and b. From the left free-body diagram in Fig. 5.5a the following expression is obtained by summing forces in the vertical direction:

$$V_c = -1 + R_a = -1 + \frac{L - x}{L} = \frac{-x}{L} \quad 0 \le x < \frac{L}{4} \tag{5.3}$$

Alternatively, using the right-hand free-body diagram and summing forces in the vertical direction gives

$$V_c = -R_b = \frac{-x}{L} \quad 0 \le x < \frac{L}{4} \tag{5.3a}$$

Thus, either Eq. (5.3) or (5.3a) can be used to construct the influence line for V_c for the segment from a to c (Fig. 5.4). Observing the usual convention for beams, we see that the shear in this case should be plotted as negative. As the unit load traverses the segment from points c to b, either of the free-body diagrams shown in Fig. 5.5b can be used to investigate the shear at section c. Using the left free-body diagram and summing forces in the vertical direction gives

$$V_c = R_a = \frac{L - x}{L} \quad \frac{L}{4} < x \le L \tag{5.4}$$

The right free-body diagram gives

$$V_c = 1 - R_b = 1 - \frac{x}{L} = \frac{L - x}{L} \quad \frac{L}{4} < x \le L \tag{5.4a}$$

If we use either Eq. (5.4) or (5.4a) and observe that V_c is positive (according to the sign convention for shear), the influence line can be completed as shown in Fig. 5.4. The influence line can be plotted by using the expressions developed for V_c in terms of x, or in a slightly different manner. For example, if the unit load is between points a and c and Eq. (5.3a) is used, we can observe that V_c is numerically equal to $-R_b$. Thus, since the influence line for R_b is already known, its shape can simply be replicated with

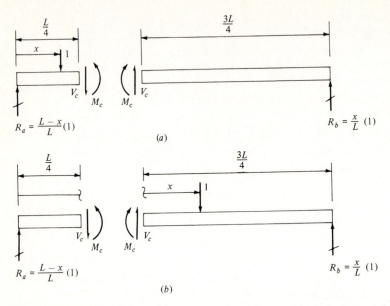

Figure 5.5 Free-body diagrams for calculating V_c and M_c: (a) diagram for $0 \leq x < L/4$, (b) diagram for $L/4 < x \leq L$.

a change in sign to obtain V_c. Similarly, as the unit load moves from point c to point b, the expression shown in Eq. (5.4) can be used to construct the rest of the influence line for V_c. Since the shape of R_a is already known, this can simply be plotted for V_c with a positive sign in this case. Since both approaches to plotting the influence line are valid, the reader is urged to experiment with these two construction techniques.

To obtain the moment influence line for this beam it is necessary to write expressions for the moment at point c as the unit concentrated load is positioned at all locations on the span. Thus, the free-body diagrams in Fig. 5.5 can be used to construct this influence line as well. For the load positioned between points a and c using the left and right free-body diagrams (Fig. 5.5a) respectively, we obtain

$$M_c = R_a \frac{L}{4} - \left(\frac{L}{4} - x\right)(1) = \frac{L}{4} - \frac{x}{4} - \frac{L}{4} + x$$

$$= \frac{3x}{4} \qquad 0 \leq x \leq \frac{L}{4} \tag{5.5}$$

and $$M_c = R_b \frac{3L}{4} = \frac{3x}{4} \qquad 0 \leq x \leq \frac{L}{4} \tag{5.5a}$$

Both equations give the same expression for M_c, but Eq. (5.5) uses the value of R_a and Eq. (5.5a) uses R_b. Plotting the final values of M_c and noting that the sign of the quantity is positive gives the left portion of M_c, shown in Fig. 5.4. As the load goes from point c to b, the free-body diagrams in Fig. 5.5b must be used. The left and right free-body diagrams give, respectively,

$$M_c = R_a \frac{L}{4} = \frac{L - x}{4} \qquad \frac{L}{4} \le x \le L \tag{5.6}$$

and
$$M_c = R_b \frac{3L}{4} - \left(x - \frac{L}{4} \right)(1) = \frac{3x}{4} - x + \frac{L}{4}$$

$$= \frac{L - x}{4} \qquad \frac{L}{4} \le x \le L \tag{5.6a}$$

Plotting this function for $L/4 \le x \le L$ gives the remainder of the influence line for M_c, shown in Fig. 5.4. This influence line could have been constructed differently by using the following expressions from Eqs. (5.5a) and (5.6):

$$M_c = \begin{cases} R_b \dfrac{3L}{4} & 0 \le x \le \dfrac{L}{4} \\ R_a \dfrac{L}{4} & \dfrac{L}{4} \le x \le L \end{cases}$$

Hence, multiplying the influence line ordinates for R_b by $3L/4$ when the load is between a and c and the R_a influence-line ordinates by $L/4$ for the interval between c and b gives the same influence line for M_c as that obtained by plotting the algebraic expressions of Eqs. (5.5) and (5.6).

Although the process for constructing influence lines for all beams is the same as that demonstrated above for the simply supported structure, the expressions required to describe the various quantities will change with the geometry and support conditions of the structure. Examples 5.1 to 5.3 illustrate the construction of influence lines for several beams.

Example 5.1 Draw the influence lines for R_a, M_a, V_b, and M_b for this cantilever beam.

SOLUTION From Fig. E.5.1a

$$\sum F_y = 0: \qquad\qquad R_a = 1$$

$$\sum M_a = 0: \qquad\qquad M_a = -1x$$

From Fig. E5.1b for load from a to b

$$V_b = 0 \qquad M_b = 0$$

From Fig. E5.1c for load from b to c

$$V_b = 1 \qquad M_b = 1(12 - x)$$

DISCUSSION The influence lines for the reactions R_a and M_a are obtained by applying the equations of statics to the total beam as shown in Fig. E5.1a. Note that R_a acts upward and must be plotted as a positive quantity, while M_a, which acts in the counterclockwise direction, distorting the beam into a negative curvature, is plotted as negative. V_b and M_b are obtained using either of the free-body diagrams in Fig. E5.1b when the unit load is between points a and b. When the unit load is between points b and c, the free-body diagrams in Fig. E5.1c are applicable. If the right-hand free-body diagram in Fig. E5.1b is used, it can quickly be seen that both V_b and M_b are zero when the load is between points a

From Fig. E5.2c (applied unit load not shown) for the load between a and c

$$V_{c-} = R_a - 1 = \frac{-x}{10} \qquad V_{c+} = 0$$

$$M_c = 10R_a - (10 - x) = 0$$

and for load between c and d

$$V_{c-} = R_a = \frac{10 - x}{10} \qquad V_{c+} = 1 \qquad M_c = 10R_a = 10 - x$$

DISCUSSION The influence lines for the reactions at points a and c are obtained using the entire structure as a free-body diagram and allowing the unit concentrated loading to move across the beam from point a to d. Note that the resulting influence lines are both linear functions of the location of the loading x and intuitively have the correct properties; i.e., they are unity when the unit load is at the reaction and zero when the load is located at the other reaction. The influence line for V_b is obtained by taking an imaginary cut through the beam at that point and writing the equations of statics for each of the individual free-body diagrams as the unit load moves across the entire structure. It is necessary to write an expression for V_b when the loading is between points a and b and another when it moves between points b and d (Fig. E5.2b). Note that the influence line can be plotted using either the expressions in terms of x or those in terms of the reaction influence lines R_a and R_c. The influence line for M_b is also obtained using the free-body diagrams in Fig. E5.2b and writing the appropriate algebraic expressions for the bending moment as the unit load moves across the beam from point a to point d. Figure E5.2c shows two sections through the beam, one just to the left and another just to the right of support c. Thus, V_{c-} and V_{c+} are not the same, but they will add algebraically to give the value of R_c. Figure E5.2c is used to write appropriate expressions for V_{c-}, V_{c+}, and M_c, which in turn are used to construct the influence lines. Note that in all cases the sign of the shear and bending moment are plotted observing the sign convention established in Sec. 4.3, i.e., according to how the beam is distorted.

Example 5.3 Draw the influence lines for R_a, R_d, R_f, V_b, M_b, V_e, and M_e for the beam illustrated.

SOLUTION From Fig. E5.3b (applied unit load not shown) for the load between a and c

$$R_a = \frac{4 - x}{4} \qquad V_c = \frac{-x}{4} \qquad R_d = \frac{x}{3} \qquad R_f = \frac{-x}{12}$$

and for the load between c and f

$$R_a = V_c = 0 \qquad R_d = \frac{20 - x}{12} \qquad R_f = \frac{x - 8}{12}$$

From Fig. E5.3c (applied unit load not shown) for the load between a and b

$$V_b = R_a - 1 = -(R_d + R_f) = \frac{-x}{4} \qquad M_b = 2R_a - (2 - x) = 6R_d + 18R_f = \frac{x}{2}$$

and for the load between b and f

$$V_b = R_a = 1 - (R_d + R_f) \qquad M_b = 2R_a = 6R_d + 18R_f - (x - 2)$$

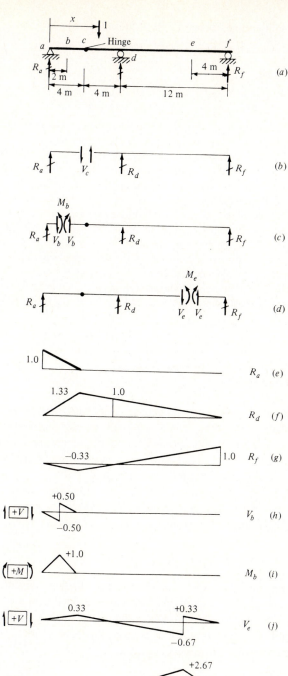

Figure E5.3

From Fig. E5.3d (applied unit load not shown) for the load between a and e

$$V_e = R_a + R_d - 1 = -R_f \qquad M_e = 16R_a + 8R_d - (16 - x) = 4R_f$$

and for the load between e and f

$$V_e = R_a + R_d = 1 - R_f \qquad M_e = 16R_a + 8R_d = 4R_f - (x - 16)$$

DISCUSSION The fact that the moment is zero at point c gives the necessary equation of condition to obtain V_c. Note that when the load has moved off of segment ac, there is no reaction at point a; however, there are nonzero values of R_d and R_f for all locations of the unit load. This behavior is due to the hinge construction at point c. The influence lines for the shear and moment at point b are obtained by using the appropriate free-body diagrams shown in Fig. E5.3c. When the unit load is located to the right of point b, the expressions for V_b and M_b are written in terms of the reaction R_a. The influence lines can be constructed by using the influence lines for R_a directly (for example, M_b is obtained by multiplying the influence ordinates for R_a by 2), or, alternatively, the algebraic expressions for R_a can be substituted into the equations for the shear and moment. If the latter approach is used, note that R_a is described with one expression when the unit load is between points b and c and another when the load traverses the structure to the right of point c. Figure E5.3d shows the free-body diagrams used in constructing the influence lines for the shear and moment at point e.

5.4 RELATIONSHIP OF INFLUENCE LINES AND STRUCTURAL LOADING

Influence lines are typically constructed using a unit concentrated load. These diagrams can be used to investigate the effect of the actual load moving across the structure. A vehicle can be characterized as a set of concentrated forces with a specified spacing between them. Similarly, a lane of traffic can be idealized as a uniformly distributed load, with the length of the load dependent upon the volume of traffic.

If a single concentrated force of magnitude P moves across the beam in Fig. 5.4, the effect of the load is obtained by simply placing it at a given location x_1 and multiplying the influence-line ordinate $IL(x_1)$ at that point by the magnitude of the load P

$$F = IL(x_1)P \qquad (5.7)$$

where F is the value of the function of interest — reaction, shear, bending moment, etc. This process can be used to obtain the value of any function associated with the beam as long as an influence line has been constructed. For example, if a single concentrated load of 50 kN is placed on the beam of Fig. 5.4 at point c, the value of the reaction at point a is obtained by multiplying the influence ordinate at that point (0.75) by the magnitude of the load, giving

$$R_a\left(x = \frac{L}{4}\right) = 50(0.75) = 37.5 \text{ kN}$$

Since the units of the load are kilonewtons and the influence ordinate is dimensionless, the units of the reaction are kilonewtons. To obtain the maximum value of the reaction

R_a from the influence line it is necessary to place the concentrated load ($P = 50$ kN) at the maximum influence ordinate for the reaction (at point a in this case). Thus,

$$(R_a)_{max} = 50(1.0) = 50 \text{ kN}$$

To obtain the value of the shear in the beam due to the concentrated force of 50 kN positioned at a given location, the influence ordinate at the location is multipled by the value of the applied load. The following calculation gives the shear at point c when the load is positioned slightly to the left of point c (designated as $x = [L/4]-$):

$$V_c(x = [L/4]-) = 50(-0.25) = -12.5 \text{ kN}$$

Also, the moment at c when the load is positioned at $x = L/2$ is

$$M_c = 50(0.125L) = 6.25L \text{ kN} \cdot \text{m}$$

if the length L is expressed in meters.

Assume that a concentrated load P moves across a structure having the influence line shown in Fig. 5.6a. The load must be positioned at point b to yield the maximum positive value of the function and at point c to give the negative value with the greatest

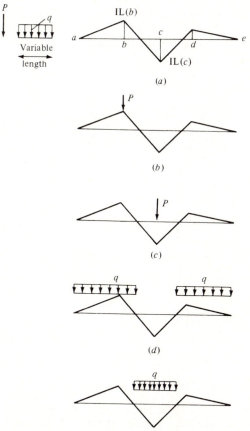

(a)

(b)

(c)

(d)

(e)

Figure 5.6 A typical influence line with the loads positioned to give maximum values: (a) influence line with loading, (b) maximum positive value using P only, (c) largest negative value using P only, (d) maximum positive value using q only; (e) largest negative value using q only.

magnitude. If a distributed load $q(x)$ is applied over a portion of a structure as shown in Fig. 5.7, its effect can also be calculated using the influence ordinates. A portion of the influence line for the quantity of interest, e.g., reaction, is shown in Fig. 5.7a. Multiplying the differential concentrated load $q(x)\, dx$ by the influence ordinate at that point gives the corresponding value for the quantity (reaction)

$$dF = IL(x)q(x)\, dx \tag{5.8}$$

This is analogous to the relationship developed for a concentrated load shown in Eq. (5.7). The effect of the loading situated between points x_a and x_b on the beam can be obtained by integrating Eq. (5.8) between those two points to give

$$F = \int_{x_a}^{x_b} dF = \int_{x_a}^{x_b} IL(x)q(x)\, dx \tag{5.9}$$

If the loading is uniformly distributed ($q = \text{const}$), the value of the function according to Eq. (5.9) is

$$F = \int_{x_a}^{x_b} IL(x)q\, dx = q \int_{x_a}^{x_b} IL(x)\, dx \tag{5.10}$$

The integral in Eq. (5.10) represents the area under the influence line between points x_a and x_b (shown crosshatched in Fig. 5.7a). Thus, for example, if the beam in Fig. 5.4 is loaded over its entire length with a load of 4 kN/m and the length of the beam L is expressed in meters, the value of the reaction at point a is

$$R_a = \tfrac{1}{2}L(1.0)\,(4) = 2.0L \qquad \text{kN}$$

Furthermore, the value of the moment at point c due to 4 kN/m is

$$M_c = \tfrac{1}{2}L(0.1875L)\,(4) = 0.375L^2 \qquad \text{kN} \cdot \text{m}$$

If it is required to obtain the maximum positive and negative magnitudes of the function with an influence line as shown in Fig. 5.6 due to the variable-length uniform

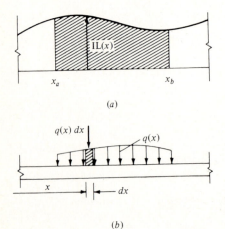

Figure 5.7 An influence line with a distributed load on the beam: (a) influence line ordinates, (b) loaded beam segment.

load q, the loading would be positioned as shown in Fig. 5.6d and e. That is, to obtain the maximum positive effect, the loading would be placed over all the positive areas of the influence line; furthermore, the load would be placed at locations of the beam where the influence ordinates are negative if it is required to obtain the greatest magnitude of the negative function. For example, if the beam of Fig. 5.4 has a uniformly distributed load of 4 kN/m distributed over the beam, the maximum positive shear is obtained by positioning the load between points c and b. This gives

$$V_{max+} = \tfrac{1}{2}(0.75L)(0.75)(4) = 1.125L \qquad \text{kN}$$

Similarly, if the negative shear for the 4-kN/m loading is required, the loading would be positioned between points a and c and the value of the function would be

$$V_{max-} = \tfrac{1}{2}(0.25L)(-0.25)(4) = -0.125L \qquad \text{kN}$$

From the above discussion and Eqs. (5.7) and (5.10) it is possible to make several general statements about the relationships between influence lines and structural loading:

1. The effect of a concentrated load can be obtained by multiplying the value of the load by the influence ordinate where the load is positioned.
2. The greatest magnitude of a function, e.g., reaction, due to a concentrated load exists when the load is positioned on the structure at the point where the influence line has the largest ordinate.
3. The effect of a uniformly distributed load is obtained by multiplying the area under the influence line (between the points where the load is distributed) by the value of the distributed loading.
4. The greatest magnitude of a function, e.g., reaction, due to a uniformly distributed load of constant value and variable length is obtained by placing the loading over those portions of the influence line which have ordinates of the same sign.

Examples 5.4 to 5.6 illustrate the process of loading the influence lines obtained for Examples 5.1 to 5.3, respectively.

Example 5.4 The beam in Example 5.1 is loaded with a single moving concentrated load of 32 kips. Using the influence lines developed previously, calculate the maximum values of V_b and M_b due to this loading.

SOLUTION It can be seen from Fig. E5.4b that to obtain the maximum shear at point b, the 32-kip load can be positioned anywhere between points b and c on the cantilever beam of Example 5.1. From Eq. (5.7), the value of the maximum shear is

$$(V_b)_{max} = 32.0(1.0) = 32.0 \text{ kips}$$

The influence line for M_b indicates that the moment will be the largest when the 32-kip load is situated at point c; therefore,

$$(M_b)_{max-} = 32.0(-8.0) = -256.0 \text{ kip} \cdot \text{ft}$$

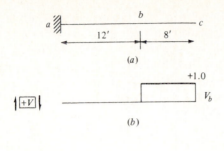

(a)

(b)

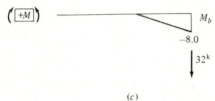

(c)

Figure E5.4

Example 5.5 The beam in Example 5.2 has the illustrated loading applied to the structure. Note that the uniformly distributed part of the load is a variable length. Calculate the largest positive and negative values of V_b and M_b due to this loading.

SOLUTION From Fig. E5.5b

$$(V_b)_{max+} = 100(0.4) + \tfrac{1}{2}(0.4)(4)(10) = 48 \text{ kN}$$

From Fig. E5.5c and the shear influence line

$$(V_b)_{max-} = 100(-0.6) + \tfrac{1}{2}(-0.4)(4)(10) + \tfrac{1}{2}(-0.6)(6)(10) = -86 \text{ kN}$$

From Fig. E5.5d

$$(M_b)_{max+} = 100(2.4) + \tfrac{1}{2}(2.4)(10)(10) = 360 \text{ kN} \cdot \text{m}$$

From Fig. E5.5e and the moment influence line

$$(M_b)_{max-} = 100(-2.4) + \tfrac{1}{2}(-2.4)(4)(10) = -288 \text{ kN} \cdot \text{m}$$

DISCUSSION Both the concentrated load and the uniformly distributed loading must be applied to the structure simultaneously to yield the maximum shear and moment in the beam. The variability of the length of the uniform loading implies that it represents a line of vehicles whereas the concentrated load is a single heavy vehicle within this line. Thus, to obtain the maximum positive value of the shear at point b, we locate the uniform loading between points b and c since the influence ordinates are positive in this region and negative elsewhere. In addition, the 100-kN concentrated load is situated over the +0.4 ordinate at point b. Similarly, the largest negative shear at point b is obtained by placing the uniform loading on the structure wherever the influence ordinates for V_b are negative, and the

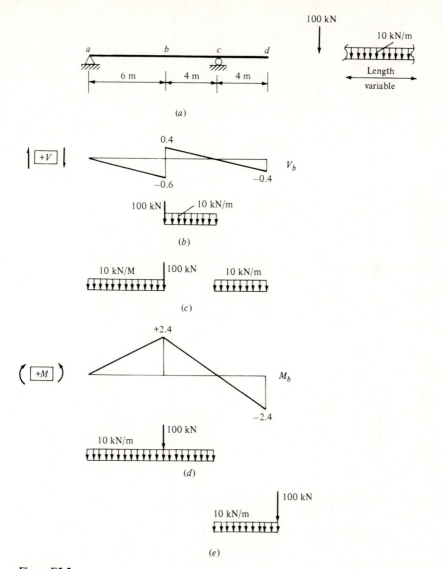

Figure E5.5

100-kN load is located over the -0.6 ordinate at point b. Likewise, the values of $(M_b)_{\text{max}+}$ and $(M_b)_{\text{max}-}$ are obtained by loading the positive and negative portions, respectively, of the influence line for M_b.

Example 5.6 The beam in Example 5.3 is loaded with a standard H 20 (M 18) highway wheel loading as shown. Using the influence lines developed previously, calculate the largest values of R_d, V_e (negative), and M_e (positive).

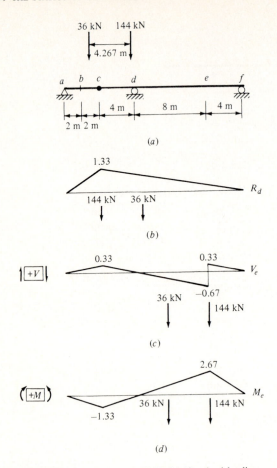

(a)

(b)

(c)

(d)

Figure E5.6 (a) Beam with H 20 (M 18) wheel loading.

SOLUTION From Fig. E5.6b

$$(R_d)_{max} = 144(1.33) + 36[\tfrac{1}{12}(11.733)] = 192 + 35 = 227 \text{ kN}$$

From Fig. E5.6c

$$(V_e)_{max-} = 144(-0.67) + 36[-\tfrac{1}{12}(3.733)] = -96 - 11 = -107 \text{ kN}$$

From Fig. E5.6d

$$(M_e)_{max+} = 144(2.67) + 36[\tfrac{1}{3}(3.733)] = 384 + 45 = 429 \text{ kN} \cdot \text{m}$$

DISCUSSION The pair of concentrated loads illustrated represents a relatively heavy vehicle that can traverse the beam in either direction; i.e., the 36-kN load can either be to the left or to the right of the 144-kN load. This is the standard AASHTO H 20 (M 18) wheel loading, discussed in Sec. 5.7. To obtain the maximum effect, the 144-kN load is placed at the largest influence ordinate, and the other load is positioned 4.267 m away on the side that places it over the largest influence ordinate. It can be observed from the calculations for $(V_e)_{max-}$ that the 36-kN load must be to the left of point e since it is the only location with negative influence ordinates, whereas if the order of the loads were reversed for the calculations for either $(R_d)_{max}$ or $(M_e)_{max}$, the 36-kN load would be off the beam and would not have any effect.

5.5 INFLUENCE LINES FOR GIRDERS WITH FLOOR BEAMS

The small bridge shown in Fig. 5.8 consists of various flexural members arranged to transfer the applied loads from the bridge deck to the piers. The primary supporting members in this array are the girders. The loading applied to the roadway slab by a vehicle will be transmitted to the stringers, and the reactions from the stringers will be transferred to the floor beams. The floor-beam reactions are transmitted in turn to the girder at discrete points along the length of this member. This loading process differs from that for beams just discussed. The loading on beams is applied directly to the member, but the bridge girder is loaded only at a number of specific points along its length. Figure 5.9 shows a bridge in which the stringers and floor beams can be observed.

An idealized drawing of a bridge girder is shown in Fig. 5.10a. The stringers are drawn as a set of simply supported beams that transfer their reactions to the floor beams.

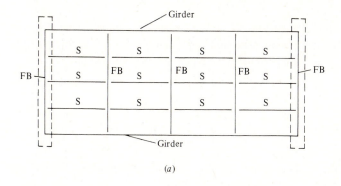

(a)

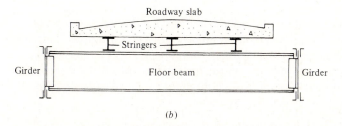

(b)

Figure 5.8 Girder bridge: (a) plan view (FB = floor beam, S = stringer), (b) cross section.

Figure 5.9 View showing floor beams and stringers of a bridge. (*Bridge Design Section Alaska DOT & PF*)

In reality, a stringer usually consists of a single continuous beam with a length equal to that of the girder. Such a continuous stringer would be indeterminate to the third degree and would require that the reactions be obtained by augmenting the equations of statics with deflection considerations. Alternatively, this member has been idealized as four individual beams in order to simplify the calculations for the reactions. For most cases, this assumption of simply supported stringers has a minor effect on the results obtained for the girder.

The influence lines for the reactions for the girder can be obtained using the same procedures as for beams; i.e., placing the unit load at some general position on the system, as shown in Fig. 5.10a, and using the equations of statics gives

$$R_a = \frac{20 - x}{20} \tag{5.11a}$$

$$R_e = \frac{x}{20} \tag{5.11b}$$

The influence lines for these two quantities, constructed using the equations above, are shown in Fig. 5.10c. Note from these influence lines that

$$R_a + R_e = 1 \tag{5.11c}$$

Sometimes it is convenient to use the influence lines for the floor-beam reactions in constructing the various shear and moment influence lines for the girder. Since the stringers are idealized as simply supported beams, the influence lines for the floor-beam reactions R_a', R_b', R_c', R_d', and R_e', are obtained using the free-body diagrams in Fig. 5.10b. It can be observed that these beam reactions are the numerical sum of the shear forces applied to the stringers. For example,

$$R_b' = V_{b-} + V_{b+}$$

where V_{b-} and V_{b+} denote the values of the shear immediately to the left and right of

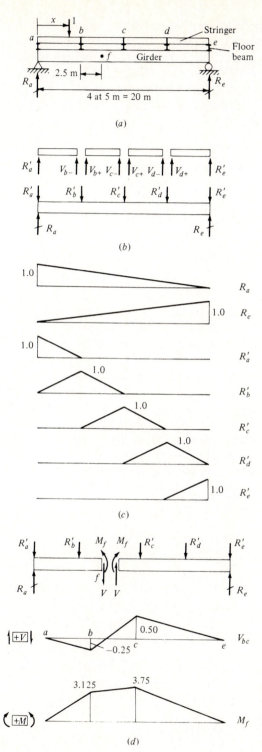

(a)

(b)

(c)

(d)

Figure 5.10 A bridge girder with floor beams and stringers: (a) girder showing stringers and floor beams, (b) free-body diagram of stringers and girder with beam reactions, (c) influence lines for floor-beam reactions and girder reactions, (d) free-body diagrams plus shear and moment influence lines at point f.

point b, respectively. As a concentrated load moves across the structure, the only stringer that will experience any reactions is the one on which the load is positioned. Hence, if the load is located between points c and d, V_{c+} and V_{d-} are the only nonzero stringer shears; consequently, R'_a, R'_b, and R'_e are all zero, while R'_c and R'_d have nonzero values. The influence lines for these five floor-beam reactions are shown in Fig. 5.10c. Note from these influence lines that for all positions of the unit concentrated load the following equation of statics is satisfied:

$$R'_a + R'_b + R'_c + R'_d + R'_e = 1 \qquad (5.11d)$$

The various shear and moment influence lines can be constructed the same as they were for the beams in the previous section. For example, if the shear and moment at point f are to be investigated, the free-body diagrams in Fig. 5.10d must be used. Summing forces in the vertical direction for the left free-body diagram gives

$$V = R_a - R'_a - R'_b \qquad (5.11e)$$

Or from the right free-body diagram in Fig. 5.10d

$$V = -R_e + R'_c + R'_d + R'_e \qquad (5.11f)$$

Either Eq. (5.11e) or (5.11f) can be used to construct the influence line for the shear at this location. The former involves the combination of three influence lines whereas the latter implies that four influence lines must be combined. Note that the unit concentrated load does not appear explicitly in either Eq. (5.11e) or (5.11f) but its location and magnitude are contained implicitly in the influence lines that must be used to obtain V. Thus, Eqs. (5.11e) and (5.11f) are valid for all locations of the unit concentrated load. Either of these two equations can be used to obtain the influence line in Fig. 5.10d. These two expressions for shear are valid if point f is located anywhere between points b and c; thus, the illustrated influence line can properly be labeled V_{bc}. The influence line has been plotted using the customary sign convention for shear.

The moment (M_f) at point f can be obtained using the free-body diagrams in Fig. 5.10d. The left and right free-body diagrams yield, respectively

$$M_f = \begin{cases} (R_a - R'_a)(7.5) - 2.5R'_b & (5.11g) \\ 12.5(R_e - R'_e) - 7.5R'_d - 2.5R'_c & (5.11h) \end{cases}$$

Using either of the above two equations and observing the usual sign convention for plotting the moment gives the influence line for the moment at f shown in Fig. 5.10d.

Since the influence lines in Fig. 5.10d for the shear and moment in the girder vary linearly between the floor-beam reaction points, they could have been obtained in another, slightly different manner. The influence lines can be constructed by placing the unit concentrated load alternately at the beam-reaction points a to e, calculating V_{bc} and M_f corresponding to these five locations of the loading, and plotting these values. For example, when the unit load is positioned at point d, the free-body diagram in Fig. 5.11 can be used to write the following for the shear and moment:

$$V_{bc} = R_a = 0.25 \qquad (5.11i)$$

$$M_f = 7.5R_a = 7.5(0.25) = 1.875 \qquad (5.11j)$$

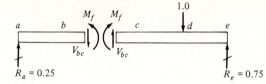

Figure 5.11 Free-body diagram of girder with the unit concentrated load at point d.

These are the same values that were obtained previously using, for example, Eqs. (5.11*e*) and (5.11*g*), respectively. The influence lines could have been constructed by continuing to calculate V_{bc} and M_f when the load is placed at the other four floor-beam reaction points, and the results would have been identical to those obtained previously. Examples 5.7 and 5.8 illustrate the construction of influence lines for other types of girder–floor-beam systems.

Example 5.7 This girder–floor-beam structure is loaded through the stringers, as shown. Construct the influence lines for the reactions plus the shear and moment in the girder at points e and f.

SOLUTION From Fig. E5.7*b*

$$V_{bc} = R_a - (R_a' + R_b') = (R_c' + R_d') - R_c$$

$$M_e = 14(R_a - R_a') - 4R_b' = 6(R_c - R_c') - 16R_d'$$

From Fig. E5.7*c*

$$V_{cd} = R_a + R_c - (R_a' + R_b' + R_c') = R_d'$$

$$M_f = 22(R_a - R_a') - 12R_b' + 2(R_c - R_c') = -8R_d'$$

DISCUSSION The influence lines for the girder reactions and the floor-beam reactions are used to construct influence lines for the shears and moments. These reaction influence lines have been constructed using a common base line to save space. From Fig. E5.7*b* it can be seen that the shear at all points between the floor beams at b and c will be equal. The expressions for both V_{bc} and M_e involve three reaction influence lines; therefore, the equations obtained from either the left or right free-body diagrams can be used to construct these two influence lines. From Fig. E5.7*c*, however, it is clear that the right free-body diagram yields the most convenient equations for constructing the influence lines for point f. Alternatively, the influence lines could have been obtained by calculating the shears and moments when the unit concentrated load is placed at points a to d and then noting that there is a linear relationship between these points.

Example 5.8 The loads applied to this structure are transmitted to the bridge girder at the floor-beam reaction points. Draw influence lines for the reactions R_a and R_d and those for V_{bc}, M_f, and M_g.

SOLUTION From Fig. E5.8*b*

$$V_{bc} = R_a - (R_a' + R_b') = (R_c' + R_d') - R_d$$

$$M_f = 20(R_a - R_a') - 5R_b' = 25(R_d - R_d') - 10R_c'$$

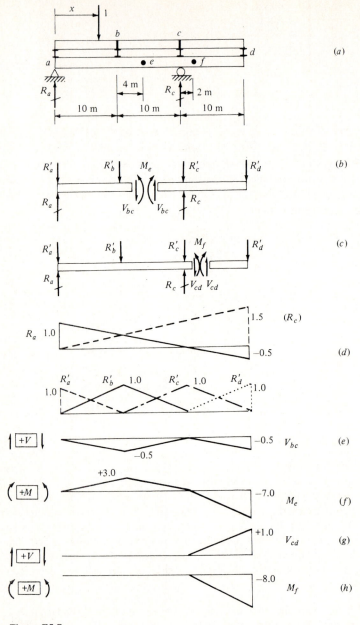

Figure E5.7

From Fig. E5.8c

$$M_g = 35(R_a - R'_a) - 20R'_b - 5R'_c = 10(R_d - R'_d)$$

DISCUSSION The influence lines for the shear and moments in this structure are obtained using the floor-beam reaction influence lines in conjunction with the influence lines for

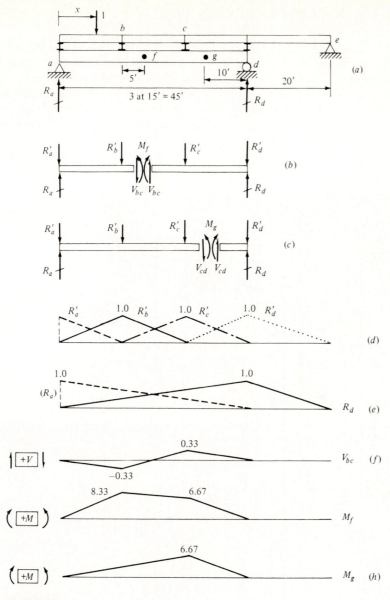

Figure E5.8

the reactions at a and d. The two equations for the shear at point f are written using both the right and left free-body diagrams in Fig. E5.8b. Note that either expression can be used to construct the influence line and both involve the same number of reaction influence lines. Therefore, neither equation presents a superior method for constructing the influence line. This is also true for the moment at point f. However, the right free-body diagram in Fig. E5.8c yields the most convenient expression for the moment at point g.

5.6 INFLUENCE LINES FOR TRUSSES

The trusses shown in Fig. 5.12 are loaded with moving loads. In order to design each of the individual members it is necessary to know the largest tensile or compressive force they must sustain as the loading moves across the structure. This information can usually be obtained most conveniently by using the influence lines for the members. The maximum force sustained by the members can be calculated by applying the service loads to the structure and using the principles for loading influence lines developed in Sec. 5.4.

Figure 5.12 (*a*) Fairview–St. Mary's skyway system, Minneapolis (*American Institute of Steel Construction*); (*b*) Outerbridge Crossing spanning the Arthur Kill (*Port Authority of New York and New Jersey*).

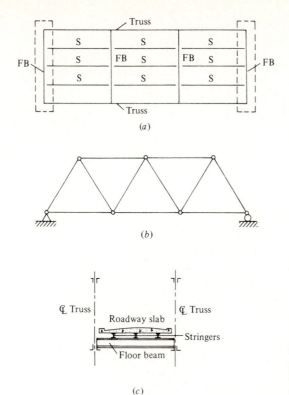

Figure 5.13 Through truss bridge showing primary structural members: (*a*) plan view (FB = floor beam, S = stringer), (*b*) side elevation, (*c*) cross section.

For a typical through truss bridge (Fig. 5.13) the loading is transferred to the trusses in the same way it is delivered to the girders of the structure in Fig. 5.8. That is, the load is first transmitted to the roadway slab and then to the stringers, which in turn deliver their reactions to the floor beams. Since the floor beams are attached to the trusses at discrete points, their reactions will constitute the loading that must be sustained by the truss. Thus the trusses in this case will be loaded only at the points where the floor beams attach to the trusses. These points are termed the *joints* or *panel points*.

If we consider the stringers to be simply supported beams spanning between the panel points, the influence lines for the truss will be linear functions between these points. This linearity was demonstrated for the girder–floor-beam system. For truss problems it is possible first to construct influence lines for the floor-beam reactions and then to write the expressions for the member forces as functions of these quantities. An alternative approach can be used because (1) loads can be applied only to the panel points of the loaded chord and (2) the influence lines are linear functions between the panel points. Thus, the unit load can be successively applied to each of the panel points, and the influence ordinates at these points are connected with straight lines.

The procedure is demonstrated by obtaining several typical influence lines for the members of the truss shown in Fig. 5.14. First, the reaction influence lines are obtained, and these are both shown in Fig. 5.15 plotted on a common base line. If the

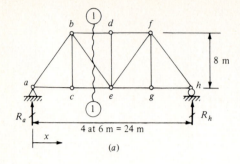

(a)

(b)

(c)

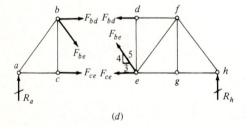

(d)

Figure 5.14 Truss with the free-body diagrams for constructing influence lines: (b) joint a, (c) joint c, (d) free-body diagrams resulting from a cut at section 1-1.

infuence line for member force F_{ab} is to be obtained, the free-body diagram for joint a (Fig. 5.14b) can be used. This diagram is correct if the load moves between points c and h and

$$F_{ab} = -\tfrac{5}{4}R_a \qquad 6 \le x \le 24 \qquad (5.12a)$$

If the unit concentrated load is located at point a, this load must be included in the free-body diagram. The value of R_a is 1 when the load is at this position; therefore,

$$F_{ab} = \tfrac{5}{4}(1 - R_a) = 0 \qquad x = 0 \qquad (5.12b)$$

Thus, the influence line can be constructed by multiplying the ordinates of the influence line for R_a as prescribed in Eq. (5.12a) for all locations from panel point c to panel

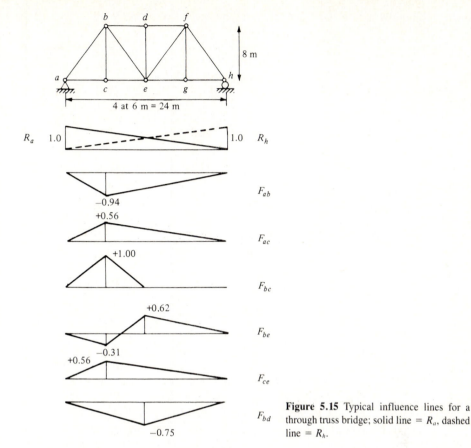

Figure 5.15 Typical influence lines for a through truss bridge; solid line = R_a, dashed line = R_h.

point h and then joining the zero ordinate at point a to the value obtained at point c. This gives the influence line shown in Fig. 5.15.

The influence line for the chord member ac can also be obtained by using the free-body diagram of joint a and summing forces in the horizontal direction to give

$$F_{ac} = -\tfrac{3}{5}F_{ab} \qquad 0 \le x \le 24 \qquad (5.12c)$$

This expression is valid for all positions of the load since the presence of the unit load at joint a would alter the free-body diagram of that joint and this vertical force would not affect the summation of forces in the horizontal direction. This influence line is displayed in Fig. 5.15. The free-body diagram shown in Fig. 5.14c for the joint c can be used to obtain the influence line for F_{bc}. This diagram is valid if the load is either at point a, e, g, or h, and in this case $F_{bc} = 0$. If the load is positioned at point c, then $F_{bc} = 1$. Plotting these five values obtained for this member and connecting them with straight lines gives the influence line shown in Fig. 5.15.

If the member forces in panel ce are to be obtained, the truss must be cut at section 1-1, which gives the two free-body diagrams shown in Fig. 5.14d. As the load moves from point a to c ($0 \leq x \leq 6$), using the right free-body diagram gives

$$\sum F_y = 0: \qquad\qquad F_{be} = -\tfrac{5}{4}R_h \qquad\qquad (5.12d)$$

$$\sum M_b = 0: \qquad\qquad F_{ce} = \tfrac{18}{8}R_h \qquad\qquad (5.12e)$$

$$\sum M_e = 0: \qquad\qquad F_{bd} = -\tfrac{12}{8}R_h \qquad\qquad (5.12f)$$

Similarly, when the unit concentrated load is positioned between points e and h, the following equations for $12 \leq x \leq 24$ can be obtained most conveniently by using the left free-body diagram:

$$\sum F_y = 0: \qquad\qquad F_{be} = \tfrac{5}{4}R_a \qquad\qquad (5.12g)$$

$$\sum M_b = 0: \qquad\qquad F_{ce} = \tfrac{6}{8}R_d \qquad\qquad (5.12h)$$

$$\sum M_e = 0: \qquad\qquad F_{bd} = -\tfrac{12}{8}R_a \qquad\qquad (5.12i)$$

Thus, by using the equation pairs shown in the above six equations, the influence lines for F_{be}, F_{ce}, and F_{bd} can be constructed. They are shown in Fig. 5.15. Note that Eq. (5.12f) is actually applicable for the load moving between points a and e and that Eq. (5.12h) is applicable for the load positioned between points c and h. In both cases, if the load had encroached upon the panel in consideration (panel ce), the floor-beam reaction introduced would have passed through the moment center for the equation being written. There are, of course, a multitude of ways to write the expressions for the member forces. For example, when the load is moving between points a and c, the left free-body diagram in Fig. 5.14 could have been used to write the appropriate equations, which would have been expressed in terms of R_a instead of R_h. Alternatively, the unit load could have been successively placed at the lower panel points and the member forces calculated in terms of the load explicitly. Examples 5.9 and 5.10 illustrate the construction of influence lines for different types of truss structures.

Example 5.9 This Warren through truss has the vehicle loads applied to the bottom panel points. Draw the influence lines for reactions R_a and R_g and those for the marked members.

SOLUTION Considering joint a for the load at a gives

$$F_{ab} = F_{ac} = 0$$

and with the load between c and g joint a gives

$$F_{ab} = -1.12R_a = \frac{-1.12(60 - x)}{60}$$

$$F_{ac} = \frac{-F_{ab}}{2.24} = \frac{R_a}{2} = \frac{60 - x}{120}$$

From Fig. E5.9i with the load at a gives

$$F_{bc} = F_{bd} = 0$$

and with the load between c and g from Fig. E5.9i we get

$$\sum F_y = 0: \qquad F_{bc} = 1.12R_a = 1.12(1 - R_g) = \frac{1.12(60 - x)}{60}$$

$$\sum M_c = 0: \qquad F_{bd} = -R_a = \frac{(x - 20) - 40R_g}{20} = \frac{-60 + x}{60}$$

From Fig. E5.9j with the load between a and c we obtain

$$\sum F_y = 0: \qquad F_{cd} = 1.12(1 - R_a) = 1.12R_g = \frac{1.12x}{60}$$

$$\sum M_d = 0: \qquad F_{ce} = 1.5R_g = \frac{30R_a - (30 - x)}{20} = \frac{1.5x}{60}$$

and with the load between c and g from Fig. E5.9j we get

$$\sum F_y = 0: \qquad F_{cd} = -1.12R_a = 1.12(R_g - 1) = \frac{-1.12(60 - x)}{60}$$

$$\sum M_d = 0: \qquad F_{ce} = 1.5R_a = \frac{30R_g - (x - 30)}{20} = \frac{1.5(60 - x)}{60}$$

DISCUSSION The two reaction influence lines are plotted on the same base line for convenience. The influence lines for members ab and ac are obtained by using a free-body diagram of joint a and writing the equations of equilibrium in terms of the influence line for R_a. When the load is at point a, $R_a = 1$ and the values for both members are zero. Plotting the values of the member forces for all joints and noting that the functions are linear between panel points gives the influence lines. Since members bd and bc carry no load when the unit applied load is at point a, their influence lines can be constructed using the equations obtained for the loading between points c and g. The influence lines for these two members could also have been obtained by isolating joint b and writing

$$F_{bc} = -F_{ab} \qquad \text{and} \qquad F_{bd} = \frac{F_{ab} - F_{bc}}{2.24}$$

Since these expressions are valid for all locations of the unit load, the influence lines could also be constructed from them. However, the first method is preferable since the influence lines are calculated independently of those for all other members. Finally, members cd and ce are investigated using the free-body diagrams obtained using section 2-2 (Fig. E5.9a). In the above analysis, the equations were developed using both the left and right free-body diagrams. This practice is not necessary but was done for this example to point out that the free-body diagram without the unit applied load usually yields the simplest equations to work with. The influence lines for the remaining members could be obtained by noting that the structure is symmetric about the centerline; i.e., the influence lines for corresponding members can be obtained by constructing the mirror image of the existing influence line.

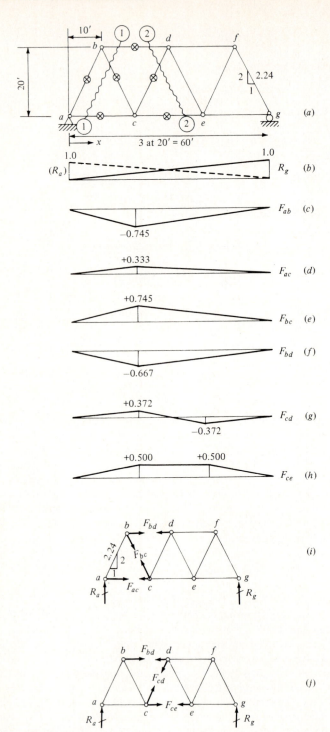

Figure E5.9

Example 5.10 This Pratt bridge truss is loaded at the panel points of the top chord. Draw the influence lines for the two reactions and for the marked members.

SOLUTION Using Fig. E5.10k we note from joint a that for all locations of the load

$$F_{ab} = -R_a = \frac{-80 + x}{80} \qquad F_{ac} = 0$$

Using Fig. E5.10k with the load at b we have

$$F_{bc} = F_{bd} = 0$$

and with the load between d and k

$$F_{bc} = \tfrac{5}{3} R_a = \tfrac{5}{3}(1 - R_j) = \frac{80 - x}{48}$$

$$F_{bd} = -\tfrac{4}{3} R_a = \frac{(x - 20) - 60 R_j}{15} = \frac{-80 + x}{60}$$

From Fig. E5.10l with the load between b and d

$$F_{de} = -\tfrac{5}{3} R_j = \tfrac{5}{3}(R_a - 1) = \frac{-x}{48}$$

$$F_{ce} = 4 R_j = \frac{20 R_a - (20 - x)}{15} = \frac{x}{20}$$

and with the load between f and k from Fig. E5.10l we have

$$F_{de} = \tfrac{5}{3} R_a = \tfrac{5}{3}(1 - R_j) = \frac{80 - x}{48}$$

$$F_{ce} = \tfrac{4}{3} R_a = \frac{60 R_j - (x - 20)}{15} = \frac{80 - x}{60}$$

From Fig. E5.10m with the load at b

$$F_{cd} = 0$$

and with the load between d and k from Fig. E5.10l we get

$$F_{cd} = -R_a = R_j - 1 = \frac{-80 + x}{80}$$

DISCUSSION Influence lines for members ab and ac are obtained by isolating joint a and writing the equilibrium equations, which are valid for all positions of the unit applied loading. The equations for the other member forces are obtained using the proper free-body diagrams, and the influence lines are constructed using the fact that the diagrams are linear functions between the panel points. Joint f can be isolated to investigate member ef. When the unit load is between points b and d, and also h and k,

$$F_{ef} = 0$$

but when the load is at point f,

$$F_{ef} = -1$$

This influence line is constructed using these equations.

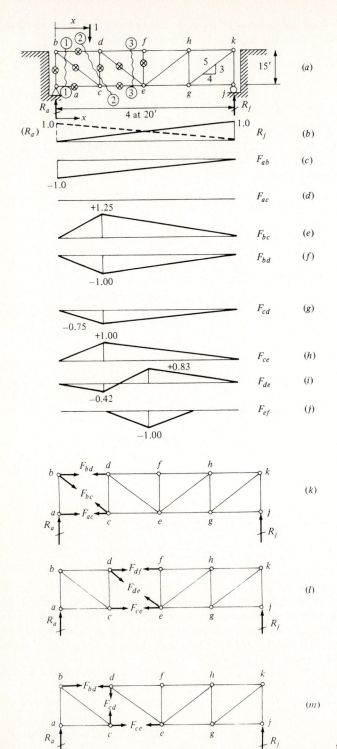

Figure E5.10

5.7 HIGHWAY LOADING

A bridge on a state or federal highway system must be designed using the specifications of the American Association of State Highway and Transportation Officials (AASH-TO).[1] The bridge live loads declared by AASHTO consist of either a truck load or a lane load. The truck load represents a single heavy vehicle on the span consisting of either a two-axle truck or a two-axle truck plus a one-axle semitrailer. The two-axle vehicles are designated by the letter H (or M if specified in SI units) followed by the weight of the truck in tons (or approximately 10^4 newtons). The three vehicles in this series are the H 10, H 15, and H 20 (M 9, M 13.5, and M 18) trucks.† These H (M) series loads have 20 percent of the vehicle weight on the front axle and 80 percent on the rear. The HS 15 and HS 20 (MS 13.5 and MS 18) trucks are composed of the corresponding H (M) series truck plus a semitrailer with a single-axle load equal to that of the rear axle of the H (M) series load. A lane load corresponding to each truck loading constitutes another possible design loading. The lane loads are made up of a line of moderate-weight vehicles plus a single heavy load in their midst. An example of the truck loads and the corresponding lane loading is shown in Fig. 5.16. Both the truck and lane loads must be used to determine the maximums, but usually the lane loads will produce the largest moments in simple spans greater than 145 ft and the largest shears in simple spans greater than 128 ft. The standard loads used to design a given bridge will depend upon the type and density of traffic on the highway.

There are also standard loads for the interstate highway system that consist of a pair of axle loads of 24 kips each, spaced 4 ft apart. This loading generally will be the critical design condition for relatively short spans up to 40 ft.

The live loads occupy a standard lane of traffic that is 10 ft (3.048 m) wide. For a multilane bridge it is necessary to find the amount of loading transmitted to each girder. This can be accomplished by calculating the reactive forces delivered from the floor beam (or bridge deck) to the various individual girders. If there are more than two girders, this will require an indeterminate analysis; therefore, AASHTO provides tables for multiple-girder bridges that can be used to calculate the load per girder.

In addition to the live loads, impact loading caused by such effects as vertical misalignment of the highway and bridge must be calculated. AASHTO specifications treat this effect using a static loading which is a percentage of the live load. The impact percentage depends upon the span length of the bridge. Since various other design criteria must be met, the structural engineer must understand the specification in its entirety when designing a bridge. The structure in Example 5.11 is investigated for the maximum shear and moment at point b for a standard HS 20 (M 18) loading.

Example 5.11 This beam is loaded with one lane of a standard HS 20 (MS 18) highway loading. Find the values of $(V_b)_{max+}$, $(V_b)_{max-}$, and $(M_b)_{max+}$ due to this live load.

SOLUTION The bridge is supported by two identical beams, each supporting one standard 10-ft lane of traffic. According to AASHTO, both lane and truck loading must be investigated and the maximum of the two values used to design the member. The loadings are

---

†The metric designations are approximate conversions, derived as follows: 10 tons = 20,000 lb × 4.448 N/lb = 88.96 kN ≈ 9×10^4 N.

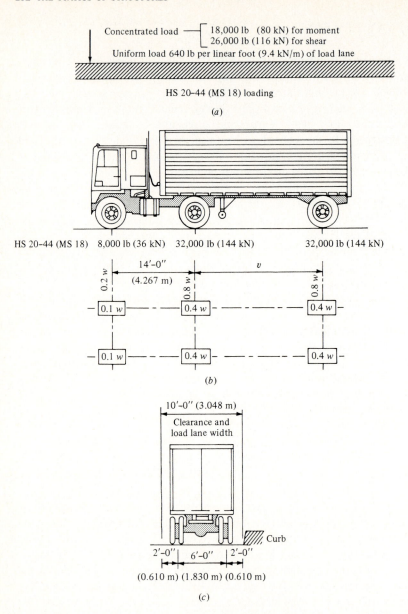

Concentrated load — 18,000 lb (80 kN) for moment
26,000 lb (116 kN) for shear
Uniform load 640 lb per linear foot (9.4 kN/m) of load lane

HS 20–44 (MS 18) loading

(a)

HS 20–44 (MS 18)   8,000 lb (36 kN)   32,000 lb (144 kN)   32,000 lb (144 kN)

(b)

10'-0" (3.048 m)

Clearance and
load lane width

Curb

2'-0"   6'-0"   2'-0"
(0.610 m) (1.830 m) (0.610 m)

(c)

Figure 5.16 Standard highway loads: (*a*) lane loading; (*b*) truck loading; w = combined weight on the first two axles, which is the same as for the corresponding H (M) truck, v = variable spacing 14 to 30 ft (4.267 to 9.144 m) inclusive; spacing to be used is that which produces maximum stresses; (*c*) lane arrangement.[1]

positioned as shown in the sketches. Generally, the truck loading is positioned so that the wheels are over the largest ordinates; the uniform lane loading is positioned over the areas of the influence line with all positive (or negative) ordinates. Since it is not apparent where the truck loading should be positioned for the maximum moment, both cases 1 and 2 (as illustrated) must be evaluated and the larger moment selected. The calculations for these

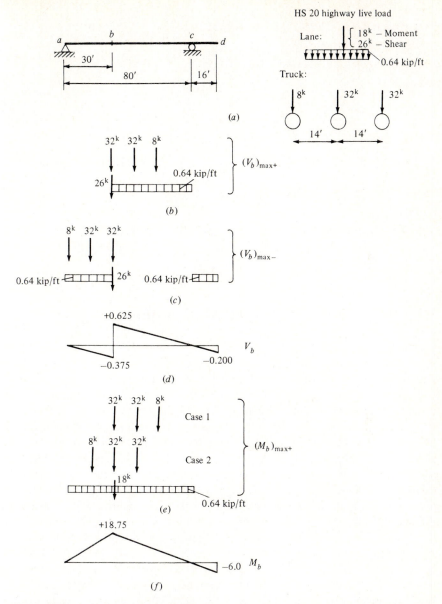

Figure E5.11 (a) HS 20 (MS 18) highway live load, (b) and (c) loading for V_b, (d) influence line for V_b, (e) loading for M_b, (f) influence line for M_b.

three effects are as follows:

$$V_{b+} = \begin{cases} 32(0.625) + \dfrac{0.625}{50}[36(32) + 22(8)] = 36.60 \text{ kips} \qquad \text{truck} \\[4mm] \tfrac{1}{2}(0.625)(50)(0.64) + 26(0.625) = 26.25 \text{ kips} \qquad \text{lane} \end{cases}$$

$$V_{b-} = \begin{cases} 32(-0.375) - \dfrac{0.375}{30}[16(32) + 2(8)] = -18.60 \text{ kips} & \text{truck} \\[2em] \tfrac{1}{2}(-0.375)(30)(0.64) + \tfrac{1}{2}(-0.20)(16)(0.64) \\[1em] \qquad\qquad\qquad + 26(-0.375) = -14.37 \text{ kips} & \text{lane} \end{cases}$$

$$M_{b+} = \begin{cases} 32(18.75) + \dfrac{18.75}{50}[36(32) + 22(8)] = 1098 \text{ kip} \cdot \text{ft} & \text{truck, case 1} \\[2em] (8)\dfrac{18.75}{30}(16) + 32(18.75) + (32)\dfrac{18.75}{50}(36) = 1112 \text{ kip} \cdot \text{ft} & \text{truck, case 2} \\[2em] \tfrac{1}{2}(18.75)(80)(0.64) + 18(18.75) = 818 \text{ kip} \cdot \text{ft} & \text{lane} \end{cases}$$

Thus, the largest values given in all cases by the truck loading are

$$(V_b)_{\text{max}+} = 36.60 \text{ kips} \qquad (V_b)_{\text{max}-} = -18.60 \text{ kips}$$
$$(M_b)_{\text{max}+} = 1112 \text{ kip} \cdot \text{ft} \qquad \text{case 2}$$

5.8 ABSOLUTE MAXIMUM SHEAR IN A BEAM

In order to design a structure that is subjected to moving loads, it is imperative to know the absolute maximum shear the loads can induce. This is possible only if it is known where the loading should be placed on the structure in order to cause this maximum effect. The maximums calculated in the previous sections were obtained for a specific point in the structure without considering whether the maximums obtained by loading the influence line corresponded to the absolute maximum shear for the structure. For a single concentrated moving load the maximum shear in the beam will generally occur in the vicinity of the reactions when the load is located as close as possible to the support point, but for some beams, such as a cantilever, the maximum shear is obtained if the loading is positioned at any point on the structure. For a uniformly distributed loading of variable length the maximum shear is obtained in the vicinity of one of the supports if the load is placed over the entire beam. For a series of moving concentrated forces the problem solution becomes less well defined. Generally, to obtain the maximum shear in the span it is necessary to locate the center of gravity of the resultant of the concentrated loads as close as possible to one of the supports while all the loads remain on the structure. Unfortunately, the maximum shear can occur when one or more of the loads has moved off the structure and the resultant of the remaining loads is situated close to the reaction. Thus, the maximum must be calculated using a trial-and-error approach; i.e., the value of the shear must be examined for various combinations and locations of the loads positioned on the beam.

5.9 ABSOLUTE MAXIMUM MOMENT IN A BEAM

Since the maximum moment must usually be known in order to design a beam, when a structure is to be designed for moving loads, it must be established where the loading must be placed and where the maximum moment in the beam will occur. This problem resembles that of determining the maximum shear in a beam. In the usual process the influence line is known and the applied load is situated to induce the largest effect in the beam, thus establishing a maximum for a given location on the structure; however, it is necessary to identify the position that will experience the absolute maximum moment in the span. For example, if a simply supported beam is subjected to a single concentrated load, the maximum moment will occur at the centerline of the beam when the load is situated at this position. For a simply supported girder the maximum will occur when the load is situated at the floor-beam reaction point (panel point) nearest the center of the girder. Also, a moving uniformly distributed load will induce the maximum moment at the center of a simply supported beam if it is applied over the entire span. Generally, for a cantilever beam, the moment of largest magnitude will occur at the fixed end if the moving single concentrated load is located at the tip of the beam. Unfortunately, for most beams subjected to a set of concentrated moving loads there is no direct way to establish the location of the load and the position in the structure where the greatest moment occurs. Typically, the task of finding the maximum moment must be accomplished by some type of trial-and-error solution.

It is possible, however, to establish the maximum bending moment in a simply supported beam that is loaded with a series of moving concentrated loads. This can be done in a direct systematic manner. The moment diagram introduced by these loads will consist of a series of straight lines that change slope at the locations of the concentrated loads. The maximum value of a function can be attained at the endpoints of its domain, where its derivative is zero or is undefined. Therefore, the maximum moment will occur under one of the wheels. The question of which wheel will be associated with the maximum moment must be answered first. Usually the heaviest wheel load adjacent to the resultant of all applied loads is the wheel under which the maximum moment in the span occurs. Then it is necessary to establish where the loads must be positioned on the structure to attain this maximum moment. Assume for the wheel loads shown in Fig. 5.17a that wheel 2 is the largest wheel load closest to the resultant; thus the maximum moment will occur under this wheel. If d is the distance from wheel 2 to the resultant of the set of loads $(\overline{P})$, and x is the unknown distance from wheel 2 to the centerline of the beam,

$$R_a = \frac{\overline{P}[(L/2) - (d - x)]}{L} \qquad (5.13)$$

From the free-body diagram in Fig. 5.17b the moment under wheel 2 is

$$M = R_a\left(\frac{L}{2} - x\right) - P_1 l_1 = \frac{\overline{P}}{L}\left(\frac{L^2}{4} - \frac{dL}{2} + dx - x^2\right) - P_1 l_1 \qquad (5.14)$$

The maximum value of the moment will occur when x is such that the first derivative of Eq. (5.14) with respect to x is zero. Thus,

$$\frac{dM}{dx} = \frac{\overline{P}}{L}(d - 2x) = 0 \qquad (5.15)$$

giving

$$x = \frac{d}{2} \qquad (5.16)$$

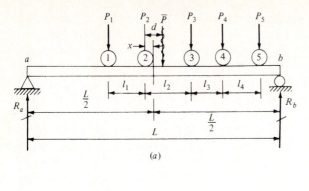

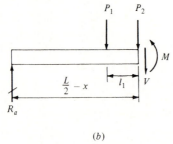

Figure 5.17 A set of concentrated loads on a simply supported beam.

Thus, the maximum moment in a simply supported beam that is loaded with a set of moving concentrated loads will occur under the heaviest wheel nearest the resultant whose distance from the centerline of the beam equals half its distance from the resultant of all of the loads. Sometimes it is necessary to review the moment under several wheels using this criterion in order to establish the maximum value. Example 5.12 illustrates the use of this maximum-moment criterion for a set of three wheels on a simply supported beam.

Example 5.12 Calculate the absolute maximum moment and shear in this beam induced by the set of wheel loads shown.

SOLUTION The resultant of the applied loads is between wheels 2 and 3; therefore, according to the criterion described in Eq. (5.16), wheel 2 and the resultant should be placed equidistant from the centerline of the beam, as shown in Fig. E5.12b. The maximum moment in the beam will then occur under wheel 2; that is,

$$M_{max} = 14R_a - 5(16) = 443 \text{ kN} \cdot \text{m}$$

This criterion could also have been applied to wheel 3, but in this position, the moment under wheel 3 is less than 443 kN · m. The maximum shear will occur near a reaction and is obtained by positioning the wheels as shown in Fig. E5.12d. Thus with the resultant located as close as possible to one support and all wheels on the structure

$$V_{max} = R_a = \tfrac{23}{30}(80) = 61.33 \text{ kN}$$

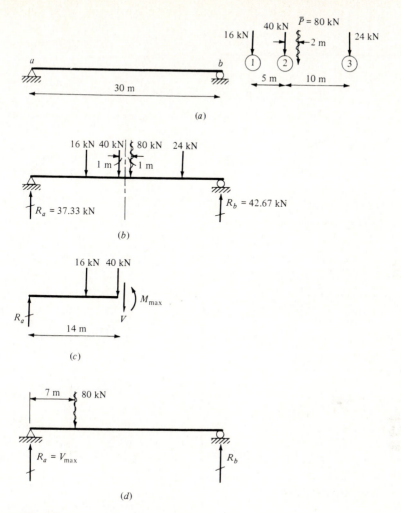

Figure E5.12

5.10 ENVELOPE OF MAXIMUM EFFECTS

An influence line gives the response at a specific point as the loading moves across the structure; therefore, it is impossible to use a single influence line to obtain the maximum effect in a beam unless the critical location is known a priori. However, by constructing a family of influence lines for all points on the beam and selecting the maximums from them an envelope of maximums can be constructed. A structural engineer can consult such an envelope to determine the maximum effect that must be used to establish the cross-sectional properties for the member.

Figure 5.18 illustrates the envelopes for the shear and bending moment in a simple span for an applied single concentrated load. These two envelopes verify the anticipated

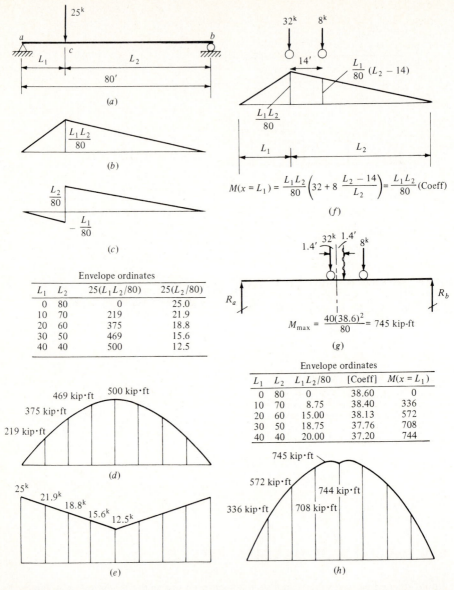

Figure 5.18 Envelopes of maximums for a simply supported beam: (*b*) influence line for the bending moment at a general point *c*; (*c*) influence line for the shear at a general point *c*; (*d*) envelope of maximum bending moments (single load); (*e*) envelope of maximum shears (single load); (*f*) influence line for the bending moment at $x = L_1$ loaded with a H 20 (M 18) wheel load; (*h*) envelope of maximum bending moments [H 20 (M 18) load].

results — that the maximum shear occurs when the load is adjacent to the supports and the maximum moment occurs when the load is at the centerline of the structure. If this same beam is loaded with a standard H 20 (M 18) truck, the shape of the bending-moment envelope is slightly altered because the maximum moment occurs 1.4 ft from the centerline (see Fig. 5.18g). The envelope of maximum bending moment is constructed in Example 5.13 for the beam of Example 5.2 with a 75-kN load applied.

Example 5.13 Construct the envelope of maximum moments for the beam of Example 5.2 if the structure is loaded with a moving concentrated load of 75 kN.

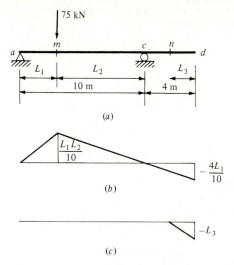

Envelope ordinates

L_1	L_2	$75\dfrac{L_1 L_2}{10}$	$-75\dfrac{4L_1}{10}$	L_3	$-75L_3$
0	10	0	0	0	0
2	8	120	−60	2	−150
4	6	180	−120	4	−300
5	5	188	−150		
6	4	180	−180		
8	2	120	−240		
10	0	0	−300		

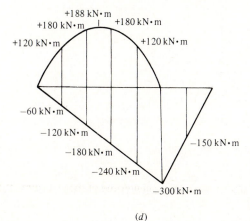

Figure E5.13 Influence line for the bending moment at (b) a general point m between a and c and (c) a general point n between c and d; (d) envelope of maximum positive and negative bending moments.

DISCUSSION This structure will have both positive and negative bending moments induced as the loading moves across, as illustrated by the influence line for a point between the supports a and c. However, it can be seen that for all positions between c and d negative moments occur in the overhanging beam segment only when the loading occupies span cd. Thus, the complete envelope, encompassing both positive and negative moments, indicates that the maximum positive moment occurs at midspan between points a and c and that the largest negative moment is introduced at point c.

PROBLEMS

5.1 Draw the influence lines for the reactions at b and d and the shear and moment at c.

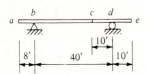

Figure P5.1

5.2 The beam of Prob. 5.1 is subjected to moving loads consisting of a uniform load of 450 lb/ft (variable length) and a concentrated load of 9000 lb. For live load calculate (a) the maximum upward reaction at b; (b) the maximum positive and negative shear at c; (c) the maximum positive and negative moments at c; and (d) the maximum moment at c due to dead plus live load if the dead load is 600 lb/ft.

5.3 Draw influence lines for the horizontal reaction at a, the shear at b, and the moment at b.

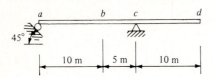

Figure P5.3

5.4 The beam of Prob. 5.3 is subjected to moving loads consisting of a uniform load of 9 kN/m (variable length) and a concentrated load of 100 kN. For live load calculate (a) the maximum upward reaction at c; (b) the maximum positive and negative shear at b; (c) the maximum positive and negative moment at b; and (d) the maximum moment at b due to dead plus live load if the dead load is 6 kN/m.

5.5 Compute the maximum (positive and negative) shear that can occur at the left support due to the illustrated wheel loads.

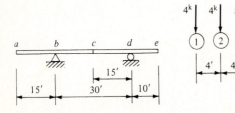

Figure P5.5

5.6 Draw the influence lines for the shear and moment at point c for the beam of Prob. 5.5 and calculate their maximum (positive and negative) values due to the illustrated wheel loads.

5.7 Loads normal to the beam move between points b and c on this rigid frame. Draw the influence lines for (a) the horizontal and vertical reactions at a; (b) the reaction at c; and (c) the moment, shear, and axial force at d.

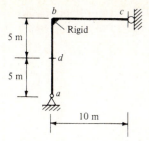

Figure P5.7

5.8 Moving loads are applied normal to members *ad* and *ac* of this rigid frame. Draw influence lines for (*a*) shear at points *a* and *b* and (*b*) bending moment at points *a* and *b*.

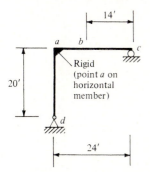

Figure P5.8

5.9 Moving loads traverse along member *abc* of this rigid frame. Draw influence lines for (*a*) the reaction R_a; (*b*) the horizontal and vertical reactions at *e*; and (*c*) the shear and moment at *d*.

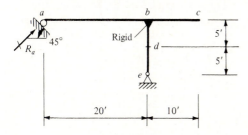

Figure P5.9

5.10 Vertical loads moving from *a* to *c* are transmitted to this structure. Draw influence lines for the moment, shear (internal force component normal to the member axis), and axial force at point *d*.

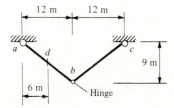

Figure P5.10

5.11 Vertical loads moving over the structure from a to c are transmitted to this three-hinge arch. Draw influence lines for the moment, shear (internal force component normal to the member axis), and axial force at point d.

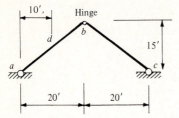

Figure P5.11

5.12 Draw influence lines for the reactions at points a and b and the shear and moment at point d.

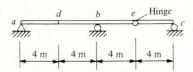

Figure P5.12

5.13 Draw influence lines for the shear at point b and the moment at point c.

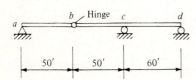

Figure P5.13

5.14 Draw influence lines for (a) the reactions at a and f; (b) the shear and moment at b; and (c) the shear and moment at d.

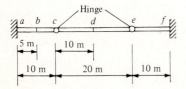

Figure P5.14

5.15 Draw influence lines for (a) the reactions at a and b; (b) the shear at c; and (c) the shear at g.

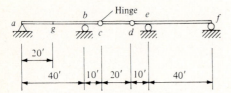

Figure P5.15

5.16 Draw influence lines for (a) the reaction at a; (b) the shear at c and e; and (c) the moment at d.

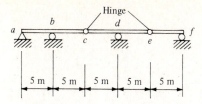

Figure P5.16

5.17 Loads (normal to the member) move along the bottom horizontal beam from a to c. Draw the influence lines for (a) the force in cable db; (b) the moment in the beam at b; and (c) the shear in the beam at e.

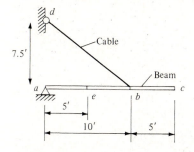

Figure P5.17

5.18 A standard highway live loading designated as H 20-44 (M 18) is shown. The structure is to be loaded individually with the truck wheel load and also with the lane loading, consisting of the uniformly distributed load plus a single concentrate load. Using this loading, find (a) the maximum (positive and negative) shear and moment at c for the beam of Prob. 5.1; (b) the maximum (positive and negative) shear and the maximum positive moment at d for the beam of Prob. 5.14.

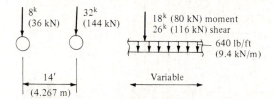

Figure P5.18

5.19 The load is transmitted to this three-hinge arch from the bridge deck through the struts. Draw the influence lines for (a) the vertical and horizontal reactions at a and e and (b) the moment at f.

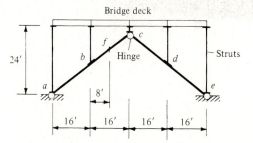

Figure P5.19

5.20 Draw the following influence lines for the girder: (a) the shear between c and d and (b) the bending moment at d.

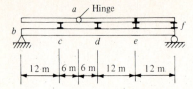

Figure P5.20

5.21 Draw the following influence lines for the girder: (*a*) the shear between *b* and *d* and (*b*) the moment at *c*.

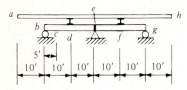

Figure P5.21

5.22 Draw the influence lines for the shear and moment at *g* in the girder.

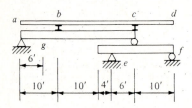

Figure P5.22

5.23 Draw the influence lines for the shear and moment at *e* in the girder.

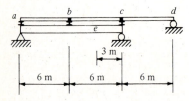

Figure P5.23

5.24 Draw the following influence lines for the girder: (*a*) the shear between *d* and *f* and (*b*) the moment at *e*.

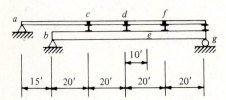

Figure P5.24

5.25 Draw the following influence lines for the girder: (*a*) the shear between *b* and *c*; (*b*) the moment at *b*; and (*c*) the moment at *c*.

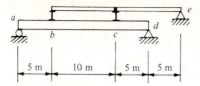

Figure P5.25

5.26 Draw the following influence lines for the girder: (a) the reaction at c; (b) the shear between c and d; and (c) the moment at d.

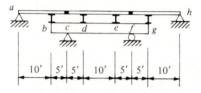

Figure P5.26

5.27 Draw the following influence lines for the girder: (a) the reactions at b and e; (b) the shear between d and e; and (c) the moment at e.

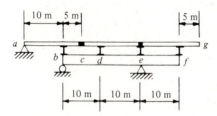

Figure P5.27

5.28 Sketch the beam (or beam-girder system) that can be identified with each of these five influence lines and indicate what the influence line characterizes (reaction, shear, moment, etc.).

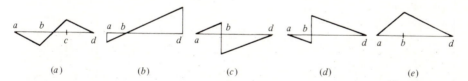

(a)   (b)   (c)   (d)   (e)

Figure P5.28

5.29 Calculate the maximum bending moment introduced in the girder at e by the moving 10-kip load.

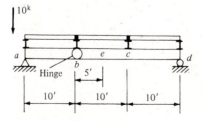

Figure P5.29

5.30 Loads move along the bottom chord *ace* of the truss of Fig. P3.5. Draw the influence lines for all the members.

5.31 Loads move along the top chord *bdfh* of the truss of Fig. P3.6. Draw the influence lines for members *df*, *ce*, *de*, *cd*, and *fe*.

5.32 Loads move along the bottom chord from *a* to *k* on this truss. Draw influence lines for the marked members *fg*, *dg*, and *eg*.

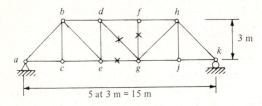

5 at 3 m = 15 m

Figure P5.32

5.33 Loads move along the bottom chord from *a* to *r* on this K truss. Draw influence lines for the marked members *lm*, *mp*, and *mq*.

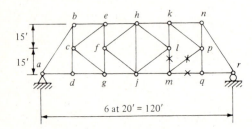

6 at 20' = 120'

Figure P5.33

5.34 Loads move along the top chord *adfgj* on the truss of Fig. P3.11. Draw influence lines for members *ce*, *ed*, and *df*.

5.35 Loads move along the bottom chord *aceg* on the truss of Fig. P3.8. Draw influence lines for members *bd*, *cd*, and *ce*.

5.36 Loads move along the bottom chord *acf* on the truss of Fig. P3.7. Draw influence lines for members *ab*, *ac*, and *cb*.

5.37 Loads move along the bottom chord from *a* to *h* on the truss of Fig. P3.9. Draw influence lines for members *bc*, *be*, and *ce*.

5.38 Loads move along the top chord from *a* to *j* on the truss of Fig. P3.10. Draw influence lines for members *eg*, *ef*, and *df*.

5.39 Loads move along the bottom chord from *a* to *r* on the truss of Fig. P3.14. Draw influence lines for members *dh*, *de*, *df*, *ef*, *eg*, *gj*, *fj*, *fg*, and *hj*.

5.40 Loads move along the bottom chord from *a* to *r* on the truss of Fig. P3.15. Draw influence lines for members *hm*, *hk*, and *kl*.

5.41 Loads move along the bottom chord from *a* to *m* on the truss of Fig. P3.16. Draw influence lines for members *eg* and *gk*.

5.42 Loads move along the top chord from *a* to *f* on the truss of Fig. P3.12. Draw influence lines for members *ce*, *de*, *df*, and *fh*.

5.43 Loads move along the bottom chord from *a* to *j* on the truss of Fig. P3.13. Draw influence lines for members *bd*, *ef*, *fg*, and *gk*.

5.44 Loads move along the top chord from *a* to *f* on this truss. Draw influence lines for members *ac*, *bd*, and *cd*.

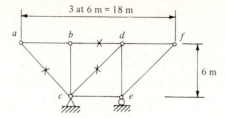

Figure P5.44

5.45 Loads move along the top chord on stringers as shown for this truss. Draw influence lines for members *df*, *de*, and *ef*.

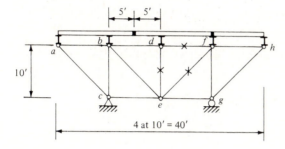

Figure P5.45

5.46 Calculate the maximum (positive and negative) shear and moment for the beam of Prob. 5.5 for the illustrated set of moving wheel loads.

5.47 Calculate the maximum (positive and negative) shear and moment for the beam of Prob. 5.1 if it is loaded with the set of moving wheel loads shown.

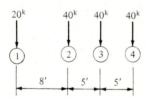

Figure P5.47

5.48 A simply supported beam has a length of 70 ft and supports a dead load of 2 kips/ft. Calculate the maximum moment for the dead load plus a standard highway HS 20-44 (MS 18) truck. See Fig. 5.16 for the wheel loads and spacing.

5.49 Construct the envelopes of maximum shear and bending moment for the beam of Prob. 5.3 for an applied concentrated load of 120 kN.

5.50 Construct the envelopes of maximum shear and bending moment for the beam of Prob. 5.13 for an applied concentrated load of 40 kips.

REFERENCE

1. The American Association of State Highway and Transportation Officials: *Standard Specifications for Highway Bridges,* 12th ed., secs. 2 and 3, AASHTO, Washington, 1977.

ADDITIONAL READING

Gerstle, K. H.: *Basic Structural Analysis,* chap. 6, Prentice-Hall, Englewood Cliffs, N.J., 1974.

Hsieh, Yuan-Yu: *Elementary Theory of Structures,* 2d ed., chaps. 6 and 7, Prentice-Hall, Englewood Cliffs, N.J., 1982.

Ketter, R. L., G. C. Lee, and S. P. Prawel, Jr.: *Structural Analysis and Design,* chap. 3, McGraw-Hill, New York, 1979.

Laursen, H. I.: *Structural Analysis,* 2d ed., chap. 10, McGraw-Hill, New York, 1978.

McCormac, J. C.: *Structural Analysis,* 3d ed., chaps. 11 to 13, Intext, New York, 1975.

Norris, C. H., J. B. Wilbur, and S. Utku: *Elementary Structural Analysis,* 3d ed., chap. 5, McGraw-Hill, New York, 1976.

Timoshenko, S. P., and D. H. Young: *Theory of Structures,* 2d ed., chap. 3, McGraw-Hill, New York, 1965.

West, H. H.: *Analysis of Structures,* chap. 9, Wiley, New York, 1980.

Willems, N., and W. M. Lucas, Jr.: *Structural Analysis for Engineers,* chaps. 4 and 5, McGraw-Hill, New York, 1978.

ROOF AND BRIDGE TRUSS STRUCTURES

6.1 INTRODUCTION

A building or a bridge is composed of a number of individual structural elements that must function together to resist the applied loads. Throughout the design process the structural engineer must be able to visualize the various individual load-carrying elements and to understand the functional interrelationships necessary for transmitting the applied loads through the structure. Usually the structure will be subjected to several distinct types of loading (e.g., lateral as well as vertical), and the effective load-transfer mechanisms typically vary with the applied loads. An example of a structure in which the load-carrying components are visible is shown in Fig. 6.1.

The variety of structural loads and how they are prescribed by the various building codes and standards were discussed in Chap. 1, the static analysis of trusses was described in Chap. 3, and Chap. 5 treated the special problem of moving loads on bridge structures and in some buildings (such as mill structures with overhead cranes). In this chapter we integrate the information from these three earlier chapters and demonstrate the process of analyzing statically determinate roof and bridge truss structural systems to obtain the maximal load that must be resisted by individual structural members. Since several different loads can be applied to a structure simultaneously, e.g., dead, live, and wind, their combined effect must be considered and it is usually necessary to appraise several combinations of loading to obtain the maximum force in each of the structural members. Therefore, one of the most important functions of structural analysis is to establish the critical load that must be resisted by each individual element during the useful life of the structure.

6.2 IDEALIZATION OF STRUCTURES WITH ROOF TRUSSES

The roof trusses in a building constitute the *primary structural components* for resisting applied vertical loads such as dead, live, and various environmental loads. Lateral forces are resisted by the primary structural system, but in addition, considerable

Figure 6.1 Structural system using wooden beams and steel trusses. *(Setter Leach & Lindstrom, Minneapolis.)*

secondary bracing is required to transmit these loads through the structure and into the foundations. For example, the load-carrying elements for the building in Fig. 6.2 consist of several distinct systems, as shown. The intermediate frames will partially resist lateral forces applied parallel to the short direction of the building, but the resistance of the building as a whole is considerably enhanced by the end frames and the associated diagonal bracing. Since the roof trusses cannot resist forces applied parallel to the long direction of the building, considerable secondary bracing is required to transmit these loads. For the structure in Fig. 6.2 the diagonal bracing is incorporated into the planes of the exterior side walls, the plane of the lower chord of the trusses, and the planes of the top truss chords. In this example, alternate bays are braced, and each of the planar trusses in the braced bays constitutes an effective element for resisting longitudinal forces. The girts and purlins are flexural members that transmit forces from the exterior walls and the roof, respectively, to the primary structure.

Sketches showing the various components for an unbraced bay of a structure composed of Fink roof trusses are shown in Fig. 6.3. The exterior cladding on the roof is attached to the sheathing, which transmits the loads to the joists. Thus, the purlins, which serve as the supports for the joists, are subjected to forces both normal and tangential to the plane of the roof. In Fig. 6.3a we observe that the purlins are connected to the panel points of the truss by the small clip angles. Typically, the purlins are inherently weak in resisting applied forces in the plane of the roof; therefore, they

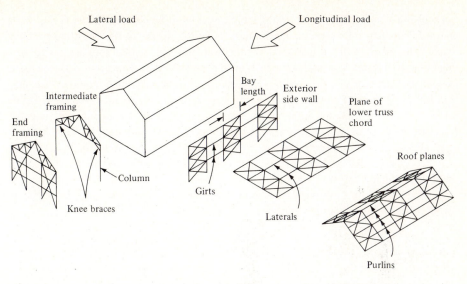

Figure 6.2 A typical structure with roof trusses showing elements transmitting horizontal forces.

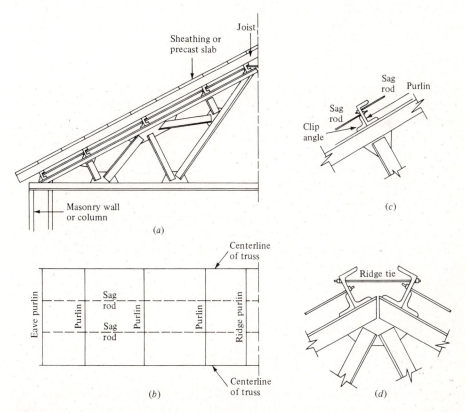

Figure 6.3 (*a*) Half of a Fink roof truss; (*b*) plan view in plane of the truss; (*c*) typical upper joint; (*d*) ridge detail.

must be braced at various points along their length with sag rods, as shown in Fig. 6.3a and b. These sag rods serve as intermittent supports for the purlins and transmit their loads to the ridge purlin, where the sag rods from both sides of the roof are joined with the ridge tie. Figure 6.3c and d shows typical details for the connections between these secondary members and the purlins. Roof trusses can be supported by either masonry bearing walls or steel columns, depending upon the type of structural framing. Uplift of the truss, imparted by wind, must be resisted either by an anchor bolt into the masonry wall or by an attachment to the column. With steel columns a knee brace is typically connected between the truss and the column, as shown in Fig. 6.2.

A complete analysis of this structure requires each primary and secondary member to be investigated for the effect of the applied loads to obtain the maximum load that must be resisted. In this chapter we shall focus our attention only upon the members of the truss and establish the required resisting forces.

6.3 ROOF LOADS

The roof of a structure can be subjected to dead and live loads plus environmental loads induced by snow, rain, wind, and earthquake. Snow and rain loads are applied statically, while wind and earthquake constitute dynamic loads. The dead loads are composed of the weights of (1) the roof, e.g., the impervious external membrane and sheathing; (2) the purlins; (3) the truss members; (4) the ceiling plus insulation; and (5) any fixed service equipment. Roof live loads include the weight of personnel and equipment for maintenance plus loads induced by suspended objects attached to either the panel points of the lower truss chord or the ceiling. The environmental loads are established by the engineer using the applicable standard or code (see Chap. 1 for a detailed discussion of the calculation of these loads).

The individual loadings must be combined to give the maximal forces to be resisted by the truss members. The procedure for combining the loads depends to a great extent upon the material used and also the standard or code being followed. Generally, two basic approaches[1] can be used in the design of structural members, *allowable-stress design* and *limit-states design*.

Allowable-stress design is a method for proportioning structural members in which the computed elastic stress must not exceed a specified limiting value. Generally, for this approach the design of a member is based on the combination of loads that induces the largest load in the member. For example, some of the basic load combinations that should be considered are

AS1: Dead
AS2: Dead + live + snow (or rain)
AS3: Dead + wind (or earthquake)
AS4: Dead + live + snow (or rain) + wind (or earthquake)

Typically, the most unfavorable effects from both wind and earthquake loads are considered, where appropriate, but they are not assumed to act simultaneously. Most loads (other than dead loads) can vary significantly, but the design should be carried out in such a way that the risk of unsatisfactory performance is reduced to an acceptably low level under the effect of combined loading. When several variable loads are assumed to act at once, it is extremely unlikely that all will attain their maximum effect simultaneously; therefore, the total of the combined load effects is usually reduced using *load-combination factors,* for example, 0.75 for combination AS4 and 0.66 for load combination AS4 plus temperature effects. Some material specifications permit the allowable stresses to be increased by 33 percent when wind or earthquake effects are considered; however, this practice is somewhat questionable if the combined load effects are also reduced by the appropriate load-combination factor.

The objective of limit-states design is to ensure against a structure's or component's becoming unfit for use, i.e., being no longer useful for its intended function (serviceability limit state) or unsafe (strength limit state). In this approach, nominal load effects are multiplied by *load factors* to account for (1) unavoidable deviations of the actual load from the nominal value and (2) uncertainties in the analysis for the member forces. Also, *resistance factors* (multipliers on the maximum load-carrying capacity of a member) account for (1) unavoidable deviations of the actual member strengths from the nominal value and (2) the manner of failure and its consequences. In the development stages of this method, load factors were typically obtained by surveys of the reliabilities inherent in existing design practice for allowable-stress design. In limit-states design it is assumed that, in addition to the dead load, one of the variable loads takes on its maximum lifetime value while the other variable loads assume values that could probably be measured at that instant. Some typical load combinations and the associated load factors are as follows:

LS1: $1.4 \times$ dead
LS2: $1.2 \times$ dead $+ 1.6 \times$ live $+ 0.5 \times$ snow (or rain)
LS3: $1.2 \times$ dead $+ 1.6 \times$ snow (or rain) $+ 0.5 \times$ live (or $0.8 \times$ wind)
LS4: $1.2 \times$ dead $+ 1.3 \times$ wind $+ 0.5 \times$ live $+ 0.5 \times$ snow (or rain)
LS5: $1.2 \times$ dead $+ 1.5 \times$ earthquake $+ 0.5 \times$ live (or $0.2 \times$ snow)
LS6: $0.9 \times$ dead $- 1.3 \times$ wind (or $1.5 \times$ earthquake)

Note that some of the specified load factors are less than unity for the various load combinations since the usual prescribed loads are in excess of the arbitrary values. Most design standards for typical buildings permit a one-third increase in allowable stress (which corresponds to a 25 percent reduction in total factored load) for wind, and these adjustments are reflected in the load factor of 1.3 for wind load in combinations LS4 and LS6.

6.4 ANALYSIS OF A ROOF TRUSS

A complete analysis of a roof truss includes (1) the calculation of member forces for each separate load case and (2) a systematic combination of these results to obtain the critical forces that must be resisted by each individual member. The truss in Fig. 6.4

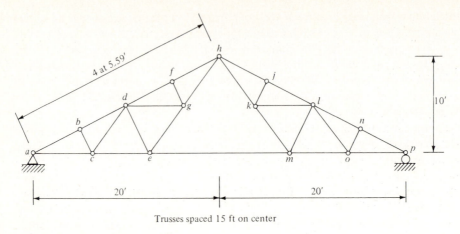

Trusses spaced 15 ft on center

Figure 6.4 Design example of a Fink roof truss.

will be investigated to illustrate the general procedure for obtaining the member design loads. For this structure, the trusses are spaced 15 ft on center, and this investigation focuses upon the analysis of a typical intermediate truss.

The dead load consists of the roofing, decking, insulation, and purlins, in addition to the weight of the truss members. The total weight of the truss, including the members, gusset plates, and other miscellaneous steel, is estimated to be 1080 lb, and it is assumed that a solution of sufficient accuracy will be obtained if this weight is considered to be divided between the panel points on the upper and lower chords; i.e., this results in a load of 90 lb per intermediate panel point and 45 lb at points a and p. The other loads are

Asphalt shingles	2 lb/ft^2
Decking, 2-in	6
Insulation	1
Purlins	1.73
Total	10.73 lb/ft^2

Note that for convenience the purlins have been assumed to consist of a load distributed uniformly over the roof surface. The distance between panel points on the upper chord measured along the roof is 5.59 ft; therefore, the total dead load per intermediate upper panel point is $10.73(5.59)(15) + 90 = 990$ lb, and lower panel points are loaded with only the dead weight of the truss (90 lb). The member forces induced by this symmetrical loading are listed in Table 6.1 under the heading D.

Since there is no ceiling in this building, the panel points on the lower chord will be accessible for suspending loads. In this case, some codes stipulate that the structure be capable of supporting a specified live load from any single panel point on the lower chord of the roof truss. For this example, the concentrated load has been chosen to be 2000 lb, and the member forces introduced by suspending this load from panel points c and e are tabulated in Table 6.1 as L1 and L2, respectively. Load cases L3 and L4

Table 6.1 Fink roof truss member forces for individual load cases

Member	D	L1	L2	L3	L4	SB	SUR	SUL	W
Upper chord, left:									
ab	−8150	−3773	−3074	−698	−1398	−15,849	−6,708	−16,771	+6195
bd	−7707	−3773	−3074	−698	−1398	−14,943	−6,708	−15,429	+6195
df	−7164	−1538	−3074	−698	−1398	−14,038	−6,708	−14,087	+6195
fh	−6721	−1538	−3074	−698	−1398	−13,132	−6,708	−12,745	+6195
Upper chord, right:									
hj	−6721	−698	−1398	−1538	−3074	−13,132	−12,745	−6,708	+6195
jl	−7164	−698	−1398	−1538	−3074	−14,038	−14,087	−6,708	+6195
ln	−7707	−698	−1398	−3773	−3074	−14,943	−15,429	−6,708	+6195
np	−8150	−698	−1398	−3773	−3074	−15,849	−16,771	−6,708	+6195
Lower chord:									
ac	+7290	+3376	+2750	+624	+1250	+14,176	+6,000	+15,000	−5541
ce	+6232	+1876	+2750	+624	+1250	+12,151	+6,000	+12,000	−4552
em	+4129	+624	+1250	+624	+1250	+8,101	+6,000	+6,000	−2573
mo	+6232	+624	+1250	+1876	+2750	+12,151	+12,000	+6,000	−4552
op	+7290	+624	+1250	+3376	+2750	+14,176	+15,000	+6,000	−5541
Web left:									
bc	−886	0	0	0	0	−1,812	0	−2,683	+885
cd	+1102	+2500	0	0	0	+2,025	0	+3,000	−990
de	−1821	−1118	0	0	0	−3,622	0	−5,367	+1770
dg	+990	0	0	0	0	+2,025	0	+3,000	−990
fg	−886	0	0	0	0	−1,812	0	−2,683	+885
eg	+2148	+1250	+2500	0	0	+4,050	0	+6,000	−1979
gh	+3139	+1250	+2500	0	0	+6,075	0	+9,000	−2968
Web right:									
hk	+3139	0	0	+1250	+2500	+6,075	+9,000	0	−2968
km	+2148	0	0	+1250	+2500	+4,050	+6,000	0	−1979
jk	−886	0	0	0	0	−1,812	−2,683	0	+885
kl	+990	0	0	0	0	+2,025	+3,000	0	−990
lm	−1821	0	0	−1118	0	−3,622	−5,367	0	+1770
lo	+1102	0	0	+2500	0	+2,025	+3,000	0	−990
no	−886	0	0	0	0	−1,812	−2,683	0	+885

correspond to suspending the 2000-lb load from panel points *o* and *m*, respectively, and are also shown in Table 6.1. The member forces for L3 and L4 are derived from those of L1 and L2, respectively. That is, the forces in the members on the right side of the truss for L1 correspond to the member forces on the left side of the truss for L3, etc.

The building is located in a region where the ground snow load p_g is 39 lb/ft^2, and the roof snow loads will be calculated using the ANSI A58.1-1982 procedure[2] described in Chap. 1. That is the flat-roof design snow load p_f is

$$p_f = 0.70 C_e C_t p_g$$

In this case, the exposure factor C_e is 1.0 since there are other structures which surround the building and snow will not necessarily be removed by the wind; the thermal factor C_t is 1.0 for this heated structure. The sloped-roof design snow load p_s is

$$p_s = C_s p_f$$

where the slope factor C_s is 1.0 since the roof slope is not sufficient to shed the snow. From these two equations the snow load is calculated to be 27 lb/ft² (measured on the horizontal projection of roof area). The member forces introduced by this uniform load are listed in Table 6.1 in the column labeled SB (snow load balanced). The unbalanced snow load results from the wind blowing the snow from the windward side of the roof onto the leeward slope, and it is calculated using the relationship $1.5p_s/C_e$, which gives 40 lb/ft² for this structure. The member forces for the unbalanced snow load on the right side of the truss are shown in Table 6.1 under SUR (snow load unbalanced on the right). The results for snow blown off the right slope and onto the left side are obtained as a mirror image of the SUR results and are shown in Table 6.1 under SUL (snow load unbalanced on the left).

The wind load on this structure is calculated using an approach similar to that of ANSI A58.1-1982,[2] discussed in Chap. 1. The equation for the velocity pressure at a height h above the ground in pounds per square foot q_h can be obtained using basic fluid mechanics and noting the proper roughness coefficient for a structure with a specific siting, which gives

$$q_h = 0.00256 K_z V^2$$

where K_z is the velocity-pressure exposure coefficient, V is the basic wind speed in miles per hour, and the constant 0.00256 for standard air is found from

$$0.50 \frac{0.0765 \text{ lb/ft}^3}{32.2 \text{ ft/s}^2} \left(\frac{1 \text{ mi}}{h} \times \frac{5280 \text{ ft}}{\text{mi}} \times \frac{1 \text{ h}}{3600 \text{ s}} \right)^2 = 0.00256$$

Standard air is at 59°F (15°C) under a pressure of 29.92 inches of mercury, which corresponds to 14.7 lb/in² (101.36 kPa). The structure is located in an urban area with numerous closely spaced obstructions having the size of single-family dwellings extending in the upwind direction for a considerable distance, and the basic wind speed is 70 mi/h. The roof trusses are supported upon walls that are 20 ft high; the velocity-pressure exposure coefficient at a mean roof height of 25 ft above the ground is 0.46. These conditions result in a velocity pressure of 5.77 lb/ft².

For this low structure the recommended design pressure p is calculated from

$$p = q_h G_h C_p - q_h(GC_{pi})$$

where G_h = gust response factor
 C_p = external pressure coefficient
 q_h = velocity pressure at mean roof height
 GC_{pi} = product of internal pressure coefficient and gust response factor for building

For the exposure of this building at a mean roof height of 25 ft above ground level $G_h = 1.54$. The height-to-span ratio h/L for the building is 0.625; therefore, for this roof, in which the angle of the plane of the roof from the horizontal is 26.57°,

$$C_p = \begin{cases} -0.39 & \text{on windward side for wind normal to ridge} \\ -0.70 & \text{on leeward side for wind normal to ridge} \\ -0.70 & \text{on both sides for wind parallel to ridge} \end{cases}$$

where minus signs denote pressures acting away from the surface. The openings in the building are such that

$$GC_{pi} = +0.75 \qquad \text{and} \qquad -0.25$$

where the plus and minus signs denote pressures acting toward and away from the surface, respectively. Thus, the critical situation will arise with the wind parallel to the ridge and the internal pressure acting in the positive direction. This combination gives

$$p = 5.77(1.54)(-0.70) - 5.77(0.75) = -6.22 - 4.33 = -10.55 \text{ lb/ft}^2$$

which is directed away from, and perpendicular to, the roof surface, is measured in terms of the sloping roof surface area, and gives a load per intermediate panel point on the upper chord of $-10.55(5.59)(15) = -885$ lb. The member forces induced by these loads are listed in Table 6.1 in column W. Note that if the wind is normal to the ridge, the pressure on the windward roof is

$$p = 5.77(1.54)(-0.39) - 5.77(0.75) = -3.47 - 4.33 = -7.80 \text{ lb/ft}^2$$

and the load per intermediate panel point on the upper chord is 654 lb. Furthermore, the panel-point loads on the leeward roof will be the same as when the wind is parallel to the ridge. Thus, it can be observed that the member forces for wind perpendicular to the ridge will be less than those obtained for the wind acting parallel to the ridge.

The member forces for dead load (D) and wind load (W) are shown in Table 6.2, along with the critical member forces resulting from live load (L) and snow load (S), respectively. Results for only part of the structure are shown in Table 6.2 since the truss design will be symmetrical with respect to the center. Since allowable-stress design is to be used, the appropriate load combinations are

AS1: D
AS2: D + L + S
AS3: D + W
AS4: D + L + S + W (load-combination factor = 0.75)

These combinations are calculated in Table 6.2 for all the members, and the critical design load for each member is obtained from them. Note that all members must be designed for load combination AS2 (D + L + S) in this case. Although the wind load does not cause any member to undergo a reversal of force (from tension to compression or vice versa), this almost occurred for members bc, dg, and fg when only the dead load and wind loads were acting on the structure.

Table 6.2 Summary of critical member forces for a Fink roof truss

Member	Basic loads				Load combinations			
	D	L	S	W	D + L + S	D + W	0.75 × (D + L + S + W)	Critical design load
Upper chord:								
ab	−8150	−3773[a]	−16,771[e]	+6195	−28,694	−1955	−16,874	−28,694
bd	−7707	−3773[a]	−15,429[e]	+6195	−26,909	−1512	−15,536	−26,909
df	−7164	−3074[b]	−14,087[e]	+6195	−24,325	−969	−13,598	−24,325
fh	−6721	−3074[b]	−13,132[d]	+6195	−22,927	−526	−12,549	−22,927
Lower chord:								
ac	+7290	+3376[a]	+15,000[e]	−5541	+25,666	+1749	+15,094	+25,666
ce	+6232	+2750[b]	+12,151[d]	−4552	+21,133	+1680	+12,436	+21,133
em	+4129	+1250[b,c]	+8,101[d]	−2573	+13,480	+1556	+8,180	+13,480
Web:								
bc	−886	0	−2,683[e]	+885	−3,569	−1	−2,013	−3,569
cd	+1102	+2500[a]	+3,000[e]	−990	+6,602	+112	+4,209	+6,602
de	−1821	−1118[a]	−5,367[e]	+1770	−8,306	−51	−4,902	−8,306
dg	+990	0	+3,000[e]	−990	+3,990	0	+2,250	+3,990
fg	−886	0	−2,683[e]	+885	−3,569	−1	−2,013	−3,569
eg	+2148	+2500[b]	+6,000[e]	−1979	+10,648	+169	+6,502	+10,648
gh	+3139	+2500[b]	+9,000[e]	−2968	+14,639	+171	+8,753	+14,639

[a]Member forces induced by L1.
[b]Member forces induced by L2.
[c]Member forces induced by L4.
[d]Member forces induced by SB.
[e]Member forces induced by SUL.

6.5 BRIDGE TRUSS STRUCTURES

In addition to dead load, bridge structures must resist forces introduced by the weight of vehicles, impact, wind, earthquake, centrifugal forces, and other effects such as those caused by temperature. Loads for highway bridges were described in Chaps. 1 and 5. Longitudinal forces are the result of vehicle acceleration, earthquake, and wind acting on surfaces normal to the bridge axis, e.g., truss members, signs, and vehicles. These loads are applied to the truss directly or to the bridge deck, but in either case they are resisted by the *primary structure,* i.e., the two primary trusses, as shown in Fig. 6.5. Forces that act in the lateral direction, i.e., normal to the longitudinal bridge axis, are generally due to wind and earthquake and must be transmitted to the foundations through the *secondary bracing,* as shown in Fig. 6.5. The horizontal trusses in the planes of the upper and lower chords are composed of the diagonals and lateral struts joining the two primary trusses; in addition, the chords of the primary trusses act as the chords for the bracing trusses. The loads transmitted from the upper horizontal truss must be resisted by the intermediate sway bracing plus the portal bracing at the two ends of the structure.

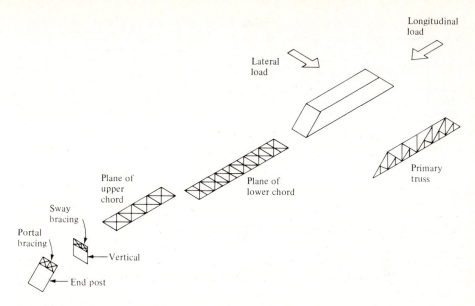

Figure 6.5 Truss bridge structure showing elements transmitting longitudinal and lateral forces.

There are two basic approaches to the design of bridge structures. *Service-load design* requires that the elastic stress not exceed a specified limiting value, while the objective of *load-factor design*[3] is to ensure that a structure or component will not become unfit for use. Service-load design and load-factor design correspond, respectively, to allowable-stress and limit-states design described in Sec. 6.3. Both methods require the individual load cases to be combined in some systematic manner to yield the proper design loads.

With the service-load approach the design of a member should be based upon the combination of loads that induces the largest force in the member. For example, some of the basic load combinations[3] that should be considered are

SL1: Dead + live + impact
SL2: Dead + wind (on structure)
SL3: Dead + live + impact + 0.3 × wind (on structure) + wind (on live load)

where the live load consists of either a standard truck or a uniform lane load accompanied by a single concentrated force, and the impact is expressed as a fraction of the live load. The basic allowable unit stresses are permitted to be multiplied by 1.25 for load combinations SL2 and SL3. Note that the reciprocal of this factor (0.80) can be interpreted as the load-combination factor discussed in Sec. 6.3. That is, since several variable loads are combined, it is extremely unlikely that the loads will attain their maximums simultaneously.

For the load-factor design approach, it is assumed that in addition to the dead load, one of the variable loads will take on its maximum life-time value while the other variable loads assume values that could probably be measured at that instant. Some typical load combinations and the associated load factors[3] are

LF1: $1.3 \times$ dead $+ 2.86 \times$ live $+ 2.86 \times$ impact
LF2: $1.3 \times$ dead $+ 1.3 \times$ wind (on structure)
LF3: $1.3 \times$ dead $+ 1.3 \times$ live $+ 1.3 \times$ impact $+ 0.39 \times$ wind (on structure)
     $+ 1.3 \times$ wind (on live load)

The loads resulting from these combinations must be used in conjunction with the appropriate resistance factors (multipliers on the maximum load-carrying capacity of a member) to arrive at the final member configurations.

6.6 ANALYSIS OF A BRIDGE TRUSS

The process of obtaining the critical member forces for a bridge is similar to that for a roof truss. That is, the various individual loadings must be considered separately and combined in a rational manner to establish the design forces for each member. For a bridge, however, the live load is introduced by moving vehicles and is accompanied by a dynamic effect called *impact*. The calculations required to give the critical member forces for various loadings will be demonstrated for the bridge truss in Fig. 6.6. This two-lane deck bridge consists of trusses spaced 24 ft on center, and the entire structure includes the deck, floor beams, stringers, and secondary bracing (see Fig. 3.2 for the location of these members in a typical through bridge).

The dead load for a highway bridge truss consists of the weight of the truss and bracing plus that contributed by the flooring (slab, wearing surface, etc.), guard rail, sidewalks, curbs, railings, stringers, and floor beams. Initially, most of these loads must be estimated, but after the first analysis and design cycle has been completed, they can be calculated accurately. For the truss in Fig. 6.6 it is estimated that the weight of the truss members, gusset plates, and other details plus the secondary bracing is 0.32 kip per foot of truss; and it is assumed that a sufficiently accurate solution will be obtained if this is divided equally between the top and bottom panel points. Thus, each of the intermediate panel points on both the bottom and top chords has a dead load contributed by the weight of the truss, which is equal to $28(0.32)/2 = 4.5$ kips. The weight of the floor system, including the deck, floor beams, stringers, and other details, is 0.75 kip per foot of truss, and this entire load is applied to the joints

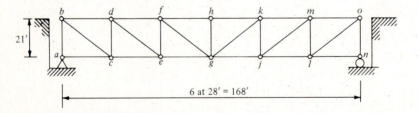

Figure 6.6 Design example of a Pratt bridge truss.

of the top chord. Thus, each of the intermediate top joints will have a dead load of $4.5 + 0.75(28) = 25.5$ kips. The member forces resulting from the dead load are listed in column D of Table 6.5.

The lane loading for a standard H 20-44 (M18) truck[3] constitutes the total live load, consisting of a uniform load of variable length with a magnitude of 0.64 kip/ft plus a single concentrated load of 18 kips (if the member force is introduced by bending) or 26 kips (if the member force results from shear). In addition, the impact, expressed as a fraction of the live load, I is[3]

$$I = \frac{50}{L + 125} \le 0.30$$

where L is the length of the structure (in feet) that is loaded to produce the maximum live-load member force.

The effects introduced by the live load can conveniently be calculated using the influence lines. The member forces resulting from a unit load placed at each upper panel point b to o are given in Table 6.3. The pertinent values from the influence lines are summarized in Table 6.4 in a form that can be used to calculate the live-load member forces. The values in the columns headed "Sum of ordinates" are used to calculate the member forces introduced by the uniform loading; those under "Maximum ordinates" are used in conjunction with the concentrated forces. The loaded lengths are summarized for convenience in calculating the impact fractions. The procedures for loading the influence lines described in Chap. 5 are used to obtain the live-load member forces, and these values are listed in Table 6.5 in column L. That is, the member forces for the

Table 6.3 Influence-line ordinates for the highway bridge truss

Member	b	d	f	h	k	m	o
				Unit vertical load at panel point			
Upper chord:							
bd	0	−1.111	−0.889	−0.667	−0.445	−0.222	0
df	0	−0.889	−1.778	−1.333	−0.889	−0.445	0
fh	0	−0.667	−1.333	−2.000	−1.333	−0.667	0
Lower chord:							
ac	0	0	0	0	0	0	0
ce	0	+1.111	+0.889	+0.667	+0.445	+0.222	0
eg	0	+0.889	+1.778	+1.333	+0.889	+0.445	0
Diagonal:							
bc	0	+1.389	+1.111	+0.833	+0.556	+0.278	0
de	0	−0.278	+1.111	+0.833	+0.556	+0.278	0
fg	0	−0.278	−0.556	+0.833	+0.556	+0.278	0
Vertical:							
ab	−1.000	−0.833	−0.667	−0.500	−0.333	−0.167	0
cd	0	−0.833	−0.667	−0.500	−0.333	−0.167	0
ef	0	+0.167	−0.667	−0.500	−0.333	−0.167	0
gh	0	0	0	−1.000	0	0	0

Table 6.4 Summary of influence lines for highway bridge truss

Member	Sum of ordinates			Maximum ordinates		Loaded length (ft) for:	
	Positive	Negative	All	Positive	Negative	Tension	Compression
Upper chord:							
bd	0	−3.334	−3.334	0	−1.111	0	168
df	0	−5.334	−5.334	0	−1.778	0	168
fh	0	−6.000	−6.000	0	−2.000	0	168
Lower chord:							
ac	0	0	0	0	0	0	0
ce	+3.334	0	+3.334	+1.111	0	168	0
eg	+5.334	0	+5.334	+1.778	0	168	0
Diagonal:							
bc	+4.167	0	+4.167	+1.389	0	168	0
de	+2.778	−0.278	+2.500	+1.111	−0.278	134	34
fg	+1.667	−0.834	+0.833	+0.833	−0.556	101	67
Vertical:							
ab	0	−3.000†	−3.000	0	−1.000	0	168
cd	0	−2.500	−2.500	0	−0.833	0	168
ef	+0.167	−1.667	−1.500	+0.167	−0.667	34	134
gh	0	−1.000	−1.000	0	−1.000	0	56

†Result obtained using half the influence ordinate at *b* to account for the shape of the influence line.

Table 6.5 Member forces (in kips) for loaded highway bridge truss

Member	D	L			$\frac{50}{L + 125}$	I	L + I	D + L + I
		Uniform	Concen-trated	Total				
bd	−100.0	−59.8	−20.0†	−79.8	0.171	−13.6	−93.4	−193.4
df	−160.0	−95.6	−32.0†	−127.6	0.171	−21.8	−149.4	−309.4
fh	−180.0	−107.5	−36.0†	−143.5	0.171	−24.5	−168.0	−348.0
ac	0	0	0	0	. . . . .	0	0	0
ce	+100.0	+59.8	+20.0†	+79.8	0.171	+13.6	+93.4	+193.4
eg	+160.0	+95.6	+32.0†	+127.6	0.171	+21.8	+149.4	+309.4
bc	+125.0	+74.7	+36.1‡	+110.8	0.171	+19.0	+129.8	+254.8
de	+75.0	+49.8¶	+28.9‡	+78.7	0.193	+15.2	+93.9	+168.9
		−5.0¶	−7.2‡	−12.2	0.300	−3.7	−15.9	
fg	+25.0	+29.9¶	+21.7‡	+51.6	0.221	+11.4	+63.0	+88.0
		−15.0¶	−14.5‡	−29.5	0.260	−7.7	−37.2	−12.2
ab	−87.8	−53.8	−26.0‡	−79.8	0.171	−13.6	−93.4	−181.2
cd	−70.5	−44.8	−21.7‡	−66.5	0.171	−11.4	−77.9	−148.4
ef		+3.0¶	+4.3‡	+7.3	0.300	+2.2	+9.6	
	−40.5	−29.9¶	−17.3‡	−47.2	0.193	−9.1	−56.3	−96.8
gh	−25.5	−17.9	−26.0‡	−43.9	0.276	−12.1	−56.0	−81.5

†Loaded with 18 kips. ‡Loaded with 26 kips.
¶Value not exact since influence line has plus and minus ordinates.

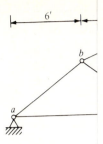

Figure P6.2

6.3 Refer to Fig. 6.6

   (*a*) Compute the
load-factor design is

   (*b*) Multiply the
member forces from
multiplier for service-
rolled-steel sections.

6.4 Calculate the me
is to be used for this
stringers, floor beam
bottom panel points)
of 20 kips, and the i

where *L* is the load

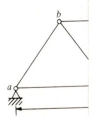

Figure P6.4

6.5 The diagonal n
(counters), as show
and the stringers,
1200 lb/ft (applied
plus a single conce

where *L* is the load
assuming that serv

uniform load of 0.64 kip/ft are obtained by multiplying the area under the influence lines by the intensity of this loading; however, this can effectively be accomplished by multiplying the sum of the influence ordinates by an equivalent panel-point load equal to 0.64(28) = 17.92 kips. Note that this approach generally will not give exact results for influence lines that have both positive and negative ordinates. The member forces due to the concentrated loads are obtained by positioning the load over the maximum influence ordinate and calculating the product of the ordinate and the magnitude of the load. Since the upper and lower chords both resist bending, the 18-kip load is used for these members whereas the 26-kip load is used for the verticals and diagonals since they resist the shear.

The impact fraction $50/(L + 125)$ is multiplied by the total live-load member force to give the member force introduced by this dynamic effect (these values are listed under column I in Table 6.5). The values tabulated under L + I are the member forces given by the sum of the total live load and impact.

Since this bridge is to be designed using the service-load design method, it is sufficient to combine the member forces due to dead, live, and impact loads to give the critical design member forces (see Table 6.5 and the column headed D + L + I). Only member *fg* experiences a reversal of force when the live load traverses the structure, i.e., from a tensile load of 88.0 kips to a compressive load of 12.2 kips. Therefore, the structural engineer must use the necessary criteria for both tensile and compressive members for the design of this member.

The function of the secondary bracing incorporated into a bridge is to resist the lateral forces imparted to the structure by such phenomena as wind, earthquake, etc. For example, the bracing in the plane of the upper chord of the bridge structure of Fig. 6.6 could be arranged as shown in Fig. 6.7. The two chords are composed of the upper chords of the two primary trusses, and the lateral struts (*bb'*, *dd'*, etc.) are compression members that transmit the loads between the two primary trusses. Note that the diagonals in each bay have been arranged in pairs; the main diagonal has been augmented with a second diagonal member called a *counter*. A cursory examination of this truss reveals that the truss has six redundant members; however, if properly designed, the truss can be regarded as statically determinate.

If the diagonals of Fig. 6.7 consist of rods with turnbuckles, they can be expected to resist tension but they are totally ineffective in compression. Similarly, tensile members fabricated from slender thin-walled steel sections will buckle when subjected to compression. Generally, the axial-load resistance of a slender compressive member is directly proportional to the moment of inertia and inversely proportional to the length

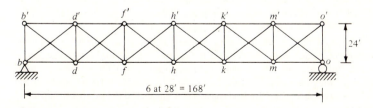

Figure 6.7 Secondary bracing in the plane of the upper chord.

squared. For

long slender r

compressive l

will take zero

directed so th

the truss, dia

remaining di

other truss c

become zero

remaining di

This behavic

wind loads a

diagonals, is

Typicall

lateral wind

must not be

chord. Since

loads, the se

points of (

loading can

analysis pro

The abo

Counters ar

similar to t

members in

panel is co

PROBLE

6.1 Refer to

   (a) Wha

to prevent up

   (b) For

and compare

loads that are

   (c) Con

that limit-sta

   (d) Mu

forces from

for allowabl

rolled-steel s

6.2 These r

critical desig

has a conce

measured o

(2) 45 lb/ft

The wind lc

(a)

(b)

Figure 7.1 (a) The Pasco-Kennewick intercity bridge (*Arvid Grant & Assoc., Olympia, Washington*); (b) Satsop River bridge (*American Institute of Steel Construction*).

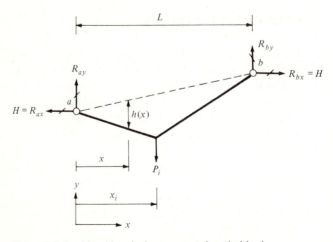

Figure 7.2 A cable with a single concentrated vertical load.

$$R_{ax} = R_{bx} = H \tag{a}$$

Summing moments about point a for the total cable gives

$$LR_{by} = P_i x_i + H(y_b - y_a) \tag{b}$$

where y_a and y_b are the y coordinates of points a and b, respectively, and x_i is the horizontal distance to the applied force P_i. Examining a typical point on the cable at point x, where $x \leq x_i$, and summing moments about this point using the free-body diagram to the right yields

$$R_{by}(L - x) = H(y_b - y) + P_i(x_i - x) \tag{c}$$

where L is the horizontal distance between the supports and y is the vertical coordinate of the generic point. Substituting Eq. (b) into Eq. (c) and rearranging terms gives

$$H\left[y_a + (y_b - y_a)\frac{x}{L} - y \right] = P_i x \left(1 - \frac{x_i}{L} \right) \tag{d}$$

Note that the terms in the square brackets on the left-hand side represent the vertical distance from the point on the cable to the line joining the two support points. This is designated $h(x)$ and referred to as the *sag* of the cable. Substituting $h(x)$ into Eq. (d) gives the simplified form

$$Hh(x) = P_i x \left(1 - \frac{x_i}{L} \right) \qquad \text{for } x \leq x_i \tag{7.1}$$

A similar expression for a point to the right of x_i can be obtained in the same manner. Thus, examining a point for which $x \geq x_i$, taking moments about the point using the free-body diagram to the right, substituting R_{by} from Eq. (b) into the resulting equation, rearranging terms, and using the cable sag $h(x)$, we have

$$Hh(x) = P_i x_i \left(1 - \frac{x}{L} \right) \qquad \text{for } x \geq x_i \tag{7.2}$$

Equations (7.1) and (7.2) can be interpreted by comparing them with the response of a simply supported beam of length L acted upon by the same loading as the cable (Fig. 7.3). The moment diagram for this beam is the illustrated bilinear shape, with

$$M(x) = \begin{cases} P_i x \left(1 - \dfrac{x_i}{L} \right) & \text{for } x \leq x_i \quad (e) \\[2ex] P_i x_i \left(1 - \dfrac{x}{L} \right) & \text{for } x \geq x_i \quad (f) \end{cases}$$

A comparison of Eqs. (7.1) and (7.2) with Eqs. (e) and (f), respectively, makes the following statement possible.

> If a cable with negligible weight and transverse stiffness is acted on by a single vertical concentrated force, the product of the horizontal component of cable tension and the sag at any point is equal to the bending moment at a corresponding section of a simply supported beam having the same

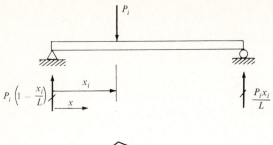

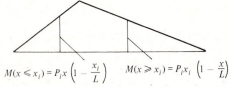

$$M(x \leqslant x_i) = P_i x \left(1 - \frac{x_i}{L}\right) \qquad M(x \geqslant x_i) = P_i x_i \left(1 - \frac{x}{L}\right)$$

Figure 7.3 Beam loaded with a single concentrated load and the associated moment diagram.

length as the horizontal span of the cable and acted upon by the load applied to the cable.

A lightweight flexible cable loaded with multiple concentrated vertical forces (Fig. 7.4) can be analyzed using the results obtained for a single vertical applied load. The left and right support points of the cable have elevations y_a and y_b, respectively, and a generic point located a distance x from the left support, with a sag $h(x)$, will be examined. By using Eqs. (7.1) and (7.2) the effects from the individual loads can be superposed to obtain the total response of this cable. Thus, for the loads to the left of x $(x \geq x_i)$, Eq. (7.2) can be used for each force and the results added to give

$$\left[Hh(x)\right]_L = \left(1 - \frac{x}{L}\right) \sum_{i=1}^{N_L} P_i x_i \qquad (g)$$

where $[Hh(x)]_L$ is the total effect of the N_L forces to the left of the point, for example, $N_L = 3$ for the cable in Fig. 7.4. Similarly, for the N_R forces to the right of x $(x \leq x_i)$ their effect $[Hh(x)]_R$ is obtained using Eq. (7.1), which yields

$$\left[Hh(x)\right]_R = \frac{x}{L} \sum_{j=1}^{N_R} P_j(L - x_j) \qquad (h)$$

Thus, the product of the total horizontal component of the cable tension H and the sag h is

$$Hh(x) = \left[Hh(x)\right]_L + \left[Hh(x)\right]_R$$

$$= \left(1 - \frac{x}{L}\right) \sum_{i=1}^{N_L} P_i x_i + \frac{x}{L} \sum_{j=1}^{N_R} P_j(L - x_j)$$

$$= \sum_{i=1}^{N_L} P_i x_i - \frac{x}{L} \sum_{i=1}^{N_L} P_i x_i + x \sum_{j=1}^{N_R} P_j - \frac{x}{L} \sum_{j=1}^{N_R} P_j x_j$$

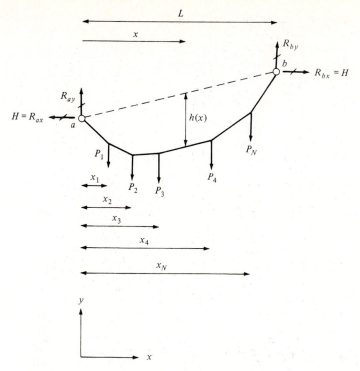

Figure 7.4 A cable with multiple concentrated vertical loads.

Noting that the total number of vertical forces $N = N_L + N_R$ and simplifying, we get

$$Hh(x) = \left(1 - \frac{x}{L}\right) \sum_1^N M_a - \sum_1^{N_R} M_R \qquad (7.3)$$

where $\sum_1^N M_a$ represents the moment of all the applied forces $(P_1, P_2, \ldots, P_i, \ldots, P_N)$ taken about point a and $\sum_1^{N_R} M_R$ is the moment of all the forces to the right of point x about point x $(P_4, P_5, \ldots, P_i, \ldots, P_N$ for the cable in Fig. 7.4). The right-hand side of Eq. (7.3) can be interpreted as the moment at a generic cross section of a simply supported beam with a span L and N applied loads:

> If a cable with negligible weight and transverse stiffness is acted on by a set of vertical concentrated forces, the product of the horizontal component of cable tension and the sag at any point is equal to the bending moment at a corresponding section of a simply supported beam having the same length as the horizontal span of the cable and acted upon by the loads applied to the cable.

The calculations for a cable supported on two compression masts are shown in Example 7.1.

Example 7.1 Calculate the sag at points c and d and the tensions in this cable if the sag at point d is 3 m.

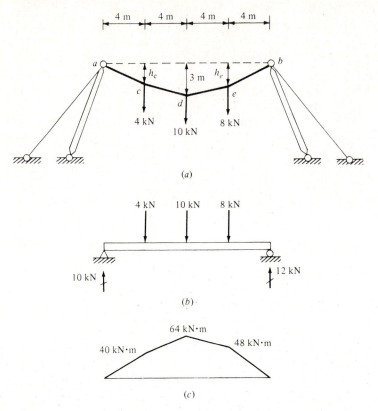

(a)

(b)

(c)

Figure E7.1 (a) Cable structure; (b) corresponding beam; (c) beam moment diagram.

SOLUTION The bending moment in the beam equals the product of H and the sag, thus at point d

$$3H = 64 \qquad H = 21.333 \text{ kN}$$

At point c

$$21.333h_c = 40 \qquad h_c = 1.875 \text{ m}$$

At point e

$$21.333h_e = 48 \qquad h_e = 2.250 \text{ m}$$

The cable tensions are calculated by dividing H by the cosine of the angle between the cable segment and the horizontal

$$T_{ac} = 23.56 \text{ kN} \qquad T_{cd} = 22.16 \text{ kN} \qquad T_{de} = 21.70 \text{ kN} \qquad T_{eb} = 24.48 \text{ kN}$$

DISCUSSION Since the horizontal component of the cable tension is constant, the tension will be a maximum where the absolute value of the slope is a maximum. For this cable it can be observed that this will occur between points e and b.

7.3 CABLES WITH UNIFORMLY DISTRIBUTED LOADS

A uniform flexible chain hanging between two supports will assume the form of a *catenary* (from the Latin word for chain). Leonardo da Vinci was the first person to sketch the form of this curve, and Galileo made the comparison between the path of a projectile and the shape of a fine chain hung between two points at the same elevation (the chain assumes the shape of a parabola only if the sag is small compared with the span). It was not until 1690 that Christian Huygens, Gottfried von Leibnitz, and Johann Bernoulli independently solved the problem of the catenary curve. The cable hanging under the effect of its own weight is a funicular structure since it resists the load in tension only.

Figure 7.5a depicts a uniform flexible cable suspended between two points with different vertical elevations and span length of L. For an incremental length Δx (Fig. 7.5b) the x components of the tension in the cable are $T_x(x)$ and $T_x(x + \Delta x)$, and those in the y direction are $T_y(x)$ and $T_y(x + \Delta x)$. Horizontal equilibrium gives

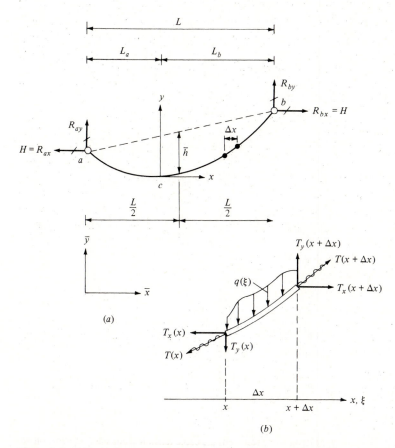

Figure 7.5 (a) The shape of a suspended cable subjected to a uniformly distributed load; (b) the free-body diagram of an incremental length of cable.

$$\sum F_x = 0 : \qquad\qquad T_x(x + \Delta x) - T_x(x) = 0$$

therefore,

$$T_x(x + \Delta x) = T_x(x) = H$$

where H is a constant. From Fig. 7.5b we note that the slope at the left end is dy/dx; thus,

$$\frac{dy}{dx} = \frac{T_y(x)}{T_x(x)} \qquad T_y(x) = H \frac{dy}{dx} \tag{a}$$

Vertical equilibrium yields

$$\sum F_y = 0 : \qquad\qquad T_y(x + \Delta x) - T_y(x) - \int_0^{\Delta x} q(\xi)\, d\xi = 0$$

From the mean-value theorem

$$q(x^*)\, \Delta x = \int_0^{\Delta x} q(\xi)\, d\xi \qquad \text{where } x \le x^* \le x + \Delta x$$

Thus

$$T_y(x + \Delta x) - T_y(x) - q(x^*)\, \Delta x = 0$$

Dividing by Δx and taking the limit as $\Delta x \to 0$ leads to

$$\lim_{\Delta x \to 0} \frac{T_y(x + \Delta x) - T_y(x)}{\Delta x} = \frac{dT_y}{dx} = q(x) \tag{b}$$

since $q(x^*) \to q(x)$ as $\Delta x \to 0$. Combining Eqs. (a) and (b) yields

$$\frac{d^2 y}{dx^2} = \frac{q}{H} \tag{7.4}$$

which is the second-order differential equation governing the shape of a free-hanging chain.

If the weight of the cable is uniformly distributed along the horizontal projection, q is a constant. For a uniform chain this is not the case, but the approximation can be made if the ratio of sag to span is $1:8$ or less. It has been demonstrated by sophisticated numerical methods that this assumption gives a close approximation to the actual behavior for ratios from $1:5$ to $1:4$. Integrating Eq. (7.4) gives

$$Hy = \tfrac{1}{2} qx^2 + C_1 x + C_2 \tag{c}$$

Taking the coordinate origin at the lowest point on the curve (c in Fig. 7.5a) with L_a and L_b as the distances to points a and b, respectively, gives

$$C_1 = C_2 = 0$$

Thus, the equation for the shape assumed by the cable is

$$y = \frac{qx^2}{2H} \tag{7.5}$$

If the load is uniformly distributed along the length of the cable, e.g., a uniform chain, the governing differential equation is

$$\frac{d^2y}{dx^2} = \frac{q}{H}\left[1 + \left(\frac{dy}{dx}\right)^2\right]^{1/2} \tag{7.4a}$$

the solution of which is

$$y = \frac{H}{q}\left(\cosh\frac{qx}{H} - 1\right) \tag{7.5a}$$

This is the equation of the catenary. The derivation and solution of Eq. (7.4a) are left as an exercise for the reader.

7.4 THE THREE-HINGED ARCH

An arch can be used to build economical and efficient long-span structures. If the proper shape is selected for a given distribution of applied load, the structure will have neither internal shear nor bending moment. For example, consider the arch shape required to support a distributed load such that the structure is totally in compression. For convenience, the origin of the coordinate system is located at the top (*crown*) of the arch, as shown in Fig. 7.6a. From examination of an incremental length of the structure (Fig. 7.6b) the *thrust* along the axis of the arch will have a value of $T(x)$ at x and will change to $T(x + \Delta x)$ at the point $x + \Delta x$. A comparison of the free-body diagrams for the cable in Fig. 7.5 with that for the arch demonstrates the similarity of behavior. It must be noted, however, that the cable is in tension, whereas the symbol T denotes compression for the arch. The derivation of the governing equation for the arch analogous to Eq. (7.4) will be identical to that for the cable if the sense of T is properly interpreted. Invoking equilibrium and using the mean-value theorem and the definition of a derivative gives the result shown in Eq. (7.4) for the arch as well. The expression for the axis of the arch if q is constant is obtained by integrating Eq. (7.4) to give

$$y = \frac{qx^2}{2H} + C_1x + C_2 \tag{a}$$

where the constants of integration, C_1 and C_2, are both zero for the illustrated location of the coordinate origin. Thus, if the structure is constructed in the form of a parabola with $y = qx^2/2H$ and a uniformly distributed loading q is applied, the arch will be exclusively in compression; i.e., neither shear nor bending moment will be present. Since H represents the horizontal reaction component at the supports (*springings*) of the structure, an entire family of parabolas can be generated to accommodate a given uniform loading. As a result of the above development, a uniformly loaded parabolic arch constructed with closely fitting stone or masonry elements and no mortar between the blocks would be both stable and functional.

From the previous discussion we can conclude that if the arch axis were not parabolic or if the loading on a parabolic shape were not uniform, bending moments and shear forces would be introduced into the structure. The distribution of internal

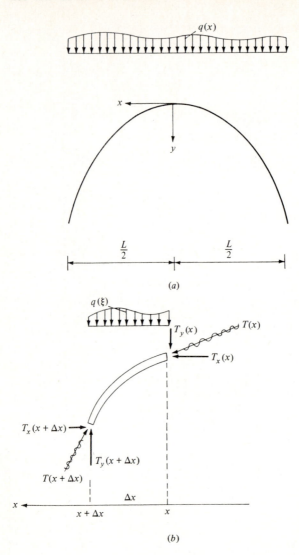

Figure 7.6 (*a*) Distributed load on an arch; (*b*) an incremental length of arch.

forces in an arch depends upon (1) the loads applied, (2) the support conditions, (3) the general shape of the arch, and (4) the details of internal construction. For example, of the three arches shown in Fig. 7.7, only the third is statically determinate. The structure in Fig. 7.7a, which is statically indeterminate to the third degree, is a *fixed arch;* the *two-hinged arch* in Fig. 7.7b has only one redundancy. Introducing an internal hinge into the latter arch gives the statically determinate *three-hinged arch* of Fig. 7.7c. The internal forces in a three-hinged arch can be investigated using only the equations of static equilibrium, but statically indeterminate arches must be analyzed using both force and deformation considerations. Examples 7.2 and 7.3 show the analysis of two different three-hinged arches.

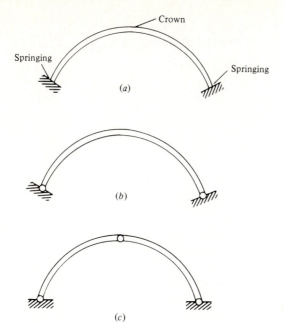

Crown

Springing

Springing

(a)

(b)

(c)

Figure 7.7 (a) Fixed arch; (b) two-hinged arch; (c) three-hinged arch.

Example 7.2 Analyze this three-hinged arch for the unbalanced snow load shown. The arch consists of a segment of a circle.

SOLUTION From Fig. E7.2a

$$\sum F_x = 0: \qquad R_{ax} - R_{cx} = 0$$

$$\sum F_y = 0: \qquad R_{ay} + R_{cy} = 0.50(200 \cos 45°) = 70.70$$

$$\sum M_c = 0: \qquad 2(200 \cos 45°)R_{ay} = 0.50(200 \cos 45°)[0.50(200 \cos 45°)]$$

From Fig. E7.2b

$$\sum M_b = 0: \qquad 58.60R_{ax} = 141.4R_{ay} \qquad \frac{R_{ax}}{R_{ay}} = 2.41$$

Solving the above four equations gives

$$R_{ax} = 42.65 \text{ kips } (\leftrightarrow) \qquad R_{ay} = 17.68 \text{ kips } (\uparrow)$$

$$R_{cx} = 42.65 \text{ kips } (\leftrightarrow) \qquad R_{cy} = 53.02 \text{ kips } (\uparrow)$$

From Fig. E7.2c, for $0° \le \phi \le 45°$

$$M(\phi) = 200[\cos 45° - \cos (45° + \phi)]R_{ay} - 200[\sin (45° + \phi) - \sin 45°]R_{ax}$$

$$= 141.4[1 - \cos \phi + \sin \phi)R_{ay} - (\cos \phi + \sin \phi - 1)R_{ax}]$$

$$T(\phi) = R_{ax} \cos (45° - \phi) + R_{ay} \sin (45° - \phi)$$

$$V(\phi) = -R_{ax} \sin (45° - \phi) + R_{ay} \cos (45° - \phi)$$

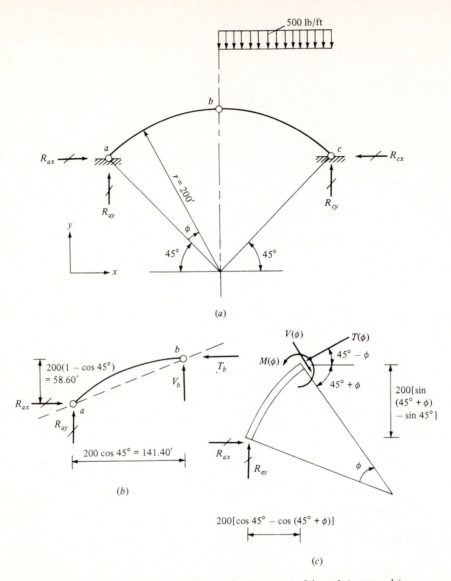

Figure E7.2 (a) Three-hinged arch; (b) half the arch; (c) segment of the arch (not to scale).

DISCUSSION The extremal point for the moment for the unloaded span (a to b) occurs at $\phi = 22.5°$, and the value there is -702.1 kip · ft. The extreme value for the thrust in the unloaded span also occurs at this point and is equal to 46.17 kips. The shear in the unloaded span has its largest values at the crown and the springing, where it is $+17.68$ and -17.68 kips, respectively.

Example 7.3 Analyze this three-hinged arch, in which the supports are at different elevations.

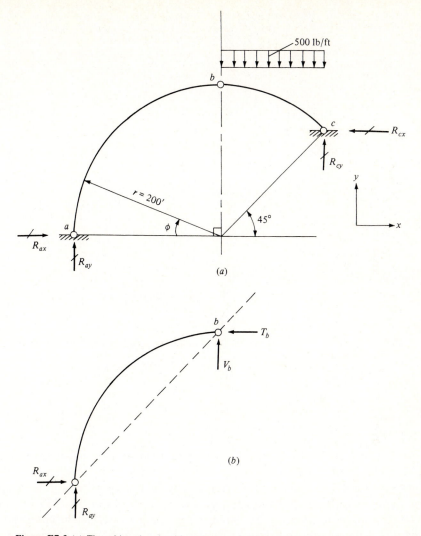

Figure E7.3 (*a*) Three-hinged arch with springings at different elevations; (*b*) free-body diagram of left portion.

SOLUTION From Fig. E7.3*a*

$$\sum F_x = 0: \qquad\qquad R_{ax} - R_{cx} = 0$$

$$\sum F_y = 0: \qquad\qquad R_{ay} + R_{cy} = 70.70$$

$$\sum M_c = 0: \qquad\qquad 4.83R_{ay} = 70.70 + 2R_{ax}$$

From Fig. E7.3*b*

$$\sum M_b = 0: \qquad\qquad R_{ax} - R_{ay} = 0$$

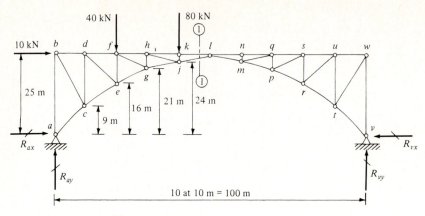

Figure 7.8 A parabolic three-hinged trussed arch.

Solving the above four equations gives

$$R_{ax} = 25.0 \text{ kips } (\rightarrow) \qquad R_{ay} = 25.0 \text{ kips } (\uparrow)$$

$$R_{cx} = 25.0 \text{ kips } (\leftarrow) \qquad R_{cy} = 45.7 \text{ kips } (\uparrow)$$

From Fig. E7.3b the internal forces at any point ϕ (as designated in Fig. E7.3a) are

$$M(\phi) = 200(1 - \cos \phi)R_{ay} - 200(\sin \phi)R_{ax}$$

$$T(\phi) = R_{ax} \sin \phi + R_{ay} \cos \phi$$

$$V(\phi) = R_{ax} \cos \phi - R_{ay} \sin \phi$$

where $0° \leq \phi \leq 90°$.

Truss structures can also be designed with an arch shape to form long-span structures in which the internal forces are minimized by virtue of the configuration. For example, the trussed arch in Fig. 7.8 attains its structural integrity through the shape of the lower chord in much the same fashion as the arch ribs in Fig. 7.7. For this three-hinged trussed arch, the lower chord has a parabolic shape, and the upper chord is typical of that used for a deck bridge. This structure has four reaction-force components similar to those of all three-hinged arches, and since it consists of two simple trusses connected with a hinge at joint l, it is statically determinate.

PROBLEMS

7.1 A light cable with negligible weight is suspended between two points spaced 300 ft apart horizontally, with the right support 20 ft lower than the left. Four vertical gravity loads of 10, 15, 20, and 25 kips (from left to right) centered between the supports and spaced 60 ft apart horizontally, are applied to the cable. If the largest sag (vertical distance between the cable and the cable chord) is 40 ft, calculate (a) the sag at each applied load; (b) the y coordinate of the cable at each load with the coordinate origin at the left support; (c) the maximum cable tension; and (d) the unstressed length of the cable.

7.2 A light cable with negligible weight is suspended between two points spaced 300 m apart horizontally, with the right support 12 m higher than the left. Four vertical gravity loads of 400, 200, 400, and 1200 kN (from left to right) centered between the supports and spaced 60 m apart horizontally, are applied to the cable. If the largest sag is 24 m, calculate the quantities requested in parts (*a*) to (*d*) of Prob. 7.1.

7.3 The equation for the shape of a cable with a uniformly distributed load along its horizontal projection can be written

$$H\bar{y} = \tfrac{1}{2}q\bar{x}^2 + H(\bar{y}_b - \bar{y}_a)\frac{\bar{x}}{L} - \tfrac{1}{2}qL\bar{x} + H\bar{y}_a$$

where the cartesian coordinates $\bar{x}$ and $\bar{y}$ shown in Fig. 7.5*a* are used, and $\bar{y}_a$ and $\bar{y}_b$ are the $\bar{y}$ coordinates of points *a* and *b*, respectively. Using the principles discussed in Sec. 7.2 for cables with concentrated forces, derive this equation. *Hint:* Either use expressions of the form shown in Eqs. (*g*) and (*h*) and integrate them over the length of the structure or use the analogy between the bending moment in an equivalent simply supported beam and the product of the horizontal cable tension and the sag.

7.4 Obtain the equation of the arch using the coordinate system illustrated if each structure is to perform as a compressive arch and support the loadings shown.

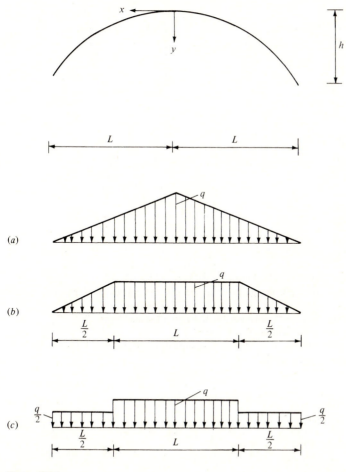

Figure P7.4

CHAPTER

EIGHT

THREE-DIMENSIONAL TRUSSES

8.1 DESCRIPTION OF BEHAVIOR

In many structures, three-dimensional systems are synthesized from a number of individual planar components; each component, functioning as an entity, performs a role in transmitting the loads. For example, for the through bridge shown in Fig. 3.2 the vertically applied loads are delivered to the primary trusses collectively by the deck, stringers, and floor beams. Furthermore, the lateral loads, imparted by wind, impact, and other dynamic effects, are resisted by various components, including the primary trusses, the trussing in the planes of both the lower and upper chords, the sway bracing, and the portal bracing (see Fig. 6.5).

For some structures it is impossible to resolve the load-transfer mechanism into a system of individual planar structural components. Many radar telescopes, transmission-line towers, and steel offshore oil-drilling platforms consist of composites of trusses arranged in multiple planes, and their structural integrity depends upon the three-dimensional interaction of the members. Early investigations of these structures were carried out by A. Föppl. The Schwedler trussed dome, the Zeiss-Dywedag trussed dome, and the geodesic domes of R. B. Fuller are all famous examples of spherical domes subdivided into triangles by members that function in tension or compression. A relatively common use of three-dimensional structures consists of the space frames used in buildings as the horizontal structure to support the gravity loads and the vertical structural system resisting lateral loads. An example of a constructed space truss is shown in Fig. 8.1.

Some three-dimensional structures are constructed using flexural members, but three-dimensional *space trusses* are generally analyzed assuming that

1. The system is composed of *two-force members;* i.e., all members transmit forces in either tension or compression.
2. The members are connected with *ideal spherical hinges.*
3. The *axes of all members* meeting at a joint *intersect at a point.*
4. All *loads* are *applied at* the points of *connection.*

214

Figure 8.1 Offshore steel-jacket platform. (*Standard Oil Company of California.*)

Note that these assumptions are analogous to those used in the analysis of planar trusses. If all these ideal conditions are satisfied, the *primary stresses* in the members are induced by axial tension or compression and any spurious stresses resulting from shear forces, bending moments, and/or torsion are termed *secondary stresses*. Assumptions 1 and 4 imply that the members are weightless, but if the engineer must include this effect, half the member weight is lumped to each end. A structure is usually analyzed for dead loads and live loads separately, and in a typical truss the secondary stresses due to the member weights are much smaller than the primary axial stresses induced by the other dead loads.

The members meeting at a joint are carefully aligned so that the centroidal axes intersect at a common point, and they are welded, bolted, or riveted so that the connections transmit primarily axial forces with negligible bending moments, resulting in a concurrent force system. The spherical hinge is an idealization analogous to the smooth frictionless pins of planar structures. In reality, the connections do not behave strictly in conformity with assumption 2 for ideal truss behavior, but if the connections are properly designed and carefully fabricated, the secondary stresses are small compared with the primary stresses and can be ignored. The analysis of secondary stresses must be accomplished by treating the structure as a semirigid to rigid frame, and the proper methods must be used for obtaining the shear, bending, and/or torsional stresses (methods for the analysis of statically indeterminate structures are presented in Chaps. 11 and 12 and also in Part Five). In the ensuing discussion, the secondary stresses will be assumed to be negligible.

The intrinsic stability characteristics of a three-dimensional space truss are determined by the arrangement of the members. For planar trusses the basic configuration which yields a rigid truss is the triangle since the three points are constrained relative to each other. For three-dimensional structures, if we start with the triangle *abc*, as shown in Fig. 8.2*a*, and add three bars all attached to the same point *d*, which does

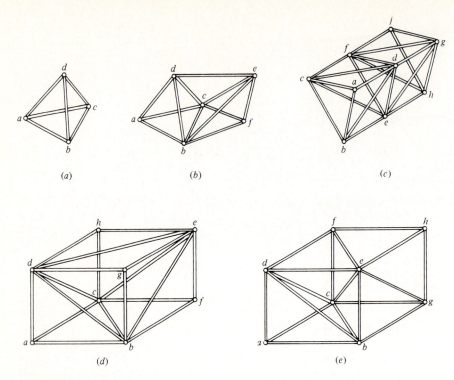

Figure 8.2 Simple space trusses.

not lie in the plane of triangle *abc,* the configuration will be stable. That is, by inspecting triangles *abc, bcd,* and *cad* we deduce that point *d* is fixed relative to points *a, b,* and *c,* and these four joints and six members form a *rigid tetrahedron.* Note that, in general, the points will displace relative to each other by small amounts because of the material deformations of the members under the effects of loading. These deformations will be ignored in the ensuing discussion.

A new rigid truss can be formed from a basic rigid tetrahedron by adding three members, which are all attached to a new joint, provided that the axes of the three new bars do not lie in one plane. This procedure can be continued to produce a structure of the desired size and shape, and the final configuration is termed a *simple space truss.* Some examples of these structures are shown in Fig. 8.2, and in all these cases the joints have been formed in alphabetical order by extending three members (not all in one plane) to a new joint from the existing simple space truss. Note that the cuboid of Fig. 8.2*d* is formed by five tetrahedra, whereas that of Fig. 8.2*e* consists of four tetrahedra plus a right pyramid, the five-sided polygon *cfhge.*

For a simple space truss with *n* joints, it requires *n* − 4 additional joints to expand the structure beyond the initial four-joint, six-bar tetrahedron. Since each of these joints is formed by the intersection of three members, the total number of bars *b* required to form any simple rigid space truss is

$$b = 3(n - 4) + 6 = 3n - 6 \qquad (8.1)$$

This is the minimum number of bars needed to form a rigid simple space truss; more bars are unnecessary to attain this objective, and fewer bars will yield a nonrigid or unstable truss.

Two or more simple space trusses can be joined together to form a rigid *compound space truss,* three examples of which are illustrated in Fig. 8.3. If the two simple rigid trusses shown shaded in Fig. 8.3*a* are connected only at point *e,* they will not form a rigid configuration because relative rotation can occur. Inserting connecting member 1 fixes points *d* and *d'* relative to each other, but not all relative motion is prevented since, for example, truss *a'b'c'd'e* can rotate about member *ed'*. Connecting member 2 situates points *c* and *c'* with reference to each other, but point *c* is not fixed relative to point *d'* (nor is *c'* fixed relative to *d*). Since member 3 fixes points *c* and *d'* relative to each other and since *d* is located relative to *c* and since *c'* is located relative to *d'*, points *d* and *c'* are also fixed relative to each other. The connection is complete to form a rigid space truss because the six components of relative translation and relative rotation of the two simple rigid trusses have been constrained. That is, the attachment at point *e* constrains the relative translations, and the three connecting bars

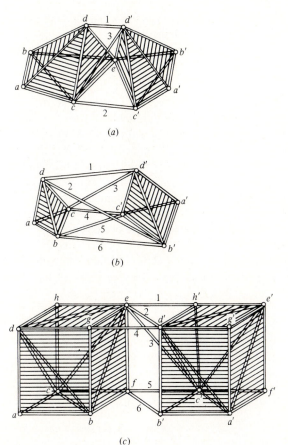

(a)

(b)

(c)

Figure 8.3 Compound space trusses.

fix the three relative rotations. This structure was formed like a compound truss, but inspection reveals that it is also a simple space truss. This can be demonstrated by (1) starting with the rigid tetrahedron *abcd*, (2) extending three members to form point *e*, and (3) expanding the structure by locating the primed points in inverse alphabetical order by simultaneously extending three members to each new point.

The above description serves to illustrate the point that to form a compound truss from two simple rigid space trusses three relative translations and three relative rotations must be constrained. For example, this can be accomplished using six connecting members that are properly arranged. Figure 8.3*b* and *c* shows two shaded simple rigid space trusses connected in this manner. Note that it is impossible to construct either of these two structures by adding three members and one joint at a time to any of the component simple rigid space trusses; furthermore, the solid polyhedra formed by the connecting members do not necessarily form tetrahedra; for example, although *egd'b'bf* in Fig. 8.3*c* is a right prism, all points are fixed relative to each other.

If two rigid simple space trusses forming a compound truss have n_1 and n_2 joints, they must have b_1 and b_2 bars, respectively, where, according to Eq. (8.1),

$$b_1 = 3n_1 - 6 \qquad \text{and} \qquad b_2 = 3n_2 - 6$$

If these are connected together with six additional members to form a rigid compound truss, the resulting truss must have b members, such that

$$b = b_1 + b_2 + 6 = (3n_1 - 6) + (3n_2 - 6) + 6$$
$$= 3(n_1 + n_2) - 6 = 3n - 6$$

where the total number of joints in the compound structure n is equal to $n_1 + n_2$. Therefore, we observe that the criterion for the minimum number of bars required to form a rigid compound space truss is identical to that for a rigid simple truss.

The configuration of some space trusses can be categorized as neither simple nor compound. For example, the structure in Fig. 8.4*a* is a simple space truss with the nodes formed in alphabetical order subsequent to forming the basic tetrahedron *abcd*. For this structure, $n = 8$, $b = 18$, and Eq. (8.1) is satisfied. If members *be* and *cd* are removed and replaced by *fg* and *ah*, respectively, as shown in Fig. 8.4*b*, a different rigid truss satisfying Eq. (8.1) is created that is neither simple nor compound; such a structure is a *complex space truss*.

The relationship of the members for space trusses and the equations of equilibrium will be discussed in subsequent sections of this chapter. Here we note that the kine-

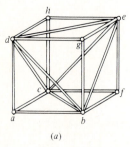

(a)

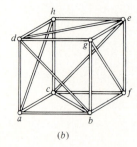

(b)

Figure 8.4 (*a*) simple space truss; (*b*) complex space truss.

matics of three-dimensional space trusses reveals that the basic rigid configuration is the tetrahedron, and a compound truss can be formed by connecting two or more simple space trusses with six conditions of connectivity created by using enough connecting conditions properly arranged to constrain the components of relative translation and relative rotation.

8.2 EQUATIONS OF EQUILIBRIUM IN THREE DIMENSIONS

The force **P** in Fig. 8.5 can be written as the vector sum

$$\mathbf{P} = P_x\mathbf{i} + P_y\mathbf{j} + P_z\mathbf{k} \qquad (a)$$

where **i**, **j**, and **k** are unit vectors having the directions of the positive x, y, and z axes, respectively. The scalar components P_x, P_y, and P_z of the vector **P** are

$$P_x = P \cos \alpha \qquad P_y = P \cos \beta \qquad \text{and} \qquad P_z = P \cos \gamma \qquad (b)$$

where P is the magnitude of the vector **P** and α, β, and γ are the angles measured to **P** from the positive x, y, and z axes, respectively (see Fig. 8.5). The numbers $\cos \alpha$, $\cos \beta$, and $\cos \gamma$ are the direction cosines of **P**, frequently denoted by λ, μ, and ν, respectively. From the geometrical relationship

$$P = \sqrt{P_x^2 + P_y^2 + P_z^2} \qquad (c)$$

it follows that

$$\lambda^2 + \mu^2 + \nu^2 = \cos^2 \alpha + \cos^2 \beta + \cos^2 \gamma = 1 \qquad (d)$$

If a force in space as shown in Fig. 8.5 is projected onto a set of mutually orthogonal planes, we obtain

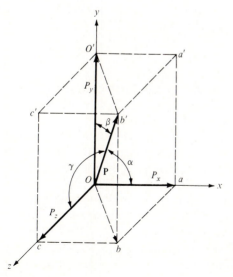

Figure 8.5 Components of a force in three dimensions.

$$\mathbf{Oa'} = P_x\mathbf{i} + P_y\mathbf{j} \qquad \mathbf{Oc'} = P_y\mathbf{j} + P_z\mathbf{k} \qquad \mathbf{Ob} = P_x\mathbf{i} + P_z\mathbf{k} \qquad (e)$$

Any system of concurrent forces in space having a generic member denoted by $\mathbf{F}$ can be replaced by a resultant $\mathbf{R}$, and if the system is in equilibrium,

$$\mathbf{R} = \sum \mathbf{F} = 0$$

These requirements can be written in scalar form as

$$\sum F_x = 0 \qquad \sum F_y = 0 \qquad \sum F_z = 0 \qquad (8.2)$$

where summation over all the force components is implied. We can use these three equations to state a number of useful *special observations* and to facilitate the analysis of member forces for space trusses:

1. *Any three concurrent forces that do not lie in a single plane will not be in equilibrium unless all three forces are zero.* Consider the vector $\mathbf{P}$ in Fig. 8.5 to be the resultant of a concurrent force system and $P_x\mathbf{i}$, $P_y\mathbf{j}$, and $P_z\mathbf{k}$ as the projections of $\mathbf{P}$ onto the mutually orthogonal coordinate axes. We conclude from Eqs. (8.2) that each of the scalar components is zero.

2. *If two forces of a system of four concurrent forces that are in equilibrium in space are collinear, the other two forces will be zero, unless these forces are themselves collinear; collinear forces will be equal in magnitude but opposite in direction.* This can be verified by summing forces normal to a plane containing the two collinear forces. Since the other two forces are not collinear, they must both be zero.

3. *If all but one force of a system of concurrent forces that are in equilibrium in space are coplanar, the out-of-plane force must be zero.* By summing forces normal to the plane containing all but the singular force it can be demonstrated that this latter force is zero.

4. *If all but two forces of a system of concurrent forces in space are in equilibrium and are coplanar and the magnitude of one of these out-of-plane forces is known, the magnitude of the other can be calculated by projecting all forces onto an axis perpendicular to the coplanar forces.*

Stabilizing a point in space with the minimum number of reactions can be accomplished using Eqs. (8.2) in conjunction with the four special observations stated above. Consider a point in space which has an applied load and which is to be constrained with the minimum number of reactions to be provided by two-force members attached to their foundations with spherical hinges. The arrangement of members in Fig. 8.6a will be in equilibrium only if P is either zero or in the plane abc, since the concurrent force system, formed by the forces in members ac and bc, and the applied load, in general, will not satisfy equilibrium (see observation 1 above). Equations (8.2) imply that there must be at least three reactions to satisfy equilibrium. The third observation above can be used to deduce that the configuration in Fig. 8.6b, with the required three reactions, will not be in equilibrium because the reaction forces are coplanar. Point c can be stabilized only if the three reactions are situated as shown in Fig. 8.6c. Note that this arrangement of the four joints (i.e., the four faces, each containing three of the

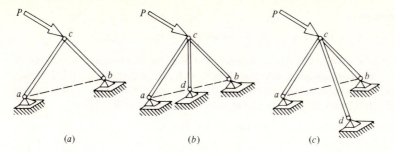

whe

mai
the
hav
and
abo
the
wer
the
resi

full
of
rea
mei
mei

Figure 8.6 Support of a point in space.

joints, form a tetrahedron) conforms to the description of a rigid simple space truss in the previous section. Example 8.1 illustrates the calculation of the reaction forces for a tripod.

Example 8.1 Calculate the reactions for this tripod.

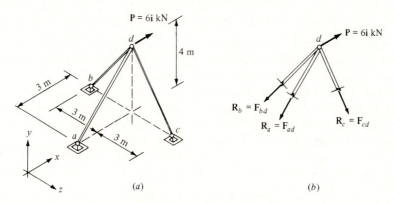

Figure E8.1 (*a*) Tripod with applied load; (*b*) free-body diagram of joint *d*.

SOLUTION Note on the free-body diagram that all member forces are assumed to be tensile, and

$$\mathbf{R}_a = R_a(-0.6\mathbf{i} - 0.8\mathbf{j}) \qquad \mathbf{R}_b = R_b(-0.8\mathbf{j} - 0.6\mathbf{k}) \qquad \mathbf{R}_c = R_c(-0.8\mathbf{j} + 0.6\mathbf{k})$$

From Eqs. (8.2)

$$\mathbf{P} + \mathbf{R}_a + \mathbf{R}_b + \mathbf{R}_c = 0$$

$$6\mathbf{i} - 0.6R_a\mathbf{i} - 0.8R_a\mathbf{j} - 0.8R_b\mathbf{j} - 0.6R_b\mathbf{k} - 0.8R_c\mathbf{j} + 0.6R_c\mathbf{k} = 0$$

Grouping terms gives

$$(6 - 0.6R_a)\mathbf{i} + (-0.8R_a - 0.8R_b - 0.8R_c)\mathbf{j} + (-0.6R_b + 0.6R_c)\mathbf{k} = 0$$

Hence, the three scalar equations of equilibrium are

$$6 - 0.6R_a = 0 \qquad -0.8R_a - 0.8R_b - 0.8R_c = 0 \qquad -0.6R_b + 0.6R_c = 0$$

determinate reactions for a rigid body. In some cases it is possible to replace the applied forces with an equivalent force system to facilitate the reaction calculations, and Example 8.3 illustrates this approach.

Example 8.2 Calculate the six reactions for this rigid body.

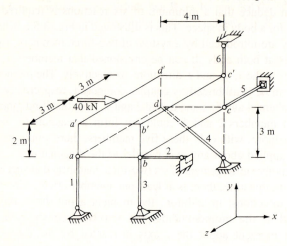

Figure E8.2

SOLUTION All reactions are equal to the forces in the numbered bars, and they have been assumed to be in tension. The equations of equilibrium in vector form are

$\mathbf{R} = \mathbf{0}$:

$$(40 + F_2 + 0.8F_4)\mathbf{i} + (-F_1 - F_3 + F_6 - 0.6F_4)\mathbf{j} + (-F_5)\mathbf{k} = \mathbf{0}$$

$\mathbf{M}_a = \mathbf{0}$:

$$[(2\mathbf{j} - 3\mathbf{k}) \times 40\mathbf{i}] + [4\mathbf{i} \times F_3(-\mathbf{j})] + [(4\mathbf{i} - 6\mathbf{k}) \times F_6\mathbf{j}] + [4\mathbf{i} \times F_5(-\mathbf{k})]$$
$$+ [-6\mathbf{k} \times (0.8F_4\mathbf{i} - 0.6F_4\mathbf{j})] = \mathbf{0}$$

Simplifying gives six scalar equilibrium equations

$$40 + F_2 + 0.8F_4 = 0 \qquad -F_1 - F_3 + F_6 - 0.6F_4 = 0$$
$$-F_5 = 0 \qquad 6F_6 - 3.6F_4 = 0$$
$$-120 + 4F_5 - 4.8F_4 = 0 \qquad -80 - 4F_3 + 4F_6 = 0$$

Solving them gives

$$F_1 = 35.0 \text{ kN (t)} \qquad F_2 = -20.0 \text{ kN (c)}$$
$$F_3 = -35.0 \text{ kN (c)} \qquad F_4 = -25.0 \text{ kN (c)}$$
$$F_5 = 0.0 \text{ kN} \qquad F_6 = -15.0 \text{ kN (c)}$$

Example 8.3 Calculate the six reactions for this rigid body.

SOLUTION By inspection (Fig. E8.3*b*)

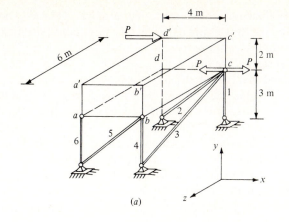

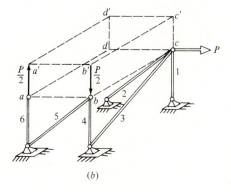

(b)

Figure E8.3 (*a*) Rigid body with applied load and reactions; (*b*) free-body diagram showing equivalent loading.

$$F_6 = +\frac{P}{2}\,\text{(t)} \qquad F_5 = 0 \qquad F_4 = -\frac{P}{2}\,\text{(c)}$$

$$F_3 = 0 \qquad F_2 = +\tfrac{5}{4}P\,\text{(t)} \qquad \text{and} \qquad F_1 = -\tfrac{3}{4}P\,\text{(c)}$$

DISCUSSION The total applied loading consists of the force $P\mathbf{i}$ applied at point d'. The forces $P\mathbf{i}$ and $-P\mathbf{i}$ have been added at point c, and it can be observed that their sum is a null vector. The forces $P\mathbf{i}$ at d' and $-P\mathbf{i}$ at c form the couple $-2P\mathbf{k}$, which can be replaced by a couple composed of the forces $(P/2)\mathbf{j}$ and $-(P/2)\mathbf{j}$ located at points a and b, respectively. By using the equation $\Sigma M_y = 0$ about point c it can be observed that the force in member 5 must be zero since there are no applied loads to create a moment about the y axis. The reaction forces F_2 and F_3 are calculated by summing forces in the x and z directions, respectively. The vertical reactions 1, 4, and 6 can all be obtained by summing forces in the y direction at points c, b, and a, respectively. Alternatively, reaction 6 could be calculated by summing moments about the z axis, and 1 and 4 obtained by summing moments about the x axis.

It is possible to arrange six restraining members in such a way that the structure is unstable; i.e., the presence of this minimum number of reactions is a necessary but not sufficient condition for stabilizing a body in space. For example, the rigid body in Fig. 8.9a with six reactions can resist applied forces in the x and y directions; however, there is no reactive force to equilibrate forces applied in the z direction. Thus, this structure is geometrically unstable. The rigid body in Fig. 8.9b is another example of an arrangement of six reactions that is *geometrically unstable*. This can be seen by first visualizing the plane formed by the line of action for reaction 5 (axis cc') and the point a which is the point of concurrency for the four reaction components 1, 2, 3, and 4. The line of action for reaction 6 intersects this plane at point Q. The axis passing through points Q and a is intersected by all six reactions; therefore, any applied loads that do not intersect this axis cannot be resisted, and the structure will not be in equilibrium. In some cases, the six reactions can be arranged to yield an unstable structure even though this fact cannot be detected by inspection. In these cases the instability will be revealed by the fact that the determinant of the coefficient matrix of the six equilibrium equations is zero.

If the six reactions are arranged to stabilize the structure in space, they form a *statically determinate* system of supports. If additional supports are added to an already stable structure, the reactions cannot be determined unless additional equations of deformation are introduced; the system of supports is *statically indeterminate*. In certain cases where there is no solution for the reactions, e.g., there are fewer than six

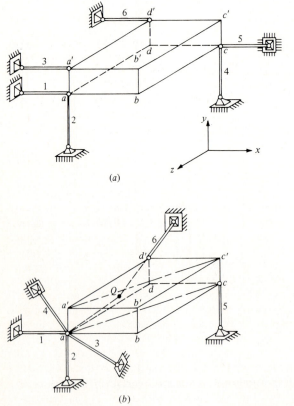

(a)

(b)

Figure 8.9 Geometrically unstable constraints of a rigid body in space: (a) reactions parallel to one plane; (b) reactions intersecting a common axis.

reactions, or the axes of all reactions intersect a common line, the determinant of the equilibrium equations vanishes and the system of supports is *unstable*.

8.3 MEMBER FORCES

The analysis for the member forces of a space truss can begin by considering the free-body diagram of the entire structure and calculating the reactions using the equations of static equilibrium. To carry out the analysis for the forces sustained by the individual members, the equations of equilibrium are applied to various portions of the structure; if every part of a structure is in equilibrium, the entire truss is in equilibrium. These calculations must be made under the assumptions of ideal truss behavior listed in Sec. 8.1.

Since neither the direction nor the magnitude of the member forces is known a priori, it is necessary at the onset of the analysis to assume a direction for each of the member forces. The options used for doing this for planar trusses are also available for space-truss analysis (Sec. 3.2). In this book it will be assumed initially that all members are in tension, i.e., the force applied to the joint by the member is directed away from the joint, and members are indicated accordingly on the free-body diagrams. After the calculations are complete, it can be concluded that the assumed direction is correct (i.e., the member is in tension) if the algebraic sign of the result is positive and that the direction of the force is opposite to that assumed if the sign is negative (i.e., the member is in compression). All subsequent free-body diagrams in an analysis will show the actual (computed) direction of the member force.

In applying the *method of joints* to a space truss the members meeting at a joint form a concurrent force system; thus, the unknown member forces can be calculated using the three equations of equilibrium shown in Eq. (8.2). The *method of sections* requires the use of the six equations of equilibrium for three-dimensional systems, as shown in Eq. (8.3). These methods are discussed individually in the next two sections.

8.4 THE METHOD OF JOINTS

The forces in the members connecting at a joint, together with the applied loads and/or reactions, form a concurrent force system; consequently, three independent equilibrium equations in the form shown in Eq. (8.2) can be constituted. Thus, the member forces in a statically determinate space truss can be calculated using the method of joints in a manner similar to that used for planar trusses. For rigid simple space trusses the member forces can be obtained by investigating the joints in a sequence ensuring that there are no more than three unknowns at the joint under consideration. Alternatively, the equilibrium equations can be formed for all the joints, giving a set of simultaneous algebraic equations with the member forces and reactions as the unknowns. This approach is relatively inefficient for hand calculations, but it is a systematic means for setting up the equations for solution on the digital computer.

For the structure in Fig 8.10*a* it is possible to start the analysis at the last joint formed, joint *j*; since there are only three unknowns at this point, they can be calcu-

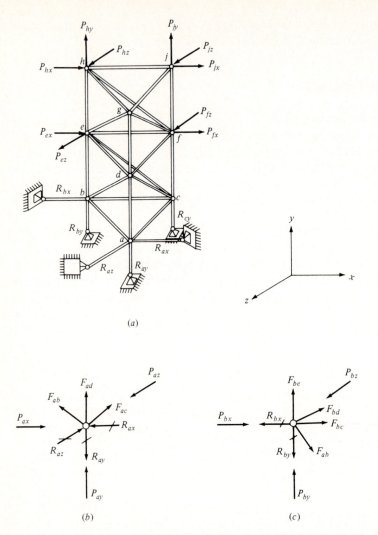

Figure 8.10 (a) Space truss; (b) free-body diagram of joint a; (c) free-body diagram of joint b.

lated. Proceeding to the joints in reverse alphabetical order, all the member forces and reactions can be obtained joint by joint since each concurrent force system has only three unknowns. Alternatively, the reactions could be calculated initially by considering the entire structure and the joints investigated starting at joint a since there are only three unknown member forces at this point. For example, the free-body diagram for joint a (Fig. 8.10b) shows all the member forces acting away from the joint and the reaction forces positive if they act in the negative coordinate direction. Note that any consistent convention can be used for the reactions, but the one adopted is useful in writing the subsequent equations since a positive reaction opposes a positive applied force. For this joint the equations of equilibrium are

$$P_{ax} = 0 = -F_{ab}\lambda_{ab} - F_{ac}\lambda_{ac} - F_{ad}\lambda_{ad} + R_{ax}$$

$$P_{ay} = 0 = -F_{ab}\mu_{ab} - F_{ac}\mu_{ac} - F_{ad}\mu_{ad} + R_{ay}$$

$$P_{az} = 0 = -F_{ab}\nu_{ab} - F_{ac}\nu_{ac} - F_{ad}\nu_{ad} + R_{az}$$

where λ, μ, and ν are the direction cosines of the members, i.e., the cosines of the angles between the member axis and the x, y, and z axes, respectively. The direction cosines can be calculated if the coordinates of the nodal points are known. Thus, for example, for member ab

$$L_{ab} = [(x_b - x_a)^2 + (y_b - y_a)^2 + (z_b - z_a)^2]^{1/2}$$

$$\lambda_{ab} = \frac{x_b - x_a}{L_{ab}} \qquad \mu_{ab} = \frac{y_b - y_a}{L_{ab}} \qquad \nu_{ab} = \frac{z_b - z_a}{L_{ab}}$$

Next, joint b (Fig. 8.10c) can be investigated and the member forces F_{bc}, F_{bd}, and F_{be} calculated since F_{ab} has been obtained from joint a. The three equations of equilibrium for the concurrent force system at joint b are

$$P_{bx} = 0 = -F_{ab}\lambda_{ab} - F_{bc}\lambda_{bc} - F_{bd}\lambda_{bd} - F_{be}\lambda_{be} + R_{bx}$$

$$P_{by} = 0 = -F_{ab}\mu_{ab} - F_{bc}\mu_{bc} - F_{bd}\mu_{bd} - F_{be}\mu_{be} + R_{by}$$

$$P_{bz} = 0 = -F_{ab}\nu_{ab} - F_{bc}\nu_{bc} - F_{bd}\nu_{bd} - F_{be}\nu_{be}$$

The analysis can be completed by proceeding to joints c, d, e, f, g, and h in sequence. When we reach the last joint j, all the member forces have already been calculated, so that the analysis of this joint merely serves as a check on the numerical accuracy of the three member forces meeting there. If the analysis were started at joint j and the joints analyzed in reverse alphabetical order, the reaction forces could be calculated; i.e., this would give the forces for equilibrium of the entire structure. Example 8.4 illustrates the calculation of member forces for a simple rigid space truss.

Example 8.4 Calculate the forces in the members of this space truss using the method of joints.

SOLUTION The equations of equilibrium give for joint a

$$F_{ab} = -12 \text{ kN (c)} \qquad F_{ac} = 24 \text{ kN (t)} \qquad F_{ae} = 16 \text{ kN (t)}$$

for joint e

$$\sum F_y = \tfrac{3}{5}F_{be} - 24 = 0 \qquad F_{be} = 40 \text{ kN (t)}$$

$$\sum F_x = -16 - \tfrac{4}{5}F_{be} - \tfrac{4}{5}F_{ec} = 0 \qquad F_{ec} = -60 \text{ kN (c)}$$

$$\sum F_z = -F_{ef} - \tfrac{3}{5}F_{ec} = 0 \qquad F_{ef} = 36 \text{ kN (t)}$$

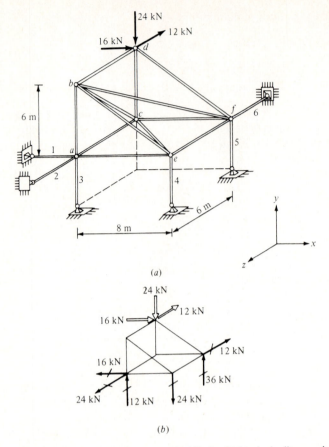

(a)

(b)

Figure E8.4 (a) Space truss with applied loads; (b) free-body diagram showing calculated reactions.

for joint b

$$\sum F_x = \frac{4}{\sqrt{34}} F_{bf} + \tfrac{4}{5}(40) = 0 \qquad F_{bf} = -8\sqrt{34} \text{ kN (c)}$$

$$\sum F_y = 12 - \tfrac{3}{5}(40) - \frac{3}{\sqrt{34}} F_{bf} - \frac{1}{\sqrt{2}} F_{bc} = 0 \qquad F_{bc} = 12\sqrt{2} \text{ kN (t)}$$

$$\sum F_z = \frac{-1}{\sqrt{2}} F_{bc} - \frac{3}{\sqrt{34}} F_{bf} - F_{bd} = 0 \qquad F_{bd} = 12 \text{ kN (t)}$$

for joint d

$$\sum F_x = \tfrac{4}{5} F_{df} + 16 = 0 \qquad F_{df} = -20 \text{ kN (c)}$$

$$\sum F_y = -\tfrac{3}{5} F_{df} - 24 - F_{cd} = 0 \qquad F_{cd} = -12 \text{ kN (c)}$$

$$\sum F_z = 12 - 12 = 0 \qquad \text{(okay)}$$

for joint c

$$\sum F_z = 24 + \frac{1}{\sqrt{2}} 12\sqrt{2} + \tfrac{3}{5}(-60) = 0 \qquad \text{(okay)}$$

$$\sum F_x = \tfrac{4}{5} F_{ce} + F_{cf} = 0 \qquad F_{cf} = 48 \text{ kN (t)}$$

$$\sum F_y = -12 + \frac{1}{\sqrt{2}} 12\sqrt{2} = 0 \qquad \text{(okay]}$$

and for joint f

$$\sum F_x = -48 + \tfrac{4}{5}(20) + \frac{4}{\sqrt{34}} 8\sqrt{34} = -48 + 16 + 32 = 0 \qquad \text{(okay)}$$

$$\sum F_y = 36 - \tfrac{3}{5}(20) - \frac{3}{\sqrt{34}} 8\sqrt{34} = 36 - 12 - 24 = 0 \qquad \text{(okay)}$$

$$\sum F_z = 36 - 12 - \frac{3}{\sqrt{34}} 8\sqrt{34} = 36 - 12 - 24 = 0 \qquad \text{(okay)}$$

MATRIX FORM OF JOINT EQUILIBRIUM EQUATIONS

The procedure of obtaining member forces joint by joint explicitly is probably the most efficient method for hand calculation, but if the solution is to be carried out in conjunction with a digital computer, it is perhaps advisable to formulate all the joint equilibrium equations as a set of simultaneous algebraic equations. Thus, for the structure illustrated in Fig. 8.10a, these equations take the matrix form

$$\mathbf{P} = \mathbf{BF} = \mathbf{B} \begin{bmatrix} \mathbf{F}_b \\ \cdots \\ \mathbf{R} \end{bmatrix}$$

where n = total number of joints in structure
  $\mathbf{P}$ = $3n \times 1$ matrix of applied forces
  $\mathbf{F}$ = $3n \times 1$ matrix of unknowns, consisting of member forces $\mathbf{F}_b$ and reactions $\mathbf{R}$
  $\mathbf{B}$ = $3n \times 3n$ global equilibrium matrix

Thus

$$\begin{bmatrix} \mathbf{F}_b \\ \cdots \\ \mathbf{R} \end{bmatrix} = \mathbf{B}^{-1}\mathbf{P}$$

For the example shown in Fig. 8.10a

$$\mathbf{F}^t = [F_{ab} \ F_{ac} \ F_{ad} \ F_{bc} \ F_{bd} \ F_{be} \ \cdots \ F_{hj} \ R_{ax} \ R_{ay} \ R_{az} \ R_{bx} \ R_{by} \ R_{cy}]$$

$$\mathbf{P}^t = [P_{ax} \ P_{ay} \ P_{az} \ P_{bx} \ P_{by} \ P_{bz} \ \cdots \ P_{jx} \ P_{jy} \ P_{jz}]$$

$$\mathbf{B} = \begin{bmatrix}
-\lambda_{ab} & -\lambda_{ac} & -\lambda_{ad} & \cdots & \cdots & \cdots & \cdots & 1 & 0 & 0 & 0 & 0 & 0 \\
-\mu_{ab} & -\mu_{ac} & -\mu_{ad} & \cdots & \cdots & \cdots & \cdots & 0 & 1 & 0 & 0 & 0 & 0 \\
-\nu_{ab} & -\nu_{ac} & -\nu_{ad} & \cdots & \cdots & \cdots & \cdots & 0 & 0 & 1 & 0 & 0 & 0 \\
-\lambda_{ab} & 0 & 0 & -\lambda_{bc} & -\lambda_{bd} & -\lambda_{be} & \cdots & 0 & 0 & 0 & 1 & 0 & 0 \\
-\mu_{ab} & 0 & 0 & -\mu_{bc} & -\mu_{bd} & -\mu_{be} & \cdots & 0 & 0 & 0 & 0 & 1 & 0 \\
-\nu_{ab} & 0 & 0 & -\nu_{bc} & -\nu_{bd} & -\nu_{be} & \cdots & 0 & 0 & 0 & 0 & 0 & 0 \\
\vdots & & & & & & & & & & & & \\
0 & 0 & 0 & 0 & 0 & 0 & \cdots & 0 & 0 & 0 & 0 & 0 & 0
\end{bmatrix}$$

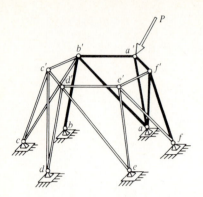

Figure 8.11 A rigid space truss.

where the inverse of **B** exists unless the space truss is inherently unstable. Note that for the structure in Fig. 8.10a, $P_{ax} = P_{ay} = P_{az} = P_{bx} = P_{by} = P_{bz} = 0$; furthermore, there are nonzero values of the applied loads only at joints e, f, h, and j.

The four special observations associated with concurrent force systems (see Sec. 8.2) can be used advantageously for some space trusses to identify the zero-force members and shorten the calculations. Consider the simple rigid space truss in Fig. 8.11, which is constructed so that each side forms a plane (members dd', ee', $d'e$, and $d'e'$ are coplanar, etc.). If this structure were to be analyzed by the method of joints, the investigation could begin at joint a' (which has three unknowns) and proceed joint by joint through the structure in a systematic manner until all member forces were calculated. A more expedient approach consists of augmenting the usual analysis with the special observations. For example, note that at joint f' all members with the exception of $e'f'$ are coplanar; therefore, we deduce that $F_{e'f'}$ must be zero. Proceeding in sequence to joints e', d', and c', we conclude in a similar manner that $F_{d'e'} = F_{c'd'} = F_{b'c'} = 0$. When we isolate joint b' and note that all the nonzero member forces except $b'c'$ and cb' are coplanar, $F_{b'c}$ must be zero since $F_{b'c'}$ is zero. For the subsequent discussion, imagine that all the zero-force members identified above are removed from the structure. Thus, at joints c', d', and e' there remain only two member forces which are not collinear; therefore, both of them must also be zero-force members. That is, $F_{cc'} = F_{c'd} = F_{dd'} = F_{d'e} = F_{ee'} = F_{e'f} = 0$. As a result, we observe that only the seven shaded members are active in resisting the applied load, and the computational effort necessary to obtain these member forces has been reduced appreciably from that required to analyze the entire structure joint by joint.

8.5 THE METHOD OF SECTIONS

Generally, if a compound truss (as in Fig. 8.3b and c) is to be analyzed by the method of joints, it is necessary to formulate all the joint equilibrium equations and solve them simultaneously. That is, it is impossible to calculate the member forces independently by progressing through the structure so that only three unknowns are encountered since eventually a joint will be found with more than three unknown member forces. An alternative procedure consists of cutting an imaginary section through the space truss and considering an entire portion of the structure as a free-body diagram. If the members cut are not coplanar, do not have a point of concurrency, and do not all intersect a single axis, six independent equations of equilibrium are available for calculating their forces. This approach, the method of sections, is similar to that used for planar trusses. With this method it is not necessary to begin at a joint with three unknowns and proceed systematically through the structure; therefore, the method is extremely useful for calculating the connecting forces in members that join the individual simple space trusses forming a compound truss.

This approach can be envisioned by studying the compound truss in Fig. 8.3b, which is composed of the two simple rigid trusses, *abcd* and *a'b'c'd'*, joined together by the six labeled connecting members. Joints *a* and *a'* each have three unknown member forces, which can be obtained using Eqs. (8.2); however, the remainder of the joints have more than three unknowns. Passing an imaginary section through the structure that cuts these six connecting members makes it possible to calculate them using Eqs. (8.3). Note that the six members that have been cut are not coplanar, do not have a point of concurrency, and do not intersect a single axis; therefore, their forces can be calculated uniquely. It is usually advantageous to use the free-body diagram that yields the simplest form of the equilibrium equations, and this is determined by the number of applied loads, etc. For the space truss of Fig. 8.3b this decision can be made only after the locations of the applied loads are known. The calculations for finding the connecting forces for a compound truss using the method of sections are presented for a cuboid space truss in Example 8.5.

Example 8.5 This compound truss, formed from simple space trusses *abcc'* and *a'b'd'd*, is joined together with the six numbered members to form a cube with side length L. Find the six connecting forces.

SOLUTION
$$\mathbf{P}_{d'} = P\frac{\sqrt{3}}{3}(\mathbf{i} - \mathbf{j} + \mathbf{k})$$

From Fig. E8.5b

$$\sum F_x = \frac{\sqrt{3}}{3}P + F_3 + F_4 = 0 \qquad (a)$$

$$\sum F_y = -\frac{\sqrt{3}}{3}P - F_1 - F_6 = 0 \qquad (b)$$

$$\sum F_z = \frac{\sqrt{3}}{3}P + F_2 + F_5 = 0 \qquad (c)$$

$$\sum M_{aa'} = \frac{PL\sqrt{3}}{3} + F_4L + F_5L = 0 \qquad (d)$$

$$\sum M_{a'c'} = \frac{PL\sqrt{3}}{3} + F_2L + F_6L = 0 \qquad (e)$$

$$\sum M_{a'd'} = F_4L - F_6L = 0 \qquad (f)$$

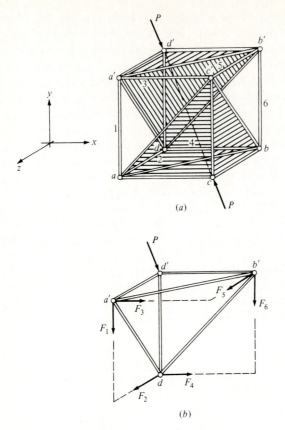

(a)

(b)

Figure E8.5 (a) Compound truss with loads; (b) free-body diagram of upper tetrahedron.

Solution of the equilibrium equations gives

$$F_1 = F_2 = F_3 = F_4 = F_5 = F_6 = -\frac{\sqrt{3}}{6}P$$

8.6 STATICAL DETERMINACY AND STABILITY

In Secs. 8.4 and 8.5 the equations of static equilibrium were used to calculate the member forces in stable space trusses that are statically determinate. The ensuing discussion will be devoted to methods that can be used to establish whether these structures are statically determinate, indeterminate, and/or unstable.

In Sec. 8.1 the tetrahedron was identified as the basic closed polyhedron used to form simple rigid space trusses. The structure is expanded beyond this primary form by adding three bars, which are not coplanar, meeting at a new joint to form a new tetrahedron that is contiguous with the existing truss configuration. Using arguments of kinematics exclusively, a simple test for statical determinacy of simple rigid space trusses was devised [see Eq. (8.1)].

A similar criterion can be obtained for supported rigid space trusses from equilibrium considerations. At each joint of a space truss there is a concurrent force system composed of the member forces plus any applied loads and reactions. Since the moment equilibrium equations are trivially satisfied in this case, the three scalar equations (8.2) can be used to ensure equilibrium at each joint in the truss. For a structure with n joints, each joint can be individually investigated and $3n$ independent equations of equilibrium formulated. The total number of unknowns, which should simultaneously satisfy these equations, is equal to the sum of the bars in the structure b and the number of independent reaction forces r. By comparing the number of unknowns and equations it is possible to draw some conclusions about the structural behavior of the system. If the number of unknowns is equal to the number of equations, the bar forces and reactions can be uniquely calculated and the structure is *statically determinate* unless the truss is inherently unstable. When the number of unknowns exceeds the number of equations, the bar forces and/or the reactions cannot be determined independently unless additional equations of deformation are introduced, and the structure is *statically indeterminate* unless it is inherently unstable. If the number of unknowns is less than the number of equations, the system is overdetermined, which implies that the structure is a *mechanism;* the truss can sustain some types of loadings but will be unstable for most loads. These criteria for classification, initially advanced by Möbius, can be summarized as follows:

$$(b + r) - 3n \begin{cases} = 0 & \text{space truss is statically determinate} \\ > 0 & \text{space truss is statically indeterminate} \\ < 0 & \text{space truss is a mechanism} \end{cases} \qquad (8.4)$$

Since six independent support forces are required to prevent rigid-body motions in three-dimensional space ($r = 6$), we can observe that the first criterion is identical to that described in Eq. (8.1).

The criteria stated above are *necessary conditions* for establishing whether a structure is statically indeterminate or determinate, but even when these conditions are met, they are *not sufficient* to state that the structure is stable. Generally, if the basic rules for constructing rigid simple and compound space trusses (see Sec. 8.1) are observed, the structure will be stable. Unfortunately, with complex space trusses, it is often difficult to visualize the kinematic behavior of the structure; furthermore it is possible to arrange the minimum number of supports for a rigid body to give an unstable condition (Sec. 8.2). For example, the complex space truss in Fig. 8.12 has $r = 12$, $b = 12$, and $n = 8$; therefore,

$$(b + r) - 3n = (12 + 12) - 3(8) = 0$$

According to Eq. (8.4), the structure is statically determinate; however, it can be envisioned that the square *efgh* is not rigid since the lengths of the two diagonal distances $\overline{fh}$ and $\overline{eg}$ can change, which implies that the space truss is unstable. Thus, a structure can be *geometrically unstable* by virtue of an improper arrangement of the supports and/or the members. That is, a space truss is geometrically stable if

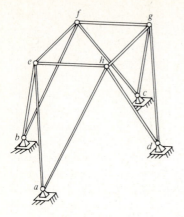

Figure 8.12 A geometrically unstable space truss.

(1) There are six or more reactions that are not parallel, are not concurrent, are not coplanar, and do not intersect a common axis and (2) there are a sufficient number of bars so arranged that internal instability does not occur.

The so-called *zero-load test* can be used to detect unstable behavior. That is, if it can be demonstrated that the members sustain loads for zero applied loads, the structure is unstable. For example, for the structure in Fig. 8.12 we hypothesize that there are no applied loads and that member *ah* is loaded in tension. To maintain equilibrium in the vertical direction at joint *h* member *hd* must sustain a compressive force equal in magnitude to F_{ah}. These two member forces will each have a component in the plane of the top horizontal square *efgh*, and by investigating joint *h* we conclude that members *eh* and *hg* must act in compression and tension, respectively, to maintain equilibrium at the joint. By investigating joints *e*, *f*, and *g* it can be demonstrated that $F_{eh} = F_{fg} = -F_{hg} = -F_{ef} \neq 0$. Therefore, it must be concluded that the solution is ambiguous and the structure is unstable since zero applied loads can result in a situation with nonzero loads in the members.

It is sometimes necessary to investigate unstable structural behavior by determining whether the set of equilibrium equations has a unique solution; thus, a design is judged unstable if the coefficient matrix of the unknown member forces and reactions (matrix **B** in Sec. 8.4) has no inverse. That is, if this matrix is singular or near singular, the structure is inherently unstable. This approach will not be discussed in this book, but it should be noted that it can be a tenuous basis for judging the acceptability of a structure since the detection of matrix singularity depends upon both the hardware and the software of the computer system being used. It is therefore suggested that all designs be carefully inspected using the principles discussed in this chapter to discover whether the structure has a propensity for displaying unstable behavior.

Equations of condition can be introduced into space trusses just as they are in planar trusses, i.e., by virtue of conditions of design and construction. Thus, for three-dimensional systems the spherical hinge gives three conditional equations; the link yields two conditional equations (as it does in planar systems); and parallel members in a given panel introduce up to two equations of condition since no forces can be transmitted normal to the direction of these bars.

8.7 SPECIAL COMPLEX SPACE TRUSSES

Complex space trusses can be analyzed in much the same fashion as the other structures described in this chapter. The framed tower and the Schwedler dome, both examples of closed-envelope bar structures, will be discussed in this section.

The tower in Fig. 8.13 is a complex space truss forming a closed envelope and having four to six unknowns per joint. The structure can be visualized as a simple truss formed joint by joint from the foundation upward, with the directions of the diagonals changed in opposite side panels; for example, c_2d_1 and a_2b_1 replace c_1d_2 and a_1b_2, respectively, in the first tier of the simple rigid truss. Note that $b = 36$, $r = 12$, and $n = 16$; therefore, by the criteria in Eq. (8.4) the structure is statically determinate. Since this tower has straight legs, it can be analyzed by treating each side as an individual planar truss. The single applied load is resolved into components that are aligned with the planes of the two contiguous sides of the tower. The component P_b is parallel to the leg b_1b_4, and it will induce loads only into the members of this leg. The force P_{ab} is parallel to member b_4a_4; hence, it will load the members in the plane formed by the joints designated with the letters a and b, while P_{bc}, which is parallel to member b_4c_4, will introduce forces in members in the plane containing all the joints denoted with the letters b and c. For this particular loading, trusses forming the other two planes will not be loaded. The total forces in the members are obtained by superposing the results obtained from the individual planar-truss investigations.

If a tower does not have straight legs, the response of the structure will differ from that described above. For example, since the legs of the tower in Fig. 8.14 have

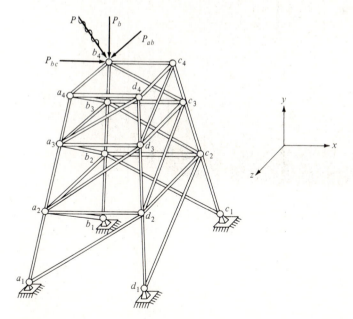

Figure 8.13 Tower with straight legs.

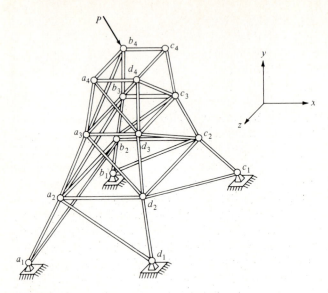

Figure 8.14 Tower having legs with variable batter.

different slopes for each tier this structure cannot be analyzed as four independent planar trusses. The transfer of loads within the tower is such that the forces in the members on one side are introduced into adjacent sides as the result of the tower geometry. For example, if the analysis is begun at the top, the load shown can be resolved into three components in a manner similar to that used for the straight-leg tower of Fig. 8.13 and the bar forces in the top tier calculated. Upon analyzing joint a_3 we note that the force in member a_3a_4 can be resolved into three components (one each parallel to planes $a_2b_2b_3a_3$ and $a_2d_2d_3a_3$ plus one parallel to member a_2a_3) in the same way that the applied load was treated. Consequently, a component of force is introduced into plane $a_2d_2d_3a_3$. This same effect will also occur at joint c_3, and as a result all four sides of this tower will transmit forces from the single applied load. This behavior is in contrast with the simpler load-transfer mechanism displayed by the straight-leg tower.

For many designs it is common to have secondary bracing arranged in the horizontal planes at the various tier levels of the tower. The presence of these members renders the structure statically indeterminate, but they are typically assumed to be zero-force members for the purpose of analyzing the forces in the main members of the tower.

Another type of complex space truss constructed in the form of a closed envelope of framed members is the Schwedler dome. An example of a two-tier hexagonal Schwedler dome is shown in Fig. 8.15. The criteria of Eq. (8.4) indicate that the structure is statically determinate ($b = 42, r = 12$, and $n = 18$). The external stability of this structure depends upon the arrangement of the horizontal support reactions as measured by the angle α_n. This can be demonstrated by writing the equations of equilibrium for the bottom ring of the dome. The determinant of the coefficient matrix vanishes if: (1) each reaction is normal to the radius of the circumscribing circle

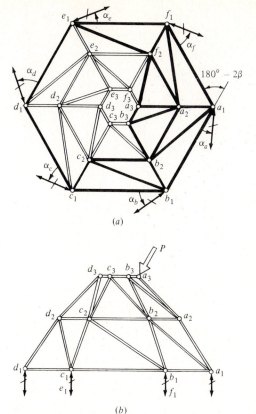

(a)

180° − 2β

(b)

Figure 8.15 A hexagonal two-tier Schwedler dome: (*a*) plan; (*b*) elevation.

(for even-sided polygons); or (2) the reactions at two adjacent joints are directed along the members that are not contiguous with these joints (for even- and odd-sided polygons); for example, $\alpha_b = 0°$ and $\alpha_a = 180° - 2\beta$ (the internal angles of the base polygon are 2β).

The dome in Fig. 8.15 is examined under the effect of a single applied load to illustrate its behavior in transferring forces. For the force applied at joint a_3 there are a number of zero-force members, which can be identified using the special observations from Sec. 8.2 for concurrent force systems. Starting at joint b_3, we note that b_3c_3 is the only member that is not contained in the plane of joints $a_2a_3b_3b_2$; therefore, $F_{b_3c_3} = 0$. Similarly, by investigating joints c_3, d_3, e_3, and f_3, we conclude that $F_{c_3d_3} = F_{d_3e_3} = F_{e_3f_3} = F_{f_3a_3} = 0$. When all these zero-force members are removed, joints c_3, d_3, e_3, and f_3 each have only two members that are not collinear; therefore, they must all be zero ($F_{c_3b_2} = F_{c_3c_2} = F_{d_3d_2} = F_{d_3d_2} = F_{e_3d_2} = F_{e_3e_2} = F_{f_3e_2} = F_{f_3f_2} = 0$). Proceeding in this manner to the first tier of the dome, we can identify the remaining zero-force members. The members that are effective in resisting the applied load are shown shaded in Fig. 8.15*a*. The analysis can now be conducted by the method of joints in the usual fashion; e.g., since joint a_3 has three unknowns, it would be a logical starting point.

PROBLEMS

8.1 to 8.3 Calculate the forces in the members of these space trusses.

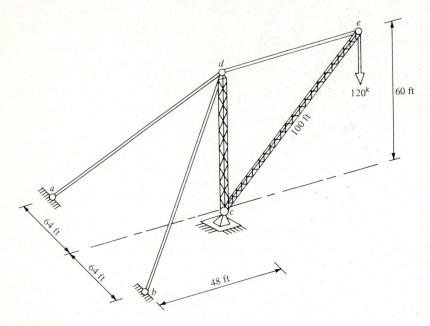

Figure P8.1

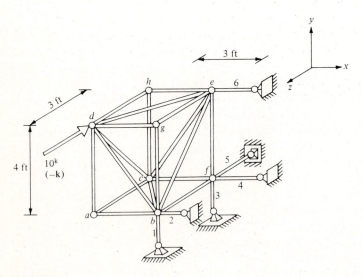

Figure P8.2

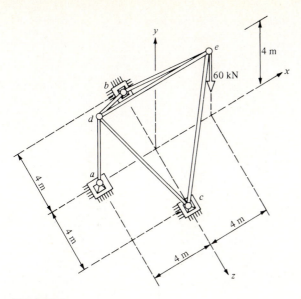

Figure P8.3

8.4 The solid plate *dfe* and the triangle formed by points *bac* are both right triangles oriented parallel to the *xz* plane. The forces at *d* act in the *x* and *z* directions, and those at *e* and *f* are directed in the $-x$ and $-z$ directions, respectively. Calculate the forces in the six supporting members if $P = 5$ kN.

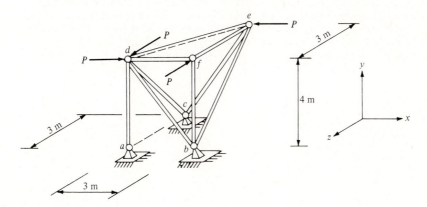

Figure P8.4

8.5 This rigid space truss consists of a square pyramid $abcdO$ supported by six members. Calculate the forces in all members.

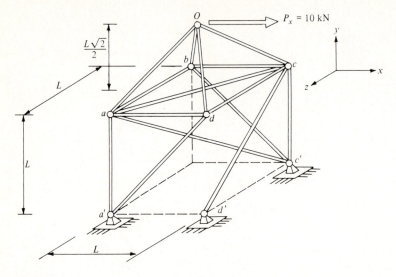

Figure P8.5

8.6 The two parallel equilateral triangles abc and def with sides L are located a distance L apart. The structure is subjected to two separate load cases, LC1, the illustrated loading (note that all loads are parallel to the xz plane and perpendicular to members ef and ac) and LC2, in which two collinear forces of magnitude P directed toward each other are applied at joints a and d. Is the space truss simple, compound, or complex? Calculate the forces in all members for (a) the illustrated structure, (b) the structure obtained if member dc is removed and replaced by a member connecting joints a and f, and (c) the structure obtained if the vertical members are removed and replaced by three members connecting joints b and f, a and e, and a and f.

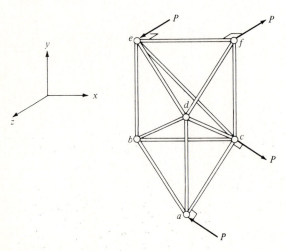

Figure P8.6

8.7 The triangles *abc* and *def* are equilateral with sides of length *L*; the distance between these two parallel planes is *L*. Is this space truss simple, compound, or complex? Calculate the forces in all members for (*a*) the illustrated structure and (*b*) the structure obtained by removing member *bd* and replacing it with a member connecting joints *b* and *f*.

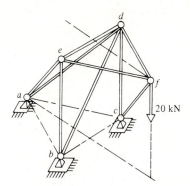

Figure P8.7

8.8 The radii of the circumscribed circles of the regular hexagons forming this structure are 6.75, 11.00, and 15.25 m. Is this a simple, compound, or complex truss? (This is a Schwedler dome.) Use Eq. (8.4) to establish that it is statically determinate. Calculate the member forces for (*a*) the single downward load of 10 kN at joint d_3 (*Hint:* Identify the zero-force members first) and for (*b*) a downward 10-kN load applied to each of the joints of the inner ring.

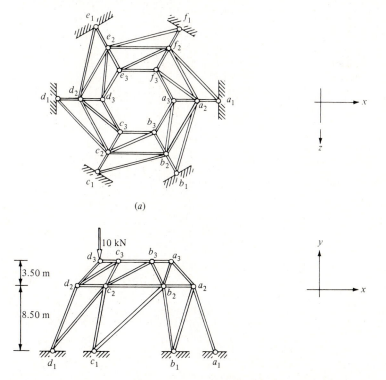

(*a*)

(*b*)   **Figure P8.8** A Schwedler dome: (*a*) plan; (*b*) elevation.

8.9 The design of this structure is attributed to Föppl. Is this a simple, compound, or complex space truss? Use Eq. (8.4) to establish that it is statically determinate. Calculate the member forces for (a) the single downward load of 10 kN at joint c_3 and (b) a downward 10-kN load applied to each of the joints of the inner square.

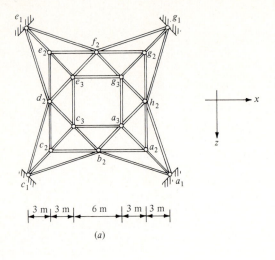

(a)

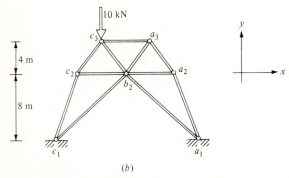

(b)

Figure P8.9 (a) Plan; (b) elevation.

8.10 This tower is subjected to three separate load cases each involving a 10-kN load applied only to joint a_3 as follows:

LC1: Load in the positive x direction
LC2: Load in the positive y direction
LC3: Load in the positive z direction

Calculate the forces in all members for the following configurations:

(a) The legs are all vertical and the square formed by the four joints at each level measures 5 m on a side.

(b) All the legs have an equal and constant batter such that the top square $a_3b_3c_3d_3$ is 5 m on each side and the bottom square $a_1b_1c_1d_1$ is 25 m on each side.

(c) The sides of the squares at the top, middle, and bottom measure 5, 10, and 20 m, respectively (the tower is symmetrical about the vertical centerline, and all legs in a given level have the same batter).

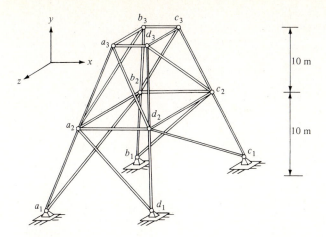

Figure P8.10

ADDITIONAL READING

Andersen, P., and G. M. Nordby: *Introduction to Structural Mechanics,* chap. 12, Ronald, New York, 1960.

Gerstle, K. H.: *Basic Structural Analysis,* chap. 4, Prentice-Hall, Englewood Cliffs, N.J., 1974.

Laursen, H. I.: *Structural Analysis,* chap. 4, McGraw-Hill, New York, 1978.

McCormac, J. C.: *Structural Analysis,* 3d ed., chap. 14, Intext, New York, 1975.

Norris, C. H., J. B. Wilbur, and S. Utku: *Elementary Structural Analysis,* 3d ed., chap. 4, McGraw-Hill, New York, 1976.

Timoshenko, S. P., and D. H. Young: *Theory of Structures,* 2d ed., chap. 4, McGraw-Hill, New York, 1965.

West, H. H.: *Analysis of Structures,* chap. 10, Wiley, New York, 1980.

PART
THREE

THE DEFLECTIONS OF STRUCTURES

NINE

DEFLECTIONS: SOLUTIONS TO THE GOVERNING EQUATIONS

9.1 DESCRIPTION OF DEFLECTION BEHAVIOR

The deflections displayed by various structures supporting their design loads can vary greatly. For example, the wing of a high-speed jet aircraft during flight can experience deflections easy to observe; whereas a high-rise building in a windstorm undergoes horizontal displacements that are not apparent to an observer on the street. The aircraft passengers may be having a normal flight while the occupants of the upper stories in the high rise experience great discomfort from the movement of the building. Most people readily perceive and disapprove of a leaning wall, a sagging beam, or a floor that bounces under normal foot traffic. The flexible floor is a deflection problem associated with the dynamic behavior of the structure; a similar problem arises when a large group of people walks across a bridge in cadence. There are many less dramatic, but equally significant, reasons why the deflection behavior of structures is important. For example, excessive structural deflections can cause plaster to crack, sensitive equipment to malfunction, doors to stick, glass to crack, and many other objectional phenomena that can render a structure unacceptable for use. Allowable deflections vary from several feet in the case of the wing tip of an airplane to a small fraction of the span, e.g., length/360, for a member supporting a plaster ceiling. The criteria for deflection behavior for different types of structures vary widely, but they are usually described explicitly in an appropriate code that is applicable to the structure. See, for example, Ref. 1.

Structures can experience both elastic and inelastic deflections. Inelastic deflections, which result in a permanent distortion of the structure, can be a realistic design criterion but require analysis methods that are beyond the scope of this text. Generally, the investigation is confined to linear elastic displacements, where the structure returns to its undeflected position after the loads are removed.

Deflection behavior of a structure is the visible evidence of how it is performing. The deflections of a statically indeterminate structure can influence the distribution of internal resisting forces and moments; however, this is not the case if the structure is statically determinate. For example, in a bridge consisting of a continuous beam over three or more piers, large bending moments can be induced in the beam by virtue of differential settlement of the supports. Although deflections are always calculated as a routine part of any structural analysis, for statically indeterminate structures this must be done simply to begin the solution process. That is, the equations of deflection augment the equations of static equilibrium to give a solution for the reactions, internal forces, and moments.

Since the maximum deflection most typical structures experience under maximum loading is usually much less than 1 percent of the member length, it is not useful to show the deflections and the structure on a drawing using the same scale for both. For clarity, the deflected shape of the structure is typically drawn using a greatly exaggerated scale for the deflections. These sketches can give insight into the behavior of a structure where deflections must be considered. For this reason, the reader is urged to construct these deflection sketches routinely and to make extensive use of them wherever possible. Figure 9.1 shows the deflected shape of a structure subjected to extreme loads.

Deflections can be calculated using many different methods, which can be divided into two categories. In this chapter the governing equations (differential and/or algebraic) will be solved by an analytical, graphical, or combined procedure. In Chap. 10 energy methods will be used to formulate and solve the problems of deflection behavior. Although it may be confusing to be exposed to so many approaches the structural engineer must be aware of a wide spectrum of methods since there are usually several ways to solve a given problem and often there is an optimal method.

9.2 LOAD-DEFORMATION EQUATIONS

Forces and deformations can be functionally related if the material properties and the cross-sectional properties of the member are known. The specific response of a member to an applied load will depend upon how the member transmits the loading; therefore, the uniaxial truss member and the beam are discussed separately.

Figure 9.1 Prestressed concrete beam under test loading. (*ABAM Engineers, Inc., Tacoma, Washington.*)

The Truss Member

A truss member is capable of resisting only uniaxial forces, i.e., forces that are compressive or tensile. Typically, a tensile test is performed on a specimen which has a uniform cross-sectional area and which is made from a linear elastic isotropic material. By graphing force against deformation or stress against strain the modulus of elasticity can be obtained using Hooke's law, i.e.,

$$\frac{F}{A} = E\frac{\Delta L}{L} \qquad \text{or} \qquad \Delta L = \frac{FL}{AE} \qquad\qquad (9.1a)$$

where A = cross-sectional area
         E = modulus of elasticity
         F = applied load
         L = undeformed length
       ΔL = deformation (see Fig. 9.2)

The linear constant of proportionality relating force and deformation in this case is L/AE, which is the flexibility constant for the element. If a truss member is subjected to a uniform temperature increase ΔT above ambient temperature, it experiences a deformation of

$$\Delta L = \alpha\,\Delta T\,L \qquad\qquad (9.1b)$$

where α is the coefficient of linear thermal expansion.

The Beam Member

In general, a loaded beam has an internal resisting shear force and a bending moment, and its behavior is more complex than that of a truss member. The uniform beam in Fig. 9.3 is originally straight and has a symmetrical cross section with respect to the plane of the applied loads. If only a moment M is applied with $V = 0$, it is in a state of pure bending and we assume that stresses exist only in the xy plane. That is, this is plane stress behavior, and it implies that the stress σ_z and the shear stresses τ_{xz} and τ_{yz} are all zero. From Hooke's law we conclude that the shear strains γ_{xz} and γ_{yz} are also zero. When the beam deforms, it bends into an arc, as shown in Fig. 9.4, such that cross sections parallel to the yz plane remain plane and one plane parallel to the zx plane remains stress free; the longitudinal line in this plane is the neutral axis. Thus, the displacements in the x (u) and y (v) directions can be written

$$u = -y\theta(x) \qquad v = v(x)$$

where $\theta(x)$ is the slope of the cross section, and y is measured from the neutral axis. We conclude that $\varepsilon_y = 0$ from the strain-displacement relation, i.e.,

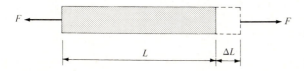

Figure 9.2 The truss member.

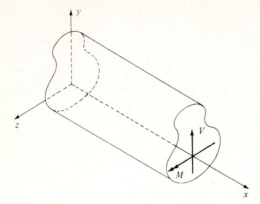

Figure 9.3 The beam member.

$$\varepsilon = \frac{\partial v}{\partial y} = 0$$

From symmetry of deformation $\gamma_{xy} = 0$, and the strain-displacement equation for this shear strain gives

$$\gamma_{xy} = \frac{\partial u}{\partial y} + \frac{\partial v}{\partial x} = -\theta(x) + \frac{dv}{dx} = 0$$

or
$$\frac{dv}{dx} = \theta(x)$$

The third strain-displacement relation, together with Hooke's law, gives

$$\varepsilon_x = \frac{\partial u}{\partial x} = -y\frac{d^2v}{dx^2} = \frac{1}{E}(\sigma_x - \nu\sigma_y)$$

Assuming that $\sigma_y = 0$ (true at the top and bottom surface) yields

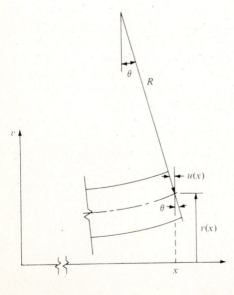

Figure 9.4 The elastic curve of a deflected beam.

$$\sigma_x = -Ey\frac{d^2v}{dx^2}$$

Note that Hooke's law, the fact that $\varepsilon_y = 0$, and the assumption that $\sigma_y = 0$ imply that σ_x should be zero, but this contradiction is ignored. Hooke's law also gives $\tau_{xy} = 0$ because $\gamma_{xy} = 0$. Enforcing equilibrium of moments about the z axis gives

$$\sum M_z = 0: \qquad M = -\int_{\text{Area}} y\sigma_x \, dA = E\frac{d^2v}{dx^2}\int_{\text{Area}} y^2 \, dA = EI\frac{d^2v}{dx^2}$$

or

$$\frac{d^2v}{dx^2} = \frac{M}{EI} \tag{9.2}$$

where the moment of inertia of the section is

$$I = \int_{\text{Area}} y^2 \, dA \tag{9.3}$$

In the subsequent discussion the beam will be oriented with its length along the x axis, and bending will be in the xy plane. Therefore, no ambiguity occurs in describing the moment of inertia without subscripts. Substituting Eq. (9.2) into the expression for σ_x gives

$$\sigma_x = -\frac{My}{I} \tag{9.4}$$

Enforcing equilibrium along the axis of the member and using Eqs. (9.2) and (9.4) yields

$$\sum F_x = 0: \qquad 0 = \int_{\text{Area}} \sigma_x \, dA = -E\frac{d^2v}{dx^2}\int_{\text{Area}} y \, dA$$

thus,

$$\int_{\text{Area}} y \, dA = 0$$

That is, the neutral surface must pass through the centroid of the cross-sectional area. Finally, equilibrium of moments about the y axis combined with Eqs. (9.2) and (9.4) gives

$$\sum M_y = 0: \qquad 0 = \int_{\text{Area}} z\sigma_x \, dA = -E\frac{d^2v}{dx^2}\int_{\text{Area}} yz \, dA$$

therefore,

$$\int_{\text{Area}} yz \, dA = 0$$

which is satisfied because the cross section is symmetric with respect to the xy plane.

The relationship between the curvature κ and the radius of curvature R can be expressed using basic concepts from the calculus

$$\kappa = \frac{d^2v/dx^2}{[1 + (dv/dx)^2]^{3/2}} \qquad \frac{1}{R} = |\kappa|$$

If $(dv/dx)^2 \ll 1$, this relationship can be approximated as

Fig. E9.1*b* the locus of all points for the deflected position of end *c* associated with member *ac* is along the straight line *AA,* and the locus of all points for the deflected position of point *c* for member *bc* is along the straight tangent line *BB*. Therefore the final deflected position for point *c* must be at point *c'*. The horizontal and vertical deflections are calculated using elementary geometry.

Beam Deflections

Equation (9.2) can be used to obtain the deflected shape of the elastic curve of a beam by integrating the equations for the bending moment. In this process it is important to observe the strength-of-material sign convention outlined in Chap. 4. That is, a positive bending moment compresses the top fibers of the beam, puts the bottom fibers in tension, and distorts the member into a shape that is concave upward. The constants of integration are evaluated by observing the boundary conditions. For example, the displacement is zero at the support of a simply supported beam, and for a fixed support both the deflection and the slope of the member are zero. Examples 9.2 to 9.4 illustrate the procedures for obtaining beam deflections using this direct calculation method.

Example 9.2 Calculate the expression for the deflection of this beam by integrating the moment-curvature relations.

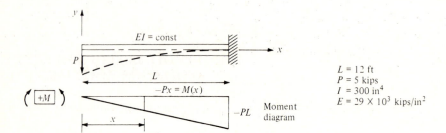

$$L = 12 \text{ ft}$$
$$P = 5 \text{ kips}$$
$$I = 300 \text{ in}^4$$
$$E = 29 \times 10^3 \text{ kips/in}^2$$

Figure E9.2

SOLUTION

$$EI\frac{d^2v}{dx^2} = -Px$$

$$EI\frac{dv}{dx} = -\tfrac{1}{2}Px^2 + C_1$$

$$EIv = -\tfrac{1}{6}Px^3 + C_1x + C_2$$

$$v(L) = 0 \qquad \frac{dv}{dx}(L) = 0$$

Therefore,
$$C_1 = \tfrac{1}{2}PL^2 \qquad C_2 = -\tfrac{1}{3}PL^3$$

$$EIv = -\tfrac{1}{6}P(x^3 - 3L^2x + 2L^3)$$

$$EI\frac{dv}{dx} = \tfrac{1}{2}P(L^2 - x^2)$$

Maximum slope and deflection occur at $x = 0$.

$$v_{max} = \frac{-PL^3}{3EI} = -\frac{(5\ \text{kips})(12^3\ \text{ft}^3)(12^3\ \text{in}^3/\text{ft}^3)}{3(29 \times 10^3\ \text{kips/in}^2)(300\ \text{in}^4)} = 0.57\ \text{in}\ (\downarrow)$$

$$\left(\frac{dv}{dx}\right)_{max} = \frac{PL^2}{2EI} = \frac{(5\ \text{kips})(12^2\ \text{ft}^2)(12^2\ \text{in}^2/\text{ft}^2)}{2(29 \times 10^3\ \text{kips/in}^2)(300\ \text{in}^4)} = 0.0060\ \text{rad}\ (\frown)$$

$$= 0.34°$$

DISCUSSION The deflected shape illustrates that the moment will be negative over the entire length of the beam. Since the moment can be described using a single equation for the entire beam, the deflection and slope equations can be conveniently obtained. The constants of integration are evaluated by observing that both the deflection and the slope at the fixed support must be zero. The general expression for deflection is used to obtain the maximum deflection at the free end of an example steel beam. The downward deflection of 0.57 in is accompanied by a positive slope at this point of 0.34°; therefore, the assumption associated with the second derivative shown in Eq. (9.5) is applicable.

Example 9.3 Use the moment-curvature relations to obtain the expression for the deflection and slope of this cantilever beam.

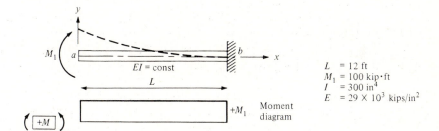

Figure E9.3

L = 12 ft
M_1 = 100 kip·ft
I = 300 in⁴
E = 29 × 10³ kips/in²

SOLUTION

$$EI\frac{d^2v}{dx^2} = M_1$$

$$EI\frac{dv}{dx} = M_1x + C_1$$

$$EIv = \tfrac{1}{2}M_1x^2 + C_1x + C_2$$

$$v(L) = 0 \qquad \frac{dv}{dx}(L) = 0$$

Therefore,
$$C_1 = -M_1 L \qquad C_2 = \tfrac{1}{2}M_1 L^2$$

$$EIv = \tfrac{1}{2}M_1(x^2 - 2Lx + L^2) \qquad EI\frac{dv}{dx} = M_1(x - L)$$

At the free end

$$v_{max} = 0.12 \text{ in } (\uparrow)$$

$$\left(\frac{dv}{dx}\right)_{max} = -0.0016 \text{ rad} = -0.09° \ (\frown)$$

DISCUSSION This is the same beam as in Example 9.2, but it is loaded with a concentrated moment at the free end. In this case, the moment deforms the beam into a positive curvature over the entire length. The same geometric-compatibility conditions used in Example 9.2 were used here to evaluate the integration constants. The maximum deflection of 0.12 in occurs at the free end, where the maximum slope of -0.0016 also occurs.

Example 9.4 Find the maximum deflection and slope for this simply supported beam.

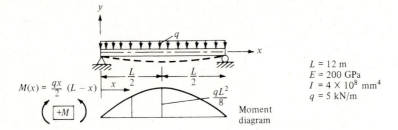

$$M(x) = \frac{qx}{2}(L - x)$$

$$L = 12 \text{ m}$$
$$E = 200 \text{ GPa}$$
$$I = 4 \times 10^8 \text{ mm}^4$$
$$q = 5 \text{ kN/m}$$

Moment diagram

Figure E9.4

SOLUTION

$$EI\frac{d^2v}{dx^2} = M$$

$$EI\frac{dv}{dx} = -\tfrac{1}{6}qx^3 + \tfrac{1}{4}qLx^2 + C_1$$

$$EIv = -\tfrac{1}{24}qx^4 + \tfrac{1}{12}qLx^3 + C_1 x + C_2$$

$$v(0) = 0 \qquad v(L) = 0$$

Hence,
$$C_1 = -\tfrac{1}{24}qL^3 \qquad C_2 = 0$$

$$EIv = \tfrac{1}{24}qx(-x^3 + 2Lx^2 - L^3)$$

$$EI\frac{dv}{dx} = \tfrac{1}{24}q(-4x^3 + 6Lx^2 - L^3)$$

Since $M = 0$ at $x = 0$ and $x = L$, the slope has the maximum magnitudes at these points. Since $dv/dx = 0$ at $x = \tfrac{1}{2}L$, v_{max} occurs there.

$$v_{max} = \frac{5}{384} \frac{qL^4}{EI} = \frac{5}{384} \frac{5(12^4)}{(200 \times 10^6)(4 \times 10^{-4})} = 16.88 \times 10^{-3} \text{ m}$$

$$\left(\frac{dv}{dx}\right)_{max} = \frac{qL^3}{24EI} = \frac{5(12^3)}{24(200 \times 10^6)(4 \times 10^{-4})} = 4.5 \times 10^{-3} \text{ rad} = 0.26°$$

DISCUSSION This symmetric beam is bent into positive curvature over its entire length by the uniform loading. The zero deflections at the two supports are used to evaluate the constants of integration. The deflection expression turns out to be a quartic polynomial with the maximum deflection occurring at $x = L/2$; the slope has its maximum magnitude at both supports and a zero value at $x = L/2$. The dimensions and size of the steel beam shown result in a maximum deflection of 16.88 mm with a maximum magnitude of slope equal to 0.26° at the two supports.

This method is more complex for a beam with a loading like that shown in Fig. 9.5. In this case the moment expression for the moment between points a and b differs from that for the moment between points b and c; therefore, the moment-curvature relation must be integrated individually for each segment. This process gives four constants of integration to describe the beam displacements. Two of these constants can be evaluated by enforcing the conditions of zero displacement at the supports, and the other two must be evaluated by matching the expressions at point b for the slope and deflection. Since there are more elegant mathematical ways of obtaining the expressions for displacement and slope for these types of problems, the reader who intends to use this method extensively would be wise to study the application of singularity functions to these problems (see for example, Ref. 2).

9.4 THE PRINCIPLE OF SUPERPOSITION

If a structure is acted upon by a number of individual forces and/or moments, the total deflection of the structure can be obtained by calculating the deflections induced by each of the individual forces and adding the results. This process of combining solutions for a given problem is called *superposition,* and the validity of the approach depends upon the linearity of the structure. Figure 9.6a and 9.6b shows load-deformation graphs for a linear and a nonlinear structural response, respectively. We can see that for the linear structure an applied load P will give the deformation Δ regardless of where in the loading process it is applied. For the nonlinear situation the same load P will give different deformations, Δ_1 and Δ_2, which depend upon the amount of loading previously applied to the structure. Nonlinearities can be introduced

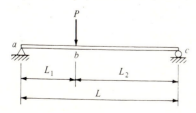

Figure 9.5 Beam with an applied concentrated force.

At the free end

$$v = -\frac{PL^3}{3EI} + \frac{M_1 L^2}{2EI}$$

$$\frac{dv}{dx} = \frac{PL^2}{2EI} - \frac{M_1 L}{EI}$$

DISCUSSION The beam shown in Fig. E9.5a is investigated for the two load cases shown in Fig. E9.5b and c. According to the principle of superposition, the moment diagrams for the two separate cases are added to give the moments for the case involving P and M_1. The two solutions have been taken directly from Examples 9.2 and 9.3. Also, the complete deflection curve can be constructed by superposing the two individual curves from the two cases. Thus the loads and moments for the individual cases combine to give their respective totals, and the separate deflection curves are superposed, giving a composite deflection curve that satisfies the geometric-compatibility conditions of the original problem.

9.5 THE MOMENT-AREA METHOD

For most structures composed of flexural members it is usually sufficient to calculate the deflections only at a number of specific points; therefore, it is not necessary to develop the general equation for the deflected shape of the elastic curve by formally integrating the moment-curvature equation (9.2). The moment-area, elastic-weights, and conjugate-beam methods all use the governing differential equation in conjunction with the geometry of the elastic curve; thus, the structural analyst can visually perceive how the member is responding to its loading. [The moment-area method is attributed to C. E. Greene (1873); the method of elastic weights to O. Mohr (1870); and the conjugate beam to H. Müller-Breslau (1885).] These are semigraphical methods for carrying out the necessary integrations and evaluating the resulting constants by using analogies relating the internal shears and moments to the deflection behavior of the beam.

Consider a segment of the elastic curve of a beam shown in Fig. 9.7 in a general deformed shape. This member was initially straight and continuous, with the x axis located at the undeflected position. The elastic curve is shown in an exaggerated deflection position for clarity. In reality, both the deflections and slopes are extremely small, and it is difficult to observe the deflected shape with the naked eye.

The differential element of length dx for an initially straight beam can be measured along the neutral axis if the radius of curvature R is large. Assume that (1) the radius of curvature is constant over the length dx and (2) $d\theta$ is small enough for the angle and its sine and tangent all to be approximately equal; therefore,

$$R\, d\theta = dx \qquad \text{or} \qquad d\theta = \frac{1}{R}\, dx = \kappa\, dx \qquad (9.6)$$

Integrating Eq. (9.6) between the two points x_a and x_b gives

$$\theta_{ba} = \theta_b - \theta_a = \int_{x_a}^{x_b} d\theta = \int_{x_a}^{x_b} \kappa\, dx \qquad (9.7)$$

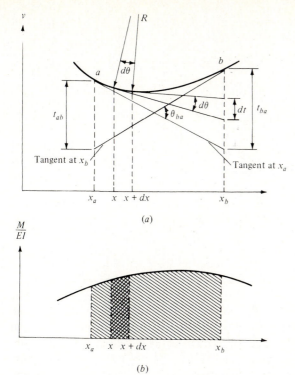

(a)

(b)

Figure 9.7 (a) Segment of the elastic curve of a beam; (b) the associated M/EI diagram.

Equation (9.7) can be used to evaluate the change in slope of any line in a beam that is straight before it is deformed. For an elastic beam, $\kappa = M/EI$ from Eq. (9.5); substituting this into Eq. (9.7) yields

$$\theta_{ba} = \int_{x_a}^{x_b} \frac{M}{EI}\, dx \qquad (9.8)$$

The right-hand side of this equation represents the area under the M/EI diagram between x_a and x_b, and the left-hand side is the change in slope of the tangents at points x_a and x_b. Equation (9.8) is the statement of the first moment-area theorem.

First moment-area theorem For an initially straight continuous beam, the change in the slope of the elastic curve between any two points is equal to the area under the M/EI diagram between these two points.

Consider the tangents to the elastic curve in Fig. 9.7a at two points along the beam located at x and $x + dx$. The change in slope of the deflected curve over this small length of beam is $d\theta$. The line segment intercepted by the two tangents on the line perpendicular to the undeflected beam passing through x_b is dt. If all angles are sufficiently small,

$$dt = (x_b - x)\, d\theta \qquad (9.9)$$

That is, the increase in t over an increment dx is dt. Substituting the definition of $d\theta$ from Eq. (9.6) into (9.9) gives

$$dt = (x_b - x)\kappa\,dx \qquad (9.9a)$$

Integrating this equation between x_a and x_b yields

$$t_{ba} = \int_{x_a}^{x_b} dt = \int_{x_a}^{x_b} (x_b - x)\kappa\,dx \qquad (9.10)$$

where the tangential deviation t_{ba} is the deflection of the elastic curve at x_b measured (perpendicular to the longitudinal axis of the undeflected beam) relative to the tangent constructed at x_a (see Fig. 9.7a). For an elastic beam the curvature κ is defined by Eq. (9.5). Substituting this into Eq. (9.10), we have

$$t_{ba} = \int_{x_a}^{x_b} (x_b - x)\frac{M}{EI}\,dx \qquad (9.11)$$

The right side of this equation is the first moment about x_b of the area under the M/EI diagram between points x_a and x_b. It is an expression of the second moment-area theorem.

> **Second moment-area theorem** Consider two points a and b on the elastic curve of a continuous initially straight beam. The tangential deviation of point b (perpendicular to the undeflected longitudinal axis) measured from the tangent to the elastic curve at point a is equal to the moment of the area under the M/EI diagram between a and b taken about point b.

From the fundamental theorem of calculus we note that these two theorems are valid only if θ is continuous between the points considered, i.e., the beam must bend into a smooth continuous curve. Therefore, the theorems cannot be used in the vicinity of a discontinuity in slope of the elastic curve introduced by a mechanism such as a hinge. The elementary calculations necessary to use these theorems can be conveniently carried out if the M/EI diagram is composed of straight lines. If the diagram is quadratic, the method can be applied, but the advantages of the method can best be realized if the properties for a parabola (Fig. 9.8) are used. If the degree of the M/EI diagram is cubic or higher, it is questionable whether this method should be applied without numerical integration. If the flexural rigidity EI of the beam changes along its length, each of the moment ordinates must be divided by the appropriate value of EI.

The algebraic signs of the slopes and tangent deviation distances obtained from the moment-area method must be carefully interpreted. Figure 9.7 illustrates the direction of positive slope change θ_{ba} and tangent deviation distance t_{ba} at the right end of the beam segment with a positive moment and the associated positive deflected shape of the elastic curve. Thus, a positive slope change implies an increase in the slope between the two points, and a tangent deviation is positive if the elastic curve lies above the tangent reference line; i.e., in the positive coordinate direction. It is usually helpful to construct a sketch of the approximate deflected shape from the bending-moment diagram before the moment-area calculations are executed. The geometry of the elastic

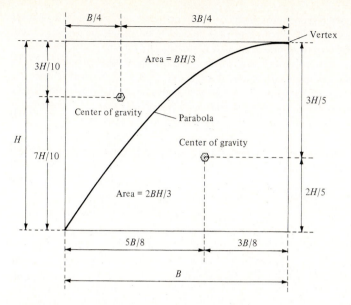

Figure 9.8 Some properties of areas defined by a parabola.

curve can be used to write equations relating deflections and slopes, which are combined with the moment-area results. The sign of the moment-area quantities is used to interpret whether the slope between two points increases or decreases and whether the tangent deviation lies above or below the elastic curve. The calculations can be used to confirm or disprove the deflected shape initially sketched. A consistent sign convention becomes extremely important when both positive and negative moments are encountered in the same beam.

Examples 9.6 to 9.9 illustrate the application of this method to four different types of beams.

Example 9.6 Using the moment-area method, calculate the slope and deflection at the free end of the beam of Example 9.3.

SOLUTION (See Fig. E9.3.) From the first moment-area theorem

$$\theta_{ba} = \frac{M_1}{EI}(L)$$

Since $\theta_b = 0$,

$$\theta_a = -\frac{M_1 L}{EI} \quad (\curvearrowleft)$$

From the second moment-area theorem

$$t_{ab} = v_a = \frac{M_1 L}{EI}\frac{L}{2} = \frac{M_1 L^2}{2EI} \quad (\uparrow)$$

DISCUSSION At the fixed support of a cantilever beam both the slope and the deflection are zero. Thus, by referencing the moment-area calculations to this point, the deflections and rotations are obtained directly. The beam is loaded with a positive moment such that it bends into positive curvature over its entire length. The change in slope between points a and b is given by applying the first moment-area theorem. $\theta_{ba} = \theta_b - \theta_a$ is positive since the M/EI area is positive; thus $\theta_a = -\theta_{ba}$. The tangent deviation at point a is positive since the M/EI ordinates are all positive. From the sketch of the deflected shape it is apparent that the deflection at point a is upward and the slope is negative (clockwise).

Example 9.7 Calculate the slopes at points a and c and the deflection at point b using the moment-area method.

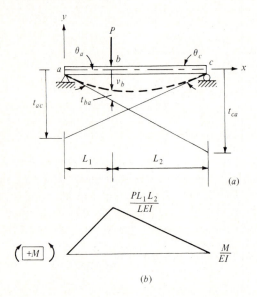

(a)

(b)   **Figure E9.7**

SOLUTION From the geometry of the elastic curve

$$\theta_a = \frac{t_{ca}}{L}$$

Applying the second moment-area theorem gives

$$t_{ca} = \tfrac{1}{2} L_1 \frac{PL_1L_2}{LEI}(L_2 + \tfrac{1}{3}L_1) + \tfrac{1}{2} L_2 \frac{PL_1L_2}{LEI}(\tfrac{2}{3}L_2)$$

$$= \frac{PL_1L_2}{2LEI}\left(L_1L_2 + \frac{L_1^2}{3} + \frac{2L_2^2}{3}\right) = \frac{PL_1L_2}{6LEI}(L_1 + 2L_2)(L_1 + L_2)$$

$$= \frac{PL_1L_2}{6EI}(L_1 + 2L_2)$$

$$\theta_a = \frac{PL_1L_2(L_1 + 2L_2)}{6LEI} \quad (\frown; \text{ negative slope})$$

From the geometry of the elastic curve

$$v_b = \theta_a(L_1) - t_{ba}$$

Applying the second moment-area theorem gives

$$t_{ba} = \frac{1}{2} \frac{PL_1L_2}{LEI} (L_1) \frac{L_1}{3} = \frac{PL_1^3L_2}{6LEI}$$

$$v_b = \frac{PL_1L_2}{6LEI} (L_1^2 + 2L_1L_2 - L_1^2) = \frac{PL_1^2L_2^2}{3LEI} \quad (\downarrow)$$

Also, from geometry,

$$\theta_c = \frac{1}{L} t_{ac}$$

or alternatively

$$\theta_c = \theta_{ac} - \theta_a$$

where

$$\theta_{ac} = \frac{1}{2} L \frac{PL_1L_2}{LEI} = \frac{PL_1L_2}{2EI}$$

$$\theta_c = \frac{PL_1L_2}{6EIL} (3L - L_1 - 2L_2) = \frac{PL_1L_2(2L_1 + L_2)}{6LEI} \quad (\frown; \text{ positive slope})$$

DISCUSSION Unlike those for the cantilever beam in Example 9.6, the slopes and deflections for this beam are not obtained directly by applying the moment-area theorems. For example, the rotation of point a is obtained by first applying the second moment-area theorem to obtain t_{ca} and dividing this deflection by the beam length. This calculation is accurate if the angle is small enough to ensure

$$\sin \theta_a \approx \tan \theta_a \approx \theta_a$$

The actual deflection at a point on the beam must also be obtained by combining geometric considerations with the results from the moment-area method. The tangent deviation t_{ba} is obtained by taking the moment about point b of the area under the M/EI diagram between a and b. This is not the deflection at point b. To obtain the deflection at b it is necessary to incorporate both the rotation at point a and the quantity t_{ba}, as shown in the figure. Note that for most structures this is how the moment-area results must be used to obtain the actual deflections of the elastic curve; i.e., the cantilever beam is one of the rare cases in which the rotations and deflections are given directly by the moment-area theorems. The magnitudes of the various quantities were calculated and used with an accurate sketch of the deflected elastic curve to interpret the sense of the deflections and slopes.

Example 9.8 Calculate the rotations at points b, c, and e and the deflections at points a and d using the moment-area method. $EI = 3 \times 10^6$ kip $\cdot$ in^2.

SOLUTION From geometry (Fig. E9.8b)

$$\theta_b = \frac{t_{eb}}{18}$$

From the second moment-area theorem and Fig. E9.8c

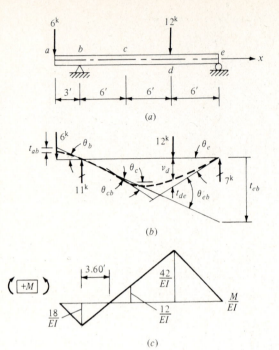

(a)

(b)

(c)

Figure E9.8

$$EIt_{eb} = \tfrac{1}{2}(6)(42)(\tfrac{2}{3})(6) + \tfrac{1}{2}(8.40)(42)(6 + 2.80) - \tfrac{1}{2}(3.60)(18)(14.40 + 2.40)$$

$$= 504.0 + 1552.3 - 544.3$$

$$= +1512.0 \text{ kip} \cdot \text{ft}^3$$

$$EI\theta_b = 84 \text{ kip} \cdot \text{ft}^2 \ (\frown) \qquad \theta_b = \frac{84(12^2)}{3 \times 10^6} = 0.0040 \text{ rad} \ (\frown)$$

i.e., negative slope. From the second moment-area theorem and Fig. E9.8c

$$EIt_{ab} = \tfrac{1}{2}(-18)(3)(2) = -54 \text{ kip} \cdot \text{ft}^3$$

i.e., the elastic curve is below the tangent, as shown. From geometry (Fig. E9.8b)

$$v_a = 3\theta_b - |t_{ab}| \qquad EIv_a = 3(84) - 54 = 198 \text{ kip} \cdot \text{ft}^3 \ (\uparrow)$$

$$v_a = \frac{198(12^3)}{3 \times 10^6} = 0.11 \text{ in} \ (\uparrow)$$

From the first moment-area theorem and Fig. E9.8c

$$EI\theta_{eb} = \tfrac{1}{2}(42)(14.4) + \tfrac{1}{2}(-18)(3.6) = 270 \text{ kip} \cdot \text{ft}^2$$

From geometry (Fig. E9.8b)

$$\theta_e = \theta_{eb} - \theta_b$$

$$EI\theta_e = 270 - 84 = 186 \text{ kip} \cdot \text{ft}^2 \; (\frown) \qquad \theta_e = \frac{186(12^2)}{3 \times 10^6} = 0.0089 \text{ rad} \; (\frown)$$

From geometry and the first moment-area theorem

$$v_d = 6\theta_e - t_{de} \qquad 1116 - \tfrac{1}{2}(6)\,(42)\,(2) = 864 \text{ kip} \cdot \text{ft}^3 \; (\downarrow) = EIv_d$$

$$v_d = \frac{864(12^3)}{3 \times 10^6} = 0.50 \text{ in} \; (\downarrow)$$

$$EI\theta_{cb} = \tfrac{1}{2}(12)\,(2.40) + \tfrac{1}{2}(-18)\,(3.60) = -18.0 \text{ kip} \cdot \text{ft}^2$$

i.e., the slope decreases from b to c

$$EI\theta_c = -84 - 18 = -102 \text{ kip} \cdot \text{ft}^2 \; (\frown) \qquad \theta_c = \frac{-102(12^2)}{3 \times 10^6} = -0.0050 \text{ rad} \; (\frown)$$

DISCUSSION The algebraic signs of the various computed moment-area quantities must be carefully interpreted for this example since the bending-moment diagram involves both positive and negative ordinates. The deflection of the elastic diagram was sketched after the moment diagram was drawn. Since only the correct curvatures are known at this point, it is not possible to determine whether the deflections are positive or negative. Thus, it has been assumed that the beam deflects downward throughout the segment be because of the heavy loading in this vicinity; however, it is not possible to guess with any certainty whether point a deflects up or down. Since t_{eb} is positive, the elastic curve at e is above the tangent, and θ_b is clockwise as shown; i.e., the beam has a negative slope at b. t_{ab} has a negative sign, indicating that the elastic curve lies below the tangent; thus from Fig. E9.8b we note that the magnitude of t_{ab} must be subtracted from $3\theta_b$ to give v_a. The area of the M/EI diagram between points b and c is negative, which indicates that the slope is less at c than at b.

The calculations could be simplified by replacing the moment diagram between b and d with two triangles, each with a base equal to the length of the entire segment bd. One of these triangles would have only positive ordinates with a maximum of $+42/EI$, while the other would have only negative ordinates with the peak value of $-18/EI$. When this approach of equivalent areas is used, neither the point of zero moment nor the areas and locations of the centroids of the original triangles need be calculated. Unfortunately, while simplifying the calculations this method may obfuscate the principles being illustrated. The reader may wish to try this approach after acquiring some experience with the moment-area method.

Example 9.9 Use the moment-area method to find the deflection and slope at point d for this beam with an internal hinge.

SOLUTION

$$\theta_b = \frac{t_{ab}}{20}$$

$$EIt_{ab} = \tfrac{1}{2}(175)\,(10)\,(\tfrac{20}{3}) + \tfrac{1}{2}(175)\,(7.78)\,(10 + 2.59) - \tfrac{1}{2}(50)\,(2.22)\,(17.78 + 1.48)$$

$$= 5833.33 + 8569.96 - 1069.96 = 13,333.33 \text{ kN} \cdot \text{m}^3$$

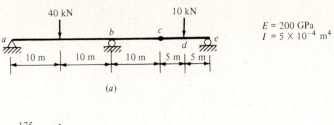

(a)

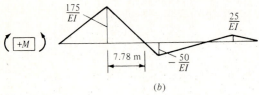

(b)

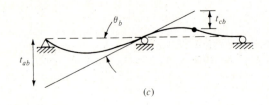

(c)

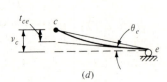

(d)

Figure E9.9

$$EI\theta_b = 666.67 \text{ kN} \cdot \text{m}^2 \qquad \theta_b = \frac{666.67}{(200 \times 10^6)(5 \times 10^{-4})} = 0.0067 \text{ rad } (\curvearrowright)$$

$$v_c = 10\theta_b - t_{cb} \qquad EIt_{cb} = \tfrac{1}{2}(50)(10)(\tfrac{20}{3}) = 1666.67 \text{ kN} \cdot \text{m}^3$$

$$EIv_c = 6666.67 - 1666.67 = 5000 \text{ kN} \cdot \text{m}^3$$

$$v_c = \frac{5000}{(200 \times 10^6)(5 \times 10^{-4})} = 0.05 \text{ m}$$

Rotation of chord $ce = v_c/10 = 0.005$ rad

$$EIt_{ce} = \tfrac{1}{2}(25)(10)(5) = 625 \text{ kN} \cdot \text{m}^3 \qquad EI\theta_e = 500 - \tfrac{625}{10} = 437.5 \text{ kN} \cdot \text{m}^2$$

$$EI\theta_d = 500 \text{ kN} \cdot \text{m}^2 \qquad \theta_d = 0.005 \text{ rad } (\curvearrowright)$$

$$EIt_{de} = \tfrac{1}{2}(25)(5)(\tfrac{5}{3}) = 104.17 \text{ kN} \cdot \text{m}^3$$

$$EIv_d = \tfrac{5000}{10}(5) - 5(\tfrac{625}{10}) + 104.17 = 2291.67 \text{ kN} \cdot \text{m}^3$$

$$v_d = 0.023 \text{ m } (\uparrow)$$

DISCUSSION The moment-area theorems cannot be applied to the beam ae because of the discontinuity in the elastic curve due to the hinge at c. Since the chord of the elastic curve between a and b does not rotate, the deflections and rotations for the span ac can be obtained using the moment-area theorems. The rotation of chord ce is established using the deflection at c; thus, the moment-area theorems for this segment give rotations and deflections relative to this rotated chord. The deflection and slope of point d are established by combining the moment-area results with the geometry of the rotated chord. Since only a single concentrated force is applied at the center of segment ce, the slope of the elastic curve at point d will be equal to the rotation of chord ce.

The location and magnitude of the maximum deflection for a beam can be calculated by applying the moment-area theorems. The unknown location of the maximum deflection for the beam in Fig. 9.9 occurs at point m, and the slope of the elastic curve at this point is zero. The slope at point a can be obtained from the relationship

$$\theta_a = \frac{t_{ba}}{L}$$

The first moment-area theorem can be used to yield θ_{am}; furthermore,

$$\theta_{am} = \theta_a - \theta_m = \theta_a$$

Since the unknown length L_m is implicit in this equation, it can be used to obtain the location of the maximum deflection. The fact that θ_m is zero implies that the maximum deflection can be calculated from the second moment-area theorem since

$$v_m = t_{am}$$

Example 9.10 illustrates this procedure for a beam, and Example 9.11 demonstrates how the moment-area method is used to obtain deflections for a rigid frame.

Example 9.10 Calculate the maximum deflection for the beam of Example 9.8 using the moment-area method.

SOLUTION Note that

$$\theta_{de} = \frac{1}{2}\frac{42}{EI}(6) = \frac{126}{EI} < \theta_e$$

therefore, m lies to the left of d. Since $\theta_m = 0$, it follows that $\theta_{me} = \theta_e$ and

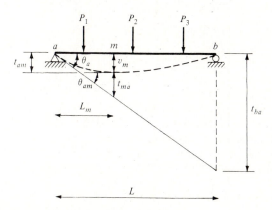

Figure 9.9 The maximum deflection of a beam.

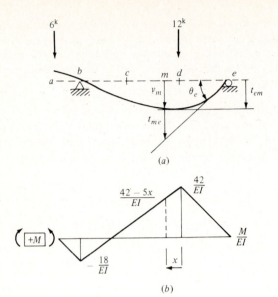

(a)

(b)   **Figure E9.10**

$$EI\theta_e = 186 = 126 + \tfrac{1}{2}(42 + 42 - 5x)x$$

$$2.5x^2 - 42x + 60 = 0$$

$$x = 1.576 \text{ ft}$$

$$EIt_{me} = 126(2 + 1.576) + \tfrac{1}{2}(42)(1.576)(\tfrac{2}{3})(1.576) + \tfrac{1}{2}(34.12)(1.576)(\tfrac{1}{3})(1.576)$$

$$= 450.64 + 34.80 + 14.13 = 499.57 \text{ kip} \cdot \text{ft}^3$$

$$v_m = 7.576\theta_e - t_{me} \qquad EIv_m = 1409.23 - 499.57 = 909.66 \text{ kip} \cdot \text{ft}^3$$

Alternatively, since $\theta_m = 0$,

$$v_m = t_{em}$$

$$EIt_{em} = 126(4) + \tfrac{1}{2}(42)(1.576)[6 + \tfrac{1}{3}(1.576)] + \tfrac{1}{2}(34.12)(1.576)[6 + \tfrac{2}{3}(1.576)]$$

$$= 504.00 + 216.03 + 189.63 = 909.66 \text{ kip} \cdot \text{ft}^3$$

$$v_m = \frac{909.66(12^3)}{3 \times 10^6} = 0.524 \text{ in} \ (\downarrow)$$

DISCUSSION The largest deflection will either occur at point a or some other point in span be (indicated as point m in Fig. E9.10a). The calculations reveal that θ_e is greater than the area under the M/EI diagram between d and e; therefore, point m lies slightly to the left of d. After finding the location of point m, the calculations for finding v_m are similar to any of those for finding a deflection. Note that the final result is the deflection with the largest magnitude for the beam.

Example 9.11 Calculate the deflection and rotation for this rigid frame using the moment-area method. $E = 200$ GPa and $I = 10^{-3}$ m^4.

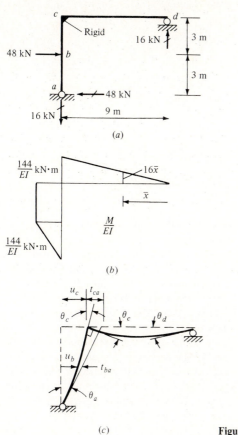

(a)

(b)

(c)

Figure E9.11

SOLUTION

(a) $\theta_d = \frac{1}{9}t_{cd}$ $EI\theta_d = \frac{1}{9}(\frac{1}{2})(144)(9)(3) = 216$ kN · m^2 ($\frown$)

$\theta_c = \frac{1}{9}t_{dc}$ $EI\theta_c = \frac{1}{9}(\frac{1}{2})(144)(9)(6) = 432$ kN · m^2 ($\frown$)

$EI\theta_{ac} = \frac{1}{2}(3)(144) + 3(144) = 648$ kN · m^2

$EI\theta_a = 432 + 648 = 1080$ kN · m^2 ($\frown$)

(b) $EIt_{ca} = \frac{1}{2}(144)(3)(3 + 1) + 144(3)(1.5) = 1512$ kN · m^3

$EIu_c = 6\theta_a - t_{ca} = 4968$ kN · m^3

(c)
$$u_c = \frac{4968}{2 \times 10^5} = 0.0248 \text{ m } (\rightarrow)$$

$$EIu_b = 3\theta_a - t_{ba} = 3240 - \tfrac{1}{2}(144)(3)(1) = 3024 \text{ kN} \cdot \text{m}^3$$

$$u_b = 0.0151 \text{ m } (\rightarrow)$$

From Fig. E9.11b the location of v_{max} in cd is

$$\theta_d = \theta_{xd}$$

$$216 = \tfrac{1}{2}(16\bar{x}^2)$$

$$\bar{x}^2 = 27$$

$$\bar{x} = 5.20 \text{ m}$$

$$v_{max} = \frac{16\bar{x}^3}{3EI} = 0.0037 \text{ m } (\downarrow)$$

DISCUSSION Assume that (1) the change in member lengths caused by axial loads will not affect the calculations and (2) that the rigid joint at point c is constructed so that moments can be transferred from the column (member ac) to the beam (member cd) with no change in geometry of the joint. The moment diagram for the beam was constructed using the strength-of-materials sign convention, but this can be ambiguous for a vertical member. The moments are therefore plotted on the column on the same side of the member that is in compression; i.e., this is also a consequence of using the usual sign convention for horizontal members. The deflected shape of the frame was constructed using the moment diagram plus the moment-curvature relations. The right-angle orientation of the rigid joint at point c was preserved after deformation. The total angle of rotation (θ_a) at point a was found by adding θ_{ac} to θ_c. This was necessary because point c deflected horizontally; thus, dividing the deviation distance t_{ca} by the length of the column gives only a part of the total rotation of point a relative to its vertical undeflected position. The rest of the deflection calculations are typical of those for a horizontal member. The location of the maximum deflection on the beam was obtained by calculating the point where the slope of the elastic curve is zero; therefore, the change in angle between the point of maximum deflection and point d equals θ_d.

9.6 METHOD OF ELASTIC LOADS

In the previous section it was demonstrated that the moment-area method yields deflections at specific points for flexural members without requiring formal integration of the moment-curvature relations or evaluation of the constants of integration. The deflections and slopes obtained from the moment-area method must be further interpreted if the beam under consideration is simply supported. However, if certain analogies are drawn between the structural deflections and the equilibrium behavior of beams on simple unyielding supports, these complications can be circumvented. These analogies free the structural analyst from having to perceive the deflected geometry and then manipulate the moment-area rotations and tangent offset distances to obtain the desired rotations and deflections of the beam.

The beam illustrated in Fig. 9.10 has simple unyielding supports at points a and b and is subjected to a loading that results in the positive moment diagram shown in the figure. It is required to calculate the rotation and slope for an arbitrary point on the beam designated as point j with a coordinate of x_j. First, it is necessary to calculate the slope at point b, but this can be accomplished only after the tangent offset distance at point a has been found. The second moment-area theorem gives

$$t_{ab} = \int_{x_b}^{0} \frac{Mx}{EI} \, dx$$

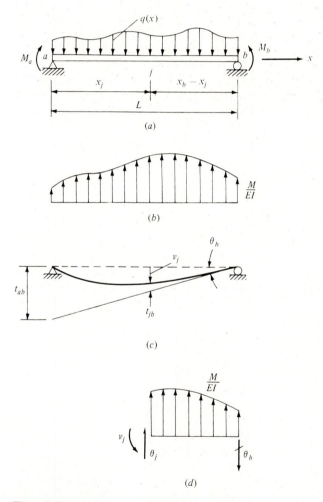

(a)

(b)

(c)

(d)

Figure 9.10 (a) Actual beam with loading; (b) elastic loads; (c) deflected shape; (d) free-body diagram showing elastic loads, elastic shears, and elastic moment.

The slope at the right end is calculated under the usual assumption that the slope is small enough to give

$$\theta_b = \frac{t_{ab}}{L}$$

The first moment-area theorem gives the change in slope between points j and b as

$$\theta_{bj} = \theta_b - \theta_j = \int_{x_j}^{x_b} \frac{M}{EI} \, dx \qquad \text{or} \qquad \theta_j = \theta_b - \int_{x_j}^{x_b} \frac{M}{EI} \, dx$$

Several analogies can be made between these calculations and those typically made in constructing shear and bending-moment diagrams. First, the actual beam is loaded with an imaginary loading consisting of the M/EI diagram, and this constitutes the *elastic load*. The quantity t_{ab} is the first moment of the M/EI diagram about point a, and θ_b is obtained by dividing this first moment by the length of the beam. That is, the calculations to obtain the slope of the actual beam at the right end are equivalent to those required for the shear at the right end of the beam loaded with the elastic loads. From Fig. 9.10d we can observe that θ_{bj} is the sum of the elastic loads between points j and b and θ_j is the difference between the elastic shear at b and the sum of the elastic loads between these two points. Thus, the slope at j corresponds to the *elastic shear* at this point.

This analogy can be extended further by using the moment-area method to calculate the deflection at point j. From the geometry of the elastic curve

$$v_j = (x_b - x_j)\theta_b - t_{jb}$$

and from the second moment-area theorem

$$t_{jb} = \int_{x_b}^{x_j} (x - x_j)\frac{M}{EI} \, dx$$

From Fig. 9.10d we observe that v_j is the moment about point j of the elastic load and the elastic reaction; thus, the deflections for the actual beam correspond to the *elastic moments*.

Generally, this approach is used for simply supported beams with unyielding supports; i.e., the slopes and deflections are referenced to the horizontal undeflected beam. However, if the chord ab in Fig. 9.10a were not horizontal, all the computed quantities would be relative to this chord. Thus, if point a in Fig. 9.10a experienced a downward deflection, the slopes and deflections would be measured with respect to the sloping chord passing through a and b. This approach is the method of elastic loads.

Method of elastic loads   The slopes and deflections of a continuous elastic curve, measured with respect to the chord drawn between two points on the curve, can be found by loading the beam with an imaginary distributed elastic load (M/EI diagram) between these two points and calculating the corresponding imaginary elastic shears and elastic moments. The slopes and deflections for the actual beam are equal to the elastic shears and elastic moments, respectively.

The strength-of-materials sign convention summarized in Fig. 4.16 for plotting shear and bending-moment diagrams should also be used for the elastic-shear and elastic-moment diagrams. Correspondingly, a positive elastic load (M/EI) must be interpreted as acting in the positive coordinate direction. Thus, a positive elastic load will yield a negative elastic-moment diagram, which implies negative deflections. The signs of the elastic shears are interpreted as positive or negative slopes for the actual beam. Example 9.12 illustrates the use of the method of elastic loads.

Example 9.12 Calculate the location and magnitude of the maximum deflection using the method of elastic loads.

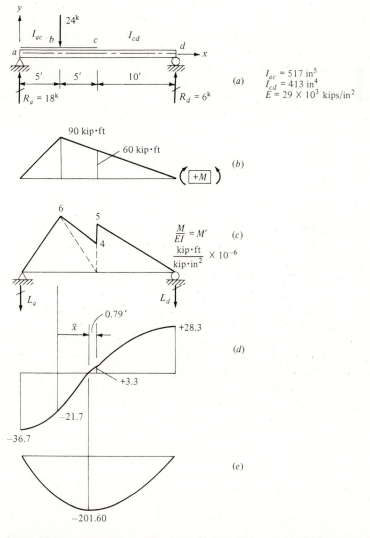

(a) $I_{ac} = 517 \ \text{in}^5$
$I_{cd} = 413 \ \text{in}^4$
$E = 29 \times 10^3 \ \text{kips/in}^2$

(b)

$$\frac{M}{EI} = M' \qquad (c)$$
$$\frac{\text{kip·ft}}{\text{kip·in}^2} \times 10^{-6}$$

(d)

(e)

Figure E9.12 (*a*) Actual beam; (*b*) moment diagram; (*c*) beam with elastic loads; (*d*) elastic shears (slopes); (*e*) elastic moments (deflections).

SOLUTION From Fig. E9.12c

$$\Sigma M_d' = 0 \curvearrowright: \qquad -\tfrac{1}{2}(10)(5)(\tfrac{2}{3})(10) - \tfrac{1}{2}(5)(4)(10 + \tfrac{5}{3})$$

$$-\tfrac{1}{2}(10)(6)(10 + 5) + 20L_a = 0$$

$$L_a = 36.7 \ (\updownarrow)$$

$$\Sigma F_y' = 0 + \uparrow: \qquad 25 + 10 + 30 - 36.7 = L_d$$

$$L_d = 28.3 \ (\updownarrow)$$

From Fig. E9.12d

$$-21.7 + \tfrac{1}{2}[6 + (6 - 0.4\bar{x})]\bar{x} = 0$$

$$-21.7 + 6\bar{x} - 0.2\bar{x}^2 = 0$$

$$\bar{x} = 4.21 \text{ ft} \quad \text{(point of maximum deflection)}$$

From Fig. E9.12c

$$v_{max} = -36.7(9.21) + 15(4.21 + 1.67) + \tfrac{1}{2}(6)(4.21)(\tfrac{2}{3})(4.21) + \tfrac{1}{2}(4.32)(4.21)(\tfrac{1}{3})(4.21)$$

$$= -201.60 \frac{\text{kip} \cdot \text{ft} \times 10^{-6}}{\text{kip} \cdot \text{in}^2}(\text{ft}^2)$$

$$= -201.60(12^3 \times 10^{-6}) = -0.348 \text{ in} \ (\downarrow)$$

DISCUSSION Because of the change in moment of inertia of the beam at point c, the simple shape of the moment diagram yields the more complex sawtooth elastic-load (M/EI diagram) distribution shown in Fig. E9.12c. The double-valued ordinate of M/EI at point c must be interpreted by noting that the value is 4 for a location slightly to the left of the point and 5 to the right of the point. These elastic loads acting in the positive coordinate direction were placed on the original simply supported beam and the internal resisting shears and moments calculated, giving the slopes and deflections, respectively, for the original beam. The trapezoid between points b and c in the M/EI diagram was divided into two triangles for convenience in calculating the reactions for the elastic loads. The elastic shears and moments were plotted using the usual sign conventions.

It is necessary to analyze the dimensions of each quantity carefully. Since the slopes are obtained using the area under the M/EI diagram, they have units of kip $\cdot$ ft^2/kip $\cdot$ in^2 $\times 10^{-6}$. Since deflections are associated with the area under the slope diagram, the numerator will contain an additional unit of feet. Therefore, the numerical results for slopes obtained from the elastic-shear diagram must be multiplied by $12^2 \times 10^{-6}$, and those for deflections taken from the elastic-moment diagram by $12^3 \times 10^{-6}$ to give results in radians and inches, respectively.

9.7 CONJUGATE-BEAM METHOD

The method of elastic loads described in Sec. 9.6 can be applied most conveniently to simply supported beams, but the analogies developed can be extended to beams with various support conditions. The conjugate-beam method uses the analogies between internal elastic generalized forces and actual generalized deflections, but it avoids the difficulties of interpreting the computed quantities for beams that are not simply supported. To predict true deflection behavior using this method it is necessary to make additional comparisons between real beam behavior and that of an imaginary conjugate beam.

Since the differential equations for beams are fundamental to the conjugate-beam method, they will be reviewed. From Chap. 4, the differential equations of equilibrium are

$$\frac{dV}{dx} = q \quad \text{and} \quad \frac{dM}{dx} = V$$

thus

$$\frac{d^2M}{dx^2} = q$$

Substituting this result into the moment-curvature equation (9.2) gives

$$\frac{d^2}{dx^2}\left(EI\frac{d^2v}{dx^2}\right) = q \tag{9.12}$$

If EI is constant over the region being considered, the equations of beam equilibrium and deformation can be summarized as follows:

Deflection $\qquad v \qquad\qquad$ elastic bending moment

Slope $\qquad \dfrac{dv}{dx} \qquad\qquad$ elastic shear

Curvature $\qquad \dfrac{d^2v}{dx^2} = \dfrac{M}{EI} \qquad$ bending moment (elastic load) $\qquad$ (9.13)

$\qquad\qquad \dfrac{d^3v}{dx^3} = \dfrac{V}{EI} \qquad$ shear force

$\qquad\qquad \dfrac{d^4v}{dx^4} = \dfrac{q}{EI} \qquad$ loading

Thus, the process of going from deflections to loading involves differentiation, while the reverse requires integration. Figure 9.11a illustrates the relations of a function $y(x)$ and its derivatives, and Fig. 9.11b shows the shapes of the various diagrams according to Eq. (9.13) for a simply supported beam that is uniformly loaded. The strength-of-materials sign convention presented in Fig. 4.16 is used to plot the various diagrams for both real and elastic internal generalized forces in Fig. 9.11b. Note that a positive deflection corresponds to a negative moment (second derivative) and a positive load

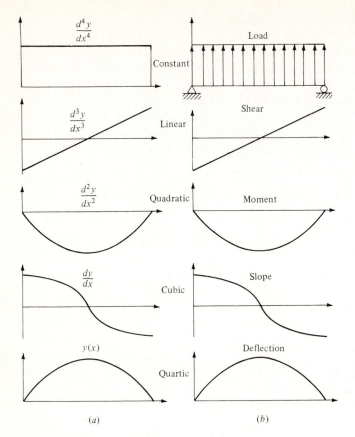

Figure 9.11 Relations of function shapes and the behavior of a beam: (*a*) function $y(x)$; (*b*) beam deflection $v(x)$.

(fourth derivative). Furthermore, a positive slope (first derivative) corresponds to a negative shear (third derivative) and vice versa.

In Sec. 9.6 it was shown that for a simply supported beam loaded with its appropriate elastic load the elastic shears and elastic moments correspond, respectively, to the slope and the deflection of the elastic curve for the actual beam; the validity of this statement can be observed in Eq. (9.13). To extend this argument for other support conditions the elastic loads at the supports of the substitute, or conjugate beam, must duplicate the kinematic behavior of the actual support conditions. For a simply supported beam there is no elastic moment (deflection) at the supports, but elastic shear (slope) can occur at this point. The various real support conditions and their corresponding conjugate substitutes are shown in Fig. 9.12. Their validity can be verified by comparing the actual deflection v and slope θ, respectively, with the elastic moment M' and shear V' for the conjugate beam. For example, the fixed-support condition must have zero slope and deflection; hence, the conjugate beam must support neither elastic shear nor elastic moment; i.e., the conjugate beam must have a free end.

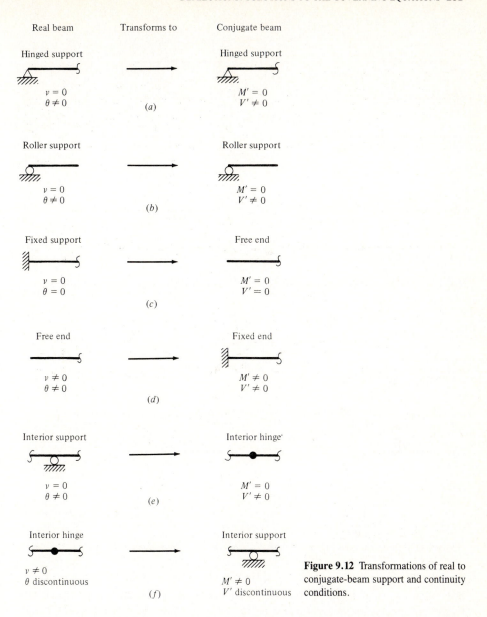

Real beam	Transforms to	Conjugate beam

Hinged support

$v = 0$
$\theta \neq 0$

(a)

Hinged support

$M' = 0$
$V' \neq 0$

Roller support

$v = 0$
$\theta \neq 0$

(b)

Roller support

$M' = 0$
$V' \neq 0$

Fixed support

$v = 0$
$\theta = 0$

(c)

Free end

$M' = 0$
$V' = 0$

Free end

$v \neq 0$
$\theta \neq 0$

(d)

Fixed end

$M' \neq 0$
$V' \neq 0$

Interior support

$v = 0$
$\theta \neq 0$

(e)

Interior hinge

$M' = 0$
$V' \neq 0$

Interior hinge

$v \neq 0$
θ discontinuous

(f)

Interior support

$M' \neq 0$
V' discontinuous

Figure 9.12 Transformations of real to conjugate-beam support and continuity conditions.

Figure 9.13 shows examples of various real beams and their conjugate counterparts for both statically determinate and indeterminate members. These conjugate beams were constructed using the corresponding conditions for beam support details in Fig. 9.12. Note that the conjugate beam for a statically determinate structure is also statically determinate. Although the conjugate beam for a statically indeterminate beam turns out to be unstable, it is stabilized by the presence of the elastic loading. This is true since the real beam is in equilibrium and thus yields a moment diagram that reflects

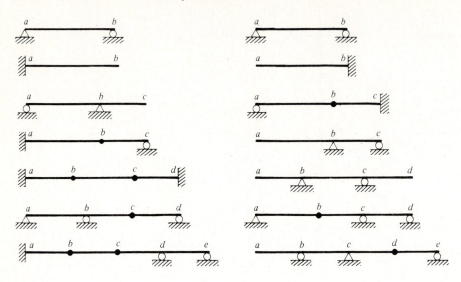

Statically determinate

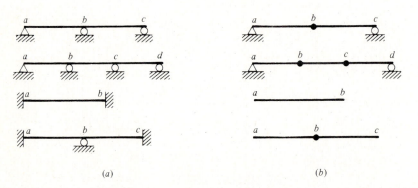

Statically indeterminate

(a)   (b)

Figure 9.13 Examples of real beams (a) and the corresponding conjugate beams (b).

this condition. For example, a conjugate beam that is free at both ends and consequently totally unsupported represents a real beam that is fixed at both ends (see Fig. 9.13).

Example 9.13 Use the conjugate-beam method to calculate the slopes and deflections for the beam of Example 9.8.

SOLUTION See Fig. E9.13d and e.

DISCUSSION The M/EI diagram obtained for Example 9.8 is the distributed elastic load applied to the conjugate beam in Fig. E9.13b. In Fig. E9.13c the conjugate beam is cut at the hinge at point b, and two free-body diagrams are used to calculate the reactions at

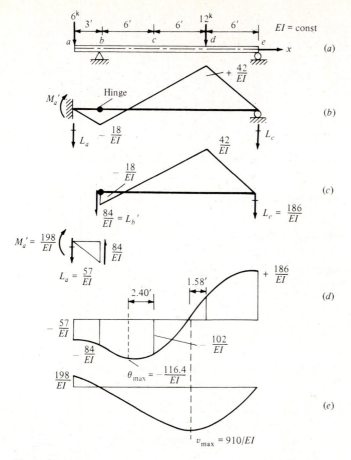

Figure E9.13 (*a*) Real beam; (*b*) conjugate beam; (*c*) free-body diagrams of conjugate beam; (*d*) diagram of slopes; (*e*) diagram of deflections.

points *a* and *e*. The elastic shear and moment diagrams were constructed using the usual sign conventions. The slopes of the various diagrams can be conveniently perceived by using the differential equilibrium equations; furthermore, the ordinates can also be calculated in this way. The values of the elastic shear and elastic moment are interpreted as the slope and deflection, respectively, of the elastic curve for the beam.

Example 9.14 Calculate the slopes and deflections for this beam using the conjugate-beam method.

SOLUTION From Fig. E9.14*d*

$$v_c = \frac{-1440 \text{ kN} \cdot \text{m}^3}{(200 \times 10^6 \text{ kN/m}^2)(175 \times 10^{-6} \text{ m}^4)} = -0.0411 \text{ m} (\downarrow)$$

DISCUSSION Since this beam is statically determinate, it has a statically determinate conjugate beam, as shown in Fig. E9.14*b*. The elastic shear and moment diagrams were

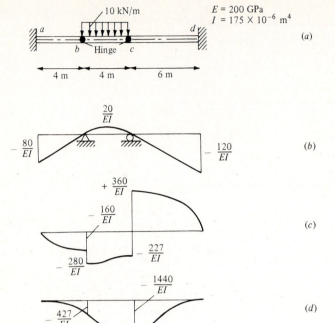

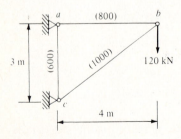

Figure E9.14 (*a*) Real beam; (*b*) conjugate beam; (*c*) slope diagram; (*d*) deflection diagram.

constructed in the usual manner. A consistent set of units must be used throughout, and it is advantageous to perform a simple dimensional analysis when calculating deflections and slopes. Note the calculations for v_c for such a check on the units.

PROBLEMS

Calculate the vertical and horizontal deflections at b for the trusses in Probs. 9.1 to 9.3 using the direct method:

9.1 Member areas in square millimeters shown in parentheses; $E = 200$ GPa.

9.2 All members have $A = 2500$ mm^2 and $E = 200$ GPa.

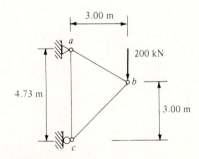

Figure P9.1

Figure P9.2

9.3 Member areas in square inches shown in parentheses; $E = 29 \times 10^3$ kips/in². Calculate deflections due to (a) the 20-kip load and (b) an increase in temperature for the entire truss of 100 °F with 20-kip load removed. $\alpha = 6.5 \times 10^{-6}$ in/in · °F.

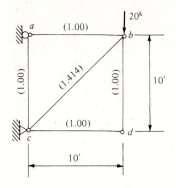

Figure P9.3

9.4 to 9.7 Find the expression(s) for the deflection by integrating Eq. (9.2). Sketch the deflected beam.

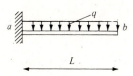

Figure P9.4

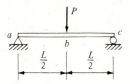

Figure P9.5

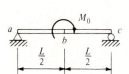

Figure P9.6

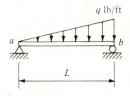

Figure P9.7

9.8 Calculate the deflection at the center of this beam by integrating Eq. (9.2).

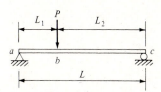

Figure P9.8

9.9 to 9.11 Using the result from Prob. 9.8 and the principle of superposition, calculate the deflection at the center of these beams. Sketch the deflected beam.

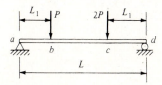

Figure P9.9

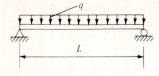

Figure P9.10

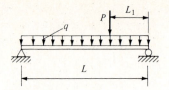

Figure P9.11

In Probs. 9.12 to 9.34 use the moment-area method. Sketch the deflected structure.

9.12 Calculate the deflection and rotation at *b* for this uniform cantilever beam; $I = 200 \times 10^6$ mm^4; $E = 200$ GPa.

9.13 Calculate the position and magnitude of the maximum deflection and the maximum rotation; $I = 900$ in^4; $E = 29 \times 10^3$ kips/in^2.

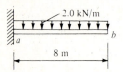

Figure P9.12

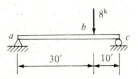

Figure P9.13

9.14 Calculate the deflection and angle of rotation at point *b* for this uniform steel beam; $I = 400 \times 10^{-6}$ m^4; $E = 200$ GPa.

9.15 Calculate the deflection and rotation at points *c* and *f* for this uniform steel beam; $L = 30$ ft; $I = 1200$ in^4; $E = 29 \times 10^3$ kips/in^2.

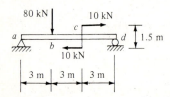

Figure P9.14

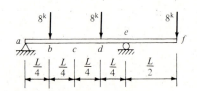

Figure P9.15

9.16 Calculate the deflection under the 60-kip load; $I = 576$ in^4; $E = 29 \times 10^3$ kips/in^2.

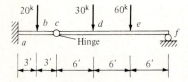

Figure P9.16

9.17 Calculate the deflection at points *b*, *c*, and *e*; $I = 500$ in^4; $E = 29 \times 10^3$ kips/in^2.

9.18 The following tangential deviations were calculated for this beam:

$$t_{ba} = \frac{4000}{EI} \qquad t_{bc} = \frac{6000}{EI} \qquad t_{cd} = \frac{2000}{EI} \qquad t_{cb} = -\frac{4000}{EI}$$

Sketch the approximate elastic curve.

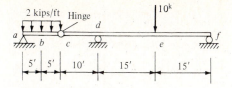

Figure P9.17

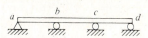

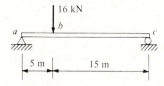

Figure P9.18

9.19 The following tangential deviations were calculated for this beam:

$$t_{ba} = \frac{200}{EI} \quad t_{cb} = -\frac{100}{EI} \quad t_{bd} = -\frac{200}{EI} \quad t_{dc} = -\frac{50}{EI} \quad t_{ce} = \frac{150}{EI}$$

Sketch the approximate elastic curve.

9.20 Calculate the location and magnitude of the maximum deflection; EI = const.

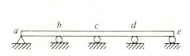

Figure P9.19

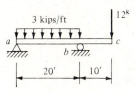

Figure P9.20

9.21 Calculate the location and magnitude of the maximum upward deflection; EI = const.

9.22 Calculate the vertical deflection at c for this uniform beam. $I = 600 \ \text{in}^4$; $E = 29 \times 10^3 \ \text{kips/in}^2$.

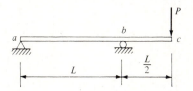

Figure P9.21

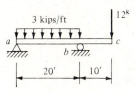

Figure P9.22

9.23 Calculate the ratio P_2/P_1 such that point c does not deflect; EI = const; $P_2 \neq 0$.

9.24 Calculate the ratio P_2/P_1 in Fig. P9.23 such that $|v_a| = |v_c| = |v_e|$; EI = const; $P_2 \neq 0$.

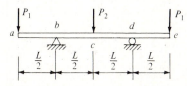

Figure P9.23

9.25 Calculate the location of the support points b and d such that the tip deflections at points a and e equal the magnitude of the centerline deflection at point c; EI = const.

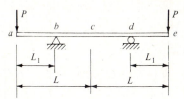

Figure P9.25

9.26 Calculate the value of the loads P such that the maximum deflection of this uniform steel beam is 20 mm; $I = 300 \times 10^{-6}$ m⁴; $E = 200$ GPa.

9.27 If $EI = $ const, calculate the magnitude of the upward load P such that: (a) the centerline deflection is zero; and (b) the rotations at the ends are zero.

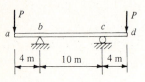

Figure P9.26

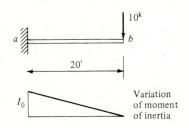

Figure P9.27

9.28 Three uniform cantilever beams all made from the same material have the same length and width, but their depths are 0.5d, 1.0d, and 2.0d, respectively. If the tip deflection of the beam of depth 1.0d is unity due to a tip load P, what are the tip deflections of the other two beams under the same type of loading?

9.29 This uniform timber beam ($E = 1.6 \times 10^3$ kips/in²) has a width of 11.5 in. Calculate the depth of beam required so that the maximum deflection at point a will be 0.67 in (neglect the weight of the beam).

9.30 This cantilever timber beam ($E = 1.6 \times 10^3$ kips/in²) has a width of 11.5 in. It is to be designed so that the moment of inertia will vary as shown. Calculate the beam cross section required at the fixed end if the maximum deflection must equal $L/360$ (neglect the weight of the beam).

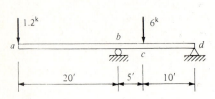

Figure P9.29

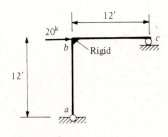

Figure P9.30

9.31 Calculate the maximum vertical deflection of beam ab for this rigid frame; $I_{ab} = 200 \times 10^6$ mm⁴; $I_{bc} = 50 \times 10^6$ mm⁴; $E = 200$ GPa.

9.32 Calculate the position and magnitude of the maximum vertical deflection of beam bc for this rigid frame; $I_{ab} = I_{bc} = 1000$ in⁴; $E = 29 \times 10^3$ kips/in².

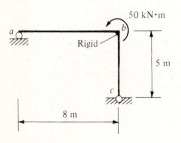

Figure P9.31

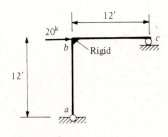

Figure P9.32

9.33 Calculate the vertical component of the deflection of point a on the bracket attached to the beam; $I = 180 \times 10^{-6}$ m^4; $E = 200$ GPa.

9.34 Compute the change in slope of the cross section on the left side of the hinge at point c. Neglect any axial deformations of the members; $I = 2500$ in^4; $E = 29 \times 10^3$ kips/in^2.

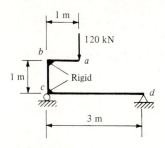

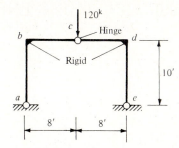

Figure P9.33   **Figure P9.34**

9.35 to 9.38 Use the method of elastic loads to solve Probs. 9.13, 9.14, 9.15 (v_c and θ_c only), and 9.17. Sketch the deflected beam.

9.39 to 9.44 Use the conjugate-beam method to solve Probs. 9.12 to 9.17. Sketch the deflected beam.

9.45 Compute the maximum vertical deflection for this beam. $I = 432$ in^4; $E = 29 \times 10^3$ kips/in^2 for all members. Sketch the deflected beam.

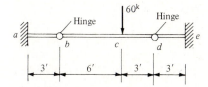

Figure P9.45

REFERENCES

1. International Conference of Building Officials, *Uniform Building Code*, sec. 2307, Whittier, Calif., 1982 Ed.
2. Crandall, S. H., N. C. Dahl, and T. J. Lardner: *An Introduction to the Mechanics of Solids*, 2d ed., chap. 7, McGraw-Hill, New York, 1978.

ADDITIONAL READING

Hsieh, Yuan-Yu: *Elementary Theory of Structures*, 2d ed., chap. 8, Prentice-Hall, Englewood Cliffs, N.J., 1982.
Ketter, R. L., G. C. Lee, and S. P. Prawel, Jr.: *Structural Analysis and Design*, chap. 4, McGraw-Hill, New York, 1979.
Laursen, H. I.: *Structural Analysis*, 2d ed., chap. 7, McGraw-Hill, New York, 1978.
McCormac, J. C.: *Structural Analysis*, 3d ed., chaps. 17 and 18, Intext, New York, 1975.
Norris, C. H., J. B. Wilbur, and S. Utku: *Elementary Structural Analysis*, 3d ed., chap. 8, McGraw-Hill, New York, 1976.
West, H. H.: *Analysis of Structures*, chaps. 7 and 8, Wiley, New York, 1980.

TEN

DEFLECTIONS USING ENERGY METHODS

10.1 GENERAL DISCUSSION

In the previous chapter the basic equations of equilibrium, compatibility, boundary conditions, and the material stress-strain relations (constitutive laws) were used to generate the governing equations for the structural deflections. These equations can be solved by an analytic, graphic, or combined method. The deflections can also be calculated using the energy of the system.

This dual approach was first pointed out by Euler while investigating the deflection of a caternary. He called them (1) the *method of effective causes* or the *direct method* and (2) the *means of final causes*. In applying the second approach to the deflection of a beam he used the expression for the strain energy that was suggested by Daniel Bernoulli. Using the principle of virtual displacements, Lagrange introduced the ideas of *generalized coordinates* and *generalized forces*. These early beginnings of the energy approach were soon followed by a number of significant developments.

Clapeyron noted in his famous theorem that the internal strain energy of an elastic solid is equal to the work done by the external forces acting during deformation (see Sec. 10.2). The theorems of Castigliano for linear elastic structures, discussed in Sec. 10.7, were generalized by Engesser in 1889 for cases in which the forces and displacements are not linearly related. Engesser introduced the concept of *complementary energy* for these structures. Since engineers work primarily with linear structures, Engesser's work did not receive much attention until the idea was further developed by Westergaard in 1941.

The energy methods of analysis have been used to obtain approximate solutions for structures. This approach was advanced by Lord Rayleigh and subsequently refined by Ritz in 1909. The Rayleigh-Ritz method is the basis for structural analysis that is commonly implemented on the digital computer, and it is used extensively today for analyzing complex structures. An introduction to this method as applied to trusses, beams, and frames is contained in Part Five.

The bridge in Fig. 10.1 is an example of a structure for which the analysis and design depend upon energy methods.

Figure 10.1 Free-cantilever segmental prestressed concrete box girder structure: Gastineau Channel Bridge, Juneau, Alaska. (*Bridge Design Section Alaska DOT & PF.*)

Energy methods have been used from their inception to investigate indeterminate structures, and recent advances have led to a better understanding of the associated fundamental concepts. It has been demonstrated[1] that all energy theorems stem from one of the two general principles: (1) virtual work (virtual displacements) or (2) complementary virtual work (virtual forces). Thus, each virtual-work theorem has a corresponding theorem based upon complementary virtual work.

The discussion in this chapter is organized into three major divisions: (1) definition of the basic terms and concepts, (2) development and classification of several energy theorems, and (3) applications of the theorems commonly used for structural investigations. The reader is encouraged to consult Tables 10.1 and 10.3 periodically for summaries of definitions and the energy theorems.

10.2 WORK AND STRAIN ENERGY; COMPLEMENTARY WORK AND COMPLEMENTARY STRAIN ENERGY

Understanding the basic definitions contained in this section will help the reader in studying the subsequent energy theorems. Generally, the ideas can be visualized by considering the truss member in Fig. 10.2 subjected to a tensile loading as illustrated. If the material is linear, the response is characterized by the plot of the applied loading versus the associated deflection shown in Fig. 10.3a. Furthermore, if the stress and strain are monitored and plotted for any typical cross section of the member, the graph shown in Fig. 10.3b is obtained. The area under the force-deflection curve, denoted by horizontal lines, is the *work* done by the external forces W_e, and the area above the curve, denoted by vertical lines, is the *complementary work* W_e^*. If the load is increased

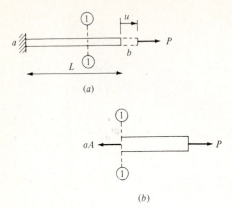

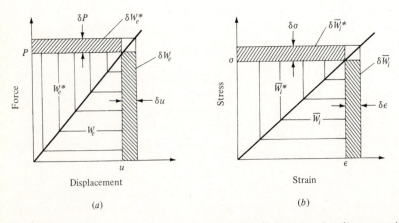

Figure 10.2 Truss member: (a) with applied load; (b) free-body diagram.

from zero,

$$W_e = W_e^* = \tfrac{1}{2}Pu \tag{10.1}$$

where P and u are the final values of the force and deflection, respectively.

If the material is nonlinear, the load-deflection and stress-strain curves appear as shown in Fig. 10.4a and b, respectively. In this case, the two quantities W_e and W_e^* are unequal.

When the displacement u is increased to $u + \delta u$ the corresponding first variation δW_e in W_e (Fig. 10.3a) becomes

$$\delta W_e = P \, \delta u \tag{10.2}$$

Referring again to Fig. 10.3a, we see that if the force P is increased to $P + \delta P$,

$$\delta W_e^* = u \, \delta P \tag{10.3}$$

Hence the quantity δW_e is obtained by varying the displacement; whereas δW_e^* requires the force to be varied.

Figure 10.3 Response of a linear truss member: (a) force-displacement plot; (b) stress-strain plot.

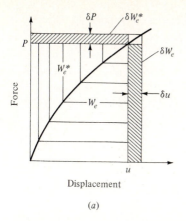 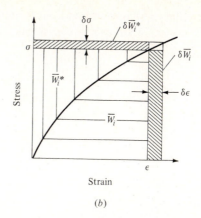

Figure 10.4 Response of a nonlinear truss member: (a) force-displacement curve; (b) stress-strain curve.

Note that if the displacement u is increased to $u + \delta u$, the corresponding total increment ΔW_e in W_e is

$$\Delta W_e = P \, \delta u + \tfrac{1}{2} \delta P \, \delta u = \delta W_e + \tfrac{1}{2} \delta^2 W_e \qquad \text{where } \delta W_e = P \, \delta u$$

Hence, with δu and δP considered as small quantities, Eq. (10.2) represents only the first-order quantities of the total increment in W_e. For nonlinear systems, terms of order higher than $\delta P \, \delta u$ would also be present in ΔW_e. In general, it is not necessary to consider terms beyond the first variation (δW_e in this case) for the derivation of strain-energy theorems; however, the higher-order terms must be retained for the investigation of such phenomena as plasticity and stability behavior.

The *internal strain energy* W_i and the corresponding *complementary strain energy* W_i^* for linear elastic structures are obtained by integrating the appropriate area in Fig. 10.3b over the volume of the member. (The areas $\overline{W}_i$ and $\overline{W}_i^*$ shown in Figs. 10.3b and 10.4b are referred to as the *density of strain energy* and the *density of complementary strain energy*, respectively, since they give the energy per unit volume. These terms must be integrated over the volume of the member in order to give the total of W_i and W_i^*.) Hence

$$W_i = W_i^* = \frac{1}{2} \int_{\text{vol}} \sigma \varepsilon \, dV \tag{10.4}$$

where σ and ε refer to their final values at equilibrium. If the material is nonlinear, $W_i \neq W_i^*$ (see Fig. 10.4b). Equations (10.1) and (10.4) are used for calculating work and strain energy in linear elastic structures in which the forces and temperature distribution are all increased from zero to their final values (the fact that $W_e = W_i$ for structures in which all the forces are increased from zero was first advanced by Clapeyron).

If the displacement is increased from u to $u + \delta u$, there will be a corresponding increase in the strains from ε to $\varepsilon + \delta \varepsilon$. Consequently, the variation in the strain energy due to this change in the strain is

$$\delta W_i = \int_{\text{vol}} \sigma \, \delta \varepsilon \, dV \tag{10.5}$$

Similarly, if the stress is increased from σ to $\sigma + \delta\sigma$, the corresponding variation in the complementary strain energy can be calculated to give

$$\delta W_i^* = \int_{\text{vol}} \varepsilon \, \delta\sigma \, dV \tag{10.6}$$

For the uniform (prismatic) truss member shown in Fig. 10.2, with cross-sectional area A and length L, Eqs. (10.5) and (10.6) give, respectively,

$$\delta W_i = \sigma \delta\varepsilon AL \tag{10.5a}$$

and
$$\delta W_i^* = \varepsilon \, \delta\sigma \, AL \tag{10.6a}$$

Table 10.1 summarizes the basic definitions established in this section.

The concepts of work and strain energy, together with their respective complements, discussed above for the simple one-dimensional case, can be extended to more general types of structures in two and three dimensions. For example, if a structure is loaded with a set of N forces P_j with corresponding displacements u_j, Eqs. (10.2) and (10.3) become, respectively,

$$\delta W_e = \sum_{j=1}^{N} P_j \, \delta u_j \qquad \text{and} \qquad \delta W_e^* = \sum_{j=1}^{N} u_j \, \delta P_j$$

These equations are valid if the concept of force is expanded to include moments and that of displacement to encompass angular displacements (rotations). That is, P_j and u_j are referred to as a *generalized force* and *generalized displacement*, respectively. Assume that the loading shown in Fig. 10.5 results in an internal state of stress in the member involving tension, shear, and bending; thus, the stresses in the member at a particular point must be described by σ_x, σ_y, and τ_{xy}. If ε_x, ε_y, and γ_{xy} are the strains corresponding to their respective stresses, the variation in strain energy [analogous to Eq. (10.5)] is

$$\delta W_i = \int_{\text{vol}} (\sigma_x \, \delta\varepsilon_x + \sigma_y \, \delta\varepsilon_y + \tau_{xy} \, \delta\gamma_{xy}) \, dV$$

That is, the change in the displacements from u_i to $u_i + \delta u_i$ results in a change in the strains from ε_x to $\varepsilon_x + \delta\varepsilon_x$, ε_y to $\varepsilon_y + \delta\varepsilon_y$, and γ_{xy} to $\gamma_{xy} + \delta\gamma_{xy}$. On the other hand, if the loads are increased from P_i to $P_i + \delta P_i$, resulting in a change in the stresses from σ_x, σ_y, and τ_{xy} to $\sigma_x + \delta\sigma_x$, $\sigma_y + \delta\sigma_y$, and $\tau_{xy} + \delta\tau_{xy}$, respectively, the variation of the complementary strain energy for the two-dimensional member is

$$\delta W_i^* = \int_{\text{vol}} (\varepsilon_x \, \delta\sigma_x + \varepsilon_y \, \delta\sigma_y + \tau_{xy} \, \delta\gamma_{xy}) \, dV$$

This equation is analogous to Eq. (10.6) for the two-force member.

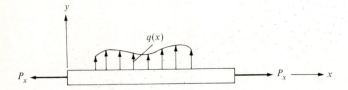

Figure 10.5 General two-dimensional member with loading.

Table 10.1 Summary of basic energy definitions

Symbol	Definition
$W_e = \int P\,du$	Work (area below the P-u curve)
$W_e^* = \int u\,dP$	Complementary work (area above the P-u curve)
$\overline{W}_i = \int \sigma\,d\varepsilon$	Density of strain energy (area below the σ-ε curve)
$\overline{W}_i^* = \int \varepsilon\,d\sigma$	Density of complementary strain energy (area above the σ-ε curve)
$W_i = \int_{vol}\left(\int \sigma\,d\varepsilon\right) dV$	Strain energy
$W_i^* = \int_{vol}\left(\int \varepsilon\,d\sigma\right) dV$	Complementary strain energy
δP	Virtual force
δu	Virtual displacement
$\delta \sigma$	Virtual stress
$\delta \overline{\sigma}$	Virtual stress due to a unit virtual force
$\delta \varepsilon$	Virtual strain
$\delta \overline{\varepsilon}$	Virtual strain due to a unit virtual displacement
$\delta W_e = P\,\delta u$	Virtual work
$\delta W_e^* = u\,\delta P$	Complementary virtual work
$\delta W_i = \int_{vol} \sigma\,\delta\varepsilon\,dV$	Virtual strain energy
$\delta W_i^* = \int_{vol} \varepsilon\,\delta\sigma\,dV$	Complementary virtual strain energy

For linear elastic behavior in which the forces and temperature distribution are all increased from zero

$$W_e = W_e^* = \tfrac{1}{2}Pu \qquad W_i = W_i^* = \tfrac{1}{2}\int_{vol} \sigma\varepsilon\,dV$$

10.3 THE PRINCIPLES OF VIRTUAL WORK AND COMPLEMENTARY VIRTUAL WORK

The infinitesimal displacements δu are referred to as *virtual displacements,* a term which comes from the principle of virtual displacements which was formulated by

Johann Bernoulli. Correspondingly, the variations δW_e and δW_i (associated with the virtual strains) are called the *virtual work* and *virtual strain energy,* respectively. The principle of virtual work is also called the principle of virtual displacements.

> **Principle of virtual work** If an elastic structure is in equilibrium under a given system of loads and temperature distribution, then for any virtual displacement away from a compatible state of deformation the virtual work is equal to the virtual strain energy; $\delta W_e = \delta W_i$.

Note that the virtual strains must be compatible with the virtual displacement.

The total increment of work ΔW_e is associated with a change in displacement δu which is assumed to be accompanied by a corresponding change in the stress and force. That is, an increment in work and strain energy is obtained from two adjacent equilibrium states in which the strain $\delta \varepsilon$ derived from the displacement δu satisfies the equations of both equilibrium and compatibility. However, since the first variations δW_e and δW_i are independent of δP and $\delta \sigma$, for the purpose of computing these quantities the force and stress can be considered to be constant while the displacement is varied from u to δu. Hence, the virtual displacement δu will be associated with a strain that satisfies the equations of compatibility but not necessarily the equations of equilibrium. This means that *the virtual displacement can be any small displacement as long as it satisfies the geometrical constraints of the structure;* i.e., it must be continuous within the boundaries of the structure and must satisfy any kinematic boundary conditions. For example, in a cantilever beam, δu must have zero displacement and slope at the fixed end.

The principle of virtual work can be demonstrated for the two-force member shown in Fig. 10.2 using fundamental concepts. The equation of equilibrium (Fig. 10.2*b*) is

$$\sigma A - P = 0$$

If both sides of the equation are multiplied by a virtual displacement δu, we have

$$(\sigma A - P)\,\delta u = 0 \qquad \text{or} \qquad \sigma A\,\delta u - P\,\delta u = 0$$

Multiplying and dividing the first term of the right-hand equation by L and noting that $\delta \varepsilon = \delta u / L$ gives

$$\sigma A L\,\delta \varepsilon - P\,\delta u = 0$$

But for the uniform rod σ and $\delta \varepsilon$ are constants, so that

$$A L \sigma\,\delta \varepsilon = \int_{\text{vol}} \sigma\,\delta \varepsilon\,dV = \delta W_i$$

Therefore,

$$\delta W_i - \delta W_e = 0 \qquad \text{or} \qquad \delta W_i = \delta W_e$$

which is the statement of virtual work. It can also be derived for two- and three-dimensional structures, but this requires the use of Green's theorem.

The corresponding dual energy principle is obtained by imposing virtual forces δP, which result in virtual internal stresses $\delta \sigma$. The resulting variations δW_e^* and δW_i^* are the *complementary virtual work* and the *complementary virtual strain energy,* respectively. The principle of complementary virtual work is also called the principle of virtual forces.

> **Principle of complementary virtual work** If an elastic structure is in a compatible state of deformation under a given system of loads and temperature distribution, then for any virtual stresses and forces away from the equilibrium

state of stress the virtual complementary work is equal to the virtual complementary strain energy of virtual forces; $\delta W_e^* = \delta W_i^*$.

Note that the virtual stresses must equilibrate the virtual forces.

From Eqs. (10.3) and (10.6) it can be observed that the variations δW_e^* and δW_i^* are both independent of the variations in displacement δu and strains $\delta \varepsilon$. Therefore, in calculating the quantities δW_e^* and δW_i^* the displacement and strain can be assumed to remain constant while the force and the stress are varied. Thus, *the virtual stress $\delta \sigma$ must satisfy the equations of internal equilibrium, and the virtual force δP must satisfy the equations of boundary equilibrium;* however, the strains associated with the virtual stresses need not satisfy the equations of compatibility. Therefore, for the two-force member, subjected to a virtual force δP (which gives the virtual stress $\delta \sigma$), the incremental equation of equilibrium is

$$\delta \sigma A - \delta P = 0$$

If this statement of equilibrium is multiplied by the actual displacement u, then

$$(\delta \sigma A - \delta P)u = 0$$

The first term of the above equation becomes

$$\delta \sigma A u = \delta \sigma A L \varepsilon = \int_{\text{vol}} \varepsilon \, \delta \sigma \, dV = \delta W_i^*$$

This is valid because the rod has a uniform cross section and both the virtual stress $\delta \sigma$ and the actual strain ε are constant throughout the member. Hence,

$$\delta W_i^* = \delta W_e^*$$

This statement of complementary virtual work can also be written for general structures in two and three dimensions. The latter general formulation requires Green's theorem.

10.4 THE UNIT-DISPLACEMENT AND UNIT-LOAD THEOREMS

The unit-displacement theorem is obtained directly from the theorem of virtual work and is used to determine the force necessary to maintain equilibrium in a structure for which the distribution of true stresses is known. If the strain $\bar{\varepsilon}$ is defined as the compatible virtual strain due to a unit virtual displacement ($\delta u = 1$), the theorem of virtual work can be written using Eqs. (10.2) and (10.5) to give the *unit-displacement theorem*,

$$P = \int_{\text{vol}} \sigma \bar{\varepsilon} \, dV$$

It is important to note that since the virtual strain is assumed to be obtained from a unit virtual deflection, this theorem is limited to linear elastic structures.

For a two-dimensional structure, the theorem of virtual work is

$$P_j \, \delta u_j = \int_{\text{vol}} (\sigma_x \, \delta \varepsilon_x + \sigma_y \, \delta \varepsilon_y + \tau_{xy} \, \delta \gamma_{xy}) \, dV$$

In a linear elastic structure the virtual strains $\bar{\varepsilon}_x$, $\bar{\varepsilon}_y$, and $\bar{\gamma}_{xy}$ are proportional to the virtual displacement

$$\delta\varepsilon_x = \overline{\varepsilon}_x\,\delta u_j \qquad \delta\varepsilon_y = \overline{\varepsilon}_y\,\delta u_j \qquad \text{and} \qquad \delta\gamma_{xy} = \overline{\gamma}_{xy}\,\delta u_j$$

where the strains $\overline{\varepsilon}_x$, $\overline{\varepsilon}_y$, and $\overline{\gamma}_{xy}$ must be compatible only with themselves and with the virtual displacement δu_j. Substituting these strains into the equation of virtual work and dividing out the quantity δu_j from both sides of the resulting equation gives a representation of the unit-displacement theorem for a two-dimensional structure

$$P_j = \int_{\text{vol}} (\sigma_x\overline{\varepsilon}_x + \sigma_y\overline{\varepsilon}_y + \tau_{xy}\overline{\gamma}_{xy})\,dV$$

The unit-load theorem is used to determine the displacement in a structure for which the distribution of true total strains is known. If a unit virtual force ($\delta P = 1$) is applied in the direction of u, giving virtual stress $\overline{\sigma}$, the theorem of complementary virtual work can be written using Eqs. (10.3) and (10.6) to give the *unit-load theorem*

$$u = \int_{\text{vol}} \varepsilon\overline{\sigma}\,dV$$

Again it is important to note that the theorem is limited to linear elastic structures since the virtual stress is proportional to the virtual applied force.

The theorem of complementary virtual work for a two-dimensional structure is

$$u_j\,\delta P_j = \int_{\text{vol}} (\varepsilon_x\,\delta\sigma_x + \varepsilon_y\,\delta\sigma_y + \gamma_{xy}\,\delta\tau_{xy})\,dV$$

In a linear elastic structure the stresses are proportional to the load; therefore, the virtual stresses are related to the virtual force as follows:

$$\delta\sigma_x = \overline{\sigma}_x\,\delta P_j \qquad \delta\sigma_y = \overline{\sigma}_y\,\delta P_j \qquad \text{and} \qquad \delta\gamma_{xy} = \overline{\tau}_{xy}\,\delta P_j$$

The stresses $\overline{\sigma}_x$, $\overline{\sigma}_y$, and $\overline{\tau}_{xy}$ must be in equilibrium only with themselves and with the imposed virtual load δP_j. Substituting these stresses into the equation of complementary virtual work and dividing out δP_j from both sides of the resulting equation gives a representation of the unit-load theorem for a two-dimensional structure

$$u_j = \int_{\text{vol}} (\varepsilon_x\overline{\sigma}_x + \varepsilon_y\overline{\sigma}_y + \gamma_{xy}\overline{\tau}_{xy})\,dV$$

10.5 DEFLECTION OF TRUSSES USING COMPLEMENTARY VIRTUAL WORK

The principle of complementary virtual work was used in the previous section to obtain the unit-load theorem. Both of them are premised on the facts that the structure is (1) initially in a state of equilibrium and (2) is altered (perturbed) from this state by varying the external loading system. The change in complementary external work must equal the change in complementary internal strain energy when the virtual load is applied. It is important to note that the displacements and the associated internal strains are the actual quantities experienced by the structure whereas the virtual external force(s) and the associated virtual stresses need not satisfy all the governing equations; i.e., they need satisfy only the equilibrium conditions for the structure.

Consider a uniform two-force truss member which has an actual load F_P resulting from the true externally applied forces. The actual strain in the member is

$$\varepsilon = \frac{\sigma}{E} = \frac{F_P}{AE} \qquad (10.7)$$

If only a virtual load is placed on the structure, the force in the member is F_Q and the corresponding virtual stress is

$$\delta\sigma = \frac{F_Q}{A} \qquad (10.8)$$

Combining Eq. (10.6) with Eqs. (10.7) and (10.8) gives for the complementary virtual strain energy for an individual truss member

$$\delta W_i^* = \int_{\text{vol}} \frac{F_P}{AE} \frac{F_Q}{A} \, dV = \frac{F_P F_Q L}{AE} \qquad (10.9)$$

Equation (10.7) is the actual strain in the member if the structure is subjected to a set of externally applied loads, but the actual strain in the member can be the result of a number of effects. For example, if the member sustains a uniform increase in temperature ΔT, the induced strain ε^0 is

$$\varepsilon = \varepsilon^0 = \alpha \Delta T \qquad (10.10)$$

where α is the coefficient of linear thermal expansion. Or if there is a fit problem, the member having been fabricated with an error ΔL in the length, then

$$\varepsilon = \varepsilon^0 = \frac{\Delta L}{L} \qquad (10.11)$$

If the member is simultaneously subjected to all these effects, the actual total strain in the member is obtained by applying the principle of superposition; i.e., the strains in Eqs. (10.7), (10.10), and (10.11) are added.

For a pin-jointed truss, the total complementary internal strain energy is obtained by summing the contributions from all the individual members. Thus, for example, in order to calculate the vertical deflection of joint d for the truss illustrated in Fig. 10.6, the complementary internal strain energy is obtained as

$$\delta W_i^* = \sum_{j=1}^{9} \frac{F_{P_j} F_{Q_j} L_j}{A_j E_j}$$

The subscript j on each of the quantities indicates that the contribution from each specific member must be included one at a time in the computation and the summation taken over all the members in the truss. Since the form of this equation is cumbersome, hereafter the equation will be written with the summation index implied

$$\delta W_i^* = \sum \frac{F_P F_Q L}{AE} \qquad (10.12)$$

The actual forces acting on the truss, shown in Fig. 10.6a, are designated as the P force system and give member forces denoted by F_P. The virtual force applied to the structure shown in Fig. 10.6b is located at the point and in the direction in which the deflection

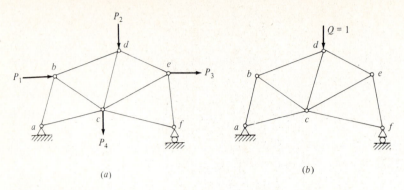

Figure 10.6 Force systems for a truss: (*a*) actual *P* force system; (*b*) virtual *Q* force system for calculating the vertical deflection at *d*.

is to be calculated. This latter force will be referred to as the *Q* force system, and the forces in the members resulting from these forces are the F_Q forces. In this case, the theorem of complementary virtual work gives

Virtual

$$v_d(1 \downarrow) = \sum \left(\frac{F_P L}{AE} \right) F_Q \tag{10.13}$$

Actual

It is important to note the relationship between the quantities on the left side of Eq. (10.13) and their counterparts on the right side. This form of the equation resembles the unit-load theorem, and the unit load has been shown in the equation to remind the reader of the presence of the *Q* force system.

Equation (10.13) is the statement of complementary virtual work for structures with unyielding supports. The complete equation is

$$\delta W_r^* + v_d(1 \downarrow) = \sum F_P F_Q \frac{L}{AE} \tag{10.14}$$

where δW_r^* is the external complementary virtual work of the reactions, i.e., the actual support movements multiplied by the reaction forces produced by the virtual load system. For unyielding supports, $\delta W_r^* = 0$, and the equation of complementary virtual work assumes the form in Eq. (10.13).

The virtual force must be applied in the direction and at the point where it is required to obtain the deflection. The *Q* force systems for various typical deflection requirements are shown in Table 10.2. Note that it is necessary to apply the *Q* force system, resulting in an expression for complementary virtual work in which the required displacement is the only unknown on the left hand side. The term δW_r^* has been included in the equations of Table 10.2 for completeness.

Table 10.2 Examples of virtual force systems for truss deflections

Generalized deflection to be investigated	Virtual force(s) Q	Complementary virtual work $\delta W_e^* = \delta W_i^*$
Vertical, joint d		$(1)(v_d\downarrow) + \delta W_r^* = \sum F_Q\,\Delta L$
Horizontal, joint d		$(1)(u_d\rightarrow) + \delta W_r^* = \sum F_Q\,\Delta L$
Inclined at angle ϕ to horizontal, joint d	$(\Delta_d\searrow)$	$(1)\cos\phi\,(u_d\rightarrow) + (1)\sin\phi\,(v_d\downarrow)$ $+ \delta W_r^* = \sum F_Q\,\Delta L$ $(1)(\Delta_d) + \delta W_r^* = \sum F_Q\,\Delta L$
Relative, joints b and e		$(1)(\Delta_b\rightarrow) + (1)(\Delta_e\leftarrow)$ $+ \delta W_r^* = \sum F_Q\,\Delta L$ $(1)(\Delta_b\rightarrow + \Delta_e\leftarrow) + \delta W_r^* = \sum F_Q\,\Delta L$ $(1)(\Delta_{be}\rightarrow\leftarrow) + \delta W_r^* = \sum F_Q\,\Delta L$ $\Delta_{be}(\rightarrow\leftarrow)$ = relative deflection of b and e
Rotation, chord bd	$(\Delta_b\searrow)$ $(\Delta_d\searrow)$	$\dfrac{1}{L}\Delta_b + \dfrac{1}{L}\Delta_d + \delta W_r^* = \sum F_Q\,\Delta L$ $\dfrac{1}{L}(\Delta_b + \Delta_d) + \delta W_r^* = \sum F_Q\,\Delta L$ $\Theta_{bd}(\frown) + \delta W_r^* = \sum F_Q\,\Delta L$ Θ_{bd} = rotation of chord bd

Examples 10.1 and 10.2 illustrate the use of this method for obtaining the deflection of trusses. It is important to observe both the algebraic signs and the units in these calculations carefully. The algebraic signs of the calculated deflections must be interpreted with respect to the direction of the applied Q force system. Both δW_e^* and δW_i^* are positive quantities. Thus, a positive deflection implies that its direction coincides with that of the applied Q force system.

Since it is imperative that a consistent set of units be used in the calculations, a careful dimensional analysis should accompany each solution. It is left to the discretion of the structural analyst whether or not to assign units to the virtual force system. That is, if the Q forces are assigned units, e.g., kips or kilonewtons, the internal member forces resulting from these forces will have the corresponding dimensions.

Example 10.1 Use complementary virtual work (virtual forces) to calculate the vertical and horizontal deflections at point b.

SOLUTION

$$\left(\frac{L}{AE}\right)_{ab} = \left(\frac{L}{AE}\right)_{ac} = \left(\frac{L}{AE}\right)_{bc} = \frac{5 \text{ m}}{(5 \times 10^{-4} \text{ m}^2)(200 \times 10^6 \text{ kPa})} = 5 \times 10^{-5} \text{ m/kN}$$

From Fig. E10.1b and c

Member	F_P, kN	F_Q, kN	$F_P F_Q$, kN²
ab	+120	$+\frac{4}{3}$	+160
ac	+90	+1	+90
bc	−150	$-\frac{5}{3}$	+250
			$\Sigma = 500$

$$(1\downarrow)v_b = \sum F_P F_Q \frac{L}{AE} = \frac{L}{AE} \sum F_P F_Q$$

$$= (5 \times 10^{-5} \text{ m/kN})(500 \text{ kN}^2)$$

$$v_b = 25 \times 10^{-3} \text{ m} = 25 \text{ mm} \ (\downarrow)$$

From Fig. E10.1b and d

Member	F_P, kN	F_Q, kN	$F_P F_Q$, kN²
ab	+120	+1	+120
ac	+90	0	0
bc	−150	0	0
			$\Sigma = 120$

$$(1\rightarrow)u_b = \sum F_P F_Q \frac{L}{AE} = \frac{L}{AE} \sum F_P F_Q$$

$$= (5 \times 10^{-5} \text{ m/kN})(120 \text{ kN}^2)$$

$$u_b = 6 \times 10^{-3} \text{ m} = 6 \text{ mm} \ (\rightarrow)$$

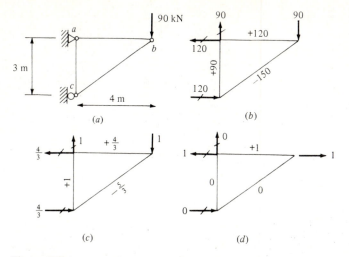

Figure E10.1 (*a*) Loaded truss; $A_{ab} = 4 \times 10^{-4}$ m², $A_{ac} = 3 \times 10^{-4}$ m², $A_{bc} = 5 \times 10^{-4}$ m², $E = 200$ GPa; (*b*) member forces F_P for the actual P force system; (*c*) member forces F_Q for the virtual Q force system for v_b; (*d*) member forces F_Q for the virtual Q force system for u_b. All forces in kilonewtons.

DISCUSSION The strength-of-materials sign convention described in Chap. 3 is used to designate the sign of the member forces. Since all members of this truss have the same value for L/AE, it is factored from under the summation sign in the statement of complementary virtual work. Note that the right-hand side of the final equations has units of kilonewton-meters, i.e., work, but since the left side contains the unit kilonewton virtual force, all units are compatible. The signs of the computed deflections are both positive, indicating that they coincide with the directions of the applied virtual forces.

Example 10.2 Using complementary virtual work (virtual forces), calculate (*a*) the relative deflection of joints c and d along the line connecting them if the truss is loaded with a downward load of 30 kips at joint d; the vertical deflection at d if (*b*) the temperature is increased by 100 Fahrenheit degrees in members ac and ce and (*c*) member be is fabricated 1 in too short. For all members $A = 1$ in², $E = 29 \times 10^3$ kip²/in² and $\alpha = 6.5 \times 10^{-6}$ in/in · °F.

SOLUTION (*a*) From Fig. E10.2*b* and *c*

Member	L, in	F_P, kips	F_Q, kips	$F_P F_Q L$, kip² · in
ac	240	-40	0.0	0
ce	240	-40	-0.8	7,680
bd	240	0	-0.8	0
bc	180	-60	-0.6	6,480
de	180	-30	-0.6	3,240
ab	300	$+50$	0.0	0
be	300	$+50$	1.0	15,000
				$\Sigma = 32,400$

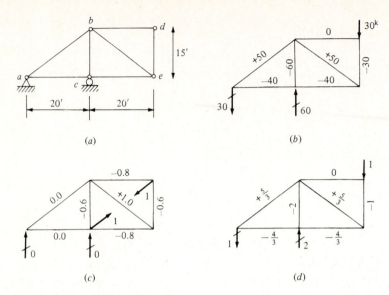

Figure E10.2 (*a*) Truss; (*b*) member forces F_P for the actual force system; (*c*) member forces F_Q for the virtual force system Q for Δ_{cd}; (*d*) member forces F_Q for the virtual force system Q for v_d. All forces in kips.

$$(1 \nearrow) \Delta_{cd} = \sum F_P F_Q \frac{L}{AE} = \frac{1}{AE} \sum F_P F_Q L = \frac{1}{(29 \times 10^3)(1)} (32{,}400)$$

$$\Delta_{cd} = 1.12 \text{ in } (\nearrow)$$

(*b*) Use Fig. E10.2*d* and

$$\delta W_i^* = \int_{\text{vol}} \varepsilon \, \delta\sigma \, dV = \int_{\text{vol}} \alpha \, \Delta T \frac{F_Q}{A} \, dV = F_Q \alpha \, \Delta T L$$

For members *ac* and *ce*

$$\alpha \, \Delta T L = (6.5 \times 10^{-6})(100)(20)(12) = 0.156 \text{ in}$$

$$(1 \downarrow) v_d = \sum F_Q \alpha \, \Delta T L = (-\tfrac{4}{3})(0.156) + (-\tfrac{4}{3})(0.156) = -0.416$$

$$v_d = 0.416 \text{ in } (\uparrow)$$

(*c*) Use Fig. E10.2*d* and

$$\delta W_i^* = \int_{\text{vol}} \varepsilon \, \delta\sigma \, dV = \int_{\text{vol}} \frac{\Delta L}{L} \frac{F_Q}{A} \, dV = F_Q \, \Delta L$$

For member *be*

$$\Delta L = -1.00$$

$$(1 \downarrow) v_d = \sum F_Q \, \Delta L = +\tfrac{5}{3}(-1) = -1.67$$

$$v_d = 1.67 \text{ in } (\uparrow)$$

DISCUSSION The strength-of-materials sign convention described in Chap. 3 is used to designate the sign of the member forces. In this case, $1/AE$ could be factored from under the summation sign. The sign of the calculated relative deflection for the 30-kip load indicates that joints c and d move closer together. The negative signs of the deflections for the thermal and fabrication situations imply that they oppose the directions of the virtual load. Note that there are no internal member forces for either parts (b) or (c). This can be observed by calculating the reactions, which must all be zero since there are no applied loads. If the reactions are zero, all member forces for this statically determinate truss must also be zero.

10.6 DEFLECTION OF BEAMS AND FRAMES USING COMPLEMENTARY VIRTUAL WORK

The theorem of complementary virtual work can also be used to calculate the deflections for linear elastic beams and frames. The expression for the internal complementary virtual work of a slender beam member with a length much greater than its depth can be obtained by considering only the bending stresses. That is, the shear strains are typically much smaller than the normal strains, so that the complementary virtual work of the shear strains is negligible relative to the complementary virtual work of the normal strains. The effect of shearing stresses generally will be neglected in subsequent text. Thus, for a slender beam only one stress and the corresponding strain need be considered, and the expression for the complementary virtual work as given in Eq. (10.6) can be used; i.e.,

$$\delta W_i^* = \int_{\text{vol}} \varepsilon \, \delta\sigma \, dV$$

where the virtual stress and the actual strain are in the longitudinal direction of the member and in general are functions of the two coordinates in the plane of the member deflections.

Figure 10.7a illustrates a beam with no axial load; it is required to obtain the vertical deflection at point b and the rotation of the beam at point a. Assume that the centroidal axis of the member is straight and that all the cross sections of the member have axes of symmetry lying in the plane of the P loads. The P load system will produce bending stresses at each cross section of the beam. The actual stress distribution at any cross section of the beam for linear elastic response is given by

$$\sigma = \frac{-M_P y}{I} \tag{10.15}$$

where M_P is the moment at a cross section produced by the P load system. M_P varies along the member length in this case, and therefore for any given beam it may be necessary to write several expressions to give a complete description of M_P for the entire member. Thus, for a linear elastic material the actual strains in the member can be obtained using Hooke's law in conjunction with Eq. (10.15), which gives

$$\varepsilon = \frac{\sigma}{E} = \frac{-M_P y}{EI} \tag{10.16}$$

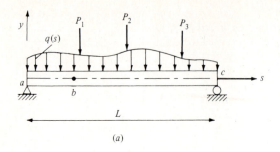

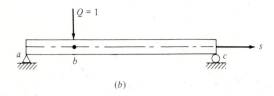

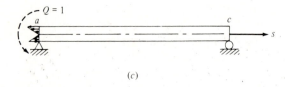

Figure 10.7 Force systems for a beam (a) actual P force system; (b) virtual Q force system for calculating the vertical deflection at b; (c) virtual Q force system for calculating the rotation at a.

The virtual force system Q shown in Fig. 10.7b lies in the xy plane and produces bending on the cross sections of the beam. Thus, the virtual bending stresses are

$$\delta\sigma = \frac{-M_Q y}{I} \tag{10.17}$$

where M_Q is the moment at any cross section of the member due to the Q load system. Substituting Eqs. (10.16) and (10.17) into the expression for complementary virtual strain energy [Eq. (10.6)] gives

$$\delta W_i^* = \int_{\text{vol}} \left(\frac{-M_P y}{EI}\right)\left(\frac{-M_Q y}{I}\right) dV \tag{10.18}$$

At a given cross section of the beam the quantities M_P, M_Q, E, and I are constant; therefore, Eq. (10.18) can be rearranged to read

$$\delta W_i^* = \int_0^L \left(\frac{M_P M_Q}{EI^2} \int_{\text{area}} y^2 \, dA\right) ds \tag{10.19}$$

Since

$$I = \int_{\text{area}} y^2 \, dA$$

we have

$$\delta W_i^* = \int_0^L \frac{M_P M_Q}{EI} \, ds \tag{10.20}$$

Note that in general E is constant but M_P, M_Q, and I vary along the member; therefore, the integration must be broken at points where the integrand changes. Often the mechanics of carrying out the integration can be simplified by selecting different origins for the variable s in various parts of the beam. This technique will be demonstrated in the examples at the end of this section.

If the vertical deflection at point b of the beam in Fig. 10.7b must be obtained, the virtual Q force system consists of a single vertical force applied at the point and directed either up or down. The Q force system induces moments M_Q in the beam; therefore, the theorem of complementary virtual work gives

$$v_b(1\downarrow) = \int_0^L \overbrace{\left(\frac{M_P}{EI}\right)}^{\text{Virtual}} M_Q\,ds \tag{10.21}$$

Actual

This statement of complementary virtual work applies to structures with unyielding supports. If the supports move, the left side of the equation must include the complementary virtual work done by the reactions δW_r^*, that is, the actual support movements multiplied by the reaction forces produced by the virtual load system.

The most satisfactory method for designating both M_P and M_Q is usually the strength-of-materials sign convention (Sec. 4.3); i.e., a positive bending moment compresses the top fibers of the beam and distorts the beam into a form that is concave upward. This can be ambiguous for members such as the columns in a rigid frame, but if the diagram is plotted in all cases on the side of the member where compression occurs, many difficulties can be avoided. Note that this convention gives the same diagrams used previously for horizontal beam members.

The method of complementary virtual work can also be used to obtain the change in slope at a point on a beam; however, the term δW_e^* must be modified accordingly. Consider the distributed virtual force system shown in Fig. 10.7c for finding the slope at point a for the beam in Fig. 10.7a. In this case, the virtual force q is a function of y, and it yields a moment that equals unity. If the beam has a constant width b, the differential force at a position y above the neutral axis is $qb\,dy$. The actual loads (P system) introduce a rotation Θ_a, and the actual horizontal displacement at y is $\Theta_a y$. Hence,

$$\delta W_e^* = \int_{-h/2}^{+h/2} (qb\,dy)(\Theta_a y) = \Theta_a \int_{-h/2}^{+h/2} qby\,dy = \Theta_a(1\frown)$$

since the moment of the virtual load is $\int qby\,dy = 1$. Therefore, in calculating rotations, it is only necessary to apply a unit couple to the beam at the point of interest, and the theorem of complementary virtual work becomes

$$\delta W_r^* + \Theta_a(1\frown) = \int_0^L \frac{M_P M_Q}{EI}\,ds$$

Examples 10.3 to 10.5 illustrate this method for beams and frames. It is important to interpret the signs of the moments and deflections with care; furthermore, each analysis should be accompanied by a dimensional analysis to ensure consistency of the units.

Example 10.3 Calculate the vertical deflections at points a and b plus the rotation at a using complementary virtual work (virtual forces); EI = constant.

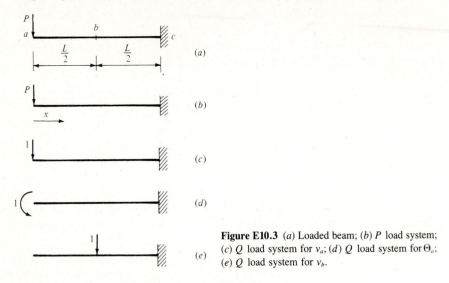

Figure E10.3 (a) Loaded beam; (b) P load system; (c) Q load system for v_a; (d) Q load system for Θ_a; (e) Q load system for v_b.

SOLUTION From Fig. E10.3b and c, respectively,

$$M_P = -Px \quad \text{and} \quad M_Q = -1x$$

$$(1 \downarrow)v_a = \int_0^L \frac{M_P M_Q}{EI} \, dx = \frac{1}{EI}\int_0^L Px^2 \, dx = \frac{PL^3}{3EI} \ (\downarrow)$$

From Fig. E3.10d

$$M_Q = -1$$

$$(1\curvearrowright)\Theta_a = \frac{1}{EI}\int_0^L (-Px)(-1)\, dx = \frac{PL^2}{2EI}\ (\curvearrowright)$$

From Fig. E3.10e

$$M_Q = \begin{cases} 0 & 0 \le x \le \dfrac{L}{2} \\[2mm] \dfrac{L}{2} - x & \dfrac{L}{2} \le x \le L \end{cases}$$

$$(1 \downarrow)v_b = \frac{1}{EI}\int_0^{L/2} (-Px)(0)\, dx + \frac{1}{EI}\int_{L/2}^L (-Px)\left(\frac{L}{2} - x\right) dx$$

$$= \frac{1}{EI}\left(-\frac{PLx^2}{4} + \frac{Px^3}{3}\right)\Bigg|_{L/2}^{L}$$

$$v_b = \frac{5PL^3}{48EI}\ (\downarrow)$$

Example 10.4 Calculate the vertical deflection at point b using complementary virtual work (virtual forces); $E = 200$ GPa, $I = 6000 \times 10^{-6}$ m^4.

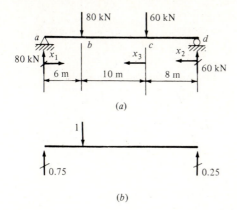

(a)

(b)

Figure E10.4 (a) Loaded beam with coordinate systems; (b) Q load system for v_b.

SOLUTION

$$EI = (200 \times 10^6 \text{ kPa}) (6 \times 10^{-3} \text{ m}^4) = 12 \times 10^5 \text{ kN} \cdot \text{m}^2$$

From a to b $(0 \le x_1 \le 6)$:

$$M_P = 80x_1 \qquad M_Q = 0.75x_1$$

From d to c $(0 \le x_2 \le 8)$:

$$M_P = 60x_2 \qquad M_Q = 0.25x_2$$

From c to b $(0 \le x_3 \le 10)$:

$$M_P = 480 \qquad M_Q = 0.25(x_3 + 8)$$

$$(1 \downarrow)v_b = \frac{1}{EI} \left[\int_0^6 (80x_1) (0.75x_1) \, dx_1 + \int_0^8 (60x_2) (0.25x_2) \, dx_2 \right.$$

$$\left. + \int_0^{10} (480) (0.25) (x_3 + 8) \, dx_3 \right]$$

$$= \frac{1}{EI} \left[20x_1^3 \Big|_0^6 + 5x_2^3 \Big|_0^8 + 120 \left(\frac{x_3^2}{2} + 8x_3 \right) \Big|_0^{10} \right]$$

$$= \frac{1}{EI} (4320 + 2560 + 15{,}600) = \frac{22{,}480}{EI} = \frac{22{,}480 \text{ kN}^2 \cdot \text{m}^3}{12 \times 10^5 \text{ kN} \cdot \text{m}^2}$$

$$v_b = 0.0187 \text{ m} = 18.7 \text{ mm} \ (\downarrow)$$

DISCUSSION Since three algebraic expressions are required to express M_P, three different variables are used to simplify the calculations. Note that they are not unique, and the reader is encouraged to try other variables. If only x_1 were used, the expressions for M_P, from b to d, would be different from those shown; furthermore, not all the integrals would have a lower limit of zero, a fact that greatly simplifies evaluation of the integrated expressions.

Example 10.5 Calculate the vertical deflection and the rotation at point c using complementary virtual work (virtual forces); $EI = 6 \times 10^6$ kips $\cdot$ in^2.

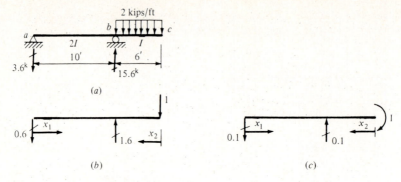

Figure E10.5 (a) Loaded beam; (b) Q load system for v_c; (c) Q load system for Θ_c.

SOLUTION Use Fig. E10.5a and b. From a to b ($0 \le x_1 \le 10$)

$$M_P = -3.6x_1 \qquad M_Q = -0.6x_1$$

From b to c ($0 \le x_2 \le 6$)

$$M_P = -x_2^2 \qquad M_Q = -x_2$$

$$(1\downarrow)v_c = \frac{1}{2EI}\int_0^{10}(-3.6x_1)(-0.6x_1)\,dx_1 + \frac{1}{EI}\int_0^6(-x_2^2)(-x_2)\,dx_2$$

$$= \frac{1}{EI}\left(0.36x_1^3\Big|_0^{10} + 0.25x_2^4\Big|_0^6\right) = \frac{684}{EI} = \frac{(684\ \text{kips}^2 \cdot \text{ft}^3)(1728\ \text{in}^3/\text{ft}^3)}{6 \times 10^6\ \text{kips} \cdot \text{in}^2}$$

$$v_c = 0.197\ \text{in} (\downarrow)$$

In Fig. E10.5c from a to b ($0 \le x_1 \le 10$)

$$M_Q = -0.1x_1$$

From b to c ($0 \le x_2 \le 6$)

$$M_Q = -1.0$$

$$(1\curvearrowright)\Theta_c = \frac{1}{2EI}\int_0^{10}(-3.6x_1)(-0.1x_1)\,dx_1 + \frac{1}{EI}\int_0^6(-x_2^2)(-1)\,dx_2$$

$$= \frac{1}{EI}\left(0.06x_1^3\Big|_0^{10} + 0.33x_2^3\Big|_0^6\right) = \frac{132}{EI} = \frac{(132\ \text{kips}^2 \cdot \text{ft}^2)(144\ \text{in}^2/\text{ft}^2)}{6 \times 10^6\ \text{kips} \cdot \text{in}^2}$$

$$\Theta_c = 0.0032\ \text{rad} (\curvearrowright)$$

The complementary virtual strain energy for a structure subjected to both axial forces and bending moments is obtained by superposing the contribution from each effect, giving

$$\delta W_i^* = \sum \frac{F_P F_Q L}{AE} + \int_0^L \frac{M_P M_Q}{EI}\,ds \qquad (10.22)$$

which assumes the two effects to be independent, so that the axial forces do not interact with the bending moments.† Additional terms must be added to the right side if strains induced by temperature and lack of fit are also present (Sec. 10.10).

If the actual shear V_P and torsion T_P are also to be considered, the right side of Eq. (10.22) must include the terms

$$\int_0^L \frac{\kappa_V V_P V_Q}{AG} \, ds + \int_0^L \frac{T_P T_Q}{\kappa_T G} \, ds \qquad (10.22a)$$

where V_Q = shear due to virtual Q forces
T_Q = torsional moment due to virtual Q forces
κ_V = shear constant, dependent upon shape of cross section
κ_T = St. Venant torsion constant, dependent upon shape of cross section
G = shear modulus of material

The complementary virtual strain energy associated with warping induced by torsion has been neglected.

Typically, axial and bending effects need not both be considered for most structures; however, it is important to realize that occasionally the combined effect can be significant. Most trusses are idealized as having pin-jointed members in order to simplify the analysis. Almost all connections in trusses will sustain some bending moment; therefore, in fact, the members are not loaded exclusively with axial forces even though the applied loads occur only at the joints. The bending stresses induced in the members by the fact that the connections are not actually pinned are termed *secondary stresses*. Typically, they are small compared with the primary axial stresses; occasionally, however, this is not the case, and the additional deflections resulting from the bending moments must be calculated. Also, both axial forces and bending moments exist in the members of rigid frames; this can be envisioned particularly for the columns. These axial effects in frames usually alter the deflections by a few percent, but for some structures the axial loads can significantly alter the deflections obtained by considering only the bending moments. Example 10.6 illustrates the use of complementary virtual work for calculating deflections when both axial and bending effects are considered.

Example 10.6 Using complementary virtual work (virtual forces), calculate the horizontal deflection of point a considering deformations due to bending and axial forces. $I_1 = 700$ in⁴, $A_1 = 12$ in², $I_2 = 950$ in⁴, $A_2 = 18$ in², and $E = 29 \times 10^3$ kips/in².

SOLUTION Use Fig. E10.6b and c. From a to b

$$L = 9 \text{ ft} \qquad 0 \le y_1 \le 9$$

$$M_P = 0 \qquad M_Q = -1y_1$$

$$F_P = -12 \qquad F_Q = -\tfrac{3}{2}$$

From b to c

$$L = 9 \text{ ft} \qquad 0 \le y_2 \le 9$$

† The interaction, called beam-column behavior, is beyond the scope of this text.

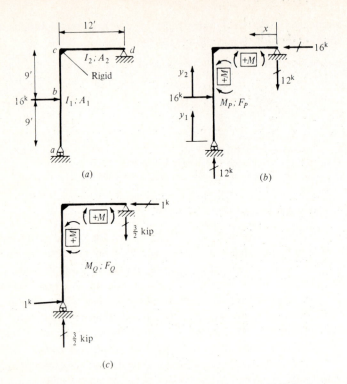

Figure E10.6 (*a*) Loaded rigid frame; (*b*) P load system; (*c*) Q load system for u_a.

$$M_P = -16y_2 \qquad M_Q = -1(9 + y_2)$$

$$F_P = -12 \qquad F_Q = -\tfrac{3}{2}$$

From d to c

$$L = 12 \text{ ft} \qquad 0 \le x \le 12$$

$$M_P = -12x \qquad M_Q = -\tfrac{3}{2}x$$

$$F_P = -16 \qquad F_Q = -1$$

$$(1\rightarrow)u_a = \frac{1}{EI_1}\left[\int_0^9 (0)(-y_1)\,dy_1 + \int_0^9 (-16y_2)(-9 - y_2)\,dy_2 \right.$$

$$\left. + \frac{1}{1.357}\int_0^{12} (-12x)(-\tfrac{3}{2}x)\,dx \right]$$

$$+ \frac{1}{EA_1}\left[(-12)(-\tfrac{3}{2})(18) + \frac{-16(-1)(12)}{1.5}\right]$$

$$(1\rightarrow)u_a = \frac{1}{EI_1}\left[(72y_2^2 + \tfrac{16}{3}y_2^3)\Big|_0^9 + \frac{6}{1.357}x^3\Big|_0^{12}\right] + \frac{1}{EA_1}(324 + 128)$$

$$(1 \rightarrow)u_a = \frac{17{,}360}{EI_1} + \frac{452}{EA_1}$$

$$= \frac{(17{,}360 \text{ kip}^2 \cdot \text{ft}^3)\,(12^3 \text{ in}^3/\text{ft}^3)}{(29 \times 10^3 \text{ kips/in}^2)\,(700 \text{ in}^4)} + \frac{(452 \text{ kip}^2 \cdot \text{ft})\,(12 \text{ in/ft})}{(29 \times 10^3 \text{ kips/in}^2)\,(12 \text{ in}^2)}$$

$$u_a = 1.4778 \text{ in} + 0.0156 \text{ in} = 1.4934 \text{ in} \ (\rightarrow)$$

DISCUSSION A careful dimensional analysis is almost mandatory for problems involving both bending and axial deformations. Note that the axial deformations account for about 1 percent of the total deflection. Since deflections for most low-rise rigid frames are usually altered by only a few percent when the axial loads are included in the calculations, this effect is generally ignored.

The method of complementary virtual work can also be used for structures in which the members are curved and/or tapered. For example, for a three-hinged arch having a cross section that varies along its length, an expression similar to Eq. (10.22) is used to obtain the deflections and slopes, since the structure is loaded with both axial forces and bending moments. Note that δW_i^* for the axial forces was obtained in Eq. (10.9) assuming that both F_P and F_Q are constant along the member length. In addition, in order to evaluate δW_i^* for the bending, it is necessary to retain the moment of inertia of the member inside the integral. The resulting expressions, which must be integrated along the axis of the curved member, will probably have to be evaluated numerically. This problem can be obviated if the arch is idealized as a series of discrete members each having a constant cross section. This procedure will give deflections and slopes that approach the true solution as the number of the member segments increases.

10.7 CASTIGLIANO'S THEOREM, PARTS I AND II

Castigliano's theorem part I is derived using the theorem of virtual work.

Castigliano's theorem part I For a linear elastic structure with a given system of loads and temperature distribution, the first partial derivative of the strain energy with respect to a particular displacement is equal to the corresponding force.

In equation form

$$\left(\frac{\partial W_i}{\partial u_j} \right)_{T=\text{const}} = P_j \tag{10.23}$$

If the loaded structure shown in Fig. 10.8 is given the virtual displacement δu_j (in the direction and at the location of the load P_j), then the virtual work is

$$\delta W_e = P_j\,\delta u_j$$

If the structure has a specified thermal distribution and the temperature is kept constant during the virtual displacement, the theorem of virtual work gives

$$\delta W_i = P_j\,\delta u_j$$

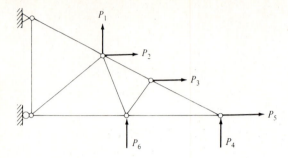

Figure 10.8 Truss with applied forces.

In the limit as $\delta u_j \to 0$,

$$\left(\frac{\partial W_i}{\partial u_j}\right)_{T=\text{const}} = P_j$$

which is Castigliano's theorem part I. If the structure is given a virtual rotation $\delta\Theta_j$ in the direction of the moment M_j, instead of the virtual displacement δu_j, Castigliano's theorem part I becomes

$$\left(\frac{\partial W_i}{\partial \Theta_j}\right)_{T=\text{const}} = M_j$$

Therefore, in Eq. (10.23), u_j and P_j can be regarded as a generalized displacement and a generalized force, respectively. Note that this theorem is valid for both linear and nonlinear elastic materials and also is applicable to structures in which the deflected geometry is dissimilar from the initial geometry (provided δW_i has been properly written to include these effects). Although a discussion of these nonlinearities is beyond the scope of this text, the reader should realize that the theorem has broad applications. It can be used for structures with induced stresses, e.g., thermal, lack of fit, provided the virtual strains (produced by the virtual displacement) do not include virtual initial strains.

The dual energy principle of Eq. (10.23) is obtained by applying the theorem of complementary virtual work.

Castigliano's theorem part II For a linear elastic structure with a given system of loads and no initial strains due to temperature, settlement, etc., the first partial derivative of the strain energy with respect to a particular force is equal to the corresponding displacement.

In equation form

$$\left(\frac{\partial W_i}{\partial P_j}\right)_{T=\text{const}} = u_j \tag{10.24}$$

For the loaded structure shown in Fig. 10.8, if only one virtual force δP_j is applied in the direction of the displacement u_j, the complementary virtual work is

$$\delta W_e^* = u_j\,\delta P_j$$

The principle of complementary virtual work gives

$$\delta W_i^* = u_j\,\delta P_j$$

In the limit as $\delta P_j \to 0$,

$$\left(\frac{\partial W^*}{\partial P_j}\right)_{T=\text{const}} = u_j$$

which is a statement of Castigliano's theorem, part II, and it can be stated as follows:

> For a linear elastic structure with a given system of loads and temperature distribution, the first partial derivative of the complementary strain energy with respect to a particular force is equal to the corresponding displacement.

The quantities P_j and u_j should be regarded as a generalized force and a generalized displacement, respectively. Therefore, the theorem can be used to obtain slopes as well as translational deflections. Note that if W_i^* is the complementary strain energy of total deformation, i.e., the strains include both the elastic strains and any induced strains such as those from temperature, the theorem is generalized.

It can be observed from Eq. (10.4) that $W_i = W_i^*$ for linear elastic materials in which the stresses are proportional to the strains and there are no induced strains. For these types of structures the general Castigliano's theorem part II simplifies to that shown in Eq. (10.24).

10.8 COMPUTATION OF DEFLECTIONS USING CASTIGLIANO'S THEOREM PART II

In order to apply Castigliano's theorem it is necessary to develop expressions for the strain energy of the structural members. For truss members and beams this can be accomplished using Eq. (10.4), which was obtained from Fig. 10.3*b* for linear elastic members. For a uniform (prismatic) truss member with a cross-sectional area A and length L subjected to an axial force F the stress and strain are, respectively,

$$\sigma = \frac{F}{A} \quad \text{and} \quad \varepsilon = \frac{\sigma}{E} = \frac{F}{AE}$$

Substituting these two expressions into the equation for strain energy (10.4) and integrating over the volume of the member gives

$$W_i = \frac{F^2 L}{2AE}$$

which is the strain energy for an individual truss member. The total strain energy for a truss is the sum of the contributions from each of the individual members; therefore,

$$W_i = \sum \frac{F^2 L}{2AE} \tag{10.25}$$

Substituting this expression into the equation for Castigliano's theorem part II and performing the partial differentiation gives

$$u_j = \sum F \frac{\partial F}{\partial P_j} \frac{L}{AE} \tag{10.26}$$

The fundamental equation for strain energy (10.4) can also be used for beams. If a slender beam is subjected to loading that produces bending moments along the member, and if the centroidal axis of the member is straight with all cross sections having an axis of symmetry lying in the plane of the loading, the stress and strain respectively due to bending are

$$\sigma = -\frac{My}{I} \quad \text{and} \quad \varepsilon = \frac{\sigma}{E} = -\frac{My}{EI}$$

Substituting these two expressions into the equation for strain energy (10.4) gives

$$W_i = \int_{\text{vol}} \frac{M^2 y^2}{2EI^2} \, dV$$

Rearranging terms gives

$$W_i = \int_0^L \left(\frac{M^2}{2EI^2} \int_{\text{area}} y^2 \, dA \right) ds$$

But since

$$I = \int_{\text{area}} y^2 \, dA$$

we have†

$$W_i = \int_0^L \frac{M^2}{2EI} \, ds \tag{10.27}$$

Substituting this expression into the equation for Castigliano's theorem part II gives

$$u_j = \frac{\partial}{\partial P_j} \left(\int_0^L \frac{M^2}{2EI} \, ds \right)$$

If the integrand is continuous, it is possible to perform the partial differentiation before the integration; thus Castigliano's theorem part II for beams is

$$u_j = \int_0^L \frac{M}{EI} \frac{\partial M}{\partial P_j} \, ds \tag{10.28}$$

Since Castigliano's theorem was derived from the theorem of complementary virtual work, the technique of applying this method yields calculations that are similar to those for the method of complementary virtual work. Examples 10.7 to 10.10 illustrate this method for trusses and beams. Note that if the deflection at a known load must be obtained, it is necessary to write the strain energy in terms of this load (taken as an independent variable) and set the load equal to its value at the end of the calculations. Similarly, if the deflection is to be calculated at a point where no load is applied, the strain energy can be written in terms of a general force at this point and its value set equal to zero at the final stage of the calculations. Since these examples were used to illustrate the theorem of complementary virtual work, it is instructional to compare the two sets of calculations and note their similarity.

---

† If shear V and torsion T are also to be considered, the right side of Eq. (10.27) must include

$$\int_0^L \frac{\kappa_V V^2}{2AG} \, ds + \int_0^L \frac{T^2}{2\kappa_T G} \, ds \tag{10.27a}$$

where the strain energy associated with warping induced by torsion has been neglected. See Eq. (10.22a) for a description of terms.

Example 10.7 Calculate the vertical and horizontal deflections at point b for the truss shown in Example 10.1. Use the method of Castigliano part II.

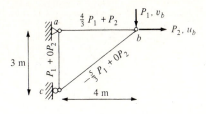

Figure E10.7

SOLUTION

$$\Delta_j = \sum F \frac{\partial F}{\partial P_j} \frac{L}{AE} = \frac{L}{AE} \sum F \frac{\partial F}{\partial P_j}.$$

See Table E10.7.

Table E.10.7

Member	F, kN	$\dfrac{\partial F}{\partial P_1}$	$F\dfrac{\partial F}{\partial P_1}$, kN	$\dfrac{\partial F}{\partial P_2}$	$F\dfrac{\partial F}{\partial P_2}$, kN
ab	$\frac{4}{3}P_1 + P_2$	$\frac{4}{3}$	$\frac{16}{9}P_1 + \frac{4}{3}P_2$	1	$\frac{4}{3}P_1 + P_2$
ac	$P_1 + (0)(P_2)$	1	P_1	0	0
bc	$-\frac{5}{3}P_1 + (0)(P_2)$	$-\frac{5}{3}$	$+\frac{25}{9}P_1$	0	0
			$\sum = \left(\frac{50}{9}P_1 + \frac{4}{3}P_2\right)$ kN		$\sum = \left(\frac{4}{3}P_1 + P_2\right)$ kN

Note that $P_1 = 90$ kN and $P_2 = 0$; hence,

$$v_b = (5 \times 10^{-5}\text{ m/kN})\left(\tfrac{50}{9}\right)(90\text{ kN}) = 0.025\text{ m} = 25\text{ mm }(\downarrow)$$

$$u_b = (5 \times 10^{-5}\text{ m/kN})\left(\tfrac{4}{3}\right)(90\text{ kN}) = 6 \times 10^{-3}\text{ m} = 6\text{ mm }(\rightarrow)$$

Example 10.8 Calculate the vertical deflection of joint d for the truss of Example 10.2 using the method of Castigliano part II.

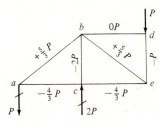

Figure E10.8

SOLUTION See Table E10.8.

Table E10.8

Member	$\dfrac{L}{A}$, in^{-1}	F, kips	$\dfrac{\partial F}{\partial P}$	$F\dfrac{\partial F}{\partial P}\dfrac{L}{A}$, kips/in
ac	240	$-\frac{4}{3}P$	$-\frac{4}{3}$	$\frac{1280}{3}P$
ce	240	$-\frac{4}{3}P$	$-\frac{4}{3}$	$\frac{1280}{3}P$
bd	240	0	0	0
bc	180	$-2P$	-2	$720P$
de	180	$-P$	-1	$180P$
ab	300	$+\frac{5}{3}P$	$+\frac{5}{3}$	$\frac{2500}{3}P$
be	300	$+\frac{5}{3}P$	$+\frac{5}{3}$	$\frac{2500}{3}P$

$$\Sigma = \frac{10,260}{3}P$$

$$v_d = \frac{10,260}{3}(30)\frac{1}{29 \times 10^3} = 3.54 \text{ in } (\downarrow)$$

Example 10.9 A uniform cantilever beam is subjected to a downward vertical load at the tip. Using the method of Castigliano part II, calculate the tip deflection and rotation.

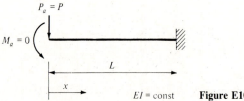

$P_a = P$

$M_a = 0$

L

x

$EI = \text{const}$ **Figure E10.9**

SOLUTION

$$M = -P_a x - M_a \qquad \frac{\partial M}{\partial P_a} = -x$$

To find v_a, let $M_a = 0$ and $P_a = P$. Then

$$v_a = \frac{1}{EI}\int_0^L (-Px)(-x)\,dx = \frac{PL^3}{3EI} \; (\downarrow)$$

To find Θ_a, note that $\partial M/\partial M_a = -1$ and let $M_a = 0$ and $P_a = P$. Then

$$\Theta_a = \frac{1}{EI}\int_0^L (-Px)(-1)\,dx = \frac{PL^2}{2EI}\;(\curvearrowright)$$

Example 10.10 Calculate the vertical deflection at point c for the beam shown in Example 10.5. Use the method of Castigliano part II.

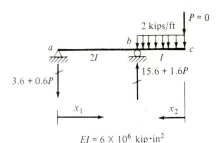

$$EI = 6 \times 10^6 \text{ kip·in}^2 \qquad\qquad \textbf{Figure E10.10}$$

SOLUTION From a to b $(0 \le x_1 \le 10)$

$$M = -(3.6 + 0.6P)x_1 \qquad \frac{\partial M}{\partial P} = -0.6x_1$$

From b to c $(0 \le x_2 \le 6)$

$$M = -x_2^2 - Px_2 \qquad \frac{\partial M}{\partial P} = -x_2$$

Now let $P = 0$ and apply Castigliano's theorem

$$v_c = \frac{1}{E2I}\int_0^{10} (-3.6x_1)(-0.6x_1)\,dx_1 + \frac{1}{EI}\int_0^6 (-x_2^2)(-x_2)\,dx_2$$

$$= \frac{684}{EI} = \frac{684 \text{ kip·ft}^3}{6 \times 10^6 \text{ kip·in}^2}(12^3 \text{ in}^3/\text{ft}^3) = 0.197 \text{ in}\;(\downarrow)$$

10.9 COMPARISON OF ENERGY THEOREMS

Table 10.3 summarizes the energy theorems presented in the previous sections. All the theorems associated with virtual work, on the left, involve a change in the actual displacements experienced by the structure, i.e., either a unit or a virtual displacement. In contrast, those associated with complementary virtual work, on the right, require that the structure be subjected to either a unit or a virtual force. That is, the left side of the table involves the area below the force-deflection and stress-strain curves, whereas the right side requires an investigation of the area above those curves. Generally the various complementary virtual-work theorems are used for calculating deflections, but those in the other category are also used for investigating structural behavior.

Table 10.3 Summary of energy theorems

Based on virtual work	Based on complementary virtual work
Principle of virtual work (principle of virtual displacements)	**Principle of complementary virtual work** (principle of virtual forces)
$\delta W_e = \delta W_i$	$\delta W_e^* = \delta W_i^*$
$\delta u, \delta \varepsilon$ virtual	$\delta P, \delta \sigma$ virtual
P, σ actual	u, ε actual
δu must be continuous and satisfy kinematic boundary conditions	δP and $\delta \sigma$ must satisfy equilibrium
Unit-displacement theorem	**Unit-load theorem**
$$P = \int \sigma \bar{\varepsilon} \, dV$$	$$u = \int \varepsilon \bar{\sigma} \, dV$$
$\bar{\varepsilon}$ = compatible strain due to unit displacement	$\bar{\sigma}$ = equilibrium stress state due to unit force
Castigliano's theorem part I	**Castigliano's theorem part II**
$$\left(\frac{\partial W_i}{\partial u_j}\right)_{T=\text{const}} = P_j$$	$$\left(\frac{\partial W_i^*}{\partial P_j}\right)_{T=\text{const}} = u_j$$
	For linear elastic structures $W_i = W_i^*$; therefore,
	$$\left(\frac{\partial W_i}{\partial P_j}\right)_{T=\text{const}} = u_j$$

10.10 ADDITIONAL EXAMPLES INVOLVING SETTLEMENT, TEMPERATURE, ETC.

The deflections resulting from externally applied forces can be altered by other effects. For example, significant deflections can be introduced into a building if one or more of the supports experiences differential settlement. An unheated structure during a cold day will display deflections resulting from the induced internal thermal strains if it was constructed in warm summer weather. Internal induced strains can also occur if the geometry of the individual members is altered during fabrication, e.g., errors in the lengths of truss members, beam members that are curved instead of straight.

All these induced deformations will alter the deflected shape of statically determinate structures but will not introduce any internal stresses. For example, it can be envisioned that if a simply supported beam experiences differential support settlement, the structure will exhibit a rigid-body rotation with no increase in the bending moments. Similarly, if this same structure is heated uniformly, the distance between supports increases but the beam will experience no additional bending moments since it is unconstrained. However, a statically indeterminate structure can sustain large stresses as the result of support settlement, thermal strains, etc. The contrasting behavior of determinate and indeterminate structures subjected to these effects will be put into perspective after the study of indeterminate structures has been completed.

Figure 10.9 illustrates the force-deflection and stress-strain plots for a linear elastic truss member with induced displacement u^0 and corresponding strain ε^0. The response plots are similar to those shown in Fig. 10.3, where the forces and deflections are simultaneously increased from zero. The discussion in Secs. 10.4 to 10.6 can be extended to situations involving induced deflections and strains. The general form of the theorem of complementary virtual work including induced deformations can be written

$$\delta P\, u \,+\, \delta W_r^* \;=\; \int_{\text{vol}} \varepsilon\, \delta\sigma\, dV \,+\, \int_{\text{vol}} \varepsilon^0\, \delta\sigma\, dV \tag{10.29}$$

where δW_r^* is the complementary virtual work due to the movement of the supports and the terms on the right side of the equation are the complementary virtual strain energy due to the linear elastic strains ε and induced strains ε^0, respectively. Using the strains

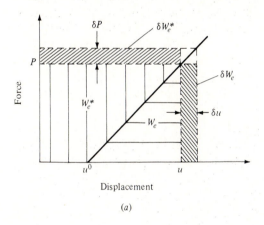

(a)

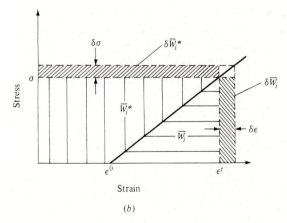

Strain

(b)

Figure 10.9 Response of a linear truss member with induced effects: (*a*) force-displacement curve; (*b*) stress-strain curve ($\varepsilon^t = \varepsilon^0 + \varepsilon$, $\varepsilon^t =$ total strain, $\varepsilon^0 =$ induced strain, $\varepsilon =$ elastic strain).

due to the thermal effects and fabrication errors, given by Eqs. (10.10) and (10.11), respectively, along with Eq. (10.8) for the virtual stress, in the second integral on the right side of Eq. (10.29) gives

$$\delta W_i^* = \begin{cases} \sum F_Q(\alpha \Delta T L) & \text{thermal strains} \qquad\qquad (10.30a) \\ \sum F_Q \Delta L & \text{fabrication errors} \qquad\qquad (10.30b) \end{cases}$$

The complementary virtual strain energy for a beam with linear elastic strains is shown in Eq. (10.20). If a beam or frame is subjected to induced strains, the corresponding δW_i^* must be calculated using the second term on the right side of Eq. (10.29). For example, if a beam (symmetric with respect to the neutral axis) is heated unevenly over the depth, with the linear thermal distribution shown in Fig. 10.10, the thermal strain is

$$\varepsilon^0 = \alpha T(y) = \alpha \left(T_m + \frac{\Delta T}{h} y \right) \qquad T_m = \tfrac{1}{2}(T_u + T_l) \qquad \Delta T = T_u - T_l$$

According to Eq. (10.29), the complementary virtual strain energy due to this thermal distribution is

$$\delta W_i^* = \int_{\text{vol}} [\alpha T(y)] \frac{M_Q y}{I} \, dV$$

$$= \int_{\text{vol}} \alpha T_m \frac{M_Q y}{I} \, dV + \int_{\text{vol}} \left(\alpha \frac{\Delta T}{h} y \right) \frac{M_Q y}{I} \, dV$$

$$= \int_0^L \left(\frac{\alpha T_m M_Q}{I} \int_{\text{area}} y \, dA \right) ds + \int_0^L \left(\frac{\alpha \Delta T M_Q}{Ih} \int_{\text{area}} y^2 \, dA \right) ds$$

Since
$$\int_{\text{area}} y \, dA = 0 \qquad \text{and} \qquad \int_{\text{area}} y^2 \, dA = I$$

We have
$$\delta W_i^* = \int_0^L \frac{\alpha}{h} \Delta T M_Q \, ds \qquad\qquad (10.31)$$

Examples 10.11 and 10.12 show the calculations for support settlement and thermal strains, respectively, using the method of complementary virtual work. Castigliano's theorem part II as shown in Eq. (10.24) is not applicable for structures with thermal or fabrication strains. This formulation is possible only if $W_i = W_i^*$; however, it can be observed from Fig. 10.9b that this is not the case if induced

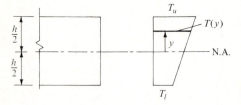

Figure 10.10 Heated beam showing thermal distribution.

deflections and strains are present. Castigliano's theorem part II is applicable for these cases only if the complementary strain energy W_i^* is used and the terms from the induced strains are properly included. This form is described in the detailed discussion following Eq. (10.24). For this reason, problems with induced deformations are illustrated in this chapter using only the method of complementary virtual work.

Example 10.11 Calculate the vertical and horizontal displacement of joint h for the following support movements:

Support a: Vertical = 5.0 mm ($\downarrow$); Horizontal = 7.5 mm ($\rightarrow$)

Support f: Vertical = 15.0 mm ($\downarrow$)

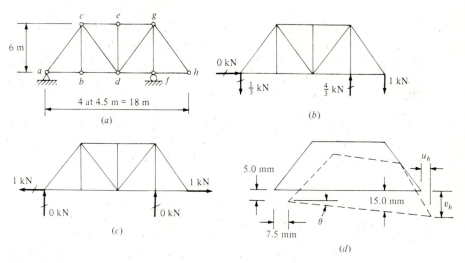

Figure E10.11 (a) The truss; (b) Q load system for v_h; (c) Q load system for u_h; (d) deflected structure.

SOLUTION From Fig. E10.11b the vertical displacement of h (v_h) is obtained as follows:

$$\delta W_i^* = 0 \qquad \delta W_e^* = (1 \downarrow) v_h$$

$$\delta W_r^* = \tfrac{1}{3}(5.0) - \tfrac{4}{3}(15.0) = -18.3$$

$$\delta W_r^* + \delta W_e^* = \delta W_i^* = 0$$

$$v_h = 18.3 \text{ mm } (\downarrow)$$

From Fig. E10.11c the horizontal displacement of h (u_h) is obtained as follows:

$$\delta W_i^* = 0 \qquad \delta W_e^* = (1 \rightarrow) u_h$$

$$\delta W_r^* = (-1)(7.5) = -7.5$$

$$\delta W_r^* + \delta W_e^* = \delta W_i^* = 0$$

$$u_h = 7.5 \text{ mm } (\rightarrow)$$

From Fig. 10.11*d* the kinematic analysis gives

$$\theta = \frac{15.0 - 5.0}{13,500} = 7.41 \times 10^{-4} \text{ rad}$$

$$v_h = 5.0 + 18,000\theta = 5.0 + 13.3 = 18.3 \text{ mm } (\downarrow)$$

$$u_h = 7.5 \text{ mm } (\rightarrow)$$

DISCUSSION Since there are no forces applied to this truss, all members have zero load $(F_P = 0)$; thus $\delta W_i^* = 0$, and only the complementary virtual work of the reactions and the applied virtual force must be considered in the equation. Note that if the displacement of the support and the virtual reaction are opposite in sign, δW_r^* is negative.

This problem can also be analyzed by observing the kinematics of the truss introduced by the support displacements. The truss has translated as a rigid body to the right 7.5 mm, and it has also rotated clockwise about point *a* through an angle θ. This angle of rotation is considered to be sufficiently small to give $\theta \approx \sin \theta \approx \tan \theta$. This is a typical assumption made for deflections if the initial and final geometries of the structure are to be considered approximately equal. If this were not the case, we would have to resort to nonlinear analysis.

Example 10.12 Compute the deflection of point *a* if the upper and lower surfaces are uniformly heated, as shown, along the length of this prismatic beam.

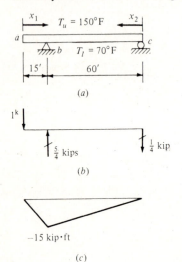

(a)

(b)

−15 kip·ft

(c)

Figure E10.12 (*a*) The beam; $h = 36$ in, $\alpha = 6.5 \times 10^{-6}$ in/in · °F; (*b*) Q load system; (*c*) M_Q diagram.

SOLUTION $\Delta T = 80$°F. From *a* to *b*

$$L = 15 \text{ ft} \qquad 0 \le x_1 \le 15$$

$$M_Q = -1x_1$$

From *c* to *b*

$$L = 60 \text{ ft} \qquad 0 \le x_2 \le 60$$

$$M_Q = -\tfrac{1}{4}x_2$$

Since the strain $\alpha \Delta T$ is negative and constant along the beam

$$v_a = \int_0^L \frac{\alpha}{h} \Delta T M_Q \, ds = \frac{\alpha}{h} \Delta T \int_0^L M_Q \, ds$$

$$= \frac{(6.5 \times 10^{-6})(80)}{36} [\tfrac{1}{2}(75)(15)(144)]$$

$$= 1.170 \text{ in } (\downarrow)$$

DISCUSSION The thermal strains are negative, which implies that the beam will be distorted into negative curvature along its length. Therefore, the sign of v_a is positive, indicating that point a deflects in the direction of the virtual load. Since $(\alpha \Delta T / h)$ is a constant, it is factored from under the integral sign. Note that $\int_0^L M_Q \, ds$ is the area under the M_Q diagram.

The problem of precambering a structure also involves the calculation of deflections produced by induced strains. In this case, the designer specifies that there should be an initial deflection in the structure that is approximately equal in magnitude but opposite in direction to that caused by live and dead loading. For example, for a truss this can be accomplished by fabricating the upper and/or lower chord members with a deviation from the planned lengths. An example of this procedure is contained in Example 10.13.

Example 10.13 The truss in Example 10.11 is to be precambered so that the initial deflection of joint h will be 30 mm upward. This is to be accomplished by uniformly shortening the upper chord members and lengthening the bottom chord members. Calculate the length change that must be fabricated into each of these six members.

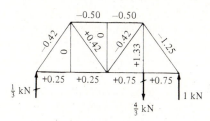

Figure E10.13 Q load system. Fabrication length changes: top chord members $= -\Delta L$, bottom chord members $= +\Delta L$.

SOLUTION Using complementary virtual work, we have

$$\sum F_Q \Delta L = 1(30)$$

$$2(-0.50)(-\Delta L) + 2(+0.25\Delta L) + 2(+0.75\Delta L) = 30$$

$$1.00\Delta L + 0.50\Delta L + 1.50\Delta L = 30$$

$$3.00\Delta L = 30$$

$$\Delta L = 10 \text{ mm}$$

DISCUSSION Fabricating the truss with each top chord member 10 mm too short and each bottom chord member 10 mm too long will give the truss an upward deflection at joint h of 30 mm when it is initially assembled. This procedure is common if there must be no deflection when live load is imposed on the structure. This is done for aesthetic and other reasons. For example, people perceive observable deflections as being unsafe, and mechanical equipment must be dead level to operate properly. Since there are no reactions, all member forces are zero when the truss is assembled.

10.11 THE RECIPROCAL THEOREM

Consider the linear elastic beam in Fig. 10.11 subjected to two individual applied loadings. First, a single concentrated load P_a is applied at point a as shown, and the deflection at point c (Fig. 10.11a) is

$$v_c = \frac{P_a L^3}{32EI}$$

Second, the load at point a is removed, and a load P_c is placed at point c (Fig. 10.11b) which produces a deflection at point a of

$$v_a = \frac{P_c L^3}{32EI}$$

If $P_a = P_c = P$, these two deflections are equal; i.e., the deflection at point c, caused by a force P at point a, is equal to the deflection at point a produced by an identical force P when applied at point c; this observation was made by Maxwell in 1864 for linear elastic trusses using the method of complementary work. Furthermore, if these two forces are not equal, the deflections v_c and v_a are not equal but

$$P_a v_a = P_c v_c$$

That is, the work done by P_a moving through the displacement at point a (produced by the force P_c) equals the work done by the force P_c moving through the displacement at point c (caused by the force P_a). This is a statement of the generalization attributed to Betti (1872).

A more general loading is shown acting on the linear elastic beam in Fig. 10.12. The applied forces are divided into two loading systems as shown. The P' loading system produces the deflections Δ_a', Δ_b', Δ_c', and Δ_d', and the P'' loading system gives the deflections Δ_a'', Δ_b'', Δ_c'', and Δ_d''. Thus, from the principle of superposition, the total deflections are the sum of those from the two loading systems, e.g., the deflection at point a is

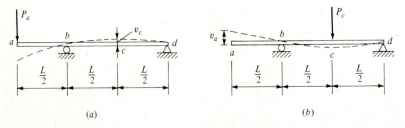

(a) (b)

Figure 10.11 Two deflection shapes of a beam: (a) with a concentrated load at point a; (b) with a concentrated load at point c.

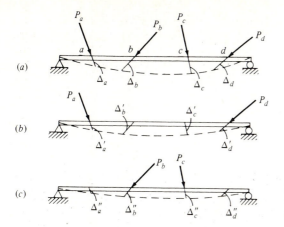

(a)

(b)

(c)

Figure 10.12 Deflections of a beam: (a) with the total loading; (b) with loading system 1 (P'); (c) with loading system 2 (P'').

$$\Delta_a = \Delta'_a + \Delta''_a$$

Assume that only the applied forces cause the deflections; i.e., the supports are unyielding and there are no induced deformations such as those due to thermal strains. If all the forces are increased from zero, the work done by all the applied loads is

$$W_e = \frac{1}{2}\sum P\Delta = \frac{1}{2}\sum P'(\Delta' + \Delta'') + \frac{1}{2}\sum P''(\Delta' + \Delta'') \qquad (10.32)$$

where P, P', and P'' are forces of the three loading systems, respectively, and Δ, Δ', and Δ'' are generic deflections of the corresponding loading system.

Alternatively, if only the P' loads are applied initially the work done by these forces is

$$\frac{1}{2}\sum P'\Delta' \qquad (10.33a)$$

If, in addition, the P'' loading system is now applied, the work done by these forces moving through their respective deflections is

$$\frac{1}{2}\sum P''\Delta'' \qquad (10.33b)$$

During the application of the P'' forces, the P' forces displace through their corresponding deflections Δ''. Therefore, this additional work of the P' forces is

$$\sum P'\Delta'' \qquad (10.33c)$$

The sum of Eqs. (10.33a) to (10.33c) is the work of the two load systems when they are applied in succession. Since the system is linear and elastic, the work done by the external forces is independent of the order in which the loads are applied; i.e., the work is path independent. Thus, the work of the total loading applied in one step [Eq. (10.32)] must equal the work when the forces are applied in two steps. Equating these two quantities gives

$$\frac{1}{2}\sum P'(\Delta' + \Delta'') + \frac{1}{2}\sum P''(\Delta' + \Delta'') = \frac{1}{2}\sum P'\Delta' + \frac{1}{2}\sum P''\Delta'' + \sum P'\Delta''$$

Simplification gives

$$\frac{1}{2}\sum P''\Delta' = \frac{1}{2}\sum P'\Delta''$$

This equation represents Betti's reciprocal theorem.

> **Betti's reciprocal theorem** For a linearly elastic structure with unyielding supports and no induced deformations, if the forces are all increased from zero, the work done by a system of forces P' moving through the corresponding displacements caused by a second set of forces P'' is equal to the work done by the forces P'' acting through the displacements produced by the forces P'.

Betti's law is a generalization of Maxwell's theorem, and the latter can be derived using Betti's law. Consider the simply supported beam loaded as shown in Fig. 10.13a. The two loading systems in this case each consist of a single force of magnitude P. The deflection Δ_{ba} at point b in the direction shown is caused by an applied force P at point a in the direction shown, and Δ_{ab} is similarly defined. The first subscript denotes the location of the displacement, and the second is the point where the force is applied. Applying Betti's law to this situation gives

$$P\Delta_{ab} = P\Delta_{ba}$$

Hence, $$\Delta_{ab} = \Delta_{ba} \tag{10.33}$$

which is a statement of Maxwell's theorem.

> **Maxwell's theorem** For a linear elastic structure with unyielding supports and no induced deformations, if the forces are increased from zero, the deflection at point a in the direction of the applied force P at this point due to a force P at point b is equal to the deflection at point b in the direction of the applied force at this point due to a force P at point a.

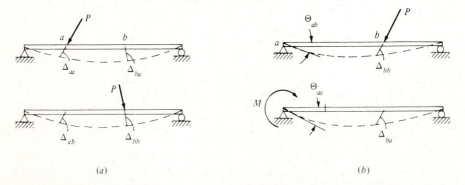

Figure 10.13 Relationship of beam deflections: (*a*) for two individual concentrated forces; (*b*) for a concentrated force and an applied moment.

Since the forces and deflections in Betti's reciprocal theorem can be considered as generalized quantities, the theorem can be applied to the beam shown in Fig. 10.13*b* to relate the rotation at point *a* with the deflection at point *b* due to the loading shown. Applying Betti's theorem gives

$$M\Theta_{ab} = P\Delta_{ba}$$

where Θ_{ab} is the slope at point *a* due to a force *P* at *b* and Δ_{ba} is the deflection at point *b* due to a moment *M* at point *a*. If $P = 1$ and $M = 1$, both with appropriate units,

$$\Theta_{ab} = \Delta_{ba}$$

Because deflections due to unit generalized forces are denoted by lowercase symbols, this will subsequently be written

$$\theta_{ab} = \delta_{ba}$$

The reciprocal theorem can also be applied to trusses. For the truss shown in Fig. 10.14 Maxwell's theorem gives

$$\Delta_{dc} = \Delta_{cd}$$

The reciprocal theorem is a useful result that will be exploited extensively in the analysis of statically indeterminate structures. It also finds broad applications in the analysis of structures using matrix methods. An interesting consequence of the reciprocal theorem is that it can be used to construct influence lines for the deflections of structures, i.e., the deflections at a specified point as a unit concentrated load moves across the structure. Assume that the influence line for the vertical deflection of point *b* for the beam of Fig. 10.15 is to be constructed. Thus, for example, the unit load would be placed at point *i* and the deflection δ_{bi} would be the corresponding influence ordinate. As the load traverses the beam, the other ordinates such as δ_{bj}, etc., must be calculated. However, thanks to the reciprocal theorem,

$$\delta_{bi} = \delta_{ib} \qquad \delta_{bj} = \delta_{jb}$$

and so on. Therefore, the total influence line for the deflection at point *b* can be obtained by placing a unit concentrated load at point *b* and calculating the deflections at all points on the beam. The reciprocal theorem can also be applied to yield the Müller-Breslau principle, which is used to construct influence lines for generalized forces in structures. This method is described in Chap. 14.

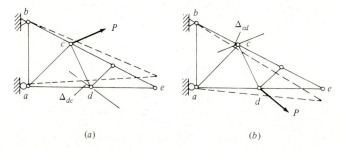

(a)   (b)

Figure 10.14 Relationship of truss deflections: (*a*) with load at joint *c*; (*b*) with load at joint *d*.

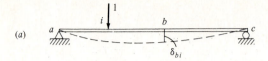

(a)

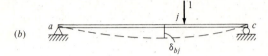

(b)

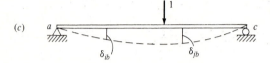

(c)

Figure 10.15 Influence lines for deflections using the reciprocal theorem: (a) unit load at general point i; (b) unit load at general point j; (c) unit load at point of interest b.

PROBLEMS

In Probs. 10.1 to 10.10 use the principle of complementary virtual work (virtual forces) to analyze the trusses.

10.1 Calculate the deflections v_c, u_c, v_d, and u_d. Member cross-sectional areas in square millimeters are shown in parentheses; $E = 200$ GPa.

10.2 Member cross-sectional areas in square inches are shown in parentheses. Calculate the horizontal deflection at joint c:

    (a) Due to the load shown ($E = 29 \times 10^3$ kips/in^2).

    (b) If member ac is fabricated 1 in too long and the 20-kip load is removed.

    (c) If the temperature in member ac increases by 154 Fahrenheit degrees ($\alpha = 6.5 \times 10^{-6}$ in/in $\cdot$ °F) and the 20-kip load is removed.

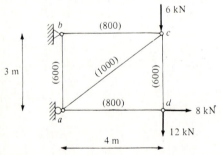

Figure P10.1

Figure P10.2

10.3 For all members, $A = 1$ in^2, $E = 29 \times 10^3$ kips/in^2. Calculate the horizontal deflection at joint d due to: (a) the illustrated forces; and (b) the fabrication errors; member bd -0.40 in; member ac $+0.50$ in (forces removed).

10.4 Member areas are shown in square millimeters in parentheses, $E = 200$ GPa. Calculate:

    (a) The horizontal deflection at joint b due to the illustrated force.

    (b) The horizontal deflection at joint d if member bd experiences a temperature increase of 50 Celsius degrees ($\alpha = 1.2 \times 10^{-5}$ mm/mm $\cdot$ °C); applied load removed.

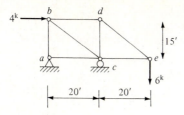

Figure P10.3

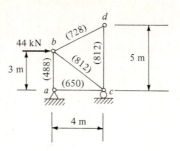

Figure P10.4

10.5 All member areas are shown in parentheses in square inches; $E = 29 \times 10^3$ kips/in². Calculate:

(*a*) The vertical deflection of joint *e* and the relative deflection of joints *d* and *g* due to an applied vertical force of 100 kips at joint *e*.

(*b*) The vertical deflection at joint *e* if there is a decrease in temperature of the bottom chord of 50 Fahrenheit degrees ($\alpha = 6.5 \times 10^{-6}$ in/in · °F); the 100-kip force is removed.

(*c*) The vertical deflection of joint *e* if members *bd* and *df* are fabricated 0.25 in too short; the 100-kip force is removed.

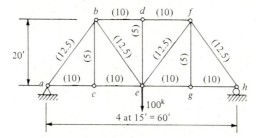

Figure P10.5

10.6 All member areas are shown in square inches in parentheses; $E = 29 \times 10^3$ kips/in². Calculate the rotation of member *ac* due to:

(*a*) The illustrated applied load.

(*b*) A temperature decrease for members *ac* and *ce* of 30 Fahrenheit degrees and 100 Fahrenheit degrees for member *ab* ($\alpha = 6.5 \times 10^{-6}$ in/in · °F); the applied force is removed.

(*c*) A fabrication error in members *ab* and *be* of +0.50 in; the applied force is removed.

10.7 Compute the horizontal deflection of joint *f* if (*a*) member *cd* is shortened 20 mm; (*b*) members *ab* and *ef* both experience a temperature increase equal to 50 Celsius degrees ($\alpha = 1.2 \times 10^{-5}$ mm/mm · °C).

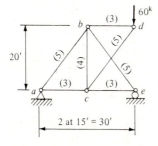

Figure P10.6

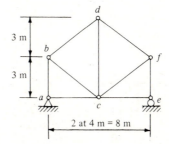

Figure P10.7

10.8 For all members $A = 1.00$ in^2, $E = 29 \times 10^3$ kips/in^2, and $\alpha = 6.5 \times 10^{-6}$ in/in · °F.

(a) Calculate the vertical deflection at joint h for the illustrated loading.

(b) Calculate the vertical deflection at joint f for the following temperature changes in Fahrenheit degrees:

$$\text{Verticals } abc \text{ and } jkl = +100$$

$$\text{Horizontal } cefhl = +100$$

$$\text{Diagonals } adf \text{ and } jgf = +120$$

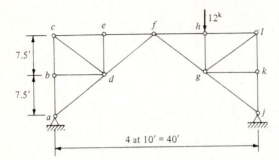

Figure P10.8

10.9 All member areas are shown in square inches in parentheses; $E = 29 \times 10^3$ kips/in^2, and $\alpha = 6.5 \times 10^{-6}$ in/in · °F. Calculate:

(a) The vertical deflection at joint b due to the illustrated loading.

(b) The vertical and horizontal deflection at joint b if member ab has a temperature change of $+110$ Fahrenheit degrees (the 3-kip load is removed).

(c) The vertical deflection at joint b if members ab and bd are fabricated 0.25 in too long (the 3-kip load is removed).

10.10 The areas of the members of the illustrated truss are shown in parentheses. Calculate the values of A_1 and A_2 if it is known that the two forces cause joints a and c to move apart 0.0345 in and joints b and d to move together 0.025 in. $E = 29 \times 10^3$ kips/in^2.

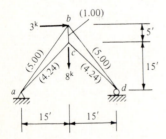

Figure P10.9

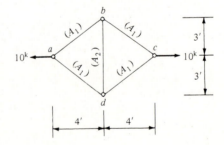

Figure P10.10

In Probs. 10.11 to 10.20 use the principle of complementary virtual work (virtual forces) to analyze the beams. Sketch the deflected beam.

10.11 Calculate Θ_a and Θ_b if $E = 29 \times 10^3$ kips/in^2 and $I = 1000$ in^4.

10.12 For the beam illustrated $I_{ab} = 100 \times 10^6$ mm^4, $I_{bc} = 300 \times 10^6$ mm^4, $I_{cd} = 200 \times 10^6$ mm^4, and $E = 200$ GPa. Calculate v_a:

(a) Due to the illustrated loading.

(b) If the upper and lower beam surfaces are 20°C and 0°C respectively, and the illustrated forces are removed ($\alpha = 1.2 \times 10^{-5}$ mm/mm · °C; $h = 250$ mm for all members).

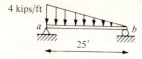

Figure P10.11

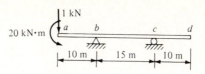

Figure P10.12

10.13 Calculate the deflection and rotation at point b; $EI = 10^7$ kip $\cdot$ in^2.

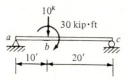

Figure P10.13

10.14 Solve Prob. 9.12.

10.15 Solve Prob. 9.14.

10.16 Calculate the magnitude of the uniformly distributed load q that will result in a deflection at point c of 1 in; $E = 29 \times 10^3$ kips/in^2; $I = 600$ in^4.

10.17 The cross section of the illustrated beam is square ($b \times b$). Calculate the dimension b if the maximum deflection for the beam is 10 mm; $E = 200$ GPa.

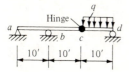

Figure P10.16

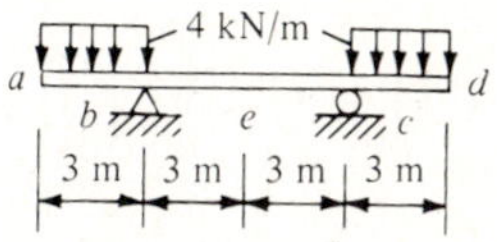

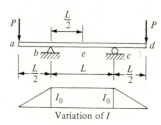

Figure P10.17

10.18 The moment of inertia varies as shown. Calculate the vertical deflection at points a and e.

Figure P10.18

10.19 Solve Prob. 9.27.

10.20 Solve Prob. 9.29.

In Probs. 10.21 to 10.24 use the principle of complementary virtual work (virtual forces) to analyze the structures. Sketch the deflected structure.

10.21 Calculate the vertical deflection at point c for this trussed beam. $E = 29 \times 10^3$ kips/in^2 for the cable and beam.

10.22 For all members $E = 29 \times 10^3$ kips/in^2. Members ab, bd, ac, bc, and cd have $A = 10$ in^2 and $I = 0$. Member df has $A = 40$ in^2 and $I = 800$ in^4. Calculate the vertical deflection at point e due to: (a) the applied load; and (b) the fact that member bc was 1 in too short when it was installed (load removed).

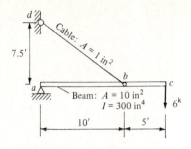

Figure P10.21

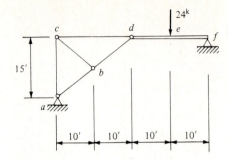

Figure P10.22

10.23 The illustrated three-hinge arch is made from timber ($E = 1.6 \times 10^6$ lb/in²) and has a uniform solid section with $I = 900$ in⁴ and $A = 60$ in². Calculate:

(a) The vertical deflection at point b due to an applied unit vertical force at point d (consider bending effects only).

(b) The vertical deflection at point b due to an applied unit vertical force at point b (consider bending effects only).

(c) The vertical deflection at point d due to an applied unit vertical force at point b (consider bending and axial effects).

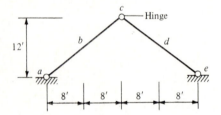

Figure P10.23

10.24 Compute the vertical deflection of point a for this rigid frame. Consider deformation due to bending only. $E = 29 \times 10^3$ kips/in².

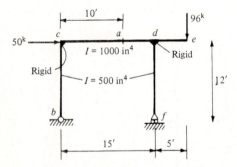

Figure P10.24

In Probs. 10.25 to 10.32 solve the original problems using Castigliano's theorem part II to analyze the trusses.

10.25 Problem 10.1

10.27 Part (a) of Prob. 10.3

10.29 Part (a) of Prob. 10.5

10.31 Part (a) of Prob. 10.8

10.26 Part (a) of Prob. 10.2

10.28 Part (a) of Prob. 10.4

10.30 Part (a) of Prob. 10.6

10.32 Part (a) of Prob. 10.9

In Probs. 10.33 to 10.42 solve the original problems, using Castigliano's theorem part II to analyze the structures. Sketch the deflected structure.

10.33 Problem 10.11   **10.34** Part (a) of Prob. 10.12
10.35 Problem 10.13   **10.36** Problem 10.16   **10.37** Problem 10.17
10.38 Problem 10.18   **10.39** Problem 10.19   **10.40** Problem 10.20
10.41 Problem 10.21   **10.42** Problem 10.24

10.43 Calculate v_{bb}, v_{bc}, v_{cb}, and v_{cc} using any method. Note that v_{ij} is the deflection at point i due to a unit load applied at point j. Using these results, calculate the forces that must be simultaneously applied at points b and c to produce the deflections $v_b = 0.50$ in and $v_c = 0.25$ in; $E = 29 \times 10^3$ kips/in^2 and $I = 250$ in^4.

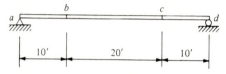

Figure P10.43

10.44 For the truss of Prob. 10.3 calculate (a) the vertical deflection at joint e due to a unit horizontal force at joint d and (b) the horizontal deflection at joint d due to a unit vertical force at joint e. How can you check the validity of your solution?

REFERENCES

1. Argyris, J. H., and S. Kelsey: *Energy Theorems and Structural Analysis*, Butterworth, London, 1960.

ADDITIONAL READING

Beaufait, F. W.: *Basic Concepts of Structural Analysis*, chap. 7, Prentice-Hall, Englewood Cliffs, N. J., 1977.
Fraser, D. J.: *Conceptual Design and Preliminary Analysis of Structures*, chap. 7, Pitman, Marshfield, Mass., 1981.
Gerstle, K. H.: *Basic Structural Analysis*, chap. 8, Prentice-Hall, Englewood Cliffs, N. J., 1974.
Ghali, A., and A. M. Neville: *Structural Analysis*, chaps. 5 and 8, Intext Educational, New York, 1972.
Gutkowski, R. M.: *Structures: Fundamental Theory and Behavior*, chap. 9, Van Nostrand Reinhold, New York, 1981.
Ketter, R. L., G. C. Lee, and S. P. Prawel, Jr.: *Structural Analysis and Design*, chap. 5, McGraw-Hill, New York, 1975.
Timoshenko, S. P., and D. H. Young: *Theory of Structures*, 2d ed., chaps. 5 and 6, McGraw-Hill, New York, 1965.
Willems, N., and W. M. Lucas, Jr.: *Structural Analysis for Engineers*, chap. 3, McGraw-Hill, New York, 1978.

PART
FOUR

ANALYSIS OF STATICALLY INDETERMINATE STRUCTURES

ELEVEN

CLASSICAL COMPATIBILITY METHODS

11.1 INTRODUCTION

A structure whose reactions and internal forces (shear, moment, axial, and/or torsional) cannot be completely established using the equations of static equilibrium is *statically indeterminate*. For example, the member forces in a two-dimensional truss cannot be obtained by statics alone if the sum of the external reactions (e.g., rectangular cartesian components) and the number of members exceeds the number of independent equations of static equilibrium. Beams, rigid frames, and other structures can also be indeterminate. Examples of indeterminate structures abound. A concrete bridge beam that is continuous over three or more supports, a Pratt truss with double diagonals, and most high-rise buildings represent this general type of construction. Figure 11.1 shows two statically indeterminate structures. This type of structure is widely accepted for various reasons:

1. Some types of construction, such as reinforced concrete, are easiest with continuous members.
2. Materials can be efficiently placed and utilized.
3. Structures display increased stiffness and the ability to redistribute stresses in the event of an overload.
4. Forces can often be resisted best with indeterminate construction, e.g., the lateral forces on a high-rise building.

Indeterminate construction also has a number of disadvantages. The analysis is generally complex and expensive because the internal forces depend upon the geometry of the structure, the elastic (or inelastic) material behavior, and the cross-sectional properties of the members. A preliminary design, complete with member sizes, etc., is required before any analysis can be attempted. The first analysis is followed by changes in the design, and subsequent cycles of analysis and design modification follow until a satisfactory structure is obtained. This process requires that the functions of analysis and design be closely intertwined. A second disadvantage of indeterminacy

Figure 11.1 (*a*) The White Bird Canyon Bridge (*American Institute of Steel Construction*); (*b*) the Jefferson National Gateway Arch, St. Louis (*American Institute of Steel Construction*).

stems from the fact that the internal stresses are affected by differential movement of the supports, nonuniform heating, fabrication errors, precambering, etc. Thus, for example, before constructing a bridge in a location with uncertain foundation conditions, careful consideration must be given to the consequences of using a continuous structure.

To analyze a linear elastic statically indeterminate structure the concepts of static equilibrium discussed in Part Two must be combined with the procedures described in Part Three for calculating displacements. In this chapter these two sets of principles are combined to give the *compatibility methods*. 1n this approach, compatibility conditions (deflection constraint conditions) are enforced throughout the structure by superposing a set of partial solutions, all of which satisfy equilibrium, force-deformation relations, and boundary conditions. This results in a set of equations with the forces as the unknowns and the flexibility quantities as the coefficients. In its general form it is termed either the *force* or the *flexibility method*. Three approaches suitable for hand calculations will be described in this chapter: (1) *the method of consistent displacements*, (2) *the theorem of least work*, and (3) *the three-moment equation*.

11.2 METHOD OF CONSISTENT DISPLACEMENTS FOR INDETERMINATE REACTIONS

The beam in Fig. 11.2*a* is statically indeterminate to the first degree since it has one reaction more than is necessary to constrain it from rigid-body translation and rotation. If the deflections are small, i.e., the geometry of the undeflected and deflected structure is the same, and the material is linearly elastic, the principle of superposition can be applied. Removing the support at point *b* while the actual loading is applied means that point *b* will deflect downward an amount equal to $5qL^4/384EI$ (Fig. 11.2*b*). By applying an upward force of sufficient magnitude, point *b* will return to its undeflected position. A unit concentrated force at point *b* gives a deflection of $L^3/48EI$; therefore, if an upward force R_b of unknown magnitude is applied at this point, it will cause a deflection equal to $R_b L^3/48EI$, as shown in Fig. 11.2*c*. Superposition of these two load cases with the condition that the displacement at point *b* equal zero gives

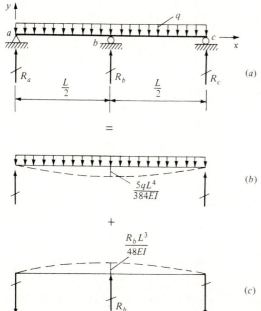

Figure 11.2 Superposition of solutions for a beam with one redundant reaction: (*a*) the beam with applied loading; (*b*) the primary structure with applied loading; (*c*) the primary structure with only the redundant force applied.

$$R_b \frac{L^3}{48EI} - \frac{5qL^4}{384EI} = 0 \tag{a}$$

Hence,
$$R_b = \frac{5qL}{8} \tag{b}$$

Thus, the reaction is obtained by enforcing compatibility (zero displacement) at point b. Since the two independent load cases both satisfy equilibrium, force-deflection relations, and boundary conditions, the final solution also satisfies these basic requirements. This method of *consistent displacements* (also known as the superposition-equation method and the Maxwell method) can also be used to investigate the reactions for various structures with multiple redundancies (see Sec. 11.5).

Basically, the method of consistent displacements involves invoking compatibility using various loading conditions for a statically determinate structure that is derived from the structure being investigated. It is possible to envision the method as consisting of five basic steps.

Step 1 Remove enough reaction forces to make the remaining structure, called the *primary structure,* statically determinate and stable. For the beam shown in Fig. 11.3 any one of the three reaction components R_a, M_a, or R_b can be removed. Although the truss in Fig. 11.4 is indeterminate to the first degree, it is not possible to remove one of the reactions arbitrarily to obtain a proper primary structure. If either of the horizontal reactions is removed, the structure will be stable, but removal of either of the vertical reactions results in an unstable structure. That is, the primary structure shown in Fig. 11.4c will rotate about point f for most applied loadings (the exception being horizontally applied forces along the bottom chord).

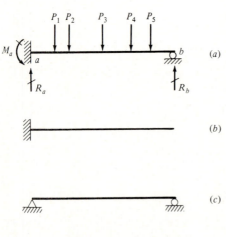

Figure 11.3 Beam with one redundant reaction showing all possible primary structures: (a) the loaded beam; (b) R_b removed; (c) M_a removed; (d) R_a removed.

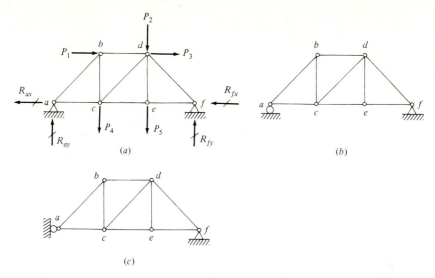

Figure 11.4 (a) Truss with one redundant reaction; (b) stable primary structure; (c) unstable primary structure.

Step 2 Calculate the displacement caused by the actual loading on the primary structure at the location and along the line of action of the reaction component that was removed. For example, for the beam in Fig. 11.5b this is the displacement designated as Δ_{b0}. This notation will be used in the subsequent discussion, the first subscript denoting the point of the displacement and the subscript 0 implying that it is the displacement of the primary structure caused by the actual applied loads.

Step 3 Apply the redundant reaction with all other loads removed. Usually it is prudent to visualize the force acting in the positive coordinate direction (see, for example, Fig. 11.5c). The displacement Δ_{bb} is the deflection at b due to the force R_b (acting at point b). For convenience, this deflection can be calculated by applying a unit force at point b, giving δ_{bb}. For linear elastic behavior

$$\Delta_{bb} = \delta_{bb} R_b \tag{11.1}$$

The quantity δ_{bb} (sometimes referred to as the *flexibility coefficient*) has the dimensions of length per unit force and represents the displacement in the positive y direction of point b due to a unit force in the positive y direction applied at point b.

Step 4 Enforce the deflection constraint equation (compatibility). The physical deflection constraint condition of the beam (see Fig. 11.5) is that the deflection Δ_b of point b be zero; hence,

$$\Delta_b = \Delta_{b0} + \delta_{bb} R_b = 0 \tag{11.2}$$

Solving for the reaction force gives

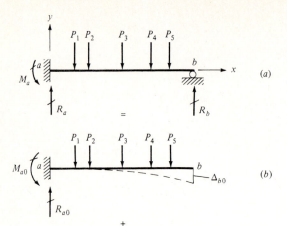

Figure 11.5 Analysis of a beam with one redundant reaction: (*a*) actual structure with applied loading; (*b*) primary structure acted upon by only the applied loading to give Δ_{b0}; (*c*) primary structure acted upon by only the redundant reaction to give $\Delta_{bb} = R_b\delta_{bb}$.

$$R_b = -\frac{\Delta_{b0}}{\delta_{bb}} \tag{11.3}$$

The deflections Δ_{b0} and δ_{bb} can both be determined using the methods developed in Chaps. 9 and 10. The signs of the deflections are to be interpreted with respect to the y axis, and the sign of R_b can then be interpreted in the same sense. For the beam in Fig. 11.5 the illustrated deflection shapes imply that Δ_{b0} is negative while δ_{bb} is positive; thus, Eq. (11.3) yields a positive sign for R_b.

Step 5 Calculate all the unknowns for the actual structure, e.g., reactions, axial forces, shears, and moments. After the value of the redundant reaction has been found, the other reaction components can be obtained using the equations of statics. Alternatively, the superposition of the forces obtained in solving for the redundant can be used. For example, in Fig. 11.5 the reaction forces at point a can be calculated as

$$M_a = M_{a0} + M_{ab} \tag{11.4}$$

$$R_a = R_{a0} + R_{ab} \tag{11.5}$$

where M_{a0} and R_{a0} are the moment and reaction, respectively, in the primary structure and M_{ab} and R_{ab} are the moment and reaction, respectively, at point a if only R_b is applied to the primary structure. The method of consistent displacements for finding reactions is illustrated for a beam, truss, and rigid frame in Examples 11.1 to 11.3.

Example 11.1 Calculate the reactions for the illustrated beam using the method of consistent displacements; $I = 200 \times 10^{-6}$ m^4, and $E = 200$ GPa.

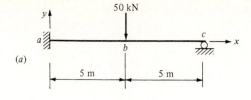

(a)

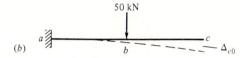

(b)

(c)

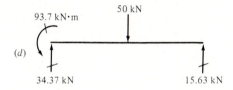

(d)

Figure E11.1 (a) Loaded beam; (b) primary structure with applied load; (c) primary structure with applied unit load; (d) total reactions.

SOLUTION From Fig. E11.1b

$$\Delta_{c0} = -\frac{5(50)(5^3)}{6EI} \quad (\downarrow; \text{ as shown})$$

From Fig. E11.1c

$$\delta_{cc} = \frac{1 \times 10^3}{3EI} \quad (\uparrow; \text{ as shown}) \qquad \Delta_{cc} = R_c\delta_{cc}$$

$$\Delta_c = \Delta_{cc} + \Delta_{c0}$$

$$0 = \frac{10^3}{3EI}R_c - \frac{5(50)(5^3)}{6EI}$$

$$R_c = 15.63 \text{ kN} \ (\uparrow)$$

DISCUSSION A primary structure could have been obtained for this beam by removing any one of the reactions M_a, R_a, or R_c. Deflections are considered positive if they correspond to a positive coordinate direction, and for consistency the unit load at the redundant has been applied in the positive direction. This yields a solution with R_c in the positive upward direction, as would be anticipated in this case. The final reactions can be obtained by applying R_c on the beam and using the equations of equilibrium. Alternatively, the reactions for the loaded primary structure and the beam loaded with only R_c can be combined

$$M_a = 50(5) - 10R_c = 93.7 \text{ kN} \cdot \text{m} \qquad R_a = 50 - R_c = 34.37 \text{ kN}$$

Example 11.2 Calculate the reactions for this truss using the method of consistent displacements; AE = const for all members.

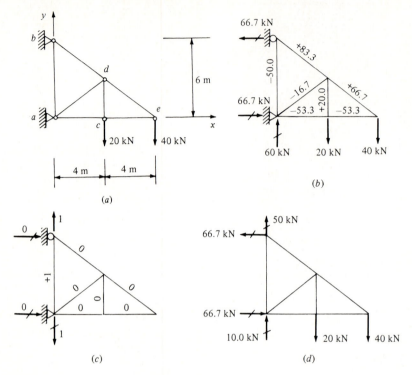

Figure E11.2 (*a*) Loaded truss; (*b*) primary structure with applied loads (F_0 forces); (*c*) primary structure with applied unit load (F_1 forces); (*d*) total reactions.

SOLUTION From Fig. E11.2*b* and *c* we have respectively the F_0 and F_1 forces

$$(\Delta_{b0})_y = \sum \frac{F_0 F_1 L}{AE} = \frac{6}{AE}(-50)(+1) = -\frac{300}{AE}\ (\downarrow)$$

$$(\delta_{bb})_y = \sum \frac{F_1^2 L}{AE} = \frac{6}{AE}\ (\uparrow)$$

$$(\Delta_b)_y = 0 = (\Delta_{b0})_y + (\Delta_{bb})_y = -\frac{300}{AE} + R_{by}\frac{6}{AE}$$

$$R_{by} = 50\ \text{kN}\ (\uparrow)$$

DISCUSSION A primary structure for this truss can be obtained by removing one of the vertical reactions, but an unstable structure results if either of the horizontal reaction components is removed. The total reactions for the truss can be calculated by combining those obtained for the two individual loadings on the primary structure, i.e., the actual loads and also R_{by} only.

Example 11.3 Use the method of consistent displacements to calculate the reactions for this rigid frame; $E = $ const.

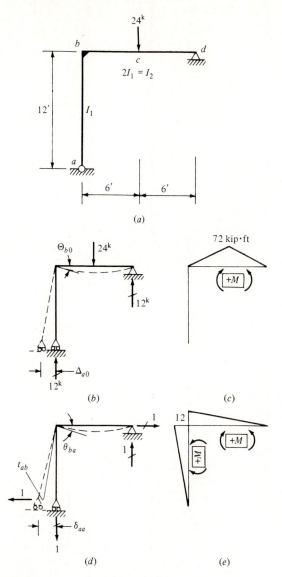

(a)

(b)

(c)

(d)

(e)

Figure E11.3 (a) Loaded rigid frame; (b) primary structure with applied load; (c) M_0 diagram; (d) primary structure with applied unit load; (e) M_{X_1} diagram.

SOLUTION From Fig. E11.3b and c

$$EI_2\Theta_{b0} = \tfrac{1}{2}(72)(6) = 216 \text{ kip} \cdot \text{ft}^2$$

$$\Delta_{a0} = 12\Theta_{b0} = \frac{1296}{EI_1} \ (\leftarrow)$$

From Fig. E11.3d and e

$$EI_2\theta_{ba} = \tfrac{1}{2}(12)(12)(\tfrac{2}{3}) = 48 \text{ ft}^2 \qquad \theta_{ba} = \frac{24}{EI_1}(\curvearrowright)$$

$$\delta_{aa} = 12\theta_{ba} + t_{ab} = \frac{288}{EI_1} + \frac{576}{EI_1} = \frac{864}{EI_1}(\leftarrow)$$

$$\Delta_a = 0 = \Delta_{a0} + X_1\delta_{aa}$$

$$X_1 = -\frac{\Delta_{a0}}{\delta_{aa}} = -\frac{1296}{EI_1}\frac{EI_1}{864} = -1.50 \text{ kips }(\rightarrow)$$

By superposition

$$R_{ax} = 1.50 \text{ kips }(\leftrightarrow) \qquad R_{ay} = 13.50 \text{ kips }(\uparrow)$$

$$R_{dx} = 1.50 \text{ kips }(\leftrightarrow) \qquad R_{dy} = 10.50 \text{ kips }(\uparrow)$$

DISCUSSION The right angle at point b will remain unchanged after the loads are applied; thus, both the beam bd and the column ab will have the same angle of rotation at point b. Since there is no bending moment in the column when the primary structure is subjected to the actual loading, the column will remain straight and from geometry $\Delta_{a0} = 12\Theta_{b0}$. The column will have a curvature when the unit load is applied; therefore both θ_{ba} and t_{ab} must be obtained in order to calculate δ_{aa}. In this case, t_{ab} is the tangential deviation of the elastic curve at point a relative to the tangent to the elastic curve drawn at point b (see Fig. E11.3d). After the horizontal reaction X_1 at point a has been calculated using the statement of consistent displacements, the other reactions are obtained by superposing the reactions from the two individual load cases on the primary structure that were investigated previously in the analysis.

11.3 METHOD OF CONSISTENT DISPLACEMENTS FOR INDETERMINATE INTERNAL FORCES

Although the reactions for the linear elastic truss in Fig. 11.6 can be obtained using the equations of static equilibrium, statics will not yield the forces in the members of the structure since there is one internal redundancy. The method of consistent displacements can also be used to solve this problem. It is necessary to cut one of the members to obtain a primary structure that is statically determinate and stable. If any one of the four members ab, ac, df, or ef in the outside panels is cut, it will be impossible for the truss to transmit the loading to the reactions; however, if any one of the other six members is cut, the structure will be statically determinate and stable. When an imaginary cut is passed through a weightless member, it will remain in its original position in the truss, but after being cut it is incapable of sustaining any loading.

If we select be as the redundant member, the cut ends will deflect an amount Δ_{10} (the subscript 1 designates the deflection of the redundant, and the 0 indicates that it is associated with the primary structure for the actual loads). Since the primary structure is statically determinate, this deflection can be computed using one of the standard deflection methods discussed in Chap. 10. The solution for the actual force in member be ($F_{be} = X_1$) must be superimposed on the solution for the primary structure. Therefore, a unit force is applied to the redundant member, and the forces in all

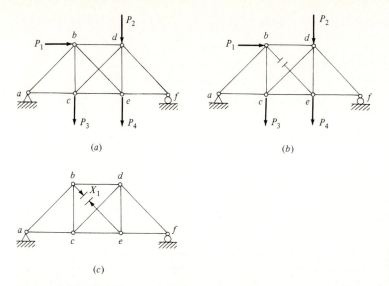

Figure 11.6 Analysis of a truss with one internal redundancy: (*a*) actual structure with applied loading; (*b*) primary structure acted upon by only the applied loading to give Δ_{10}; (*c*) primary structure acted upon by only the redundant-member force to give $\Delta_1 = X_1\delta_{11}$.

members are computed. This unit force can be envisaged as being applied in various ways such as tightening a turnbuckle that has been installed in the member. Note that with the unit applied internal load, the reactions (and many of the member forces) are zero. The unit load in the redundant member results in a deflection δ_{11} across the cut. The subscripts indicate the deflection at the redundant member 1 due to a unit load applied at the redundant member 1. The deflection Δ_{11} due to the actual value of the redundant member will be

$$\Delta_{11} = X_1\delta_{11} \tag{11.6}$$

The total deflection Δ_1 across the cut in member *be* is

$$\Delta_1 = \Delta_{10} + X_1\delta_{11} \tag{11.7}$$

Since the cut must be closed ($\Delta_1 = 0$), this gives

$$X_1 = -\frac{\Delta_{10}}{\delta_{11}} \tag{11.8}$$

The member forces for the truss can now be obtained using the equations of static equilibrium; i.e., the method of joints and/or the method of sections can be applied. Alternatively, the member forces for the primary structure can be superposed on those resulting from the application of only the redundant-member force. This gives

$$F = F_0 + X_1 F_1 \tag{11.9}$$

where F = force in member for actual structure
F_0 = member force in primary structure with actual load
F_1 = member force in primary structure with unit applied redundant force

Example calculations illustrating consistent displacements for internal redundancies are presented in Examples 11.4 to 11.6.

Example 11.4 Use the method of consistent displacements to calculate the member forces for this truss with one internal redundancy; $E = $ const, and the member areas are

$$A_{bc} = A \qquad A_{cd} = \sqrt{2}A \qquad A_{ac} = \tfrac{5}{3}A$$

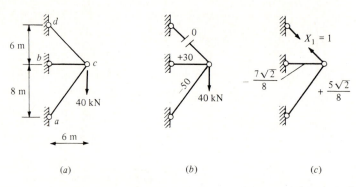

(a)   (b)   (c)

Figure E11.4 (a) Loaded truss; (b) primary structure with applied load (F_0 forces); (c) primary structure with applied unit redundant (F_1 forces).

SOLUTION From Fig. E11.4b and c

$$E \Delta_{10} = \sum \frac{F_0 F_1 L}{A} = \frac{0(1)(6\sqrt{2})}{\sqrt{2}A} + \frac{30(-7\sqrt{2}/8)(6)}{A} + \frac{-50(5\sqrt{2}/8)(10)}{\tfrac{5}{3}A}$$

$$EA\Delta_{10} = 0 - 157.5\sqrt{2} - 187.5\sqrt{2} = -345.0\sqrt{2}$$

$$E\delta_{11} = \sum \frac{F_1^2 L}{A} = \frac{1(6\sqrt{2})}{\sqrt{2}A} + \frac{\tfrac{49}{32}(6)}{A} + \frac{\tfrac{25}{32}(10)}{\tfrac{5}{3}A}$$

$$EA\delta_{11} = 6 + \tfrac{147}{16} + \tfrac{75}{16} = \tfrac{318}{16}$$

$$\Delta_1 = 0 = \Delta_{10} + X_1 \delta_{11} = \frac{-345.0\sqrt{2}}{A} + X_1 \frac{318}{16A}$$

therefore, $X_1 = 24.54$ kN. Using superposition, we find the final member forces

$$F_{cd} = +24.54 \text{ kN (t)} \qquad F_{bc} = -0.37 \text{ kN (c)} \qquad F_{ac} = -28.31 \text{ kN (c)}$$

DISCUSSION This truss can be analyzed by considering any one of the members as the redundant one. For this simple truss the calculations are sufficient in the form shown in the example, but the investigation can also be carried out using a tabular array (Table E11.4). In these calculations it was recognized that L/AE is the same for the three members; thus,

$$\Delta_1 = 0 = \frac{L}{AE} \sum F_0 F_1 + X_1 \frac{L}{AE} \sum F_1^2$$

$$0 = \sum F_0 F_1 + X_1 \sum F_1^2$$

The units for each quantity must be considered with care; they are shown in the heading for each column. For example, since the member forces F_1 are produced by a unit applied force, they are dimensionless; i.e., they have units of kilonewtons per kilonewton. This tabular array is convenient for the calculation of the final member forces by superposition in the column labeled $F_0 + X_1 F_1$.

Table E11.4

Member	F_0, kN	F_1	$F_0 F_1$, kN	F_1^2	$F_0 + X_1 F_1$, kN
ac	-50	$+\dfrac{5\sqrt{2}}{8}$	$-\dfrac{125\sqrt{2}}{4}$	$\dfrac{50}{64}$	-28.31
bc	$+30$	$-\dfrac{7\sqrt{2}}{8}$	$-\dfrac{105\sqrt{2}}{4}$	$\dfrac{98}{64}$	-0.37
dc	0	$+1$	0	1	$+24.54$
		$\Sigma = -\dfrac{230\sqrt{2}}{4}$		$\dfrac{212}{64}$	

$$0 = -\frac{230\sqrt{2}}{4} + X_1 \frac{212}{64} \qquad X_1 = 24.54 \text{ kN}$$

Example E11.5 Calculate the member forces for this truss using the method of consistent displacements; $E = $ const, and the member areas are 0.60 in^2 for verticals, 0.80 in^2 for horizontals, and 1.00 in^2 for diagonals.

SOLUTION Note that for all members $L/A = 20$ ft/in^2; see Table E11.5.

Table E11.5

Member	F_0, kips	F_1	$F_0 F_1$, kips	F_1^2	Member forces $F_1 X_1$, kips	Member forces $F_0 + F_1 X_1$, kips
bd	0	-0.8	0.0	0.64	$+16.0$	$+16.0$
ac	-32	-0.8	$+25.6$	0.64	$+16.0$	-16.0
ab	-24	-0.6	$+14.4$	0.36	$+12.0$	-12.0
cd	0	-0.6	0.0	0.36	$+12.0$	$+12.0$
bc	$+40$	$+1.0$	$+40.0$	1.00	-20.0	$+20.0$
ad	0	$+1.0$	0.0	1.00	-20.0	-20.0
			$\Sigma = 80.0$	4.00		

$$\Delta_{10} = \frac{L}{AE}80.0 \qquad \delta_{11} = \frac{L}{AE}4.00$$

$$\Delta_1 = 0 = \Delta_{10} + X_1 \delta_{11} \qquad X_1 = -\frac{80.0}{4.00} = -20.0 \text{ kips (c)}$$

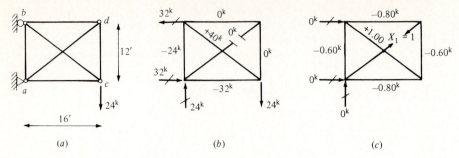

Figure E11.5 (*a*) Loaded truss; (*b*) primary structure with applied load (F_0 forces); (*c*) primary structure with applied unit redundant (F_1 forces).

Example 11.6 Calculate the reactions and the moment diagram for the beam shown using the method of consistent displacements; EI = const.

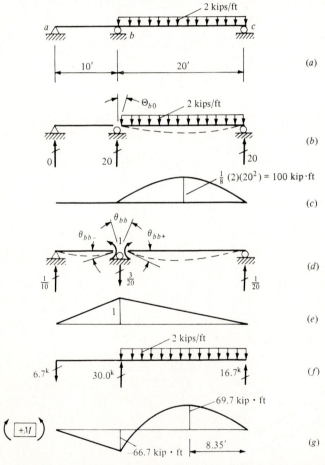

Figure E11.6 (*a*) Loaded beam; (*b*) primary structure with applied load; (*c*) M_0 diagram; (*d*) primary structure with applied unit redundant moment; (*e*) M_{x_1} diagram; (*f*) total reactions; (*g*) final moment diagram.

SOLUTION

$$EI\Theta_{b0} = \tfrac{1}{3}(100)(20) = \tfrac{2000}{3} \text{ kip} \cdot \text{ft}^2$$

$$EI\theta_{bb} = \tfrac{2}{3}(\tfrac{1}{2})(1)(20) + \tfrac{2}{3}(\tfrac{1}{2})(1)(10) = \tfrac{20}{3} + \tfrac{10}{3} = 10 \text{ ft}$$

$$\Theta_b = 0 = \Theta_{b0} + X_1\theta_{bb}$$

$$X_1 = -\frac{2000}{3EI}\frac{EI}{10} = -66.7 \text{ kip} \cdot \text{ft}$$

DISCUSSION Any one of the three reactions could have been chosen as redundant for this beam, but in this case the primary structure was obtained by removing the rotational continuity from the beam at point b. This can be accomplished physically by making an imaginary cut through the beam at this point. If the bending moment at point b is calculated, the reactions can subsequently be calculated. The compatibility condition to be satisfied for this primary structure is that the slope of the beam be continuous at point b. Θ_{b0} is the change in slope of the beam between two points located just to the left and to the right of point b, and it is obtained by using the M_0 diagram (Fig. E11.6c). Next, a unit positive bending moment is applied at point b, resulting in a slope θ_{bb-} (to the left of b) and another θ_{bb+} (to the right of b). The total change in slope at point b due to this unit applied moment is

$$\theta_{bb} = \theta_{bb-} + \theta_{bb+}$$

Note that the moments resulting from the unit internal moment M_{X_1} at b (Fig. E11.6e) are dimensionless. That is, if the applied moment has units of kilonewton-meters, the resulting moments will be described in terms of kN $\cdot$ m/kN $\cdot$ m and θ_{bb} has units of radians per kilonewton-meter.

The statement of compatibility ensuring that there will be no discontinuity in the slope at point b is

$$\Theta_b = 0 = \Theta_{b0} + X_1\theta_{bb}$$

where X_1 is the bending moment at point b. The total reactions and the final bending moments can be calculated by superposing the two individual load cases applied to the primary structure.

11.4 METHOD OF CONSISTENT DISPLACEMENTS FOR SUPPORT SETTLEMENT, TEMPERATURE, ETC.

The method of consistent displacements can also be used to analyze indeterminate structures that are subjected to displacements induced by support movement, temperature changes, fabrication errors, etc. If a structure displays small deflections and is made from linear elastic materials, each of these induced displacements can be investigated separately and all the resulting deflections, stresses, and reactions superposed to give the final results. Therefore, the final deflection of a single redundant (support or member) in a structure is given by

$$\Delta_i = \Delta_{iS} + \Delta_{iT} + \Delta_{iE} + \Delta_{i0} + X_i\delta_{ii} \tag{11.10}$$

where Δ_i = total deflection of ith redundancy in actual structure
Δ_{iS} = deflection of ith redundancy in primary structure due to support movement
Δ_{iT} = deflection of ith redundancy in primary structure due to temperature change
Δ_{iE} = deflection of ith redundancy in primary structure due to fabrication error
Δ_{i0} = deflection of ith redundancy in primary structure due to applied forces
X_i = redundant force
δ_{ii} = deflection in direction of redundant force due to unit force at the redundant support or member

Generally, the value of Δ_i is zero unless the redundancy has experienced a net deflection. For an internal truss member Δ_i is the deflection of the imaginary cut of the member, and, as such, it must be zero for the final configuration. However, a structure may experience settlement at the supports. In this case Δ_i is the total settlement for the actual structure, and Δ_{iS} is the net settlement for the primary structure. The calculations for structures with induced displacements are shown in Examples 11.7 and 11.8.

Example 11.7 The beam in Fig. E11.7 with one redundant reaction experiences support settlements of 14 mm at a, 22 mm at b, and 5 mm at c. Calculate the reactions of the beam due to this effect; $I = 300 \times 10^{-6}$ m^4, and $E = 200$ GPa.

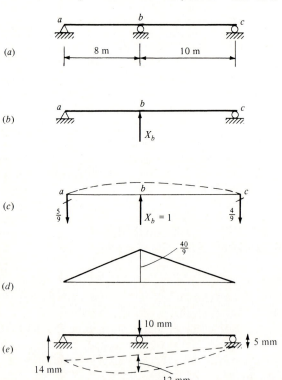

Figure E11.7 (a) The beam; (b) primary structure; (c) primary structure with applied unit redundant; (d) M_{X_b} diagram; (e) settlements.

SOLUTION From the moment-area theorems and Fig. E11.7c and d

$$\theta_{cb} = 3.21 \times 10^{-4} \text{ rad}$$

$$\delta_{bb} = 10\theta_{cb} - t_{bc} = 0.00321 - \frac{1}{2}(10)\frac{40}{9}\frac{10}{3}\frac{1}{6 \times 10^4} = 0.00198 \text{ m}$$

Since

$$\Delta_b = \Delta_{bS} + X_b\delta_{bb}$$

$$-0.022 = -0.010 + 0.00198X_b$$

We have

$$X_b = -\frac{0.012}{0.00198} = -6.08 \text{ kN}(\updownarrow)$$

Hence,

$$R_a = \tfrac{5}{9}(6.08) = 3.38 \text{ kN }(\uparrow) \qquad R_b = 6.08 \text{ kN }(\updownarrow)$$

$$R_c = \tfrac{4}{9}(6.08) = 2.70 \text{ kN }(\uparrow)$$

DISCUSSION The deflection δ_{bb} is computed here using the moment-area method, but any applicable method from Chaps. 9 and 10 can be used. In this case the statement of consistent displacements for point b must include the settlement. Thus, the net settlement for the primary structure Δ_{bS} is 10 mm, as shown in Fig. E11.7e, and the total displacement Δ_b is given as 22 mm. After the value of the redundant force has been calculated, the other reactions are obtained directly from the unit-load solution.

Example 11.8 Members bd and cd of the truss of Example 11.5 experience a rise in temperature of 154 Fahrenheit degrees. Calculate the resulting member forces if the 24-kip load is removed; $\alpha = 6.5 \times 10^{-6}$ in/in $\cdot$ °F; $E = 29 \times 10^3$ kips/in^2.

SOLUTION See Table E11.8.

Table E11.8

Member	L, ft	A, in^2	$\frac{L}{A}$, ft/in^2	F_1	$F_1^2\frac{L}{A}$, ft/in^2	ΔT, °F	$F_1\alpha \Delta T L$, ft	Member forces, kips
bd	16	0.8	20	−0.8	12.8	154	−0.0128	−5.8
ac	16	0.8	20	−0.8	12.8	0	0	−5.8
ab	12	0.6	20	−0.6	7.2	0	0	−4.4
cd	12	0.6	20	−0.6	7.2	154	−0.0072	−4.4
bc	20	1.0	20	+1.0	20.0	0	0	+7.2
ad	20	1.0	20	+1.0	20.0	0	0	+7.2
					$\Sigma = 80.0$		−0.0200	

Δ_{1T}:
$$F_Q = F_1 \qquad \Delta_{1T}(\nearrow) = \sum F_1\alpha \Delta T L = -0.0200 \text{ ft}$$

δ_{11}:
$$F_Q = F_P = F_1 \qquad \delta_{11}(\nearrow) = \frac{1}{E}\sum F_1^2\frac{L}{A} = \frac{80}{29 \times 10^3} = +0.002759 \text{ ft/kip}$$

$$\Delta_1 = 0 = \Delta_{1T} + \delta_{11}X_1$$

$$0 = -0.0200 + 0.002759 X_1 \qquad X_1 = +7.25 \text{ kips (t)}$$

DISCUSSION Member ad is considered the redundant one for this investigation; therefore, the F_1 forces are obtained from Example 11.5. The calculations indicate that the cut in member ad opens an amount equal to 0.02 ft under the effect of the heating and closes by 0.00276 ft due to the unit applied load on the member. The condition of consistent displacements requires that the cut be closed, and this gives the value of the force in the redundant member. The final member forces are obtained by the multiplication X_1F_1. Note that heating this indeterminate truss induces nonzero member forces throughout the structure.

11.5 METHOD OF CONSISTENT DISPLACEMENTS FOR MULTIPLE INDETERMINACY

The application of consistent displacements to structures having more than one redundancy results in a set of simultaneous algebraic equations that can be solved to obtain the redundant forces. The equations are generated by applying compatibility to the redundants in much the same manner as that used for single redundancies. The number of equations and the number of unknowns equal the number of redundancies involved. For example, the beam shown in Fig. 11.7a has two redundant reactions; therefore, in order to obtain a stable statically determinate structure it is necessary to remove two of the reactions to define the primary structure. The selection of the redundant reactions is arbitrary in this case since a stable statically determinate structure results by removing any two of the reactions. For example, if the reactions at points b and c are removed, the simply supported beam shown in Fig. 11.7b is the primary structure. Application of the actual forces causes the beam to deflect as shown, (deflections at points b and c are denoted Δ_{b0} and Δ_{c0}, respectively). Applying the

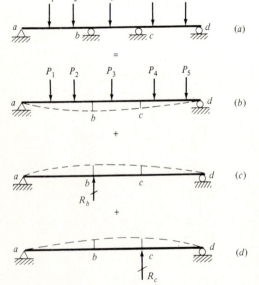

(a)

(b)

(c)

(d)

Figure 11.7 Analysis of a beam with two redundant reactions: (a) actual structure with applied loading; (b) primary structure acted upon by only the applied loading to give Δ_{b0} and Δ_{c0}; (c) primary structure acted upon by only the redundant R_b to give Δ_{bb} and Δ_{cb}; (d) primary structure acted upon by only the redundant R_c to give Δ_{bc} and Δ_{cc}.

force R_b gives the deflections Δ_{bb} and Δ_{cb} (the first subscript denotes the location where the deflection occurs, and the second subscript refers to the point where the force is applied). Since the magnitude R_b is unknown, it is convenient to apply a unit force at this point which results in the deflections δ_{bb} and δ_{cb}. If the structure behaves in a linear elastic fashion, we have

$$\Delta_{bb} = R_b\delta_{bb} \tag{11.11}$$

and
$$\Delta_{cb} = R_b\delta_{cb} \tag{11.12}$$

Similarly, for a unit force applied at point c the deflections at points b and c are denoted δ_{bc} and δ_{cc}, respectively. Thus,

$$\Delta_{bc} = R_c\delta_{bc} \tag{11.13}$$

and
$$\Delta_{cc} = R_c\delta_{cc} \tag{11.14}$$

where Δ_{bc} and Δ_{cc} are the deflections at points b and c, respectively, due to the force R_c. The total deflection Δ_b at point b and Δ_c at point c are obtained by superposing the deflections from the three individual load cases

$$\Delta_b = \Delta_{b0} + R_b\delta_{bb} + R_c\delta_{bc} \tag{11.15}$$

$$\Delta_c = \Delta_{c0} + R_b\delta_{cb} + R_c\delta_{cc} \tag{11.16}$$

The values for the redundant reactions are obtained by setting Eqs. (11.15) and (11.16) to zero. The reactions, bending moments, and shears can now be obtained either by applying the equations of static equilibrium or by superposing the values obtained for the three individual load cases shown in Fig. 11.7b to d.

Thus, the method of consistent displacements can be applied to any problem with multiple redundancies by the same five basic steps described in Sec. 11.2:

Step 1 Remove the redundancies, leaving a stable and statically determinate primary structure.

Step 2 Load the primary structure with the actual applied forces and/or the induced strains and calculate the deflections at the points and in the directions of the redundants.

Step 3 Successively apply the individual redundant forces to the primary structure and calculate the displacements at all the redundants.

Step 4 Enforce compatibility (deflection constraint conditions) at each redundant to obtain a set of simultaneous algebraic equations.

Step 5 Solve for the redundant forces and obtain the remaining reaction forces, axial forces, moments, shears, etc., either by applying the equations of static equilibrium or by superposing the results from the individual load cases used previously.

For multiple redundancies the compatibility conditions (step 4) for problems involving loads and induced strains have the general form

$$\Delta_1 = \Delta_{1S} + \Delta_{1T} + \Delta_{1E} + \Delta_{10} + X_1\delta_{11} + X_2\delta_{12} + \cdots + X_i\delta_{1i} + \cdots + X_n\delta_{1n}$$

$$\Delta_2 = \Delta_{2S} + \Delta_{2T} + \Delta_{2E} + \Delta_{20} + X_1\delta_{21} + X_2\delta_{22} + \cdots + X_i\delta_{2i} + \cdots + X_n\delta_{2n}$$

$$\cdots\cdots\cdots\cdots\cdots\cdots\cdots\cdots\cdots\cdots\cdots\cdots\cdots\cdots\cdots\cdots\cdots\cdots\cdots\cdots\cdots$$ (11.17)

$$\Delta_i = \Delta_{iS} + \Delta_{iT} + \Delta_{iE} + \Delta_{i0} + X_1\delta_{i1} + X_2\delta_{i2} + \cdots + X_i\delta_{ii} + \cdots + X_n\delta_{in}$$

$$\cdots\cdots\cdots\cdots\cdots\cdots\cdots\cdots\cdots\cdots\cdots\cdots\cdots\cdots\cdots\cdots\cdots\cdots\cdots\cdots\cdots$$

$$\Delta_n = \Delta_{nS} + \Delta_{nT} + \Delta_{nE} + \Delta_{n0} + X_1\delta_{n1} + X_2\delta_{n2} + \cdots + X_i\delta_{ni} + \cdots + X_n\delta_{nn}$$

where Δ_i = total deflection of the ith redundancy in actual structure

Δ_{iS} = deflection of ith redundancy due to support movement of primary structure

Δ_{iT} = deflection of ith redundancy in primary structure due to temperature change

Δ_{iE} = deflection of ith redundancy in primary structure due to fabrication error

Δ_{i0} = deflection of ith redundancy in primary structure due to actual applied forces

δ_{ij} = deflection of ith redundancy in primary structure due to unit value of jth redundant force ($X_j = 1$)

In general, $\delta_{ij} = \delta_{ji}$ as a consequence of the Betti-Maxwell reciprocal theorem. Not all these coefficients need be calculated explicitly, but the reader is urged initially to evaluate both δ_{ij} and δ_{ji} as a computational check.

In Eq. (11.17) the redundant forces X_i may be either reaction forces or internal member forces. For example, if a truss has one internal redundancy and is so supported that there is one redundant reaction, the total structure is indeterminate to the second degree. One of the internal members and one of the support reaction components can be removed to obtain a stable and statically determinate primary structure. In this case there are two deflection constraint conditions, but one is associated with the reaction and the other is written to ensure that the cut in the redundant member will be closed. Although the mixture of internal and external redundancies appears physically different from the situation in problems discussed previously, it is solved in exactly the same way as structures with exclusively external or internal redundancies.

For structures with multiple redundancies the selection of the primary structure can be extremely important. Any proper primary structure will yield the correct solution, but the amount of calculation effort can be minimized appreciably if the redundancies are judiciously chosen. It is usually prudent to take advantage of any symmetry of the loading and/or the structure. For example, the truss with one redundant reaction shown in Fig. 11.8a is symmetrical about the centerline; therefore, if the center support is selected as the redundant, the member forces for the unit-load case will be obtained much more conveniently than if one of the end supports were removed to obtain the primary structure. Note that for this structure the member force distribution for the actual applied forces will not be symmetrical. The symmetrical truss with a symmetrical loading in Fig. 11.8b has two internal redundancies. If one of the symmetric pairs of diagonal members (be and ef or cd and dg) is cut, a symmetrical primary structure is obtained. In this case, the member forces for the actual loading will be

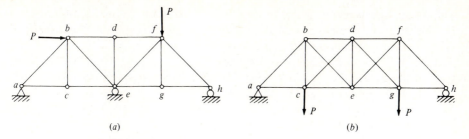

Figure 11.8 (*a*) Truss with a redundant reaction; (*b*) truss with two redundant members.

symmetrical, and the two individual unit load cases for the redundants will be mirror images of each other.

It is also important to consider the complexity of the stress distribution after the redundancies have been removed; i.e., if possible, the effects of the loadings should be localized to a small region of the structure. For example, the beam shown in Fig. 11.9*a*

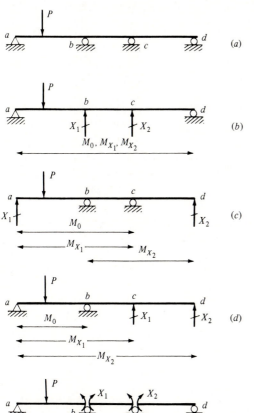

Figure 11.9 Alternate primary structures with location of moments: (*a*) actual structure with applied loading; (*b*) primary structure 1; (*c*) primary structure 2; (*d*) primary structure 3; (*e*) primary structure 4.

with two redundancies has a single concentrated load applied to the structure. If the reactions at points b and c are selected as the redundants, the load P will have an effect on the moments throughout the length of the beam (shown as M_0 in Fig. 11.9b). In addition, both the redundants will propagate throughout the beam when they are successively applied (M_{X1} and M_{X2} in Fig. 11.9b). Primary structure alternatives 2 and 3 are shown in Fig. 11.9c and d, respectively, along with the regions of the structure that will be affected by the application of the three load cases. In these two cases since the effect of each of the three forces has been somewhat localized, these two primary structures are superior to primary structure 1. It is also possible to cut the beam at points b and c in such a way that no continuity of rotation exists at these two points. The internal moments at b and c are the redundants. For this primary structure the moments due to the actual loading have been localized to span ab, while M_{X1} will be distributed from a to c and M_{X2} will exist between b and d. Primary structure 4 is the best choice since it minimizes the computational effort. Examples 11.9 and 11.10 illustrate the use of consistent deflections for structures with multiple redundancies.

Example 11.9 Calculate the reactions and moments for this beam with two redundancies. Use the method of consistent displacements; $E = 200$ GPa, and $I = 300 \times 10^{-6}$ m^4.

SOLUTION The compatibility equations are

$$\Theta_a = 0 = \Theta_{a0} + \theta_{aa} X_a + \theta_{ab} X_b \tag{1}$$

and
$$\Theta_b = 0 = \Theta_{b0} + \theta_{ba} X_a + \theta_{bb} X_b \tag{2}$$

Calculate Θ_{a0}, θ_{aa}, θ_{ab}, Θ_{b0}, θ_{ba}, and θ_{bb} by the conjugate-beam method. From Fig. E11.9c and d for Θ_{a0} and Θ_{b0} compute the shear at a and b for the conjugate beam

$$EI\Theta_{a0} = 5(200) = 1000 \text{ kN} \cdot \text{m}^2$$

$$EI\Theta_{b0} = 5(200) + 6(360) = 3160 \text{ kN} \cdot \text{m}^2$$

From Fig. E11.9e and f for θ_{aa} and θ_{ba} compute the shear at a and b for the conjugate beam

$$EI\theta_{aa} = \tfrac{2}{3}(1)(10) = \tfrac{20}{3} \text{ m} \qquad EI\theta_{ba} = \tfrac{1}{3}(1)(10) = \tfrac{10}{3} \text{ m}$$

From Fig. E11.9g and h for θ_{ab} and θ_{bb} compute the shear at a and b for the conjugate beam

$$EI\theta_{ab} = \tfrac{1}{3}(1)(10) = \tfrac{10}{3} \text{ m} \qquad EI\theta_{bb} = \tfrac{2}{3}(1)(10 + 12) = \tfrac{44}{3} \text{ m}$$

Substituting into Eqs. (1) and (2) gives

$$0 = 1000 + \tfrac{20}{3} X_a + \tfrac{10}{3} X_b$$

$$0 = 3160 + \tfrac{10}{3} X_2 + \tfrac{44}{3} X_b$$

Solving simultaneously gives

$$X_a = -47.7 \text{ kN} \cdot \text{m} \qquad X_b = -204.6 \text{ kN} \cdot \text{m}$$

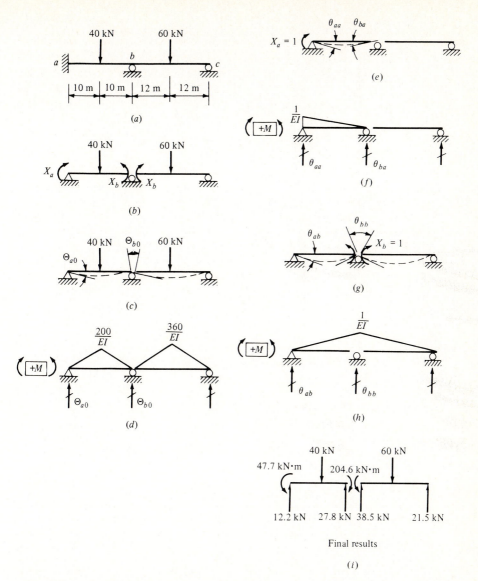

Figure E11.9 (*a*) Loaded beam; (*b*) primary structure with applied loading; (*c*) deflected primary structure; (*d*) conjugate beams with M_0/EI; (*e*) primary structure with $X_a = 1$ applied; (*f*) conjugate beams with M_{X_a}/EI; (*g*) primary structure with $X_b = 1$ applied; (*h*) conjugate beams with M_{X_b}/EI; (*i*) total reactions.

Example E11.10 Calculate the reactions and member forces for the illustrated truss using the method of consistent displacements. Note that there is one internal and one external redundancy. For all members

$$\frac{L}{A} = 10^4 \text{ m}^{-1} \quad \text{and} \quad \frac{L}{AE} = 5 \times 10^{-5} \text{ m/kN}$$

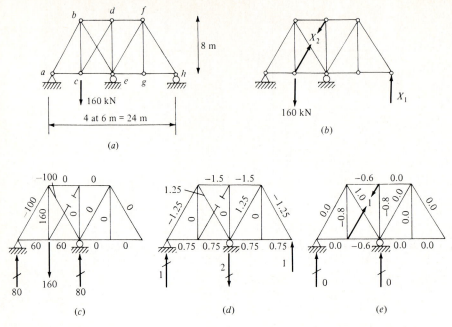

Figure E11.10 (*a*) Loaded truss; (*b*) primary structure with applied load; (*c*) primary structure with $X_1 = X_2 = 0$ (F_0 forces); (*d*) primary structure with $X_1 = 1$ (F_{X_1} forces); (*e*) primary structure with $X_2 = 1$ (F_{X_2} forces).

SOLUTION See Table E11.10.

Table E11.10‡

Member	F_0, kN	F_{X1}	F_{X2}	F_0F_{X1}	F_0F_{X2}	F_{X1}^2	F_{X2}^2	$F_{X1}F_{X2}$	X_1F_{X1}, kN	X_2F_{X2}, kN	Member forces F, kN †
ac	+60	+0.75		+45.0		0.5625			−9.4		+50.6
ce	+60	+0.75	−0.6	+45.0	−36.0	0.5625	0.36	−0.45	−9.4	−25.5	+25.1
eg		+0.75				0.5625			−9.4		−9.4
gh		+0.75				0.5625			−9.4		−9.4
bd		−1.50	−0.6			2.25	0.36	+0.90	+18.8	−25.5	−6.7
df		−1.50				2.25			+18.8		+18.8
bc	+160		−0.8		−12.8		0.64			−34.0	+126.0
de			−0.8				0.64			−34.0	−34.0
fg											0.0
ab	−100	−1.25		+125.0		1.5625			+19.5		−80.5
be	−100	+1.25	+1.0	−125.0	−100.0	1.5625	1.00	+1.25	+19.5	+42.5	−37.5
cd			+1.0				1.00			+42.5	+42.5
ef		+1.25				1.5625			+19.5		+19.5
fh		−1.25				1.5625			+19.5		+19.5
Total				+90.0	−148.8	+13.0	+4.0	+1.70			

† $F = F_0 + X_1F_{X1} + X_2F_{X2}$.
‡ Zero entries are blank.

$$\Delta_{10} = 45 \times 10^{-4} \qquad \delta_{11} = 6.5 \times 10^{-4} \qquad \delta_{12} = 0.85 \times 10^{-4}$$

$$\Delta_{20} = -74.4 \times 10^{-4} \qquad \delta_{21} = 0.85 \times 10^{-4} \qquad \delta_{22} = 2.0 \times 10^{-4}$$

$$0 = \Delta_{10} + \delta_{11} X_1 + \delta_{12} X_2$$

$$0 = \Delta_{20} + \delta_{21} X_1 + \delta_{22} X_2$$

$$X_1 = -12.5 \text{ kN}(\updownarrow) \qquad X_2 = +42.5 \text{ kN (t)}$$

11.6 CASTIGLIANO'S THEOREM PART II: LEAST WORK

In Chap. 10 Castigliano's theorem part II was described and applied to various determinate structures. The purpose of this section is to use this principle to analyze statically indeterminate structures. Recall that although the theorem stems from complementary virtual work, under certain restrictions the theorem can be used in conjunction with the strain energy W_i of the system.

For the beam shown in Fig. 11.2a the strain energy of the system (neglecting torsion and shear) can be computed using Eq. (10.27)

$$W_i = \int_0^L \frac{M^2}{2EI}\, dx$$

The following relationship is obtained by applying Castigliano's theorem part II to obtain an expression for the deflection at point b (which is zero)

$$\Delta_b = 0 = \frac{\partial W_i}{\partial R_b} = \frac{\partial}{\partial R_b} \int_0^L \frac{M^2}{2EI}\, dx \qquad (a)$$

For this structure the moment diagram is symmetric with respect to the centerline; thus, Eq. (a) can be rewritten

$$\Delta_b = 0 = \frac{\partial}{\partial R_b} \int_0^{L/2} \frac{M^2}{EI}\, dx \qquad (b)$$

Since M is continuously differentiable in this interval,

$$0 = 2 \int_0^{L/2} \frac{\partial M}{\partial R_b} \frac{M}{EI}\, dx \qquad (c)$$

The moment in $0 \le x \le L/2$ is

$$M = \frac{qL}{2} x - \frac{qx^2}{2} - \frac{R_b}{2} x \qquad (d)$$

Hence,

$$\frac{\partial M}{\partial R_b} = -\frac{x}{2} \qquad (e)$$

Substituting Eqs. (d) and (e) into (c) and performing the integration gives

$$0 = -\frac{5qL^4}{384EI} + R_b \frac{L^3}{48EI} \qquad (f)$$

which yields
$$R_b = \frac{5qL}{8} \tag{g}$$

This beam was also analyzed in Sec. 11.2 using the method of consistent displacements, and the results were shown in Eqs. (a) and (b) of that section. Comparing them with Eqs. (f) and (g) reveals that the two formulations yield the same solution. It can be also observed by recalling Eq. (11.2) that

$$\Delta_{b0} = -\frac{5qL^4}{384EI} \quad \text{and} \quad \delta_{bb} = +\frac{L^3}{48EI}$$

Thus, applying Castigliano's theorem part II to this problem gives intermediate and final results that are identical to those from the method of consistent displacements.

Since the displacement investigated in the previous calculations was zero, Eq. (a) amounts to finding the extremum for W_i. Taking the second partial derivative of W_i will reveal that this is a minimum and not a maximum. Thus, in general, if this method is used to calculate the deflection for an unyielding redundant X_i, it is referred to as the *theorem of least work* and represented as

$$\left(\frac{\partial W_i}{\partial X_i}\right)_{T=\text{const}} = 0 \tag{11.18}$$

Since W_i was substituted for W_i^* in the development of this theorem in Chap. 10, Eq. (11.18) is applicable only to linear elastic structures with no support movement, temperature effects, fabrication errors, etc. That is, Eq. (11.18) should be used for structures whose deflections are caused only by loads.

In the foregoing, the method of least work was described for a structure with a single redundant. When multiple redundancies occur, it is possible to write as many equations as there are unknowns. They are all generated using Eq. (11.18). The theorem can be stated in general as follows:

Theorem of least work For a linear elastic structure with a given system of loads and no induced displacements due to temperature, settlement, etc., the redundants are such that they minimize the strain energy.

It should be noted that the method is also applicable to types of structures other than beams. For a truss, W_i must be calculated using Eq. (10.25)

$$W_i = \sum \frac{F^2 L}{2AE}$$

And for rigid frames in which both axial forces and bending moments are to be considered the strain energy is obtained by combining the results from Eqs. (10.25) and (10.27).

Thus, the theorem of least work, which is derived from energy considerations, is an approach equivalent to that of the method of consistent displacements. In addition, least work can be applied only when there are no induced displacements. Probably the distinct advantage of the method is that it gives the structural engineer an opportunity to set up the equations in a systematic manner. That is, once the expression for the strain energy has been obtained, it is simply operated upon using Eq. (11.18) for each redundant. Although it is possible to use the theorem of least work where the structure is deformed by heating, etc., this requires the calculation of the deflections due to the induced deformations in the primary structure to be carried out by another applicable method. Such hybrid methods are not discussed in this text. Example 11.11 illustrates the arrangements of the calculations required to use the method of least work.

Example 11.11 Use the theorem of least work to calculate the axial forces in the members and the bending-moment diagram for member abd. For all members $E = 29 \times 10^3$ kips/in²; and

$$A = \begin{cases} 12 \text{ in}^2 & \text{member } abd \\ 4 \text{ in}^2 & \text{members } ac \text{ and } cd \\ 8 \text{ in}^2 & \text{member } bc \end{cases}$$

and $I = 500$ in⁴ for member abd.

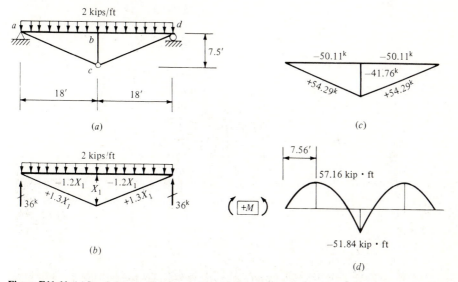

Figure E11.11 (a) Loaded truss; (b) primary structure showing loads and redundant member forces; (c) final axial forces; (d) final bending moments.

SOLUTION

$$M = 36x - x^2 - \tfrac{1}{2}X_1 x \qquad \frac{\partial M}{\partial X_1} = -\tfrac{1}{2}x \qquad 0 \le x \le 18$$

Because of symmetry, the contribution to Eq. (11.18) from the strain energy of bending is

$$\int_0^L \frac{\partial M}{\partial X_1} \frac{M}{EI}\, ds = \frac{2}{EI} \int_0^{18} \left(-\frac{x}{2}\right)\left(36x - x^2 - \frac{1}{2}X_1 x\right) dx$$

$$= \frac{1}{EI}(-43{,}740 + 972X_1) \qquad \text{kip} \cdot \text{ft}^3$$

Table E11.11

Member	$\dfrac{L}{A}$, in^{-1}	F, kips	$\dfrac{\partial F}{\partial X_1}$	$\dfrac{L}{A}F\dfrac{\partial F}{\partial X_1}$, kips/in
ab	18.00	$-1.2X_1$	-1.2	$25.920X_1$
bd	18.00	$-1.2X_1$	-1.2	$25.920X_1$
ac	58.50	$+1.3X_1$	$+1.3$	$98.865X_1$
cd	58.50	$+1.3X_1$	$+1.3$	$98.865X_1$
bc	11.25	$-1.0X_1$	-1.0	$11.250X_1$
				$\Sigma = 260.820X_1$

The contribution to Eq. (11.18) by the strain energy of axial loads is shown in Table E11.11. The theorem of least work gives

$$0 = \frac{\partial W_i}{\partial X_1} = \int_0^L \frac{\partial M}{\partial X_1}\frac{M}{EI}\,dx + \sum \frac{\partial F}{\partial X_1}\frac{FL}{AE}$$

$$0 = \frac{1}{E(500\text{ in}^4)}(-43{,}740 + 972X_1)(\text{kip}\cdot\text{ft}^3)(1728\text{ in}^3/\text{ft}^3) + \left(\frac{260.82X_1}{E}\text{ kips/in}\right)$$

$X_1 = 41.76$ kips (c)

DISCUSSION Since the deflections of this reversed king-post truss are influenced by the bending and axial deformations, the energy of both must be considered. From Castigliano's theorem part II, the criterion stating that the cut in member cb must be closed is given by

$$\frac{\partial W_i}{\partial X_1} = 0$$

The final axial forces and bending moments can be obtained by using either the equations of static equilibrium or the principle of superposition.

11.7 THE THREE-MOMENT EQUATION

Clapeyron developed a method of analysis for the reconstruction of the bridge at Asnières, near Paris, in 1849 and presented this procedure for analyzing continuous beams to the Academy of Sciences in 1857. The condition of compatibility of the angle of rotation for one point on the beam is used to obtain the governing equation. This equation relates the internal moments at three points on the beam to the applied loading, the material behavior, and the cross-sectional properties of the beam members. Typically, the points chosen for obtaining the internal moments are at the supports. This method can be used for both applied loads and support displacements; these two effects will be discussed separately for the sake of clarity.

The beam in Fig. 11.10a is composed of two uniform beam segments and is loaded with an arbitrary lateral loading. (The equation can also be derived if the moments of inertia are variable, but that is not illustrated here.) The subscripts i and j are used to denote quantities, say I and L, associated with the left and right spans, respectively.

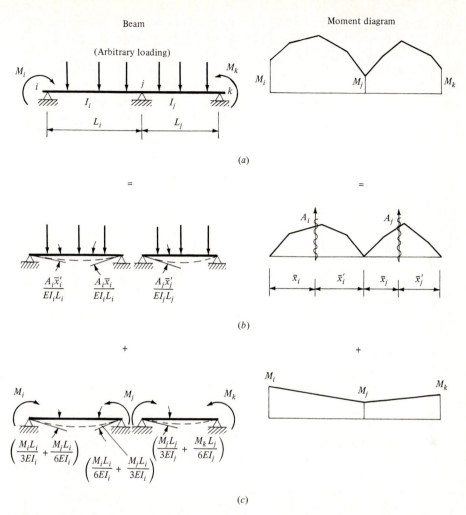

Figure 11.10 A two-span continuous beam with the moment diagram: (*a*) the beam with loading; (*b*) the beam with only lateral loads applied; (*c*) the beam with only end moments applied.

The moment diagram for the loaded beam in Fig. 11.10*a* is drawn with all the moments positive; they put the top fibers of the beam in compression. These total moments result not only from the lateral loads, but also from the action at the support of the adjacent span on the span being considered. Thus, the total moments are the superposition of these two effects, shown separately in Fig. 11.10*b* and *c*.

The areas under the moment diagrams for the lateral loads without the end moments are designated A_i and A_j for spans i and j, respectively. The centroids of these areas are located $\bar{x}_i$ and $\bar{x}'_j$ from points i and k, respectively, as shown in Fig. 11.10*b*. The angles of rotation at point j for the two spans can be found using one of the methods from Chaps. 9 and 10 and are shown in the figure. The angles of rotation at point j due to the three moments M_i, M_j, and M_k are shown in Fig. 11.10c. By superposition the angles of rotation to the left and right of point j are

$$\Theta_{ji} = \frac{M_i L_i}{6EI_i} + \frac{M_j L_i}{3EI_i} + \frac{A_i \bar{x}_i}{EI_i L_i} \qquad \Theta_{jk} = -\frac{M_j L_j}{3EI_j} - \frac{M_k L_j}{6EI_j} - \frac{A_j \bar{x}'_j}{EI_j L_j} \qquad (a)$$

where Θ_{ji} and Θ_{jk} denote the angles of rotation slightly to the left and to the right of point j, respectively. The signs of the rotations are interpreted in the mathematical sense (counterclockwise positive). For continuity of slope at point j

$$\Theta_{ji} = \Theta_{jk} \qquad (b)$$

Thus

$$M_i \frac{L_i}{I_i} + 2M_j \left(\frac{L_i}{I_i} + \frac{L_j}{I_j} \right) + M_k \frac{L_j}{I_j} = -\frac{6A_i \bar{x}_i}{I_i L_i} - \frac{6A_j \bar{x}'_j}{I_j L_j} \qquad (11.19)$$

which is the *three-moment equation*. For the special case where $I_i = I_j$ it simplifies to

$$M_i L_i + 2M_j(L_i + L_j) + M_k L_j = -\frac{6A_i \bar{x}_i}{L_i} - \frac{6A_j \bar{x}'_j}{L_j} \qquad (11.20)$$

The load terms A_i, $\bar{x}_i$, A_j, and $\bar{x}'_j$ for a concentrated and uniform load are shown in Fig. 11.11. For the beam of Fig. 11.12a, $M_a = M_c = 0$, $\bar{x}_i = \bar{x}'_j = L/4$, and $A_i = A_j = qL^3/96$. Substituting these into Eq. (11.20) gives

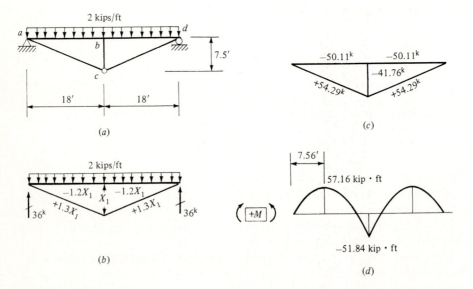

(a)

(b)

(c)

(d)

Figure 11.11 Lateral load terms for the three-moment equation: (a) a uniformly loaded beam; (b) a beam with a single concentrated load.

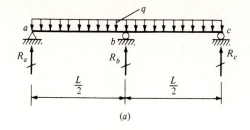

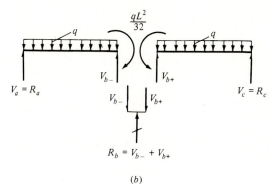

Figure 11.12 (a) A two-span uniform continuous beam; (b) free-body diagrams of the two individual spans.

$$2M_b L = -\frac{qL^3}{16} \qquad \text{or} \qquad M_b = -\frac{qL^2}{32}$$

The reaction at point b can be obtained by using free-body diagrams of the two beam segments (Fig. 11.12b). Since the shears to the left (V_{b-}) and to the right (V_{b+}) of point b are both equal to $5qL/16$ acting in the directions shown,

$$R_b = V_{b-} + V_{b+} = \frac{5qL}{8}$$

An equation like (11.19) can be written for each intermediate support of a continuous beam. If the ends are simply supported, the number of equations will equal the number of redundant reactions; see Example 11.12. If either of the end-support conditions is fixed, it is necessary to add an imaginary span with a large value of I/L such that $L/I \to 0$. This procedure is illustrated in Example 11.13.

An alternative form of Eq. (11.19) can also be derived for fixed-end conditions instead of using an imaginary span. This, of course, requires the structural engineer to work with two forms of the equation. To develop this alternate form if point i is fixed, the rotation is set equal to zero, which gives

$$2M_i + M_j = -\frac{6A_i \bar{x}_i'}{L_i^2}$$

The same formulation is obtained from Eq. (11.19) if segment i is the unloaded imaginary span with $L_i/I_i \to 0$.

Example 11.12 Use the three-moment equation to calculate the moments at points b and c. For all members $E = \text{const}$, $I_{ab} = 240 \text{ in}^4$, $I_{bc} = 180 \text{ in}^4$, and $I_{cd} = 300 \text{ in}^4$.

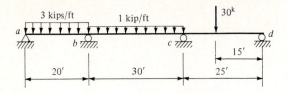

Figure E11.12

SOLUTION For span abc, $M_a = 0$, and

$$A\bar{x}_i = \frac{qL^3}{12}\frac{L}{2} = 20,000 \text{ kip} \cdot \text{ft}^3$$

$$A_j\bar{x}_j' = \frac{qL^3}{12}\frac{L}{2} = 33,750 \text{ kip} \cdot \text{ft}^3$$

$$\frac{6A_i\bar{x}_i}{I_iL_i} = \frac{6(20,000 \text{ kip} \cdot \text{ft}^3)(12 \text{ in/ft})}{(240 \text{ in}^4)(20 \text{ ft})} = 300 \text{ kip} \cdot \text{ft/in}^3$$

$$\frac{6A_j\bar{x}_j'}{I_jL_j} = \frac{6(33,750 \text{ kip} \cdot \text{ft}^3)(12 \text{ in/ft})}{(180 \text{ in}^4)(30 \text{ ft})} = 450 \text{ kip} \cdot \text{ft/in}^3$$

Eq. (11.19) yields

$$6M_b + 2M_c = -750 \tag{1}$$

For span bcd, $M_d = 0$, and

$$A_i\bar{x}_i = 33,750 \text{ kip} \cdot \text{ft}^3$$

$$A_j\bar{x}_j = \frac{PL_aL_b}{2}\left(\frac{L_a + 2L_b}{3}\right) = 30,000 \text{ kip} \cdot \text{ft}^3$$

$$\frac{6A_i\bar{x}_i}{I_iL_i} = 450 \text{ kip} \cdot \text{ft/in}^3$$

$$\frac{6A_j\bar{x}_j'}{I_jL_j} = \frac{6(30,000 \text{ kip} \cdot \text{ft}^3)(12 \text{ in/ft})}{(300 \text{ in}^4)(25 \text{ ft})} = 288 \text{ kip} \cdot \text{ft/in}^3$$

Eq. (11.19) yields

$$2M_b + 6M_c = -738 \tag{2}$$

Solving Eqs. (1) and (2) gives

$$M_b = -94.5 \text{ kip} \cdot \text{ft} \qquad M_c = -91.5 \text{ kip} \cdot \text{ft}$$

DISCUSSION Since the moments of inertia for the three spans are different, we must use the three-moment equation in the form of Eq. (11.19). In order to write Eqs. (1) and (2) as shown, the span lengths in inches are used to make the left side dimensionally consistent with the quantities on the right sides of the equations.

Example 11.13 Use the three-moment equation to compute the moments at the support points. For all members $E = 200$ GPa, $I_{ab} = 400 \times 10^{-6}$ m^4, and $I_{bc} = 640 \times 10^{-6}$ m^4.

SOLUTION For span $a'ab$

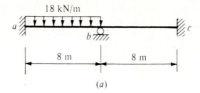

(a)

(b)

Figure E11.13 (a) Loaded beam; (b) idealized structure with $(L/I)_{a'a} \rightarrow 0$ and $(L/I)_{cc'} \rightarrow 0$.

$$M_{a'} = 0 \qquad A_i \bar{x}_i = 0 \qquad \frac{6A_j \bar{x}_j'}{I_j L_j} = 576 \times 10^4 \text{ kN/m}^2$$

$$2M_a + M_b = -288 \tag{1}$$

For span abc

$$\frac{6A_i \bar{x}_i}{I_i L_i} = 576 \times 10^4 \text{ kN/m}^2 \qquad A_j \bar{x}_j' = 0$$

$$M_a + 3.25M_b + 0.625M_c = -288 \tag{2}$$

For span bcc'

$$M_c' = 0 \qquad A_i \bar{x}_i = A_j \bar{x}_j' = 0$$

$$M_b + 2M_c = 0 \tag{3}$$

Solving Eqs. (1) to (3) gives

$$M_a = -114.46 \text{ kN} \cdot \text{m} \qquad M_b = -59.08 \text{ kN} \cdot \text{m} \qquad M_c = +29.54 \text{ kN} \cdot \text{m}$$

DISCUSSION The three-moment equation in the form of Eq. (11.19) requires the use of two fictitious end spans, each with $L/I \rightarrow 0$, as shown in Fig. E11.13b. The results in Fig. 11.11 are used to compute the loading terms. The simplified forms shown as Eqs. (1) and (2) are obtained by substituting the span lengths in meters and the moments of inertia in m⁴; these give the bending moments in kilonewton-meters.

If the supports of the loaded continuous beam of Fig. 11.10a experience displacements, the angles of rotation are obtained by combining the angles previously obtained with those introduced by the support movements. Δ_i, Δ_j, and Δ_k are the displacements of points i, j, and k, respectively, in the positive coordinate direction, as shown in Fig. 11.13. The angle of rotation of the chord between points i and j is clockwise about point j by an amount $(\Delta_i - \Delta_j)/L_i$. Similarly, the chord connecting points j and k rotates counterclockwise about j an amount $(\Delta_k - \Delta_j)/L_j$. Combining these with the results shown in Eq. (a) gives the total angles of rotation to the left and right of point j, respectively, as

$$\Theta_{ji} = \frac{M_i L_i}{6EI_i} + \frac{M_j L_i}{3EI_i} + \frac{A_i \bar{x}_i}{EI_i L_i} - \frac{\Delta_i - \Delta_j}{L_i}$$

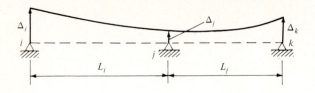

Figure 11.13 A two-span continuous beam with positive support movements for the three-moment equation.

and
$$\Theta_{jk} = -\frac{M_j L_j}{3EI_j} - \frac{M_k L_j}{6EI_j} - \frac{A_j \bar{x}_j'}{EI_j L_j} + \frac{\Delta_k - \Delta_j}{L_j} \qquad (c)$$

Again, equating Θ_{ji} and Θ_{jk} enforces compatibility at point j and leads to the following form of the three-moment equation

$$M_i \frac{L_i}{I_i} + 2M_j \left(\frac{L_i}{I_i} + \frac{L_j}{I_j} \right) + M_k \frac{L_j}{I_j} = -\frac{6A_i \bar{x}_i}{I_i L_i} - \frac{6A_j \bar{x}_j'}{I_j L_j} + 6E \left[\frac{\Delta_i}{L_i} - \Delta_j \left(\frac{1}{L_i} + \frac{1}{L_j} \right) + \frac{\Delta_k}{L_j} \right]$$

$$(11.21)$$

For the special case where $I_i = I_j = I$ the equation simplifies to

$$M_i L_i + 2M_j (L_i + L_j) + M_k L_j = -\frac{6A_i \bar{x}_i}{L_i} - \frac{6A_j \bar{x}_j'}{L_j} + 6EI \left[\frac{\Delta_i}{L_i} - \Delta_j \left(\frac{1}{L_i} + \frac{1}{L_j} \right) + \frac{\Delta_k}{L_j} \right]$$

$$(11.22)$$

It can be observed from Eqs. (11.19) and (11.21) that the three-moment equation has the same general form if support movements are involved. In using Eqs. (11.21) and (11.22) it is important to interpret the signs of the support displacements properly. The calculations in Example 11.14, carried out using the three-moment equation, illustrate the effect of support settlements upon a continuous beam.

Example 11.14 The loading is removed from the beam of Example 11.13. Using the three-moment equation, compute the moments at the support points if point a settles 8 mm and point c rotates counterclockwise 0.005 rad.

SOLUTION For span $a'ab$

$$M_a' = 0 \qquad \Delta_k = 0 \qquad \Delta_i = \Delta_j = -8 \times 10^{-3} \text{ m} \qquad A_i \bar{x}_i = A_j \bar{x}_j' = 0$$

$$2M_a \frac{8}{400 \times 10^{-6}} + M_b \frac{8}{400 \times 10^{-6}} = 6(200 \times 10^6) \left(\frac{-8 \times 10^{-3}}{8} \right)$$

$$2M_a + M_b = 60 \qquad (1)$$

For span abc

$$\Delta_j = \Delta_k = 0 \qquad \Delta_i = -8 \times 10^{-3} \text{ m} \qquad A_i \bar{x}_i = A_j \bar{x}_j = 0$$

$$M_a \frac{8}{400 \times 10^{-6}} + 2M_b \left(\frac{8}{400 \times 10^{-6}} + \frac{8}{640 \times 10^{-6}} \right) + M_c \frac{8}{640 \times 10^{-6}}$$

$$= 6(200 \times 10^6) \frac{-8 \times 10^{-3}}{8}$$

$$M_a + 3.25M_b + 0.625M_c = -60 \qquad (2)$$

For span bcc'

$$M_c' = 0 \qquad \Delta_i = \Delta_j = 0 \qquad \frac{\Delta_k}{L_j} = +0.005 \qquad A_i\bar{x}_i = A_j\bar{x}_j' = 0$$

$$M_b\frac{8}{640 \times 10^{-6}} + 2M_c\frac{8}{640 \times 10^{-6}} = 6(200 \times 10^6)(5 \times 10^{-3})$$

$$M_b + 2M_c = 480 \tag{3}$$

Solving Eqs. (1) to (3) gives

$$M_a = +79.24 \text{ kN} \cdot \text{m} \qquad M_b = -98.47 \text{ kN} \cdot \text{m} \qquad M_c = +289.24 \text{ kN} \cdot \text{m}$$

DISCUSSION Equation (11.21) must be used for this beam, and it is important to interpret the algebraic signs of the support displacements properly. For example, since support a settles, the displacement is considered to be negative. Note that the settlement of the fictitious support at a' must be considered to equal that at a ($\Delta_{a'} = \Delta_a$) since a short stiff beam is assumed to exist from a' to a. In this case we can observe that the three-moment equation (11.21) becomes

$$2M_a\frac{L_{ab}}{I_{ab}} + M_b\frac{L_{ab}}{I_{ab}} = 6EI\left[\frac{\Delta_{a'}}{L_{aa'}} - \Delta_a\left(\frac{1}{L_{aa'}} + \frac{1}{L_{ab}}\right)\right] = 6E\frac{I_{ab}}{L_{ab}}(-\Delta_a)$$

This same result can be obtained without using the analogy of a fictitious span by simplifying the second of Eqs. (c) if $\Theta_{jk} = 0$. The rotation of support c can be interpreted as though the fictitious support c' had an upward displacement such that $\Delta_{c'}/L_{cc'} = 0.005$. Except for the treatment of the displacement terms, the equations are formulated and solved as they were in the previous two examples.

11.8 DISCUSSION

The three classical methods for investigating indeterminate structures discussed in this chapter are related in that the governing equations are each derived by enforcing compatibility. The resulting equation(s) are all cast with the forces as the unknowns and the flexibility quantities as the coefficients.

The method of consistent displacements can be applied to trusses, beams, and rigid frames when the structure is subjected to applied loads, settlement, heating, fabrication errors, etc. Although this fundamental method is useful for studying the behavior of indeterminate structures, it can be tedious and inefficient if structures with multiple redundancies are to be investigated.

Since the theorem of least work presents an approach that yields the governing equations in a systematic and straightforward manner, it is more appealing than the method of consistent displacements. It is limited to structures in which the displacements are caused by loads only, since it is derived from Castigliano's theorem part II, in which the strain energy is assumed equal to the complementary strain energy.

The three-moment equation is typically used for continuous beams with applied loads and/or support movements. In using the equations, compatibility is enforced implicitly; therefore, the structural engineer need not go through the process of selecting redundancies, enforcing compatibility, etc. This advantage is offset by the danger that the physical behavior may be masked. For this reason it is important to be aware of the derivation of the equation to understand that the basic conditions of equilibrium, force-displacement, compatibility, and boundary conditions must all be satisfied.

PROBLEMS

Sketch the approximate final deflected shape for beams and rigid frames.
In Probs. 11.1 to 11.29 use the method of consistent displacements to analyze the structure.

11.1–11.2 Calculate the reactions; EI = const.

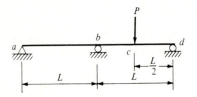

Figure P11.1

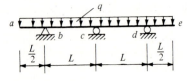

Figure P11.2

11.3 $I_{ab} = I_{bc} = I_{cd} = I_{de} = 200$ in⁴ and $E = 29 \times 10^3$ kips/in². Calculate the reactions if (a) the structure is loaded as shown; and (b) the loads are removed, and support c settles 0.50 in.

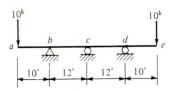

Figure P11.3

11.4 $I = 500$ in⁴ and $E = 29 \times 10^3$ kips/in². Calculate the reactions if (a) the structure is loaded as shown; (b) the loads are removed, and supports a and b each settle 0.25 in, while support f settles 1.50 in.

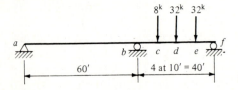

Figure P11.4

11.5 $I = 150 \times 10^{-6}$ m⁴ and $E = 200$ GPa. Calculate the reactions if (a) the structure is loaded as shown; (b) the loads are removed, and support b settles 20 mm.

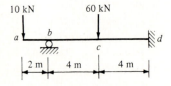

Figure P11.5

11.6–11.8 Calculate the reactions and draw the moment diagram; EI = const.

11.9 If EI = const, calculate the locations of the supports such that (a) the magnitudes of the moments at points b, c, and d are equal; (b) the reactions at points b, c, and d are equal.

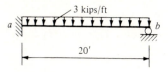

3 kips/ft

20'

Figure P11.6

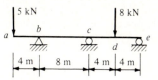

5 kN   8 kN

4 m | 8 m | 4 m | 4 m

Figure P11.7

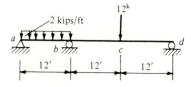

12^k

2 kips/ft

12' | 12' | 12'

Figure P11.8

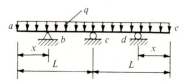

q

x

L

x

L

Figure P11.9

11.10 Calculate the member forces and reactions for this truss. For all members E = const and L/A = 48 in^{-1}.

11.11 Member areas: horizontals = 3 in^2, verticals = 4 in^2, and diagonals = 5 in^2. For all members $E = 29 \times 10^3$ kips/in^2. Calculate the member forces and reactions if (a) the truss is loaded as shown; (b) the load is removed, and support e moves horizontally to the right 0.60 in.

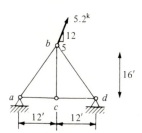

5.2^k

12
5

16'

12' | 12'

Figure P11.10

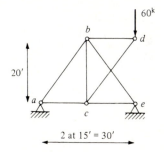

60^k

b   d

20'

a   e

c

2 at 15' = 30'

Figure P11.11

11.12 For all members A = 2000 mm^2 and E = 200 GPa. Calculate the member forces and reactions if (a) the truss is loaded as shown; (b) the load is removed, and support e moves horizontally to the right 5 mm.

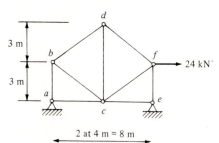

d

3 m

b   f

24 kN

3 m

a

c   e

2 at 4 m = 8 m

Figure P11.12

11.13 $A_{ab} = 30 \times 10^{-4}$ m^2, $A_{ac} = 20 \times 10^{-4}$ m^2, $A_{ad} = 15 \times 10^{-4}$ m^2, and $E = 200$ GPa. Calculate the member forces and reactions if:

(a) The truss is loaded as shown.

(b) The loads are removed, and member ab increases in temperature by 45°C ($\alpha = 1.2 \times 10^{-5}$ mm/mm · °C).

(c) The loads are removed, and member ab has been fabricated with a length error of -5 mm.

11.14 For all members $A = 1$ in^2 and $E = 29 \times 10^3$ kips/in^2. Calculate the member forces and reactions if:

(a) The truss is loaded as shown.

(b) The loads are removed, and member bd increases in temperature by 80 °F ($\alpha = 6.5 \times 10^{-6}$ in/in · °F).

(c) The loads are removed, and member bd has been fabricated with a length error of -0.40 in.

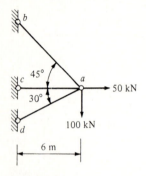

Figure P11.13

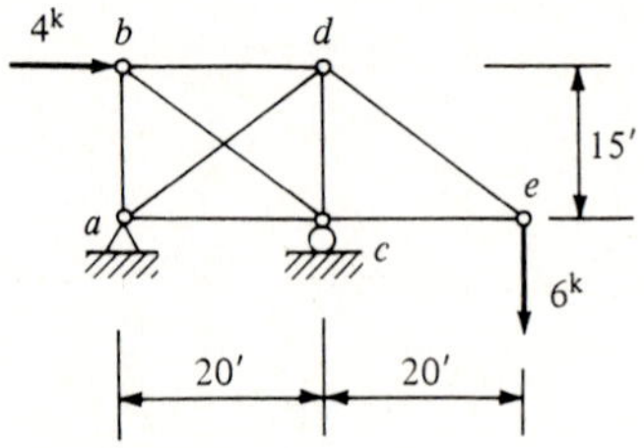

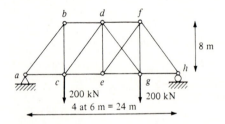

Figure P11.14

11.15 For all members $AE = $ const; calculate the member forces and reactions.

11.16 For all members $A = 1500$ mm^2 and $E = 200$ GPa. Calculate the member forces and reactions if (a) the truss is loaded as shown; (b) the loads are removed, and the top chord (members bd and df) experience an increase in temperature of 83°C ($\alpha = 1.2 \times 10^{-5}$ mm/mm · °C).

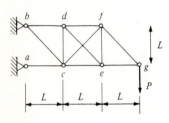

Figure P11.15

Figure P11.16

11.17 Calculate the reactions and draw the moment diagram; $EI = $ const.

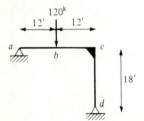

Figure P11.17

11.18 Calculate the reactions and draw the moment diagram. Ignore axial deformations of the members; $2I_{ab} = I_{bc} = I_{cd} = 2I$, and E = const.

11.19 Calculate the reactions. The rod area is 1 in², the beam area is 16 in², the beam moment of inertia is 600 in⁴, and $E = 29 \times 10^3$ kips/in².

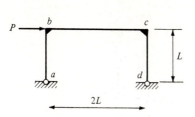

Figure P11.18

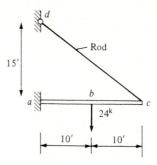

Figure P11.19

11.20 The spring, which approximates an elastic foundation, has a stiffness of 7000 kN/m. The beam has $EI = 8400$ kN · m². Calculate the reactions.

11.21 Calculate the reactions. The rod area is 1.33 in², the beam moment of inertia is 100 in⁴, and $E = 29 \times 10^3$ kips/in².

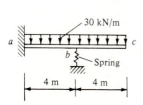

Figure P11.20

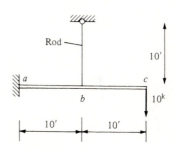

Figure P11.21

11.22 EI = const. Calculate the reactions and draw the moment diagram if (*a*) the beam is loaded as shown; (*b*) the concentrated load is removed, and the beam is loaded with a uniformly distributed load of intensity *q*.

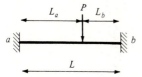

Figure P11.22

11.23 $I_{ab} = 2I_{bd} = 400$ in⁴, and $E = 29 \times 10^3$ kips/in². Calculate the reactions and draw the moment diagram.

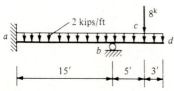

Figure P11.23

11.24 EI = const. Calculate the reactions and draw the moment diagram if (a) the beam is loaded as shown; (b) the illustrated loading is removed, and a single concentrated load of magnitude P is placed at midlength of span ab.

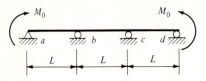

Figure P11.24

11.25 I = 746 in⁴, and E = 29 × 10³ kips/in². Calculate the reactions and draw the moment diagram if (a) the beam is loaded as shown; (b) the loads shown are removed, and support c settles 0.75 in.

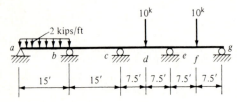

Figure P11.25

11.26 Member areas: horizontals = 4 in², verticals = 3 in², diagonals = 5 in², and E = 29 × 10³ kips/in². Calculate the member forces and reactions if (a) the truss is loaded as shown; (b) the load shown is removed, and support b moves 0.30 in to the right.

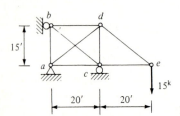

Figure P11.26

11.27 Calculate the member forces and reactions. For all members E = const and L/A = 10³ m⁻¹.

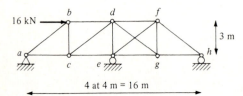

Figure P11.27

11.28 The truss of Prob. 11.16 has a member added between joints b and e with the same area and made from the same material as the other members. Calculate the member forces and reactions for the loading illustrated in Fig. P11.16.

11.29 Calculate the reactions and draw the moment diagram; EI = const.

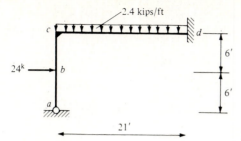

Figure P11.29

In Probs. 11.30 to 11.58 solve the original problem, using the theorem of least work to analyze the structure.

11.30 Problem 11.1
11.32 Part (a) of Prob. 11.3
11.34 Part (a) of Prob. 11.5
11.36 Problem 11.7
11.38 Problem 11.9
11.40 Part (a) of Prob. 11.11
11.42 Part (a) of Prob. 11.13
11.44 Problem 11.15
11.46 Problem 11.17
11.48 Problem 11.19
11.50 Problem 11.21
11.52 Problem 11.23
11.54 Part (a) of Prob. 11.25
11.56 Problem 11.27
11.58 Problem 11.29

11.31 Problem 11.2
11.33 Part (a) of Prob. 11.4
11.35 Problem 11.6
11.37 Problem 11.8
11.39 Problem 11.10
11.41 Part (a) of Prob. 11.12
11.43 Part (a) of Prob. 11.14
11.45 Part (a) of Prob. 11.16
11.47 Problem 11.18
11.49 Problem 11.20
11.51 Problem 11.22
11.53 Problem 11.24
11.55 Part (a) of Prob. 11.26
11.57 Problem 11.28

Use the three-moment equation to analyze the structures in Probs. 11.59 to 11.70.

11.59 Calculate the moments and draw the shear and bending-moment diagrams; EI = const.

11.60 This continuous beam has $I = 250$ in^4, and $E = 29 \times 10^3$ kips/in^2. Calculate the moments at points b and c if (a) the structure is loaded as shown; (b) the loading illustrated is removed, and supports a, b, and c settle 0.50, 0.25, and 0.10 in, respectively.

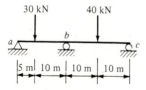

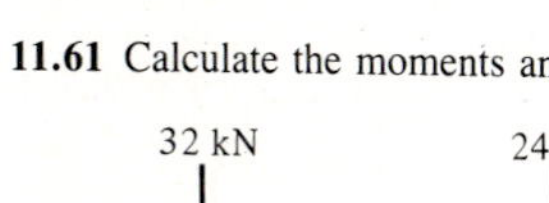

Figure P11.59

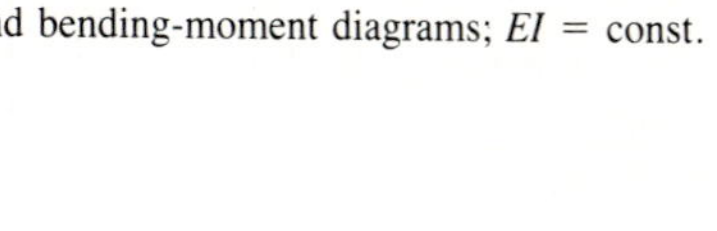

Figure P11.60

11.61 Calculate the moments and draw the shear and bending-moment diagrams; EI = const.

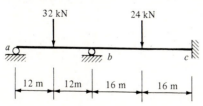

Figure P11.61

11.62 Calculate the moments and obtain the reactions; $I_{ab} = I_{bc} = 1.2I_{cd}$, and E = const.

11.63 Calculate the moments and obtain the reactions; $I_{ab} = I_{bc} = 200$ in^4, $I_{cd} = I_{de} = 300$ in^4, and E = const.

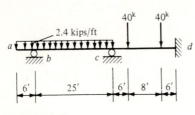

Figure P11.62

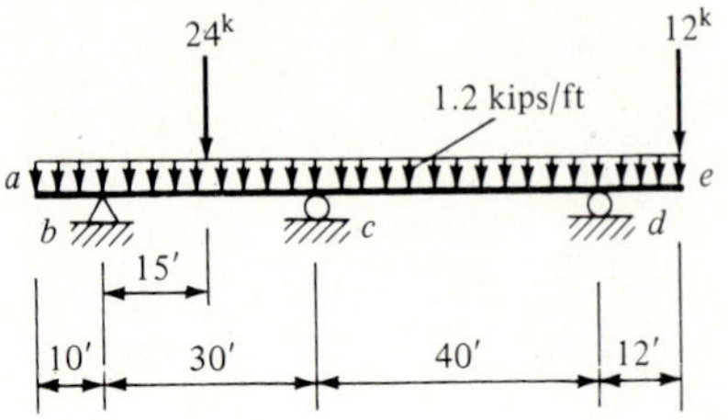

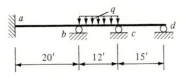

Figure P11.63

11.64 Solve parts (a) and (b) of Prob. 11.25.

11.65 Calculate the load q required to make the moment at point a equal to zero if the support at point b has settled 0.25 in. Construct the shear and moment diagrams; $I = 414$ in^4, and $E = 29 \times 10^3$ kips/in^2.

11.66 Calculate the load q required to make the moment at point a equal to zero if the support at a rotates 0.0001 rad in a counterclockwise direction. Construct the shear and moment diagrams; $I_{ab} = 2000$ in^4, $I_{bc} = 1200$ in^4, $I_{cd} = 1500$ in^4, and $E = 29 \times 10^3$ kips/in^2.

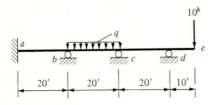

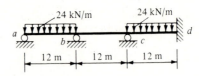

Figure P11.66

Figure P11.65

11.67 Calculate the distance x such that the magnitude of extrema of positive and negative moments are equal; EI = const.

11.68 Calculate the moments at points b, c, and d. Obtain the reactions and construct the moment diagram; $I_{ab} = I_{cd} = 600 \times 10^6$ mm^4, $I_{bc} = 1200 \times 10^6$ mm^4, and E = const.

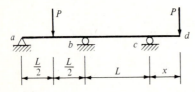

Figure P11.67

Figure P11.68

11.69 Calculate the moments at points a, b, c, and d. Obtain the reactions and construct the moment diagram; $I_{ab} = I_{cd} = 200$ in^4, $I_{bc} = 400$ in^4, and E = const.

11.70 Calculate the moments at points a, b, c, and d. Obtain the reactions and construct the moment diagram; $I_{ab} = I$, $I_{bc} = 2I$, $I_{cd} = 3I$, and E = const.

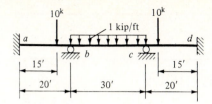

Figure P11.69

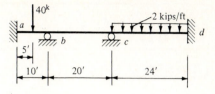

Figure P11.70

ADDITIONAL READING

Gerstle, K. H.: *Basic Structural Analysis,* chap. 10, Prentice-Hall, Englewood Cliffs, N.J., 1974.

Laursen, H. I.: *Structural Analysis,* chap. 11, McGraw-Hill, New York, 1978.

McCormac, J. C.: *Structural Analysis,* 3d ed., chaps. 19 and 21, Intext Educational, New York, 1975.

Norris, C. H., J. B. Wilbur, and S. Utku: *Elementary Structural Analysis,* 3d ed., chap. 9, McGraw-Hill, New York, 1976.

Timoshenko, S. P., and D. H. Young: *Theory of Structures,* 2d ed., chaps. 5 and 9, McGraw-Hill, New York, 1965

West, H. H.: *Analysis of Structures,* chap. 13, Wiley, New York, 1980.

TWELVE

CLASSICAL EQUILIBRIUM METHODS

12.1 INTRODUCTION

In the previous chapter the governing equations were obtained by enforcing compatibility at discrete points on the structure. These equations also implicitly satisfied the force-displacement relations, equilibrium, and boundary conditions. The emphasis in this chapter will be upon methods in which the conditions of equilibrium are explicitly enforced to give solutions for beams and rigid frames that are statically indeterminate.

The concepts of static equilibrium as discussed in Part Two are combined with the procedures described in Part Three for calculating displacements to give the so-called *equilibrium methods*. In this approach the force-displacement relations, equilibrium, and boundary conditions are imposed on individual beam members, and the equilibrium conditions are enforced at points of connectivity. Typically, this results in a set of equations with the displacements as the unknowns and the stiffness quantities as the coefficients. In its general form, it is termed either the *displacement* or the *stiffness method*. The general formulation of this method, discussed in Part Five, is commonly used in conjunction with a digital computer and the equations are cast in matrix form.

Two approaches suitable for hand calculations will be described in this chapter, *the slope-deflection method* and *the moment-distribution method*. Both use moment equilibrium at discrete points to yield the governing equations. The first method gives a set of simultaneous linear algebraic equations that must be solved explicitly, whereas the second poses an iterative solution procedure. A structure that can be analyzed by these methods is shown in Fig. 12.1.

12.2 THE SLOPE-DEFLECTION METHOD

In the slope-deflection method, equilibrium is used to formulate the governing equations for structures composed of flexural members. The resulting set of simultaneous linear algebraic equations is cast with the angles of rotation and displacements at discrete points as the unknowns and the stiffness quantities as the coefficients. Initially,

Figure 12.1 Tower Road bridge over the Toutle River, Cowlitz County, Washington. (*Arvid Grant and Associates, Olympia, Washington.*)

this basic approach was used by Manderla (1880) and Mohr (1892) to investigate secondary stresses in trusses. Later (1914) Bendixen developed the slope-deflection method for frame structures, but at approximately the same time (1915), Maney in the United States independently discovered the approach and broadened its scope. This is a systematic method for analyzing indeterminate beams and frames, but if the structure has a large number of unknowns, use of a computer is almost mandatory. For structures with relatively few unknowns the method is useful, and it can also be used to explain the moment-distribution method.

First, the equations relating the angles of rotation at the ends of the member to the bending moments produced by the applied moments and the lateral loading will be developed. These force-displacement relations can be calculated using an appropriate method from Chaps. 9 and 10. The loaded flexural member, together with the bending-moment diagram, is shown in Fig. 12.2a. The moments at the ends are prescribed using two subscripts. Together they describe the member; the first indicates the end at which the moment is acting, for example, M_{ab} is the moment at end a of member ab. Any consistent sign convention can be used to describe the various quantities, but since equilibrium equations are being written, there are substantial advantages to be realized by considering (1) clockwise rotations, for example, Θ_a and Θ_b, as positive and (2) clockwise member end moments, for example, M_{ab} and M_{ba}, as positive. Since the effects of shear and axial loading will be neglected, all deformations are considered to be introduced by the bending moments. The material is linear elastic, obeying Hooke's law, and the member is prismatic and initially straight. It is possible to derive the slope-deflection equations for a member with a variable cross section (Sec. 12.7).

For clarity, the loading on the beam in Fig. 12.2a is considered to be the super-position of two load cases (Fig. 12.2b and c). The first load case, consisting of only the applied transverse loads, has the moment diagram shown, with A denoting the area under the moment diagram and $\bar{x}$ and $\bar{x}'$ the distances from points a and b, respectively, to the centroid of the area. With either the moment-area theorems or the conjugate-beam method the angles of rotation for this case can be obtained as

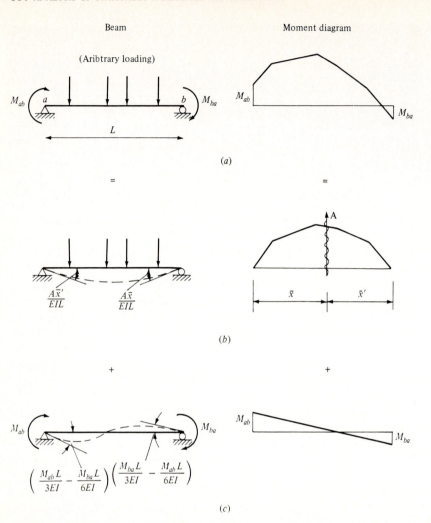

Figure 12.2 A beam member with the moment diagram: (*a*) the beam with loading and end moments shown in the positive directions; (*b*) the beam with only transverse loads applied; (*c*) the beam with only end moments applied.

$$\Theta_a = \frac{A\bar{x}'}{EIL} \qquad \Theta_b = -\frac{A\bar{x}}{EIL} \qquad (12.1)$$

If the moment diagram consists of positive and negative parts with $A = 0$ (e.g., this is the case if a couple is applied at the centerline of a beam), the angles of rotation for each part must be computed separately and Eq. (12.1) modified accordingly.

The moments applied at the ends of the member are the only loads imposed on the beam in the second load case; they give the moment diagram in Fig. 12.2*c*. The angles of rotation are

$$\Theta_a = \frac{M_{ab}L}{3EI} - \frac{M_{ba}L}{6EI} \qquad \Theta_b = \frac{M_{ba}L}{3EI} - \frac{M_{ab}L}{6EI} \qquad (12.2)$$

From superposition of the two load cases the angles of rotation for $A \neq 0$ are

$$\Theta_a = \frac{M_{ab}L}{3EI} - \frac{M_{ba}L}{6EI} + \frac{A\bar{x}'}{EIL} \qquad \Theta_b = \frac{M_{ba}L}{3EI} - \frac{M_{ab}L}{6EI} - \frac{A\bar{x}}{EIL} \qquad (12.3)$$

Solving Eq. (12.3) for the moments gives (for $A \neq 0$)

$$M_{ab} = \frac{2EI}{L}(2\Theta_a + \Theta_b) + \frac{2A}{L^2}(\bar{x} - 2\bar{x}')$$

$$M_{ba} = \frac{2EI}{L}(2\Theta_b + \Theta_a) + \frac{2A}{L^2}(2\bar{x} - \bar{x}') \qquad (12.4)$$

The terms with $2A/L^2$ in the above equations represent the effect of the applied transverse loads. They must be computed for each individual loading. If $\Theta_a = \Theta_b = 0$, the member corresponds to a fixed-end beam, and the load terms are the only effects contributing to the end moments. Thus, these two terms are the *fixed-end moments*, and for $A \neq 0$ they are denoted as

$$\text{FEM}_{ab} = \frac{2A}{L^2}(\bar{x} - 2\bar{x}') \qquad \text{FEM}_{ba} = \frac{2A}{L^2}(2\bar{x} - \bar{x}') \qquad (12.5)$$

Substituting Eqs. (12.5) into Eqs. (12.4) gives

$$M_{ab} = \frac{2EI}{L}(2\Theta_a + \Theta_b) + \text{FEM}_{ab} \qquad M_{ba} = \frac{2EI}{L}(2\Theta_b + \Theta_a) + \text{FEM}_{ba} \qquad (12.6)$$

which can be written in a convenient form by adopting the notions of a near end n and a far end f; hence,

$$M_{nf} = 2EK(2\Theta_n + \Theta_f) + \text{FEM}_{nf} \qquad (12.7)$$

where $K = I/L$ is the *stiffness factor*. Equation (12.7) is a form of the *slope-deflection equation*.

The fixed-end moments can be calculated using Eqs. (12.5) for an applied loading with $A \neq 0$. Figure 12.3 illustrates three common load cases. For the concentrated load in Fig. 12.3a

$$A = \frac{PL_aL_b}{2} \qquad \bar{x} = \tfrac{1}{3}(2L_a + L_b) \qquad \bar{x}' = \tfrac{1}{3}(L_a + 2L_b) \qquad (a)$$

Substituting these results into Eqs. (12.5) gives

$$\text{FEM}_{ab} = \frac{PL_aL_b}{L^2}\left(\frac{2L_a}{3} + \frac{L_b}{3} - \frac{2L_a}{3} - \frac{4L_b}{3}\right) = -\frac{PL_aL_b^2}{L^2}$$

$$\text{FEM}_{ba} = \frac{PL_aL_b}{L^2}\left(\frac{4L_a}{3} + \frac{2L_b}{3} - \frac{L_a}{3} - \frac{2L_b}{3}\right) = \frac{PL_a^2L_b}{L^2} \qquad (12.8)$$

For the uniform loading distributed over the span shown in Fig. 12.3b

$$A = \frac{qL^3}{12} \qquad \bar{x} = \bar{x}' = \frac{L}{2} \qquad (b)$$

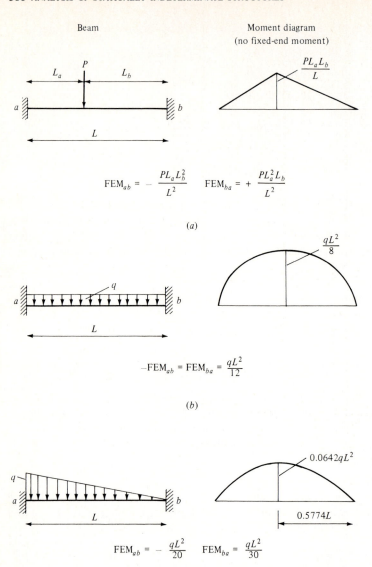

Beam

Moment diagram
(no fixed-end moment)

$$\text{FEM}_{ab} = -\frac{PL_aL_b^2}{L^2} \qquad \text{FEM}_{ba} = +\frac{PL_a^2L_b}{L^2}$$

(a)

$$-\text{FEM}_{ab} = \text{FEM}_{ba} = \frac{qL^2}{12}$$

(b)

$$\text{FEM}_{ab} = -\frac{qL^2}{20} \qquad \text{FEM}_{ba} = \frac{qL^2}{30}$$

(c)

Figure 12.3 Fixed-end moments: (a) single concentrated load; (b) uniform load; (c) linearly varying load.

$$-\text{FEM}_{ab} = \text{FEM}_{ba} = \frac{qL}{6}\left(L - \frac{L}{2}\right) = \frac{qL^2}{12} \qquad (12.9)$$

Alternatively, Eq. (12.9) could have been obtained by considering the uniform load as a series of infinitesimal loads of intensity $q\,dx$ (where dx is a differential length) and integrating the results from the concentrated load to give

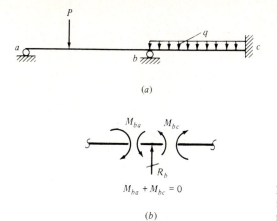

(a)

$M_{ba} + M_{bc} = 0$

(b)

Figure 12.4 (a) Continuous beam with loading; (b) free-body diagram of joint b (shears not shown for clarity).

$$\text{FEM}_{ab} = -\int_0^L \frac{x(L-x)^2}{L^2} q\, dx = -\frac{q}{L^2}\left(L^2\frac{x^2}{2} - 2L\frac{x^3}{3} + \tfrac{1}{4}x^4\right)\Bigg|_0^L = -\frac{qL^2}{12}$$

where x is measured along the beam with point a as the origin. Thus it is possible to obtain the fixed-end moments for literally any loading by judicious use of the results for a single concentrated load.

The slope-deflection equations can be used to analyze continuous beams and rigid frames. For example, the structure in Fig.12.4a is composed of segments ab and bc, which are continuously connected at point b. Initially, the moments and rotations at a, b, and c are considered nominally as unknowns. The slope at point b is single-valued (Θ_b is the same whether it is measured slightly to the right or to the left of the point). Note that $\Theta_c = 0$ implies that Θ_a and Θ_b are the only unknown rotations. Moment equilibrium at joint a gives

$$M_{ab} = 0 \tag{c}$$

It can be observed from Fig. 12.4b that moment equilibrium at point b gives

$$M_{ba} + M_{bc} = 0 \tag{d}$$

Writing Eq. (12.7) for these three moments and substituting into Eqs. (c) and (d) gives two equations with the two unknowns Θ_a and Θ_b. After they have been solved, the values of Θ_a and Θ_b can be substituted into the slope-deflection equations to give the bending moments. The calculations for a continuous beam are shown in Example 12.1.

Example 12.1 Calculate the moments at points a, b, c, and d using the slope-deflection equations. $E = $ const.

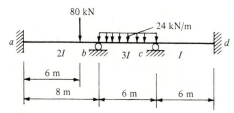

Figure E12.1

SOLUTION For span ab

$$M_{ab} = 2E\frac{2I}{8}\Theta_b - \frac{80(6)(4)}{64} = \frac{EI}{2}\Theta_b - 30$$

$$M_{ba} = 2E\frac{2I}{8}2\Theta_b + \frac{80(36)(2)}{64} = EI\Theta_b + 90$$

For span bc

$$M_{bc} = 2E\frac{3I}{6}(2\Theta_b + \Theta_c) - \frac{24(36)}{12} = EI(2\Theta_b + \Theta_c) - 72$$

$$M_{cb} = 2E\frac{3I}{6}(2\Theta_c + \Theta_b) + \frac{24(36)}{12} = EI(2\Theta_c + \Theta_b) + 72$$

For span cd

$$M_{cd} = 2E\frac{I}{6}2\Theta_c \qquad M_{dc} = 2E\frac{I}{6}\Theta_c$$

Equilibrium at joint b yields

$$M_{ba} + M_{bc} = 0$$

$$3EI\Theta_b + EI\Theta_c + 18 = 0 \tag{1}$$

Equilibrium at joint c yields

$$M_{cb} + M_{cd} = 0$$

$$8EI\Theta_c + 3EI\Theta_b + 216 = 0 \tag{2}$$

Solving Eqs. (1) and (2) gives

$$\Theta_b = \frac{24}{7EI} \qquad \Theta_c = -\frac{198}{7EI}$$

Substituting them into the expressions for the moments gives

$$M_{ab} = -\frac{198}{7} = -28.29 \text{ kN} \cdot \text{m} \qquad M_{ba} = -M_{bc} = \frac{654}{7} = 93.43 \text{ kN} \cdot \text{m}$$

$$M_{cb} = -M_{cd} = \frac{132}{7} = 18.86 \text{ kN} \cdot \text{m} \qquad M_{dc} = -\frac{44}{7} = -9.43 \text{ kN} \cdot \text{m}$$

DISCUSSION These equations can also be solved by an iteration procedure such as the Gauss-Seidel method (iterative and relaxation methods applied to structural engineering are discussed in Ref. 1). The equations have been rearranged as follows:

$$8EI\Theta_c + 3EI\Theta_b + 216 = 0 \tag{2}$$

$$EI\Theta_c + 3EI\Theta_b + 18 = 0 \tag{1}$$

They are solved by first letting $\Theta_b = 0$ and computing the value for Θ_c from Eq. (2). This value for Θ_c is substituted into Eq. (1) and the corresponding Θ_b obtained. In general, this value will not be zero, and so the calculated value of Θ_b is substituted into Eq. (2) and the process repeated until the desired accuracy is reached. These results are shown in Table E12.1. Note that the values of M_{ba} and M_{cd} were also calculated at each iteration.

Table E12.1 Iterative solution for slope-deflection equations

$EI\Theta_b$	$EI\Theta_c$ Eq. (2)	$EI\Theta_b$ Eq. (1)	M_{ba} †	M_{cd} †
0.000	−27.000	3.000	90.000	−18.000
3.000	−28.125	3.375	93.000	−18.750
3.375	−28.266	3.422	93.375	−18.844
3.422	−28.283	3.428	93.422	−18.856
3.428	−28.285	3.428	93.428	−18.857

† Computed using the initial value of Θ_b.

The beam in Fig. 12.2 can also rotate as a rigid body because of displacements of the supports, shown in Fig. 12.5. The angle of this rotation ψ_{ab} is positive if the member rotates clockwise. Equations (12.3) must be modified to give the total angles of rotation

$$\Theta_a = \frac{M_{ab}L}{3EI} - \frac{M_{ba}L}{6EI} + \frac{A\bar{x}'}{EIL} + \psi_{ab} \qquad \Theta_b = \frac{M_{ba}L}{3EI} - \frac{M_{ab}L}{6EI} - \frac{A\bar{x}}{EIL} + \psi_{ab} \qquad (12.10)$$

Solving these equations and substituting the expressions for FEM from Eq. (12.5) gives

$$M_{ab} = \frac{2EI}{L}(2\Theta_a + \Theta_b - 3\psi_{ab}) + \text{FEM}_{ab}$$

$$\tag{12.11}$$

$$M_{ba} = \frac{2EI}{L}(2\Theta_b + \Theta_a - 3\psi_{ab}) + \text{FEM}_{ba}$$

In the notation of the near and far end

$$M_{nf} = 2EK(2\Theta_n + \Theta_f - 3\psi) + \text{FEM}_{nf} \qquad (12.12)$$

These are the slope-deflection equations used in the analysis of continuous beams with support displacements and rigid-frame structures. Example 12.2 illustrates this approach for a beam with support movements.

Example 12.2 The continuous beam of Example 12.1 has the loads removed; the supports b and c settle 20 and 10 mm, respectively. Use the slope-deflection method to calculate the angles of rotation at points b and c and the moments at the ends of the members if $I = 150 \times 10^{-6}$ m^4 and $E = 200$ GPa.

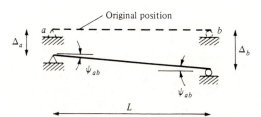

Figure 12.5 Rigid-body rotation of beam member.

SOLUTION For span ab

$$\psi = \frac{0.020}{8.0}$$

$$M_{ab} = 2E\frac{2I}{8}(\Theta_b - 3\psi) = \frac{(200 \times 10^6)(150 \times 10^{-6})}{2}(\Theta_b - 0.0075)$$

$$= (15 \times 10^3)(\Theta_b - 0.0075)$$

$$M_{ba} = (15 \times 10^3)(2\Theta_b - 0.0075)$$

For span bc

$$\psi = -\frac{0.010}{6.0}$$

$$M_{bc} = 2E\frac{3I}{6}(2\Theta_b + \Theta_c + 0.005) = (30 \times 10^3)(2\Theta_b + \Theta_c + 0.005)$$

$$M_{cb} = (30 \times 10^3)(\Theta_b + 2\Theta_c + 0.005)$$

For span cd

$$\psi = -\frac{0.010}{6.0}$$

$$M_{cd} = 2E\frac{I}{6}(2\Theta_c + 0.005) = (10 \times 10^3)(2\Theta_c + 0.005)$$

$$M_{dc} = 10 \times 10^3(\Theta_c + 0.005)$$

From equilibrium at joint b

$$M_{ba} + M_{bc} = 0 = 6\Theta_b + 2\Theta_c + 0.0025 \qquad (1)$$

and from equilibrium at joint c

$$M_{cb} + M_{cd} = 0 = 8\Theta_c + 3\Theta_b + 0.020 \qquad (2)$$

Solving Eqs. (1) and (2) gives

$$\Theta_b = 4.762 \times 10^{-4} \text{ rad} \qquad \Theta_c = -26.786 \times 10^{-4} \text{ rad}$$

Substituting these into the expressions for the moments leads to

$$M_{ab} = -105.36 \text{ kN} \cdot \text{m} \qquad M_{ba} = -M_{bc} = -98.21 \text{ kN} \cdot \text{m}$$

$$M_{cb} = -M_{cd} = 3.57 \text{ kN} \cdot \text{m} \qquad M_{dc} = 23.21 \text{ kN} \cdot \text{m}$$

12.3 APPLICATION OF THE SLOPE-DEFLECTION METHOD TO FRAMES

Frames can also be investigated using the slope-deflection equations. The rigid frame shown in Fig. 12.6a has the members fabricated at joints b and d such that the rotations of all the member ends at each joint are equal and consequently there is only one unknown angle of rotation at each point. For example, at joint b the angle of

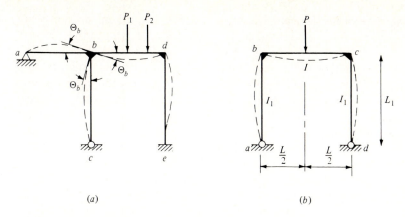

Figure 12.6 Frame structures without lateral sway: (a) constrained by supports; (b) symmetrical structure and loading.

rotation of the tangent at the top of the column is equal to the rotation of the tangent to the elastic curve for the beam. The axial deformations of the members for most frames can generally be ignored. Thus, for this particular frame there are no translations of the joints, and there are four unknown rotations (Θ_a, Θ_b, Θ_c, and Θ_d). Four independent equations can be obtained by enforcing moment equilibrium at joints a, b, c, and d. If the slope-deflection equations are substituted into these equations, four equations with four unknown rotations are obtained and the problem can be solved. Although the frame shown in Fig. 12.6b is not restrained laterally, it will not sway since the loading is symmetrically applied. This structure has four unknown angles of rotation, but thanks to symmetry, $\Theta_a = -\Theta_d$ and $\Theta_b = -\Theta_c$. Therefore the problem is reduced to two unknowns. The slope-deflection equations give

$$M_{ab} = 2EK_1(2\Theta_a + \Theta_b) \qquad M_{ba} = 2EK_1(2\Theta_b + \Theta_a)$$

$$M_{bc} = 2EK(2\Theta_b - \Theta_b) - \frac{PL}{8} \qquad (a)$$

where $K = I/L$ and $K_1 = I_1/L_1$. Moment equilibrium at joints a and b gives, respectively,

$$M_{ab} = 0 \qquad \text{and} \qquad M_{ba} + M_{bc} = 0 \qquad (b)$$

When Eqs. (a) are substituted into Eqs. (b), the angles of rotation can be calculated as

$$\Theta_a = -\tfrac{1}{2}\Theta_b \qquad \Theta_b = \frac{PL}{8E(3K_1 + 2K)} \qquad (c)$$

Hence, from Eqs. (a)

$$M_{ba} = -M_{bc} = \frac{3PL}{8(3 + 2K/K_1)} \qquad (d)$$

Thus, frames without lateral sway can be solved in a manner similar to that used for continuous-beam problems where support displacements are not involved.

If this same symmetrical frame is loaded as shown in Fig. 12.7a, the frame will displace laterally. Since it is assumed that the members do not change length because of axial deformations, the tops of both columns will displace laterally an unknown amount Δ. Thus, in addition to the unknown four angles of rotation at points a, b, c, and d there is an unknown displacement. The solution procedure for no lateral sway as described above must be augmented with an additional equation. First, the slope-deflection equations for the moments on the members at points a, b, c, and d can be expressed in terms of the unknowns Θ_a, Θ_b, Θ_c, Θ_d, and Δ. Second, the moment equilibrium equations at all the points, combined with the slope-deflection equations, give four equations in terms of the unknown displacements. One additional equation can be obtained by enforcing equilibrium in the horizontal direction for the total frame. The free-body diagrams for the three individual members are shown in Fig. 12.7b. Upon summing moments about the tops of the columns the horizontal reactions (shears in the members) at points a and d are found to be, respectively,

$$V_a = \frac{M_{ba}}{L_1} \quad \text{and} \quad V_d = \frac{M_{cd}}{L_1} \tag{e}$$

Summing forces in the horizontal direction for the total frame establishes that

$$V_a + V_d = 0 \tag{f}$$

Substituting Eqs. (e) into Eq. (f) gives the fifth equilibrium equation

$$M_{ba} + M_{cd} = 0 \tag{g}$$

By substituting the slope-deflection expressions for M_{ba} and M_{cd} this final equation can be expressed in terms of the unknown displacements. Example 12.3 gives numerical computations for a frame with sidesway.

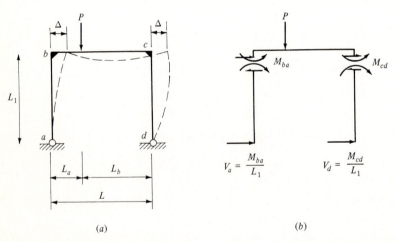

(a) (b)

Figure 12.7 Frame structure with lateral sway: (a) the frame with applied loading; (b) free-body diagrams of the three members.

Example 12.3 Calculate the moments at points a, b, c, and d for this frame structure using the slope-deflection method; $EI = $ const for all members.

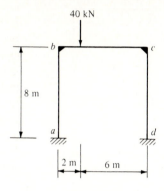

40 kN

b c

8 m

a d

2 m   6 m

Figure E12.3

SOLUTION For member ab

$$M_{ab} = 2EK(\Theta_b - 3\psi) \qquad M_{ba} = 2EK(2\Theta_b - 3\psi)$$

For member bc

$$M_{bc} = 2EK(2\Theta_b + \Theta_c) - \frac{40(2)(6^2)}{64} \qquad M_{cb} = 2EK(2\Theta_c + \Theta_b) + \frac{40(2^2)(6)}{64}$$

For member cd

$$M_{cd} = 2EK(2\Theta_c - 3\psi) \qquad M_{dc} = 2EK(\Theta_c - 3\psi)$$

From equilibrium at joint b

$$4\Theta_b + \Theta_c - 3\psi = \frac{45}{2EK} \qquad (1)$$

and from equilibrium at joint c

$$\Theta_b + 4\Theta_c - 3\psi = -\frac{15}{2EK} \qquad (2)$$

From horizontal equilibrium of the entire frame (shear equation)

$$\frac{M_{ab} + M_{ba}}{8} + \frac{M_{cd} + M_{dc}}{8} = 0$$

$$\Theta_b + \Theta_c - 4\psi = 0 \qquad (3)$$

Solving Eqs. (1) to (3) gives

$$\Theta_b = \frac{100}{14EK} \qquad \Theta_c = -\frac{40}{14EK} \qquad \psi = \frac{15}{14EK}$$

Substituting these results into the moment expressions gives

$$M_{ab} = 7.86 \text{ kN} \cdot \text{m} \qquad M_{ba} = -M_{bc} = 22.14 \text{ kN} \cdot \text{m}$$

$$M_{cb} = -M_{cd} = 17.86 \text{ kN} \cdot \text{m} \qquad M_{dc} = -12.14 \text{ kN} \cdot \text{m}$$

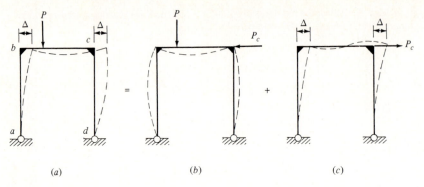

(a)   (b)   (c)

Figure 12.8 Solution strategy for a frame with lateral sway: (a) the actual frame with applied loading; (b) the frame constrained from lateral sway; (c) the frame with lateral sway only.

Frames with sway can also be analyzed using the superposition of two individual solutions. This strategy is suggested in Fig. 12.8 for the frame already discussed. The complete solution response shown in Fig. 12.8a is obtained by superimposing the solution without lateral sway (Fig. 12.8b) on that with only lateral sway (Fig. 12.8c). Since the solution suggested by Fig. 12.8b has no horizontal displacements of the column tops, only four unknown rotations are involved. This solution will yield horizontal shears at points a and d that will not sum to zero, implying that there must be a constraining force P_c applied to the top of the frame to prevent sway (Fig. 12.8b). The solution with sidesway has five unknown displacements. In this case it is convenient to assume some value for Δ and solve for the associated moments and applied constraining force at c. Generally, the assumed displacement will yield an applied force at c that does not equal P_c. Since this is a linear elastic structure, it is a straightforward procedure using proportions to calculate the values of the lateral displacement, rotations, and moments that correspond to the exact numerical value of P_c. This procedure, shown in Example 12.4, is used as an alternate solution for the frame of Example 12.3.

Example 12.4 Solve for the moments in the frame of Example 12.3 using the slope-deflection equations in two separate parts: (a) no lateral sway with the load applied and (b) lateral sway only.

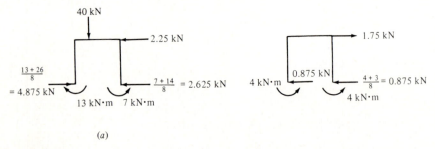

(a)

Figure E12.4 Shears and moments at the base of the columns for (a) no lateral sway; (b) lateral sidesway only.

SOLUTION (*a*) No lateral sway ($\psi = 0$):

$$4\Theta'_b + \Theta'_c = \frac{45}{2EK} \tag{1}$$

$$\Theta'_b + 4\Theta'_c = -\frac{15}{2EK} \tag{2}$$

which give
$$\Theta'_b = \frac{13}{2EK} \qquad \Theta'_c = \frac{-7}{2EK}$$

Hence,
$$M'_{ab} = 13 \text{ kN} \cdot \text{m} \qquad M'_{ba} = -M'_{bc} = 26 \text{ kN} \cdot \text{m}$$

$$M'_{cb} = -M'_{cd} = 14 \text{ kN} \cdot \text{m} \qquad M'_{dc} = -7 \text{ kN} \cdot \text{m}$$

This requires a constraining force at c of 2.25 kN (Fig. E12.4*a*).

(*b*) Lateral sway only: Assume an arbitrary value for the member rotations, say $\psi = 5/6EI$; then

$$4\Theta''_b + \Theta''_c - 3\frac{5}{6EI} = 0 \tag{3}$$

$$\Theta''_b + 4\Theta''_c - 3\frac{5}{6EI} = 0 \tag{4}$$

The solution for Eqs. (3) and (4) yields

$$\Theta''_b = \Theta''_c = \frac{1}{2EK}$$

$$M''_{ab} = -4 \text{ kN} \cdot \text{m} \qquad M''_{ba} = -M''_{bc} = -3 \text{ kN} \cdot \text{m}$$

$$M''_{cb} = -M''_{cd} = 3 \text{ kN} \cdot \text{m} \qquad M''_{dc} = -4 \text{ kN} \cdot \text{m}$$

These moments correspond to an applied force at c of 1.75 kN (Fig. E12.4*b*). Correcting this sidesway force to 2.25 kN gives

$$M''_{ab} = -5.14 \text{ kN} \cdot \text{m} \qquad M''_{ba} = -M''_{bc} = -3.86 \text{ kN} \cdot \text{m}$$

$$M''_{cb} = -M''_{cd} = 3.86 \text{ kN} \cdot \text{m} \qquad M''_{dc} = -5.14 \text{ kN} \cdot \text{m}$$

The final moments are $M' + M''$

$$M_{ab} = +7.86 \text{ kN} \cdot \text{m} \qquad M_{ba} = -M_{bc} = +22.14 \text{ kN} \cdot \text{m}$$

$$M_{cb} = -M_{cd} = +17.86 \text{ kN} \cdot \text{m} \qquad M_{dc} = -12.14 \text{ kN} \cdot \text{m}$$

12.4 THE MOMENT-DISTRIBUTION METHOD

The slope-deflection method applied to continuous beam and frame structures gives a set of simultaneous linear algebraic equations. Although these equations can be solved using a standard computer program, better ways to formulate the solution if computer methods are to be used will be discussed in Part Five. The method of successive approximations (or iteration) can be used to solve the equations, and hand calculations can be used if the equations are posed in the proper form. This approach was used by Mohr (1906) and also by Hardesty in calculating secondary stresses in trusses. The method was extended to the analysis of frame structures with and without lateral constraint by Čališev (1923). The development of the moment-distribution method in its final form is attributed to Cross (1932). The general form of this solution method, discovered by Southwell (1940) and called *relaxation methods,* is used for solving various types of engineering problems. Moment distribution consists of a series of computational cycles, each representing an iterative step in the solution of the set of governing equations.

The method has a relatively simple physical interpretation, which will be discussed using the concepts developed for the slope-deflection method. From Eq. (12.12) it can be deduced that the bending moment acting on the end of a beam is introduced by four separate effects:

1. The loads which would be applied to the member if the beam were fixed at both ends; these are the fixed-end moments.
2. The rotation Θ_n of the tangent to the elastic curve at the near end of the member.
3. The rotation Θ_f of the tangent to the elastic curve at the far end of the member.
4. The rotation ψ of the chord joining the endpoints of the member.

These independent effects are illustrated in Fig. 12.9. Each can be analyzed separately and the results combined by superposition to give the total moments.

In this section the solution procedure for structures with no joint translations ($\psi = 0$) will be developed. It is assumed that (1) the members are prismatic, (2) they are made from linear elastic materials obeying Hooke's law, and (3) axial deformations can be ignored. The method will be described using the four-member frame in Fig. 12.10a. There is only one unknown rotation for this structure since the members are joined at point a with a moment connection; i.e., the angles between members remain the same before and after the loading is applied. The necessary equation for

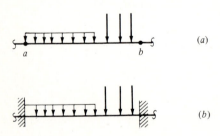

(a)

(b)

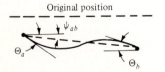

Original position

(c)

Figure 12.9 The effects introducing moments into a bending member: (a) the loaded beam; (b) the fixed-end beam; (c) the deflected beam, joint rotations and translations.

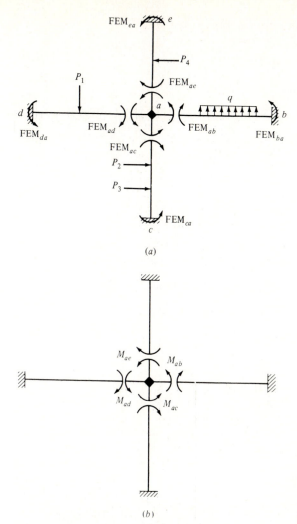

Figure 12.10 Bending moments for a frame structure without lateial sway: (a) fixed-end moments; (b) moments due to rotation of joint a.

solving the structure can be written by enforcing moment equilibrium at joint a (see Fig. 12.10b), which gives

$$M_{ab} + M_{ac} + M_{ad} + M_{ae} = 0 \qquad (a)$$

where the first and second subscripts on the moments denote the near and far ends of the member, respectively. The sign convention for moments used in the slope-deflection method is also used here; i.e., positive end moments act clockwise on the member. All the moments have been shown in Fig. 12.10b acting positive. When the slope-deflection equations (12.7) are used, the end moments at point a can be expressed as

$$M_{ab} = 4EK_{ab}\Theta_a + \text{FEM}_{ab} \qquad M_{ac} = 4EK_{ac}\Theta_a + \text{FEM}_{ac}$$
$$M_{ad} = 4EK_{ad}\Theta_a + \text{FEM}_{ad} \qquad M_{ae} = 4EK_{ae}\Theta_a + \text{FEM}_{ae} \qquad (b)$$

where the *stiffness factor* K_{am} for a typical member *am* is

$$K_{am} = \frac{I_{am}}{L_{am}} \tag{12.13}$$

Substituting these into Eq. (*a*) gives

$$4E\Theta_a(K_{ab} + K_{ac} + K_{ad} + K_{ae}) = -(\text{FEM}_{ab} + \text{FEM}_{ac} + \text{FEM}_{ad} + \text{FEM}_{ae}) = M_a \tag{c}$$

The sum of the fixed-end moments is the *unbalanced moment*, and M_a is the unbalanced moment at joint *a* with the sign reversed. Solving Eq. (*c*) gives

$$\Theta_a = \frac{M_a}{4E \sum_a K} \tag{d}$$

where

$$\sum_a K = K_{ab} + K_{ac} + K_{ad} + K_{ae} \tag{e}$$

Note that the summation is taken over all members meeting at joint *a*. Substituting Eq. (*d*) into Eq. (*b*) we get the moments at the ends of the members.

$$M_{ab} = \text{DF}_{ab} M_a + \text{FEM}_{ab} \qquad M_{ac} = \text{DF}_{ac} M_a + \text{FEM}_{ac}$$
$$M_{ad} = \text{DF}_{ad} M_a + \text{FEM}_{ad} \qquad M_{ae} = \text{DF}_{ae} M_a + \text{FEM}_{ae} \tag{f}$$

For a typical member *am* the dimensionless constant DF_{am}, the *distribution factor*, is

$$\text{DF}_{am} = \frac{K_{am}}{\sum_a K} \tag{12.14}$$

Note that all the distribution factors at a joint must sum to 1. The product of the distribution factor and the moment M_a shown in Eq. (*f*) is the *distributed moment*. From Eq. (*f*) it can be observed that the moments at the ends of the members meeting at point *a* are the sum of the distributed moment and the fixed-end moment.

The concepts introduced above in a mathematical form can be summarized for this structure in terms of physical behavior. First, joint *a* is *locked* against rotation, which requires the *fixed-end moments*. The sum of all the fixed-end moments at the joint, called the *unbalanced moment*, represents the total constraining moment necessary to lock the joint. Second, the joint is *unlocked* by applying the *balancing moment*, i.e., the unbalanced moment with the sign reversed, which causes the joint to rotate. This moment is distributed to the members connected to the joint in proportion to their distribution factors; these are the distributed moments. Thus the joint is placed in moment equilibrium, and the moments at the ends of the members (at the joint) are the sum of the fixed-end moments and the distributed moments.

Equations (*f*) give the moments at point *a* for each of the members. The moments at the far ends can be calculated using the slope-deflection equations. For example, the moment at *b* for member *ab* is

$$M_{ba} = 2EK_{ab}\Theta_a + \text{FEM}_{ba} \tag{g}$$

Substituting the calculated value for Θ_a and using the notation introduced previously gives

Figure 12.11 Moment at the fixed end due to rotation at the simply support end.

$$M_{ba} = \tfrac{1}{2} \mathrm{DF}_{ab} M_a + \mathrm{FEM}_{ba} \tag{h}$$

That is, when the distributed moment $\mathrm{DF}_{ab} M_a$ is applied to end a of member ab, $\tfrac{1}{2}\mathrm{DF}_{ab} M_a$ will be *carried over* to end b of the member. The ratio of these two end moments is given by the dimensionless constant called the *carryover factor*

$$C_{ab} = \tfrac{1}{2} \tag{12.15}$$

The moment $\tfrac{1}{2}\mathrm{DF}_{ab} M_a$ is the *carryover moment*. The distributed and carryover moments for member ab are illustrated in Fig. 12.11. Note that the carryover moment has the same sign as the distributed moment.

The physical explanation of the behavior of the four-member frame has one final basic step. When the joint a rotates under the effect of the balancing moment, the distributed moments are developed in the members meeting at the joint. As these four members deform, the carryover moments are developed at the far ends of the members. The final moments at the far ends of the members are the sum of the carryover and fixed-end moments.

In general, the moments acting on the ends of the individual members for structures without joint translation ($\psi = 0$) can be envisioned as being composed of (1) fixed-end moments, (2) distributed moments, and (3) carryover moments. The behavior of structures with only prismatic members has been discussed in this section. When members with variable I are included, these concepts must be generalized, which is done in Sec. 12.7. Carrying out the moment-distribution process numerically is illustrated in Example 12.5 for the four-member frame.

Example 12.5 Using the moment-distribution method, calculate the moments for (a) $I_{ab} = I_{ac} = I_{ad} = I_{ae} = I$; and (b) $I_{ab} = I_{ac} = I_{ae} = I$ and $I_{ad} = 3I$. For both cases, $L_{ab} = L_{ac} = L_{ad} = L_{ae} = L = 12$ ft.

SOLUTION (a)

$$\mathrm{FEM}_{da} = -\frac{PL}{8} = -\frac{64(12)}{8} = -96 \text{ kip} \cdot \text{ft} \qquad \mathrm{FEM}_{ad} = -\mathrm{FEM}_{da}$$

$$K_{ab} = K_{ac} = K_{ad} = K_{ae} = \frac{I}{12}$$

$$\sum_a K = 4\frac{I}{12} = \frac{I}{3} \qquad \mathrm{DF}_{am} = \frac{I/12}{I/3} = 0.25$$

See Fig. E12.5b.

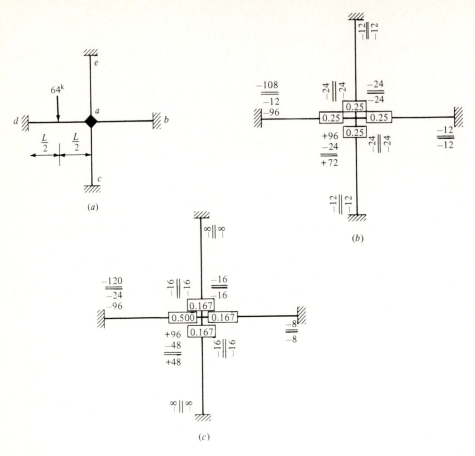

Figure E12.5 (a) Loaded rigid frame; (b) moment distribution for case a; (c) moment distribution for case b.

(b)
$$\text{FEM}_{ad} = -\text{FEM}_{da} = 96 \text{ kip} \cdot \text{ft}$$

$$K_{ad} = \frac{I}{4} \qquad K_{ab} = K_{ac} = K_{ae} = \frac{I}{12}$$

$$\sum_a K = \frac{6I}{12} \qquad \text{DF}_{ad} = 0.500 \qquad \text{DF}_{ab} = \text{DF}_{ac} = \text{DF}_{ae} = 0.167$$

See Fig. E12.5c.

DISCUSSION The distribution factors are shown on the members in square boxes, and the FEM are indicated on the ends of member ab. The second step of the solution involves applying the balancing moment (-96 kip · ft) to the members at joint a and distributing it using the four distribution factors. Half of these distributed moments must be carried over to the far ends of the members. The moments at the joints are the sum of the fixed-end moments, distributed moments, and carryover moments.

If the structure is modified so that the moment of inertia for ad is $3I$, the moments are calculated as shown in Fig. E12.5c. Note that this structural modification changes the factors — and also the distributed moments — from those obtained in the first configuration of the frame. The final moment at end d of member ad is larger than that for the structure with all members the same. This type of behavior is generally true for statically indeterminate structures; i.e., stiffer regions have a proportionately larger share of the total moment. Structural designers should be aware that the stiffness properties of the individual members in statically indeterminate structures influence the distribution of the bending moments throughout the entire structure.

12.5 MOMENT DISTRIBUTION APPLIED TO BEAMS AND FRAMES

The basic steps and concepts of the moment-distribution method will be used to investigate various structures with no joint translation ($\psi = 0$). The examples increase in complexity so that the procedure and some of its application refinements will be illustrated.

The structure shown in Fig. 12.12 is similar to the four-member frame discussed previously in that there is only one unknown angle of rotation. The stiffness factors, distribution factors, and fixed-end moments are shown in the tabular array. Since joints a and c are permanently fixed against rotation, only joint b can be unlocked; this results in the distributed moments shown. The carryover moments at joints a and c are one-half of the distributed moments at end b of the two members. The total moments are obtained by summing the fixed-end moments, the distributed moments, and the carryover moments. Thus, the solution for this structure consists of locking and unlocking joint b.

For the three-span beam in Fig. 12.13a the stiffness factors shown below each beam member give

$$\sum_b K = \frac{I}{6} \qquad DF_{ba} = \frac{I/9}{I/6} = \frac{2}{3} \qquad DF_{bc} = \frac{I/18}{I/6} = \frac{1}{3}$$

$$\sum_c K = \frac{5I}{36} \qquad DF_{cb} = \frac{I/18}{5I/36} = \frac{2}{5} \qquad DF_{cd} = \frac{I/12}{5I/36} = \frac{3}{5}$$

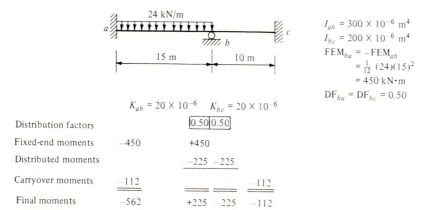

Distribution factors		0.50	0.50	
Fixed-end moments	−450	+450		
Distributed moments		−225	−225	
Carryover moments	−112			−112
Final moments	−562	+225	−225	−112

$I_{ab} = 300 \times 10^{-6}\ m^4$
$I_{bc} = 200 \times 10^{-6}\ m^4$
$FEM_{ba} = -FEM_{ab}$
$= \frac{1}{12}(24)(15)^2$
$= 450\ kN \cdot m$
$DF_{ba} = DF_{bc} = 0.50$

$K_{ab} = 20 \times 10^{-6} \qquad K_{bc} = 20 \times 10^{-6}$

Figure 12.12 Moment distribution for a beam with two spans.

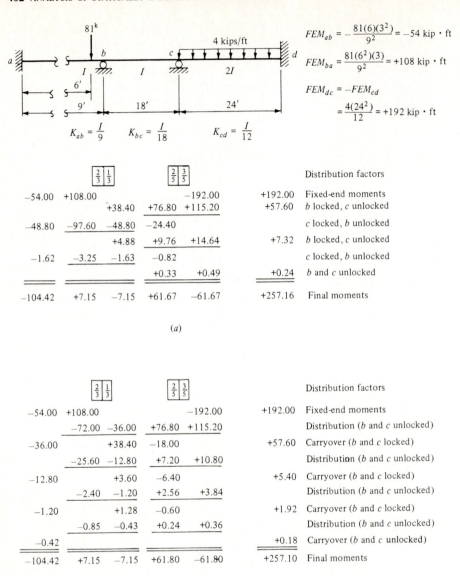

Figure 12.13 Moment distribution for a beam with three spans: (*a*) alternate locking and unlocking of joints; (*b*) simultaneous locking and unlocking of both joints.

The distribution factors are shown in the boxes in Fig. 12.13*a*, and the fixed-end moments for the two loaded spans are also shown. Although the computations need not necessarily be arranged as illustrated, the student must adopt some systematic tabular array for carrying out the calculations (see Table E12.6 for a different arrangement).

Initially, joints *b* and *c* are locked against rotation, which requires that the fixed-end moments be applied. Next, joint *c* is unlocked and allowed to rotate while all the

Table 12.1 Change in M_{cd}
(Fig. 12.13a)

M_{cd}, kip · ft (Iteration)	Change in value, kip · ft
−76.80 (After first)	
−62.16 (After second)	+14.64
−61.67 (After third)	+0.49

other joints remain fixed; this requires that a balancing moment of +192 kip · ft be applied to the members at joint c. This moment is distributed to members cb and cd in proportion to their distribution factors DF_{cb} and DF_{cd}. Since this process has allowed joint c to rotate, carryover moments will occur at the far ends of the two members which are equal to half of the distributed moments. This results in moments of +38.4 and +57.6 kip · ft at ends b and d, respectively. At this point in the computations, joint c is in equilibrium as indicated by the line drawn under the two distributed moments at that joint. Joint c is now locked in this position.

Joint b has remained locked, and as a result it is out of equilibrium. The moment required to balance this joint is

$$-(108.0 + 38.4) = -146.4 \text{ kip} \cdot \text{ft}$$

This is distributed to members ab and bc according to their distribution factors. Joint b is now in equilibrium, and it can be locked in this position. Note that as a result of this distribution at joint b, a moment was carried over to joint c (joint c is no longer in equilibrium).

Joint c must be unlocked and the new balancing moments distributed. The alternate locking and unlocking of joints b and c continues until the moments at the joints have converged to a solution of the desired accuracy. Since the moment-distribution method is an iterative solution for the equilibrium equations, it is a simple matter of comparing the moments at two succeeding iterations to determine the accuracy. For example, from Fig. 12.13a we can draw the conclusions shown in Table 12.1. There was a change of less than 1 percent in the last iteration, which is sufficiently accurate, and the process can be terminated at that point. This logic can be used to estimate the accuracy for the moments at each joint. The number of significant figures used for the calculations is based on such factors as the precision of results required, the accuracy of the loads, etc. Note that all the balancing moments should be distributed at each distribution cycle and the carryover moments are rounded to the number of significant figures. Example 12.6 compares an iterative solution of the slope-deflection equations with the solution using the moment-distribution method.

Example 12.6 Calculate the moments at points a, b, c, and d for the beam of Example 12.1 using the moment-distribution method.

SOLUTION See Fig. E12.6.

DISCUSSION The slope-deflection equations obtained in Example 12.1 were solved by Gauss-Seidel iteration. A comparison of the moment-distribution results with those obtained

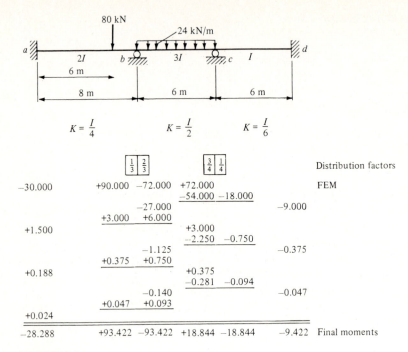

Figure E12.6

Table E12.6 Moment-distribution table for Example 12.6

Joint	a	b		c		d
Member	ab	ba	bc	cb	cd	dc
K	I/3	I/3	I/2	I/2	I/6	I/6
DF	0	1/3	2/3	3/4	1/4	0
FEM	−30.000	+90.000	−72.000	+72.000		
Distribution				−54.000	−18.000	
Carryover			−27.000			−9.000
Distribution		+3.000	+6.000			
Carryover	+1.5000			+3.000		
Distribution				−2.250	−0.750	
Carryover			−1.125			−0.375
Distribution		+0.375	+0.750			
Carryover	+0.188			+0.375		
Distribution				−.281	−0.094	
Carryover			−0.140			−0.047
Distribution		+0.047	+0.093			
Carryover	+0.024					
Total	−28.288	+93.422	−93.422	+18.844	−18.844	−9.422

by the Gauss-Seidel method (Table E12.1) shows that the two approaches give identical solutions. The Gauss-Seidel calculations were extended for one more iteration for M_{ba} and two more iterations for M_{cd}, but the results up to these points are identical. The moment-distribution method is commonly referred to as a relaxation method wherein the residual is calculated at each iteration.[1]

The format for the calculations in Fig. E12.6 will be used throughout this text; another format in use is shown in Table E12.6.

This method of successive approximations for the structure in Fig. 12.13a could have been executed in a different way, as demonstrated in Fig. 12.13b. After all the joints have been locked (which results in the fixed-end moments), joints b and c can be simultaneously unlocked. The corresponding calculations are shown on the line labeled "Distribution" in Fig. 12.13b. Next, both these interior joints are locked, and the carryover moments are computed simultaneously for all joints in the structure. The process continues until the final moments with proper accuracies are obtained. Note that a line has been drawn under the moments at each joint when it is in equilibrium. The final step in the process consists of carrying moments to joints a and d since both are physically fixed. Moments have not been carried over to joints b and c as well because that would have required these joints to be locked; i.e., it would have taken these joints out of equilibrium and required another distribution.

The calculation of the stiffness factors, distribution factors, and fixed-end moments for the beam shown in Fig. 12.14a are carried out in the usual manner. After locking all joints and applying the fixed-end moments, we allow the joints to rotate by unlocking each in turn. The iterative process is carried out, and the final moments are obtained by summing the fixed-end, distributed, and carryover moments.

This beam with a simple support at a can also be analyzed in a different manner. After applying the fixed-end moments with all joints locked, it is advantageous to unlock joint a permanently. Hence, before proceeding it is necessary to develop a stiffness factor for a member that is simply supported at one end, as shown in Fig. 12.15a. Using the slope-deflection equations and noting that the moment at a is zero, we get

$$M_{ba} = 3E \frac{I}{L} \Theta_b = 4E(\tfrac{3}{4}K_{ba})\Theta_b \qquad (12.16)$$

where the *modified stiffness factor* K_{ba}^M is

$$K_{ba}^M = \tfrac{3}{4}K_{ba} \qquad (12.17)$$

Hence Eq. (12.16) can be written

$$M_{ba} = 4EK_{ba}^M\Theta_b \qquad (12.18)$$

Since the stiffness factor for a member bm can be either K_{bm} or K_{bm}^M, depending upon whether end m is fixed or simply supported, respectively, the definition of the distribution factor (12.14) must be modified to read

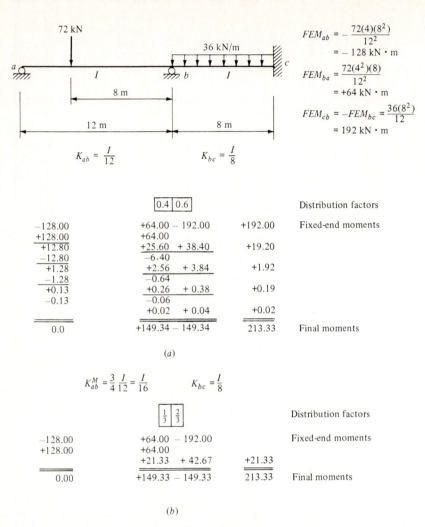

$$FEM_{ab} = -\frac{72(4)(8^2)}{12^2}$$
$$= -128 \text{ kN} \cdot \text{m}$$

$$FEM_{ba} = \frac{72(4^2)(8)}{12^2}$$
$$= +64 \text{ kN} \cdot \text{m}$$

$$FEM_{cb} = -FEM_{bc} = \frac{36(8^2)}{12}$$
$$= 192 \text{ kN} \cdot \text{m}$$

Figure 12.14 Moment distribution for a beam with a simple support: (*a*) use of standard stiffness factors; (*b*) use of modified stiffness factor.

$$DF_{bm} = \frac{(K_{bm})_{\text{eff}}}{\sum_b K_{\text{eff}}} \qquad (12.14a)$$

where K_{eff} designates the effective stiffness of the member, i.e.,

$$K_{\text{eff}} = \begin{cases} K & \text{end fixed} \\ K^M & \text{far end simply supported} \end{cases}$$

Since the modified stiffness factor implies that the far end is simply supported, the moment carried over to this end is zero.

The analysis of the beam in Fig. 12.14*a* using the modified stiffness factor is shown in Fig. 12.14*b*. After applying the fixed-end moments with all joints locked, we unlock joint *a* to allow rotation. Half of the distributed moment is carried over to

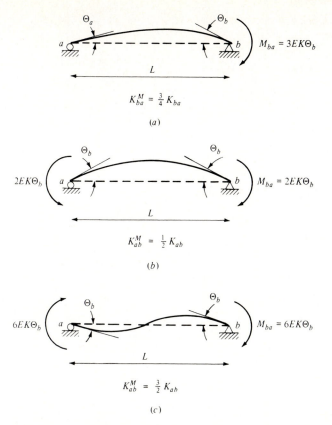

Figure 12.15 Modified stiffness factors: (*a*) beam with far end simply supported; (*b*) symmetric beam; (*c*) antisymmetric beam.

joint b since it has remained locked. After joint b has been unlocked and half the moment at end b of span bc has been carried over to c, the iteration process is complete. The moments at the three joints are summed to give the final moments. The convergence is obtained quickly by using the modified stiffness factor for member ab, and in this case the solution is the exact answer. Also, calculation and truncation errors are avoided or minimized with fewer iterations.

The modified stiffness factor for a member with one end simply supported is a very useful concept and relatively simple to use. There are other cases where modified stiffness factors are appropriate. For example, if a structure deflects symmetrically, as shown in Fig. 12.15*b*, the slope-deflection equations can be used to show that $-M_{ab} = M_{ba} = 2EK\Theta_a$ and $K_{ab}^M = \frac{1}{2}K_{ab}$. Similarly, for antisymmetrical deflections (Fig. 12.15*c*) it can be demonstrated that $K_{ab}^M = \frac{3}{2}K_{ab}$. Since these modified stiffness factors imply a known relationship between the end moments, in these cases the carryover factors are zero. Using these modified stiffness factors will reduce the number of iterations for a given structure. Since they may also confuse the initial learning process, it is advisable for students to postpone using these special stiffness factors until the basic moment-distribution method has been mastered.

The moment-distribution method applied to a frame without lateral displacement is presented in Example 12.7.

Example 12.7 Calculate the moments at all joints on this rigid-frame structure using the moment-distribution method.

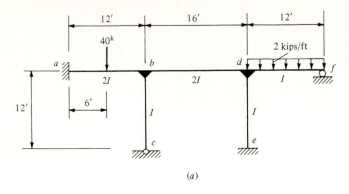

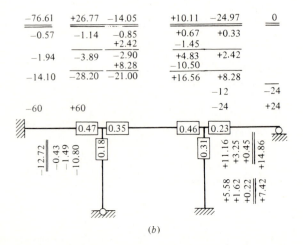

(b)

Figure E12.7 (a) Loaded rigid frame; (b) moment-distribution calculations.

SOLUTION

$$K_{ab} = \frac{I}{6} \quad K_{bc}^M = \frac{3}{4}\frac{I}{12} = \frac{I}{16} \quad K_{bd} = \frac{I}{8} \quad K_{de} = \frac{I}{12} \quad K_{df}^M = \frac{3}{4}\frac{I}{12} = \frac{I}{16}$$

$$\sum_b K = \frac{17I}{48} \qquad DF_{ba} = \frac{I/6}{17I/48} = 0.47 \qquad DF_{bd} = \frac{I/8}{17I/48} = 0.35$$

$$DF_{bc} = \frac{I/16}{17I/48} = 0.18$$

$$\sum_d K = \frac{13I}{48} \qquad DF_{db} = \frac{I/8}{13I/48} = 0.46 \qquad DF_{df} = \frac{I/16}{13I/48} = 0.23$$

$$DF_{de} = \frac{I/12}{13I/48} = 0.31$$

$$\text{FEM}_{ba} = -\text{FEM}_{ab} = \tfrac{1}{8}(40)(12) = 60 \text{ kip} \cdot \text{ft}$$

$$\text{FEM}_{fd} = -\text{FEM}_{df} = \tfrac{1}{12}(2)(12)^2 = 24 \text{ kip} \cdot \text{ft}$$

See Fig. E12.7*b*.

12.6 MOMENT DISTRIBUTION APPLIED TO STRUCTURES WITH JOINT TRANSLATIONS

The moment-distribution method will be expanded in this section to include the effects of joint translations ($\psi \neq 0$). Before proceeding with the analysis we must calculate an expression for the end moments associated with this effect. The beam member illustrated in Fig. 12.16 rotates in a clockwise direction through an angle ψ because the ends of the member are displaced differentially by an amount Δ; therefore,

$$\psi = \frac{\Delta}{L}$$

The ends of the member are constrained against rotation by the applied moments, which can be calculated using the slope-deflection equations (12.12)

$$M_{ab} = M_{ba} = 2EK(-3\psi) = -6EK\frac{\Delta}{L} = -6EI\frac{\Delta}{L^2} \qquad (12.19)$$

These are the *initial end moments* required to lock the joints against rotations. Note in Fig. 12.16 that the angle of rotation ψ for the member has been shown in the positive (clockwise) direction, resulting in negative (counterclockwise) moments at the ends to enforce the condition of no rotation at these points.

The moment-distribution process is carried out in the format used when loads are present, these initial end moments replacing the fixed-end moments. If both applied loading and joint translations are to be considered, the fixed-end moments due to the loads and initial end moments due to the translations are combined before carrying out the analysis. Alternatively, an independent distribution can be carried out for each of these effects and the two analyses combined to give the total solution. This latter approach will be used in this section for frame structures. A continuous beam with support settlements is analyzed using moment distribution in Example 12.8.

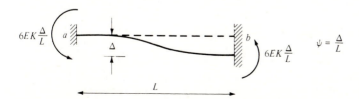

Figure 12.16 Initial end moments for a beam with joint translation.

Example 12.8 Calculate the moments at points a, b, c, and d using moment distribution if supports b and c settle 5.00 and 12.5 mm, respectively; $I = 800 \times 10^{-6}$ m⁴; $E = 200$ GPa.

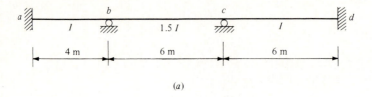

(a)

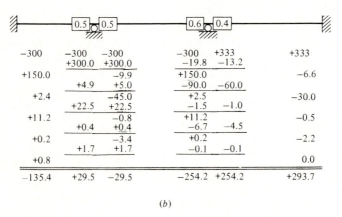

−300	−300	−300		−300	+333		+333
	+300.0	+300.0		−19.8	−13.2		
+150.0		−9.9		+150.0			−6.6
	+4.9	+5.0		−90.0	−60.0		
+2.4		−45.0		+2.5			−30.0
	+22.5	+22.5		−1.5	−1.0		
+11.2		−0.8		+11.2			−0.5
	+0.4	+0.4		−6.7	−4.5		
+0.2		−3.4		+0.2			−2.2
	+1.7	+1.7		−0.1	−0.1		
+0.8							0.0
−135.4	+29.5	−29.5		−254.2	+254.2		+293.7

(b)

Figure E12.8 (a) Continuous beam; (b) moment-distribution calculations.

SOLUTION

$$K_{ab} = K_{bc} = \frac{I}{4} \qquad K_{cd} = \frac{I}{6} \qquad \sum_b K = \frac{I}{2} \qquad \sum_c K = \frac{5I}{12}$$

$$DF_{ba} = DF_{bc} = \frac{I/4}{I/2} = 0.50 \qquad DF_{cb} = \frac{I/4}{5I/12} = 0.60 \qquad DF_{cd} = \frac{I/6}{5I/12} = 0.40$$

$$M_{ba} = M_{ab} = -\frac{6EI\Delta}{L^2} = -\frac{6(200 \times 10^6)(800 \times 10^{-6})(0.005)}{16} = -300 \text{ kN} \cdot \text{m}$$

$$M_{bc} = M_{cb} = -\frac{6(200 \times 10^6)(1200 \times 10^{-6})(0.0075)}{36} = -300 \text{ kN} \cdot \text{m}$$

$$M_{cd} = M_{dc} = \frac{6(200 \times 10^6)(800 \times 10^{-6})(0.0125)}{36} = 333 \text{ kN} \cdot \text{m}$$

See Fig. E12.8b.

DISCUSSION The initial end moments are computed using Eq. (12.19) with the differential displacements used for the values of Δ in the equation. For example,

$$\Delta_{bc} = 12.50 \text{ mm} - 5.00 \text{ mm} = 7.50 \text{ mm}$$

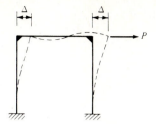

Figure 12.17 Frame structure with sidesway.

The stiffness factors and distribution factors are computed and used in the same way as in all previous moment-distribution analyses. The final moments are obtained by summing the initial end moments, the distributed moments, and the carryover moments.

All the distributed and carryover moments are rounded to one significant figure beyond the decimal place. This approach gives distributed moments at joint b for the second distribution which appear to violate the distribution factors; however, joint equilibrium is satisfied. Since moment distribution is an approximate method, numerical accuracies must be considered carefully.

The joints of rigid frames can also translate when load is applied to the structure; this is termed sway (or sidesway) behavior. The applied loading will cause joints b and c of the frame in Fig. 12.17 to displace laterally, and both these displacements will be equal if the axial deformations of the members are neglected. This structure can be analyzed in much the same way as that demonstrated for beams with support movements. In this case, however, the magnitude of the joint translation is unknown, and the initial end moments cannot be explicitly calculated. If an arbitrary value of the displacement is selected, the corresponding initial end moments can be calculated using Eq. (12.19). After the moment-distribution process is complete, an equation of equilibrium can be written relating the base shears to the applied load. In general, the base shears resulting from the assumed joint displacements will not equilibrate the applied loading; however, since the structure behaves in a linear fashion, the final moments can be obtained using a simple proportion. The solution for frames with sidesway is shown in Examples 12.9 and 12.10.

Example 12.9 Calculate the moments in this rigid frame using moment distribution; EI = const.

SOLUTION See Fig. E12.9b to d.

DISCUSSION Initially all the joints are locked and the frame is allowed to translate laterally as shown. It can be observed from this deflected shape that the moments in the columns at all four points will be equal and there will be no initial moments in member bc. Since the horizontal deflections are unknown, an arbitrary value for the column moments is assumed and the moments computed by moment distribution. These calculations have been carried out to an extreme accuracy to obtain the exact solution (the fact that the moments are exact is verified by comparing the final results with the slope-deflection solution shown in Example 12.4). The column base shears are obtained by enforcing moment equilibrium for each individual column (Fig. E12.9c). This yields

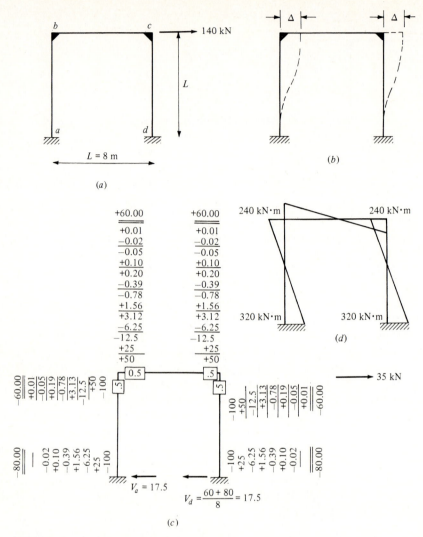

Figure E12.9 (*a*) The loaded rigid frame; (*b*) sidesway deflections with all joints locked; (*c*) moment-distribution calculations; (*d*) final moments.

$$V_a = V_d = \frac{60.0 + 80.0}{8} = 17.5 \text{ kN}$$

This obviously is not the correct solution since the sum of the shears must equal the applied horizontal load at joint *c*, or 140 kN. The initial end moments (and hence the joint displacements) were incorrectly assumed, but since the moments are linearly related to the displacement the moments can be adjusted using a multiplication factor. Thus, if *X* is the multiplier that must be applied to obtain the correct solution, equilibrium in the horizontal direction gives

$$140.0 - (17.5 + 17.5)X = 0$$

which yields $X = 4.0$. The final moments are obtained by multiplying all the moments by this factor.

Example 12.10 Calculate the moments in this rigid frame using moment distribution. $E = $ const.

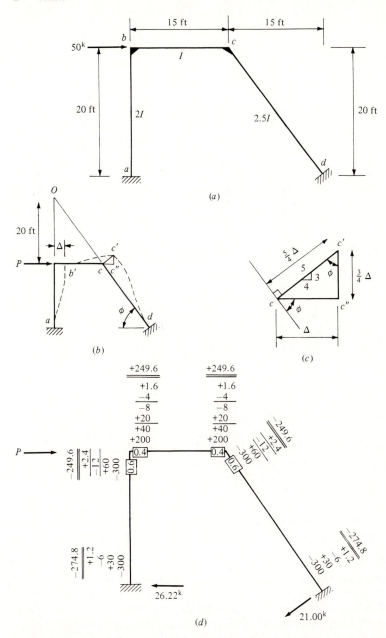

Figure E12.10 (*a*) Loaded rigid frame; (*b*) deflected structure with all joints locked; (*c*) detail of deflected joint *c*; (*d*) moment-distribution calculations.

SOLUTION From Fig. E12.10a

$$K_{ab} = \frac{2I}{20} \qquad K_{bc} = \frac{I}{15} \qquad K_{cd} = \frac{2.5I}{25} \qquad \sum_b K = \sum_c K = \frac{I}{6}$$

$$DF_{ba} = \frac{I/10}{I/6} = 0.60 \qquad DF_{bc} = \frac{I/15}{I/6} = 0.40$$

$$DF_{cb} = \frac{I/15}{I/6} = 0.40 \qquad DF_{cd} = \frac{I/10}{I/6} = 0.60$$

From Fig. E12.10b and c the initial end moments are

ab:
$$\frac{6E2I\Delta}{(20)^2} = 0.03EI\Delta \propto 300 \text{ kip} \cdot \text{ft}$$

bc:
$$\frac{6EI(\frac{3}{4}\Delta)}{(15)^2} = 0.02EI\Delta \propto 200 \text{ kip} \cdot \text{ft}$$

cd:
$$\frac{6E(2.5I)(\frac{5}{4}\Delta)}{(25)^2} = 0.03EI\Delta \propto 300 \text{ kip} \cdot \text{ft}$$

See Fig. E12.10d.
From Fig. E12.10b and d

$$\sum M_0 = 0: \qquad 20P = 26.22(40) + 21.00(50) - 274.8 - 274.8$$

$$P = 77.46 \text{ kips}$$

$$\text{Multiplier to obtain final moments} = \frac{50}{77.46} = 0.6455$$

From Fig. E12.10d the final moments are

$$M_{ab} = M_{dc} = 0.6455(-274.8) = -177.4 \text{ kip} \cdot \text{ft}$$

$$M_{bc} = M_{cb} = -M_{ba} = -M_{cd} = 0.6455(249.6) = 161.1 \text{ kip} \cdot \text{ft}$$

DISCUSSION The deflection of this slant-leg bent results in different displacements for each of the three members. The deflection behavior is more complex than that displayed by the frame in Example 12.9. It is extremely important to construct an accurate picture of the deflected shape with all the joints locked (Fig. E12.10b). To do so it is assumed that (1) the deflections are small compared with the member lengths (this implies that the length of a member in its curved configuration equals the original length); (2) the axial deformations of members are neglected; and (3) the end of a deflected member forms a circular arc which can be approximated by the tangent to the arc normal to the undeflected member orientation. From the deflected shape it can be seen that all deflections can be expressed in terms of only one unknown; a set of consistent initial end moments is calculated using Eq. (12.19).

When the moment distribution is complete, the base shears are computed. The applied load P corresponding to the initial end moments can conveniently be calculated for this structure by summing moments about point O because the reaction components parallel to members ab and cd pass through this point. The final moments are obtained by multiplying the results from the moment distribution by the calculated multiplier.

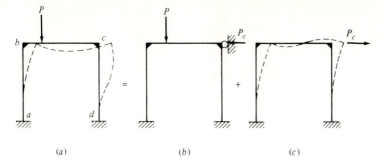

Figure 12.18 Solution procedure for a frame with vertical loading displaying sidesway: (*a*) the loaded symmetrical frame; (*b*) case 1: solution with sidesway prevented; (*c*) case 2: solution for sidesway only.

A rigid frame loaded with vertical loading can exhibit sidesway. If the symmetrical frame in Fig. 12.18 has a vertical load applied near joint *b*, the structure will translate to the right. This introduces an unknown joint displacement, and the effects of joint translation must be considered. It is convenient in this case to carry out the analysis by considering two individual load cases (Fig. 12.18). In case 1 an imaginary support is placed at joint *c* to prevent any lateral sway. Then the fixed-end moments are applied and the moment distribution carried out. The shear at the base of each column is computed using statics, and the sum of these shears is the magnitude of the force that must be applied at the artificial support. If these calculations had indicated that the resultant of the base shears was zero, the frame would not exhibit sidesway behavior. For case 2 a force P_c with the direction reversed is applied to the frame, which results in sidesway of the frame. The sidesway is analyzed in the usual way by assuming a set of consistent initial end moments and distributing them throughout the frame. The moments for case 2 are calculated by proportioning to correspond to a lateral force of P_c. The final moments are obtained by combining the results from the case 1 and case 2 solutions. Calculations for sidesway behavior of two frames with vertical loads are shown in Examples 12.11 and 12.12.

Example 12.11 Calculate the moments in this frame using moment distribution; EI = const.

SOLUTION

$$\text{FEM}_{bc} = -\frac{40(2)(36)}{64} = -45 \text{ kN} \cdot \text{m} \qquad \text{FEM}_{cb} = \frac{40(4)(6)}{64} = 15 \text{ kN} \cdot \text{m}$$

See Fig. E12.11*b* for the moment distribution; the shear at the base of the columns is

$$V_a = \frac{25.90 + 12.95}{8} = 4.86 \text{ kN} \ (\rightarrow) \qquad V_d = \frac{13.95 + 6.97}{8} = 2.61 \text{ kN} \ (\leftarrow)$$

The results from Example 12.9 are used to find the multiplier for sidesway moments

$$\frac{2.25}{35} = 0.0643$$

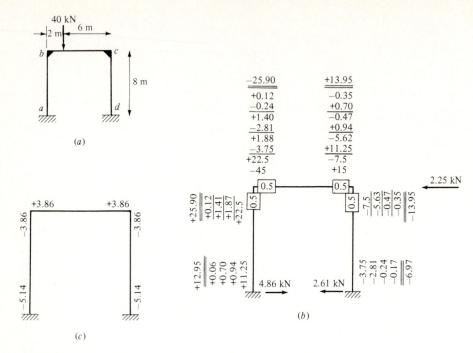

Figure E12.11 (*a*) Loaded rigid frame; (*b*) moment distribution with sidesway prevented; (*c*) moments for sidesway only.

Multiplying the final moments in Fig. E12.9*c* by 0.0643 gives the sidesway moments in Fig. E12.11*c*. From Fig. E12.11*b* and *c* the final moments are

$$M_{ab} = +12.95 - 5.14 = +7.81 \text{ kN} \cdot \text{m}$$

$$M_{ba} = -M_{bc} = +25.90 - 3.86 = +22.04 \text{ kN} \cdot \text{m}$$

$$M_{cb} = -M_{cd} = +13.95 + 3.86 = +17.81 \text{ kN} \cdot \text{m}$$

$$M_{dc} = -6.97 - 5.14 = -12.11 \text{ kN} \cdot \text{m}$$

DISCUSSION The solution with sidesway prevented is carried out by moment distribution of the fixed-end moments induced by the loading. The results of this analysis indicate that a load of 2.25 kN must be applied horizontally at joint *c* to constrain the frame from lateral sway. Since the results for sidesway of the frame have been obtained in Example 12.9, proportioning all the moments from this previous example gives the moments for an applied load (horizontal force at *c*) of 2.25 kips. The final moments are the sum of these two solutions. Note that the results are comparable to those obtained for this frame using the slope-deflection method in Examples 12.3 and 12.4.

Example 12.12 Calculate the moments in this frame using moment distribution; *E* = const.

SOLUTION

$$\text{FEM}_{bc} = -\frac{5qL^2}{192} = -\frac{5(4)(24)^2}{192} = -60 \text{ kip} \cdot \text{ft}$$

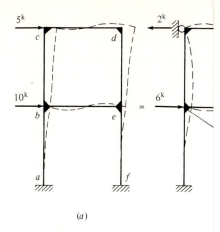

The carryover

Similarly, fixi

and

It can be obse
If the member
in general, th
member, C_{ab}

The carryove
   The ini
members usi
through an
Eq. (12.11a

For symme

For prisma

where X and Y are the multipliers fo
X and Y the moments for each of the t
yield the final solution for the actual

   The independent degrees of freedo
arbitrary. Alternatively, joints b and
case 1, with case 2 remaining the sam
end moments and total moments for ca
also have been different, but the final
regardless of how the independent de

   This same procedure can also be u
freedom with respect to sidesway. T
solutions, and hence the number of u
degrees of freedom. As the number
computations. Sometimes, an additio
conventional moment distribution to s
recalled that the slope-deflection equat
of rotation and the displacements and
to recognize that all methods have thei
probably advisable to use a method t
(Part Five).

Figure 12.19 Solution procedure for a structu
(a) the structure with actual loading; (b) case 1
Y.

12.8 SY

In using ei
symmetric
number of
solution.
symmetri
symmetri
shape is s
$\Theta_c = 0.$
rotation i
actual be

   Ant
five angle
be conclu
could als

12.7 MOMENT DISTRIBUTI

The cross section of a beam frequently
steel-plate girder construction, where the
accommodate higher moments. The m
nonprismatic members, but the elastic p
(fixed-end and initial end moments) must

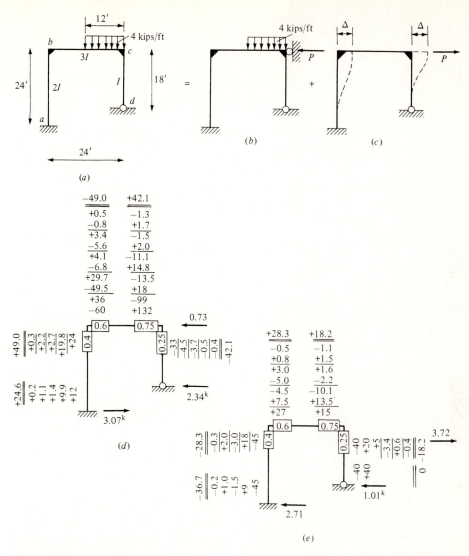

Figure E12.12 (a) Loaded rigid frame; (b) case 1: sidesway prevented; (c) case 2: sidesway only; (d) moment distribution for case 1; (e) moment distribution for case 2.

$$\text{FEM}_{cb} = \frac{11qL^2}{192} = \frac{11(4)(24)^2}{192} = 132 \text{ kip} \cdot \text{ft}$$

See Fig. E12.12d.
   Initial end moments for members are

ab:
$$-\frac{6E(2I)\Delta}{(24)^2} \propto -45 \text{ kip} \cdot \text{ft}$$

cd:
$$-\frac{6EI\Delta}{(18)^2} \propto -40 \text{ kip} \cdot \text{ft}$$

See Fig. E12.12*e*. The multip
Fig. E12.12*d* and *e* the final r

$$M_{ab} = +24.6 -$$

$$M_{ba} = -M_{bc} =$$

$$M_{cb} = -M_{cd} =$$

DISCUSSION The analysis of thi
prevented with the loads applie
for member *cd*. The fixed-end r
for a single concentrated load sh
origin is placed at *c* with the *x*

FEM$_{bc}$

and   FEM$_{cb}$

The results of the moment distrit
at point *c*. From the deflected s
that there will be initial end mor
Eq. (12.19). The final step in the
two load cases.

The sidesway of the structur
one lateral deflection. For exan
Fig. 12.17 can all be written in ter
in Fig. 12.19 has two independe
freedom with respect to sidesway
ments plus the rotations of eac
equilibrium must be written for the
implicit in the moment-distributic

The two relative horizontal di
as the independent degrees of free
for the case 1 analysis, and the
maintain this deflected shape are
the shears in the upper-story colu
total of the column shears in the
forces applied to the structure.
displacement at point *c* as shown a
to sustain this deflection.

These two independent cases
loading. Assuming that the force
analysis, the two equations of hoi

inertia.
will be
in sever
If
to calcu

where
for a u
A
are ap
slope-

where
symm
Eq. (

wher

and

Not
fact
is p

The
par

dev
pre
Sir

Th

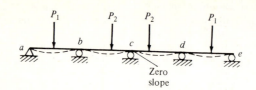
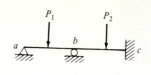

Actual structure   Alternate structure

(*a*)

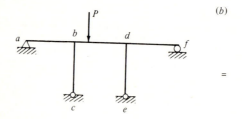
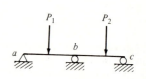

Actual structure   Alternate structure

(*b*)

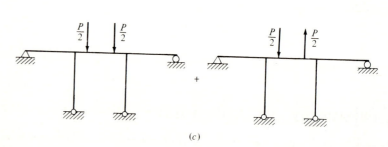

(*c*)

Figure 12.20 Symmetry and antisymmetry: (*a*) symmetrically loaded beam; (*b*) antisymmetrically loaded beam; (*c*) general loading on a symmetric frame synthesized as the sum of a symmetric and an antisymmetric load case.

It can be demonstrated mathematically that a solution for any linear problem can be represented as the sum of a symmetric and an antisymmetric part. This fact is useful for analyzing structures with a general loading. For example, envision the loading on the symmetric frame structure in Fig. 12.20*c* to be the sum of the individual symmetric and antisymmetric load cases. This structure has six unknown angles of rotation. From Fig. 12.20*c* it can be observed that the symmetric and the antisymmetric cases each have three unknowns. If this structure were being analyzed by the slope-deflection method, the solution for the actual loading would be obtained by solving a set of six simultaneous linear algebraic equations with six unknowns. An approximate estimate for solving sets of equations is that the amount of time (and hence the cost) is proportional to the number of unknowns squared; by using the symmetric and antisymmetric load cases the cost will be about half of that for analyzing the actual loading.

12.9 DEFLECTIONS OF STATICALLY INDETERMINATE STRUCTURES

The results of any indeterminate analysis, whether obtained by compatibility or equilibrium methods, contain the reactions plus the internal member forces. These can be used to calculate the member stresses and carry out the final design of the member configuration. At this point, however, the deflections of the structure are still unknown.

After the final bending moments for a continuous beam or a frame are obtained, the structural behavior is better understood by sketching the deflected shape of the members. This can be done by using the fact that the curvature of a linear elastic member is equal to the value of the M/EI ordinate at a point (Chap. 4). By making such an approximate sketch the engineer can establish points of inflection and the general shape of the deflected structure. This will not give quantitative deflections unless the moment-curvature relations are integrated and the boundary conditions on slopes and deflections are enforced. Generally, this tedious process should not be attempted unless the structure is extremely simple.

If the numerical values of the displacements are required, one of the methods discussed in Chaps. 9 and 10 must be used, i.e., either a so-called direct method or a method based upon energy considerations. For example, assume that it is required to calculate the displacement at point c for the continuous beam in Fig. 12.21a after the reactions, shears, and bending moments are known. Since the bending moments at joints b and d are known, it is convenient to carry out the calculation using the superposition of the two loading cases in Fig. 12.21b and d. The first case has only the transverse

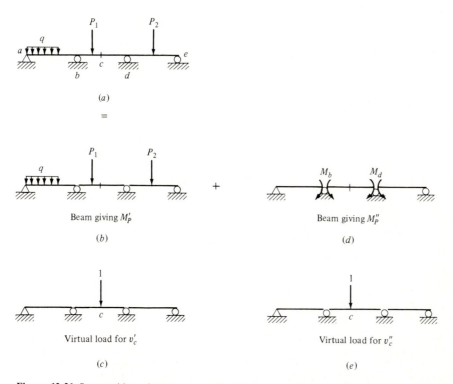

Figure 12.21 Superposition of transverse applied loads and end moments for computing deflections: (a) continuous beam with actual loading; (b) primary structure with transverse loads only; (c) virtual load for v'_c; (d) primary structure with redundant forces only; (e) virtual load for v''_c.

loads applied and results in the bending moments M_P', the second case has only the end moments on the members, which give moments M_P'' throughout the structure. If the method of complementary virtual work is to be used, a unit force is placed at point c, which gives the bending moments M_Q. The deflection introduced by the moments M_P' is calculated using Eq. (10.21) to give

$$v_c' = \int_0^L \frac{M_P' M_Q}{EI}\, dx \tag{a}$$

Similarly, to calculate the displacement at c for the end moments only (M_P''), the unit load is again placed at point c, resulting in the moments M_Q; that is, these are equivalent to those used in calculating the deflection due to M_P'. Therefore, the deflection at c for the second load case is

$$v_c'' = \int_0^L \frac{M_P'' M_Q}{EI}\, dx \tag{b}$$

By superposition the final displacement at c is the sum of the displacements from the two load cases. This gives

$$v_c = v_c' + v_c'' = \int_0^L \frac{(M_P' + M_P'')M_Q}{EI}\, dx \tag{c}$$

Note that the quantity $M_P' + M_P''$ represents the total moments in the structure as obtained previously using an indeterminate analysis method. Furthermore, the moments M_Q are those obtained by applying the unit load to a simply supported beam. In Chap. 11 this beam was referred to as the primary structure; i.e., it is the structure obtained by removing a sufficient number of redundants, giving a statically determinate structure. Therefore, the calculations can be carried out by applying the unit load to a statically determinate structure and multiplying these moments by the actual internal bending moments.

This can also be observed by recalling the discussion in Chap. 10, where the concept of complementary virtual work was developed and discussed. The form of the equation presented there was

Virtual

$$v_c(1\downarrow) = \int_0^L \left(\frac{M_P}{EI}\right) M_Q\, dx \tag{10.21}$$

Actual

From Eq. (10.21) it can also be observed that it is not necessary to load the actual structure with the unit load. Only an equilibrium solution for the applied virtual force need be obtained, and the associated physical structure need not correspond to the actual structure. From the example presented in Fig. 12.21 it is apparent that this fact can save a significant number of computations.

The statement of complementary virtual work for truss deflections is

Virtual

$$v_c(1\downarrow) = \sum \left(\frac{F_P L}{AE}\right) F_Q \tag{10.13}$$

Actual

The same arguments used above for continuous beams and frame structures can also be used for calculating the deflections for trusses. That is, it is not necessary to load the actual truss with the unit load. The F_Q represent an equilibrium solution for the virtual force applied to the less complex primary structure.

Even with the observation that the unit load can be applied to the primary structure, the task of computing displacements for indeterminate structures can be laborious. This approach can be used for relatively simple structures where only a few displacements are required, but its limitations must be recognized. For most complex structures in which both member forces and displacements are to be calculated it is prudent to use a method amenable to machine calculations (Part Five).

12.10 SECONDARY STRESSES IN TRUSSES

It is assumed for the analysis of most trusses that the members are joined together by frictionless pins allowing free rotation of the members. Trusses are generally designed with the centroids of the members meeting at the panel points, but there is usually a minimal amount of rotational restraint of the joints whether they are welded, riveted, or bolted. The truss in Fig. 12.22 illustrates the deflected configuration for a truss with moderately rigid joints. The axial forces in the members constitute the *primary stresses,* and the bending moments yield the *secondary stresses.* Significant secondary stresses occur in some trusses in spite of careful joint design and fabrication. For example, the secondary stresses can dominate the member designs in a relatively short but deep truss. Note that if the loads are applied between the panel points, bending moments are introduced; generally, however, they are not termed secondary stresses.

Secondary stresses can be analyzed using either the slope-deflection or the moment-distribution method. If the loads are applied at the joints, the moments induced by joint translations must be considered for the analysis of secondary stresses. The presence of bending moments in the members can alter the axial member forces, but this is seldom a significant effect. The analysis generally proceeds using the following basic steps:

1. Calculate the axis forces in the members assuming that the members are pinned at the joints and are free to rotate.
2. Calculate the displacements at each of the joints.
3. Calculate the rotation ψ for each member and carry out the analysis for the moments at the ends of the members using either the slope-deflection or moment-distribution method.
4. Calculate the axial forces, shears, and bending moments in the members using the results from step 3.

Since it was assumed in step 1 that the bending moments do not affect the axial forces, it is advisable to compare the axial forces obtained from step 4 with those obtained from the initial analysis. Usually there is agreement, but if there is an appreciable difference, steps 2 to 4 must be repeated. Generally, this iterative procedure is not necessary.

Therefore, for some trusses it is necessary to calculate the bending moments in addition to the axial forces. Since this procedure is extremely slow and laborious, in all but the simplest trusses it is suggested that the more systematic methods discussed in Part Five be used to investigate secondary stresses.

12.11 DISCUSSION

The two classical methods discussed in this chapter for analyzing beams and frame structures are derived by enforcing moment equilibrium at the joints. The basic slope-

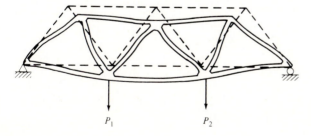

Fig. 12.22 A truss with secondary stresses.

P_1 P_2

deflection equations for each individual member are combined for the total structure, resulting in a set of simultaneous linear algebraic equations with the angles of rotation at the joints as the unknowns. This method is useful for studying relatively simple structures and forms the basis for the other method described in this chapter.

The moment-distribution method is basically an iterative solution procedure for solving the governing joint equilibrium equations derived from the slope-deflection equations. The angles of rotation are implicit in the equations, but the bending moments at the ends of the members are obtained in this procedure. This method is amenable to hand calculations for moderately complex structures if they have applied lateral loads or the joints experience translations, or both.

PROBLEMS

Using the moment diagram, sketch the approximate final deflected shape for these beams and rigid frames. Unless noted otherwise all members in each problem are made from the same material with $E = $ const.

In Probs. 12.1 to 12.18 use the slope-deflection equations to analyze these structures. Calculate the end moments and draw the moment diagram.

12.1 $I_{ab} = 60 \times 10^6$ mm^4 and $I_{bc} = 120 \times 10^6$ mm^4.
12.2 $I = $ const.

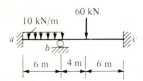

Figure P12.1

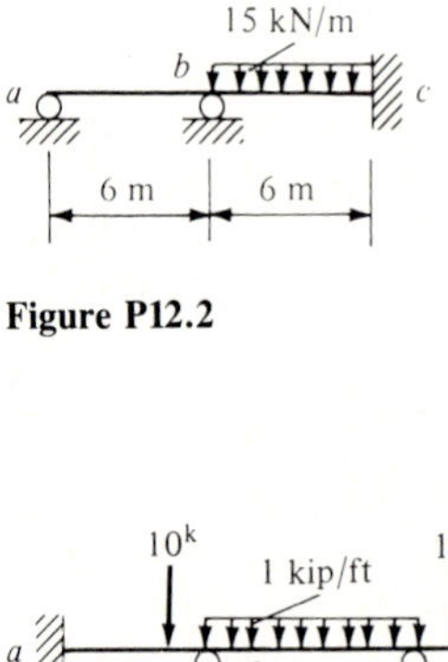

Figure P12.2

12.3 $I_{ab} = 1500$ in^4 and $I_{bc} = 2000$ in^4.
12.4 $I_{ab} = I_{cd} = 200$ in^4 and $I_{bc} = 400$ in^4.

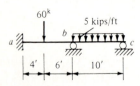

Figure P12.3

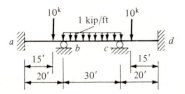

Figure P12.4

12.5 $I_{ab} = I_{cd} = 600 \times 10^6$ mm^4 and $I_{bc} = 1200 \times 10^6$ mm^4.

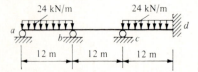

Figure P12.5

12.6 I = const.

12.7 $I_{ab} = I_{bc} = 500 \times 10^6$ mm^4 and $I_{bd} = 100 \times 10^6$ mm^4.

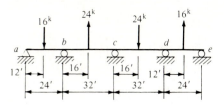

Figure P12.6

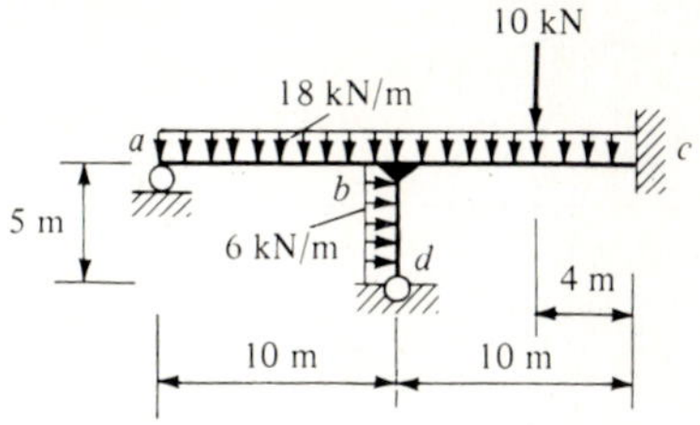

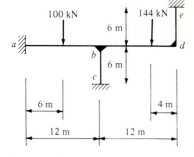

Figure P12.7

12.8 I = const.

12.9 $I_{ab} = I_{bd} = 360 \times 10^6$ mm^4 and $I_{cb} = I_{de} = 120 \times 10^6$ mm^4.

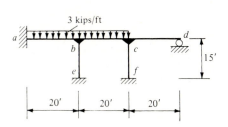

Figure P12.8

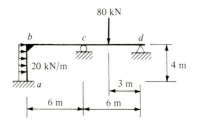

Figure P12.9

12.10 $I_{ab} = I$, $I_{bc} = 1.5I$, and $I_{cd} = 2I$.

12.11 Supports b and d settle 0.24 and 0.48 in, respectively; $I_{ab} = 1000$ in^4, $I_{bc} = 800$ in^4, $I_{cd} = 900$ in^4, $I_{de} = 1200$ in^4, and $E = 29 \times 10^3$ kips/in^2.

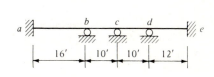

Figure P12.11

Figure P12.10

12.12 Supports b and d settle 10 and 8 mm, respectively; $I_{ab} = 600 \times 10^6$ mm^4, $I_{bc} = 600 \times 10^6$ mm^4, $I_{cd} = 900 \times 10^6$ mm^4, and $E = 200$ GPa.

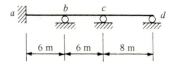

Figure P12.12

12.13 The support at *b* moves upward 1 in, and the support at *c* settles 1 in; $I = 200$ in^4 and $E = 29 \times 10^3$ kips/in^2.

12.14 In addition to the loading illustrated, support *b* settles 15 mm; $I = 1000 \times 10^6$ mm^4 and $E = 200$ GPa.

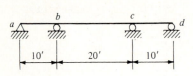

Figure P12.13

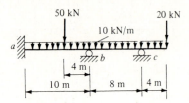

Figure P12.14

12.15 In addition to the loading illustrated, supports *b* and *c* move downward 1.0 in and upward 0.7 in, respectively; $I = 400$ in^4 and $E = 29 \times 10^3$ kips/in^2.

12.16 $I_{ab} = I_{bc} = I_{ed} = I$ and $I_{bd} = 3I$.

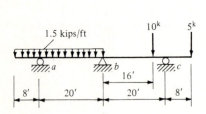

Figure P12.15

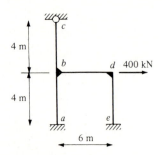

Figure P12.16

12.17 $I_{ab} = I$, $I_{bc} = 2I$, and $I_{cd} = 3I$.

12.18 $I_{ab} = I_{cd} = 1000$ in^4 and $I_{bc} = 1600$ in^4.

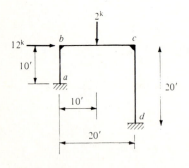

Figure P12.17

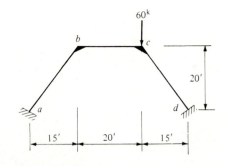

Figure P12.18

12.19 Calculate the end moments by the moment-distribution method if:

    (*a*) All member lengths are equal and *I* is the same for all members.
    (*b*) All member lengths are equal and $I_{ab} = I_{ac} = I_{ae} = I$, with $I_{ad} = 0.5I$.
    (*c*) All member lengths are equal and $I_{ab} = I_{ac} = I_{ae} = I$, with $I_{ad} = 2I$.
    (*d*) *I* is the same for all members and $L_{ab} = L_{ac} = L_{ae} = L$, with $L_{ad} = 0.5L$.
    (*e*) *I* is the same for all members and $L_{ab} = L_{ac} = L_{ae} = L$, with $L_{ad} = 2L$.

12.20 Calculate the end moments by the moment-distribution method if (*a*) the structure is loaded as shown; (*b*) the illustrated loading is removed and uniform loading *q* is applied over the entire beam length; *I* = const.

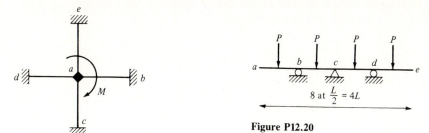

$$8 \text{ at } \frac{L}{2} = 4L$$

Figure P12.20

Figure P12.19

12.21 Calculate the end moments by the moment-distribution method if (*a*) the structure is loaded as shown; (*b*) the illustrated loading is removed and a uniform loading *q* is applied on spans *bc* and *de*; *I* = const.

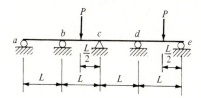

Figure P12.21

In Probs. 12.22 to 12.27 solve the original problem using the moment-distribution method to analyze the structure.

12.22 Problem 12.1	**12.23** Problem 12.2	**12.24** Problem 12.3
12.25 Problem 12.4	**12.26** Problem 12.5	**12.27** Problem 12.6

In Probs. 12.28 to 12.52 use the moment-distribution method.

12.28 (*a*) Calculate the end moments and draw the moment diagram; *I* = const.; (*b*) Calculate the moment of inertia required for span *ab* to give $|M_a| = \frac{3}{4}|M_b|$; $I_{bc} = I_{cd} = I$.

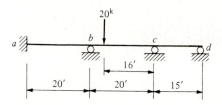

Figure P12.28

12.29 Calculate the moment of inertia required for span *bc* to give $|M_a| = 4|M_c|$; $I_{ab} = 100 \text{ in}^4$.

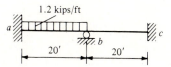

Figure P12.29

12.30 Calculate the moment of inertia required for span cd that will cause $|M_c|$ to be 25 percent greater than $|M_b|$; $I_{ab} = 400$ in^4 and $I_{bc} = 300$ in^4.

12.31 Calculate x in terms of L for this uniform beam if (a) the magnitudes of the bending moments at each of the supports are to be equal; (b) the three support reactions are to be equal.

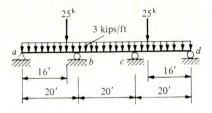

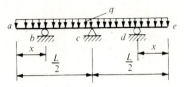

Figure P12.31

Figure P12.30

12.32 Problem 12.7	**12.33** Problem 12.8	**12.34** Problem 12.9
12.35 Problem 12.10	**12.36** Problem 12.11	**12.37** Problem 12.12
12.38 Problem 12.13	**12.39** Problem 12.14	**12.40** Problem 12.15

12.41 Calculate the end moments and draw the moment diagram; $I_{ab} = I_{bc} = 2I$ and $I_{ed} = I_{db} = I$.

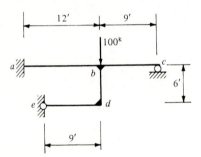

Figure P12.41

12.42 Problem 12.16.

12.43 This rigid frame has $I_{ab} = 125 \times 10^{-6}$ m^4, $I_{bc} = 100 \times 10^{-6}$ m^4, and $I_{cd} = 141 \times 10^{-6}$ m^4. Calculate the end moments if (a) the structure is loaded as shown; (b) the load illustrated is removed and a 200-kN force is placed 1 m to the left of joint c acting downward.

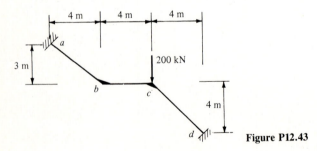

Figure P12.43

12.44 Calculate the end moments and draw the moment diagram; $I_{ab} = 600 \times 10^{-6}$ m^4, $I_{bc} = 250 \times 10^{-6}$ m^4, and $I_{cd} = 150 \times 10^{-6}$ m^4.

12.45 Problem 12.17.

12.46 Problem 12.18.

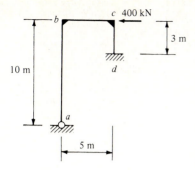

Figure P12.44

12.47 Calculate the end moments and draw the moment diagram; $I_{ab} = 1.5I$, $I_{bc} = I_{cd} = 2I$.

12.48 Calculate L_a such that this frame has no sidesway; $I_{ab} = I_{cd} = I$ and $I_{bc} = 2I$.

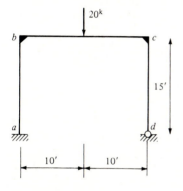

Figure P12.47

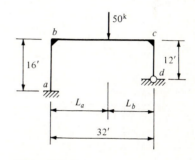

Figure P12.48

12.49 Calculate the end moments; $I_{ab} = 200$ in⁴, $I_{bc} = 400$ in⁴, and $I_{cd} = 250$ in⁴.

12.50–12.51 Calculate the end moments and draw the moment diagram. EI = const.

12.52 Calculate the end moments and draw the moment diagram; $I_{ab} = 40 \times 10^6$ mm⁴, $I_{dc} = I_{ce} = I_{fg} = 80 \times 10^6$ mm⁴, $I_{bc} = 100 \times 10^6$ mm⁴, and $I_{ef} = 120 \times 10^6$ mm⁴.

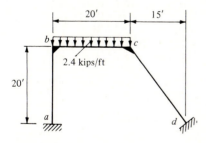

Figure P12.49

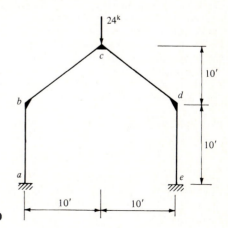

Figure P12.50

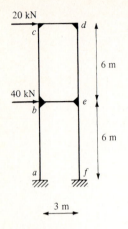

Figure P12.51

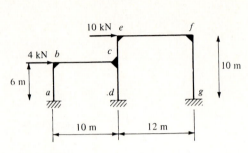

Figure P12.52

REFERENCES

1. Rubinstein, M. F.: *Matrix Computer Analysis of Structures,* chaps. 5, 10, and 11, Prentice-Hall, Englewood Cliffs, N.J., 1966.
2. Timoshenko, S. P., and D. H. Young: *Theory of Structures,* 2d ed., chap. 9, McGraw-Hill, New York, 1965.
3. Gere, J. M.: *Moment Distribution,* Van Nostrand, Princeton, N.J., 1963.
4. Portland Cement Association: *Handbook of Frame Constants,* Chicago, 1958.

ADDITIONAL READING

Beaufait, F. W.: *Basic Concepts of Structural Analysis,* chap. 9, Prentice-Hall, Englewood Cliffs, N.J., 1977.
Borg, S. F., and J. J. Gennaro: *Modern Structural Analysis,* chaps. 3 and 4, Van Nostrand Reinhold Co., New York, 1969.
Gerstle, K. H.: *Basic Structural Analysis,* chap. 14, Prentice-Hall, Englewood Cliffs, N.J., 1974.
Gutkowski, R. M.: *Structures: Fundamental Theory and Behavior,* chap. 12, Van Nostrand Reinhold, New York, 1981.
Laursen, H. I.: *Structural Analysis,* chaps. 12 and 13, McGraw-Hill, New York, 1978.
McCormac, J. C.: *Structural Analysis,* 3d ed., chaps. 22 and 23, Intext Educational, New York, 1975.
Norris, C. H., J. B. Wilbur, and S. Utku: *Elementary Structural Analysis,* 3d ed., chap. 9, McGraw-Hill, New York, 1976.
Parcel, J. I., and R. B. B. Moorman: *Analysis of Statically Indeterminate Structures,* chaps. 5 and 6, Wiley, New York, 1955.
West, H. H.: *Analysis of Structures,* chaps. 15 and 16, Wiley, New York, 1980.

THIRTEEN

APPROXIMATE METHODS

13.1 INTRODUCTION

The objective of this chapter is to present approximate methods of analysis that can be applied to certain types of indeterminate structures. It can be asserted that all solutions are approximate when examined in the light of the response of the actual structure; i.e., the first step in using the classical methods of Chaps. 11 and 12 is to develop an idealized model of a constructed facility so that its behavior can be described mathematically. For example, the rigid joints and fixed foundation supports of a high-rise building are approximations of the actual behavior. Thus, an approach such as moment distribution is used to obtain a relatively accurate solution for the idealized structure. In contrast, the methods presented in this chapter are applied to give approximate solutions for the response of the idealized structure.

The analysis of a statically indeterminate structure requires the geometry and member properties to be known a priori because the deformations affect the distribution of the forces. This presents a dilemma for the structural engineer and dramatically points out the need for approximate solutions. Thus, before performing the final solution for an indeterminate structure it is necessary to make simplifying assumptions about the response, execute a preliminary approximate analysis, and use these results to establish a tentative structural configuration that can be examined using a more exact method. Approximate analyses are also justified under other circumstances. For example, the loadings sustained by a structure over its lifetime may be undefined, and the response is rarely (if ever) linear elastic. Thus, an approximate method can be justified as consistent with such uncertainties.

In some cases an approximate analysis predicts the response of the structure adequately and the solutions need not be refined. That is, certain indeterminate structures behave predictably, so that if the proper initial simplifying assumptions about the behavior are made, the approximate solution will predict the actual response accurately.

Approximate methods are also vital for interpreting computer solutions such as those obtained using the approach in Part Five. Computer solutions should routinely be examined to detect unanticipated errant behavior. For example, if the structure was incorrectly described through the input, the correct solution for the wrong problem is obtained. By studying the results with respect to an approximation of the actual problem it is possible to obtain a minimal assurance that the computer solution is correct.

To obtain an exact solution for a statically indeterminate structure the equations of equilibrium must be augmented with equations of deflection behavior so that the total number of independent equations equals the number of independent unknowns. The classical methods of Chaps. 11 and 12 require sufficient equations to obtain the unknown generalized forces (compatibility methods) or the unknown generalized displacements (equilibrium methods). Alternatively, the system behavior can be anticipated a priori (through an understanding of the concepts of indeterminate structures presented in the two previous chapters) and rational assumptions made about the response. Thus, the deflection equations, which are necessary for classical methods, can be supplanted by an equal number of response assumptions and the structure investigated using only the equations of static equilibrium. The first task in an approximate analysis of this type is to determine the number of redundancies and make an equal number of simplifying assumptions. If too many assumptions are made, i.e., more than the degree of redundancy, the equations of equilibrium will be overdetermined. If the assumptions are inappropriate, a geometrically unstable model will result and the equations of equilibrium will not be consistent for all types of loading.

13.2 TRUSSES

A parallel-chord truss with double diagonals (Fig. 13.1a) is a statically indeterminate structure that can be analyzed approximately using several different sets of assumptions. If the diagonals are slender members that can sustain only tension, the shear will be resisted by one of the two diagonals in a given panel. For this loading, members *ad*, *cf*, *eh*, *hj*, *kl*, and *mn* are all zero-force members (shown dashed in Fig. 13.1b), and the truss can be considered to contain six members less than shown. That is, the assumptions required to analyze this structure with six redundancies have been dictated by the design. Hence, for this load case, there are 25 members, 14 joints, and 3 reactions, and the 28 unknown forces can be calculated using the independent equations of equilibrium. The results are shown in Fig. 13.1b.

If this same truss were fabricated using diagonal members capable of resisting both tension and compression, different assumptions would be required to obtain an approximate solution. By passing an imaginary vertical section through a panel we can observe that the shear must be resisted exclusively by the diagonals, and if these members have similar elastic properties (cross-sectional area and modulus of elasticity) they will share the shear equally. Since the internal forces in a statically indeterminate structure are distributed to the members in proportion to their respective stiffnesses, this assumption will be exact only for trusses with equal diagonals. Hence, an approximate solution can be obtained if the assumption is made that *the shear in each panel is divided equally between the two diagonals*. The member forces calculated using these assumptions are displayed in Fig. 13.1c.

The truss in Fig. 13.2a is the same as that in Fig. 13.1a with the intermediate vertical members removed. A count of the number of unknowns and the independent

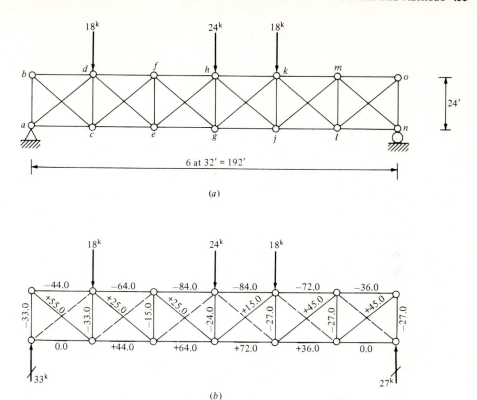

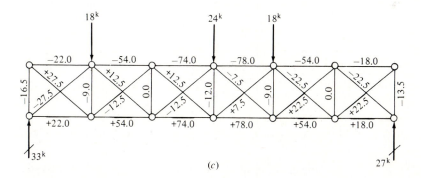

Figure 13.1 (a) Parallel-chord truss with double diagonals; (b) analysis with diagonals in tension onl;· (c) analysis with diagonals resisting half the panel shear.

equations reveals that this truss has only one redundancy. It is assumed that in any one panel the external shear is resisted by the two diagonals as though the truss functioned as two independent component trusses (Fig. 13.2b and c). This one assumption, combined with the equations of equilibrium, implies that all pairs of diagonals at every vertical section also respond in this fashion. Thus, the truss can be visualized as being

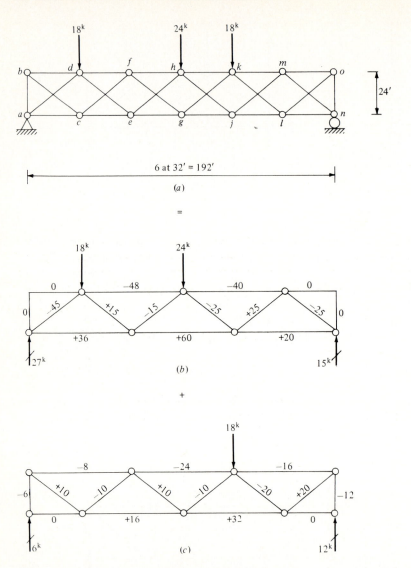

Figure 13.2 (a) Truss with one redundant member; (b) first component truss; (c) second component truss.

composed of two separate statically determinate structures. After each of these trusses has been analyzed using the equilibrium equations, the solutions are combined to give an approximate analysis for the actual structure. It is important to note that this truss must not be designed with diagonal members that cannot sustain compression — a situation that would result in an unstable structure.

Logic similar to that applied to the behavior of parallel-chord trusses can also be used for approximate analyses of towers with straight legs. In Sec. 8.7 we observed that a three-dimensional tower with straight legs can be analyzed by investigating each of

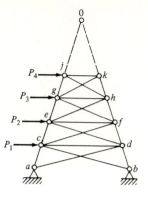

0

P_4 ⟶ j   k

P_3 ⟶ g   h

P_2 ⟶ e   f

P_1 ⟶ c   d

a   b

Figure 13.3 Tower with four redundant members.

the individual planar faces and adding the results for the four faces to give the total member forces. We see that the tower face in Fig. 13.3 has four redundancies. Extending the line of action of the two legs to their point of intersection O and taking moments about this point, we note that the component of horizontal force at a given level must be resisted by the horizontal components of the member forces in the diagonals. If this structure has slender diagonals capable of resisting only tension, members bc, de, fg, and hj are all zero-force members, and the solution can be obtained using only the equations of equilibrium.

If the diagonal members of this tower can sustain both tension and compression and it is assumed that the unbalanced horizontal component of shear at a given level is divided equally between the two diagonals, the member forces can be calculated using the equations of equilibrium. That is, one such assumption is made per panel, and they supplant the four deformation equations required for an exact analysis.

The face of the tower in Fig. 13.4a has only one redundancy. If it is assumed that at any one level the external horizontal shear is resisted by the two diagonals as though the tower functioned as two independent component towers, the equations of equilibrium can be used to calculate the member forces for the entire tower face. Since the structure can be envisioned as being composed of the two individual towers (Fig. 13.4b and c) to simplify the analysis, adding the solutions for these two statically determinate structures gives the final member forces.

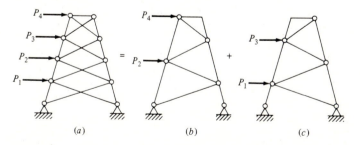

P_4⟶ P_4⟶

P_3⟶ P_3⟶

P_2⟶   = P_2⟶   +

P_1⟶ P_1⟶

(a)   (b)   (c)

Figure 13.4 (a) Tower with one redundant member; (b) first component tower; (c) second component tower.

13.3 PORTALS AND MILL BENTS SUBJECTED TO LATERAL FORCES

Portal frames of various configurations, which are efficient in resisting lateral forces, are typically incorporated into bridges (see Fig. 6.5) and other structures to transmit wind loads, etc., into the foundations. An example of a portal is the rigid frame in Fig. 13.5a. Since this structure has three redundancies, it is necessary to make an equal number of assumptions about the structural response if it is to be analyzed using only the equations of static equilibrium. This portal was analyzed in Example 12.9 using the moment-distribution method, and the results from this approach yield the moment

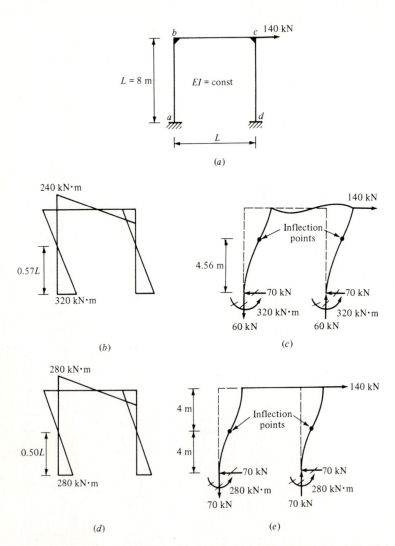

Figure 13.5 (a) Portal with applied load; (b) exact moment diagram; (c) exact deflected shape and reactions; (d) moment diagram for a rigid girder; (c) deflected shape for a rigid girder.

diagram in Fig. 13.5b. Note that the point of zero moment occurs at 0.57L above the column base. Recall from Eq. (9.5) that the curvature of a linear elastic member is proportional to the bending moment; therefore, the point of zero moment is the point where the column changes curvature: it is the inflection point. The approximate deflected frame, sketched from the moment diagram, is shown in Fig. 13.5c along with the reactions at points a and d. In this case each base column shear resists half of the applied lateral force.

The frame of Fig. 13.5a has the same cross section for the columns and the girder bc; however, if the structure is modified so that the girder is much stiffer than the columns, different results will be obtained. By modifying the calculations in Example 12.9 we observe that with a rigid girder ($I/L \rightarrow \infty$) the bending moment at the base of each column will be equal to that at the top; consequently, the inflection points occur at the midheight of the columns. The moment diagram corresponding to this structure is shown in Fig. 13.5d and the deflected configuration, along with the reactions, in Fig. 13.5e. In either case, the shear in each column is also equal to half of the applied lateral force.

A typical portal frame has a girder with a larger cross section than that of the columns; therefore, based upon these two analyses, the following assumptions can be made for the approximate behavior of all regular portal frames subjected to lateral loads:

1. An inflection point occurs near the midheight of the left column.
2. An inflection point occurs near the midheight of the right column.
3. The shear in each column is equal to half the applied lateral load.

Thus, for a portal frame thrice statically indeterminate, the solution can be obtained using these three assumptions and the equations of equilibrium.

The trussed portal or bridge portal in Fig. 13.6a is also statically indeterminate to the third degree. The columns are continuous to the top of the structure, and the trussing is attached with pinned connections. The truss is an extremely stiff component of the structure, just like the rigid girder in the frame described previously; the behavior of this structure will therefore resemble that of the rigid portal frame. The equations of equilibrium will give a solution to the structure if the same three assumptions are used, namely, each column has an inflection point at midheight ($L/2$ from the base in this case), and the horizontal reactions are equal to P for this structure. Thus, all the member forces, as well as the column shears and bending moments, can be calculated.

The mill bent in Fig. 13.6b is a typical structure used for resisting lateral forces in low-rise industrial buildings. The Fink truss is connected to the top of the columns with a pinned joint, and the knee braces are incorporated to lend lateral stability to the configuration. Again, this structure with three redundancies can be analyzed using only the equations of static equilibrium if three assumptions are made. Observing that this structure will also function like the rigid portal frame, we assume that both columns will have inflection points a distance $L/2$ from the foundations and the horizontal reactions will both be equal to 2.5P. A free-body diagram of the bent above the points of inflection can be investigated with the equations of statics to give the member forces in the truss, and the bending moments in the columns can also be calculated.

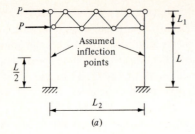

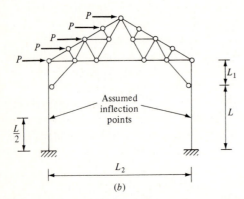

(a)

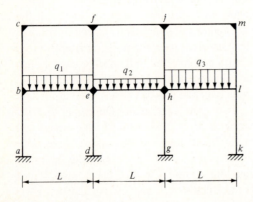

(b)

Figure 13.6 (a) Trussed portal; (b) mill bent.

13.4 BUILDING FRAMES SUBJECTED TO VERTICAL FORCES

The structure in Fig. 13.7 represents one of the types of construction used in high-rise structures. The original angles between members in a rigid frame like this are preserved when the structure deforms under loading. The cross section of each member can be subjected to axial force, shear, and bending moment. If the values of these generalized forces are known at one point in a member, the distribution of forces throughout the member can be calculated using the equations of equilibrium. If b is the number of

Figure 13.7 A rigid frame subjected to vertical forces.

members and r represents the number of independent generalized reaction forces, the total number of unknowns is $3b + r$. Three independent equations of static equilibrium must be satisfied at each joint; thus, if n is the total number of joints, the redundancy N of the structure can be expressed as

$$N = 3b + r - 3n \qquad (13.1)$$

For the structure in Fig. 13.7

$$N = 3(14) + 12 - 3(12) = 54 - 36 = 18$$

Thus, 18 assumptions are required if this rigid frame is to be analyzed using only the equations of equilibrium.

The deflected shape of a typical girder such as eh (Fig. 13.8a) depends upon the elastic end restraints contributed by the stiffnesses of the members connected at each joint. In general, the moments at the supports are negative, and those in the vicinity of the center are positive; therefore, there are inflection points at $\alpha_1 L$ and $\alpha_2 L$ from the left and right ends, respectively. The locations of these points can be calculated only if the exact solution for the entire structure is established; however, an estimate can be made if the two beams in Fig. 13.8b and c are examined. These beams represent two extremes of behavior, since the fixed-fixed beam is an approximation of the situation in which the end members are extremely stiff with respect to the stiffness of the beam under consideration, and the simply supported beam response will occur if the contiguous members have a negligible stiffness compared with that of the subject beam. The inflection points for the indeterminate fixed-fixed beam can be calculated using the methods from Chaps. 11 and 12, and they occur at $0.21L$ from each support. The simply supported beam has only positive moments between the supports, and there are no inflection points; that is, $\alpha_1 = \alpha_2 = 0$. Since the elastic supports for the beam eh will cause the member to deflect in a configuration between these two extremes, the inflection points will be located a distance less than $0.21L$ from each support; therefore, a reasonable representation is embodied in the assumption that $\alpha_1 = \alpha_2 = 0.10$. In

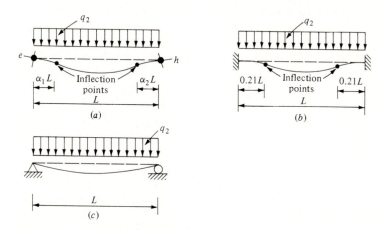

Figure 13.8 Girder with: (a) elastic supports, (b) fixed supports, and (c) simple supports.

addition, we observe that when the structure is subjected to vertical loads only, the girders will have virtually no axial force. Therefore, for a *rigid frame* like that shown in Fig. 13.7 *under vertical forces,* it can be assumed that:

1. An inflection point will occur approximately $0.10L$ from the left end of each girder.
2. An inflection point will occur approximately $0.10L$ from the right end of each girder.
3. There is no axial force in the girder.

Three assumptions of this type can be made for each girder in the structure, but they depend upon such factors as relative flexibility of the members at each joint, type of loading, etc. For the rigid frame under consideration in Fig. 13.7, the 6 girders will yield 18 assumptions, and the frame can be analyzed using only the equations of static equilibrium; i.e., the member elastic properties need not be established before investigation.

The bending moments and shears in the girders can be calculated approximately with the above analysis method; however, certain design information for the columns can also be obtained. For example, the maximum axial load in an interior column can be attained by loading the two adjacent girders with their distributed vertical design live loadings. In general, the moments in these two girders will be unequal; however, since they act in opposite directions and their difference is typically small, this influence need not be considered. Caution must be exercised, however, if the maximum axial force is to be obtained for an exterior column. The negative bending moment in the contiguous loaded girder will be introduced into the joint, and it must be distributed to the columns framing into the joint. This moment can be distributed to the columns according to their stiffness factors which will result in column shears. Since the shear in the columns induces an axial load into the girder, the third assumption used for the approximate analysis is violated.

13.5 BUILDING FRAMES SUBJECTED TO LATERAL FORCES

Structural systems must be designed to resist the lateral forces introduced by environmental effects such as wind and earthquakes. In the steel frame building in Fig. 13.9 rigid frames with moment-resisting connections provide the required strength. For example, if the floor system functions as a rigid diaphragm and torsion of the building is negligible, the total load imposed by incident wind pressure on the left side of the structure will be distributed in proportion to the frame stiffnesses. That is, a typical interior frame must resist the load applied to a width of exterior wall equal to the sum of the distances halfway between the adjacent two frames (the distance between frames if the frames are all equally spaced). If the lateral loads are significantly larger than the dead loads, the structure will deflect as in Fig. 13.9a. The deflections have been displayed on an exaggerated scale to emphasize that each member experiences a change in curvature over its length, an observation that will be used in the ensuing approximate analyses. In Sec. 13.4 we established that this type of structure has a redundancy equal to 3 times the number of girders; thus, the rigid frame under consideration has

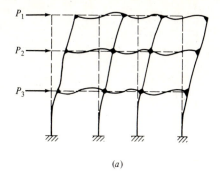

(a)

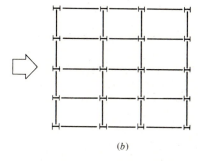

(b)

Figure 13.9 (a) Deflected multistory rigid frame subjected to lateral loads; (b) plan view.

27 redundancies, which implies that an equal number of assumptions is required if only the equations of static equilibrium are to be used for the analysis.

The portal[1] and cantilever[2] methods are representative of approximate analyses that gained prominence in the 1930s in the United States for treating high-rise structures subject to lateral forces. Additional approximate solutions such as those by Spurr and Witmer (both of which require a knowledge of the elastic properties of the members) were introduced during this period. A comprehensive study of these methods was made by the American Society of Civil Engineers.[3] If the structural framing is not regular, an exact method taking account of the elastic member properties must be used (slope-deflection, moment-distribution, or matrix structural analysis). The lateral force-resisting system of the Empire State Building was analyzed by the cantilever method. Even today, approximate analyses are as useful for developing preliminary designs as they are for checking computer solutions.

The Portal Method

The multibay rigid frame in Fig. 13.10a is statically indeterminate to the ninth degree; therefore, if it is to be analyzed using only the equations of equilibrium, an equal number of assumptions must be made. It can be asserted that this structure responds to lateral forces as three individual symmetric portals (see Fig. 13.10b), and there is an inflection point at the midlength of all members. These three portals can be investigated approximately using the assumptions developed in Sec. 13.3, but first it is necessary

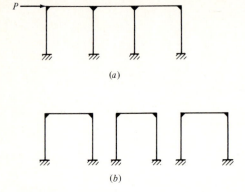

(a)

(b)

Figure 13.10 (a) Rigid frame; (b) substitute portals.

to establish the amount of the lateral applied load the various portals resist. For example, if each of the portals has the same length of girder, it is tenable to assume that an individual portal resists one-third of the applied load, since each has an equal lateral stiffness. For the more general case of unequal girder lengths this same assumption can be made, but it introduces the possibility of significant errors. This assumption implies that the total portal in Fig. 13.10a resists the applied load in such a way that the interior columns have twice the shear of the exterior columns. If the response of this one-story rigid frame is extrapolated to that of the multistory structure of Fig. 13.9, the implied approximate analysis is termed the *simple portal method,* and the following assumptions are required:

1. Inflection points occur at the midspan of the girders.
2. Inflection points occur at the midheight of the columns.
3. The horizontal shear at a given story is distributed equally to a set of equivalent symmetric portals.

The third assumption implies that the shear in the interior columns is equal to the total horizontal shear at the story level divided by the number of columns less 1; furthermore, this in turn implies that the exterior columns have half the shear of the interior columns. It is significant to note that the three assumptions for the portal method are not independent because behavior as a symmetric portal implies midlength girder inflection points as well as equal column shears. Thus, we see that the total number of consistent and independent assumptions implied by symmetric portal behavior for a multistory building equals the total number of girders and columns (one inflection point per member) plus the sum of the number of interior columns at all levels (the number of assumptions for horizontal shear distribution). Note that the shear in the exterior columns depends upon that in the interior columns. For the structure of Fig. 13.9, there are 9 girders, 12 columns, and 6 interior columns, which suggests 27 assumptions; therefore, the associated structural response can be investigated using only the equations of static equilibrium. An illustration of the use of the simple portal method is given in Example 13.1.

Example 13.1 Use the portal method to obtain the bending moments in this symmetric rigid frame. Draw the bending-moment diagram.

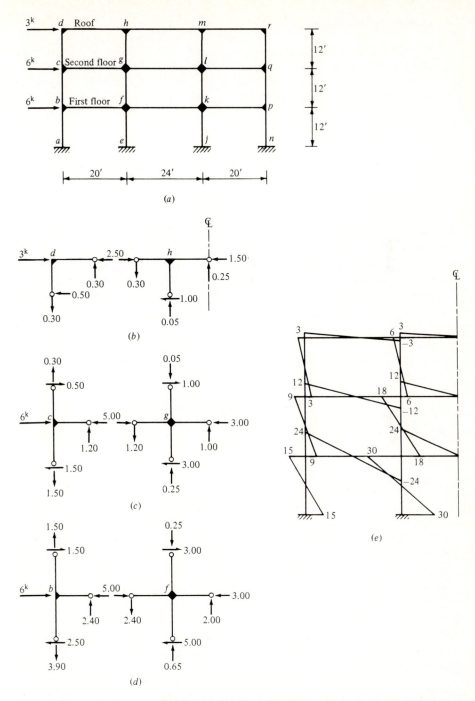

Figure E13.1 (*a*) Rigid frame with lateral forces; (*b*) roof level with inflection points at midlengths of girders and columns (forces in kips); (*c*) second-floor level with inflection points at midlengths of girders and columns (forces in kips); (*d*) first-floor level with inflection points at midlengths of girders and columns (forces in kips); (*e*) final moment diagram (moments in kip·feet).

SOLUTION See Fig. E13.1*b* to *e*.

DISCUSSION Since the response of this structure will be antisymmetric with respect to the centerline, only half of the building has been investigated. Starting with the top floor, the column shears are calculated assuming that the shear in the interior columns V_{int} is equal to twice that in the exterior columns V_{ext}; thus,

$$3.0 = 2V_{int} + 2V_{ext} = 3V_{int}$$

or $$V_{int} = 1.0 \text{ kip} \qquad V_{ext} = 0.5 \text{ kip}$$

Beginning with the left free-body diagram in Fig. E13.1*b*, we can calculate the unknown forces at the inflection point of the girder and the axial force in the column using the equations of equilibrium. The girder forces can then be used for the right free-body diagram of Fig. E13.1*b* to obtain the unknown axial force in column *gh* and the forces at the inflection point of girder *hm*. This same approach is used on each floor progressing from the top to the base of the building. Note that the total horizontal shear in the columns *bc*, *fg*, etc., is 9.0 kips and in the lower level is 15.0 kips.

The bending moments in the members are calculated using the forces in Fig. E13.1*b* to *d*, and the resulting moment diagram is presented in Fig. E13.1*e*. These calculations are carried out using static equilibrium; thus, for example, at joint *g*, the girder moments are

$$M_{gc} = (1.2 \text{ kips}) (10 \text{ ft}) = 12.0 \text{ kip} \cdot \text{ft} \ (\curvearrowright)$$

$$M_{gl} = (1.0 \text{ kip}) (12 \text{ ft}) = 12.0 \text{ kip} \cdot \text{ft} \ (\curvearrowright)$$

and the moments on the columns are

$$M_{gh} = (1.0 \text{ kip}) (6 \text{ ft}) = 6.0 \text{ kip} \cdot \text{ft} \ (\curvearrowright)$$

$$M_{gf} = (3.0 \text{ kips}) (6 \text{ ft}) = 18.0 \text{ kip} \cdot \text{ft} \ (\curvearrowright)$$

Joint *g* is in equilibrium since the sum of the clockwise moments equals the sum of the counterclockwise moments.

The Cantilever Method

Another approximate method for analyzing multistory rigid frames for lateral loads involves assuming that the structure functions as a vertical cantilever beam such that the leeward columns are in compression and the windward columns in tension. For this approach, the inflection points in the girders and the columns are also assumed to occur at the member midlengths, as in the portal method; however, the third assumption is supplanted by this analogy between the building and a cantilever beam. It is envisioned that the columns resemble the longitudinal fibers of a beam, and it is assumed that plane sections of the beam in the undeformed position remain plane after loading. This gives an approximation for the distribution of the strains and hence the stresses and axial forces in the columns. In general, the following assumptions are used for the *cantilever method*:

1. Inflection points occur at the midspans of the girders.
2. Inflection points occur at the midheights of the columns.

3. The axial stresses in the columns are proportional to their respective distances from the neutral axis of the bent, i.e., that formed by the columns at a given story level.

The third assumption implies a dependence between the axial forces in the columns, and in combination with the first two assumptions, there are enough independent assumptions to make the problem solvable by the equations of static equilibrium. Thus, as in the portal method, these three behavioral assertions give a number of independent assumptions equal to the sum of the girders and columns plus the total number of the interior columns at all levels. Example 13.2 illustrates the use of the cantilever method for the same rigid frame analyzed by the portal method in Example 13.1.

Example 13.2 Assuming the column areas at a given level to be equal, use the cantilever method to obtain the bending moments for the frame of Example 13.1. Draw the bending-moment diagram.

SOLUTION See Fig. E13.2a to d.

DISCUSSION Figure E13.2a indicates the distribution of stress in the columns for half the structure; furthermore, if the column areas are all equal, this illustrates the relationship of the column axial loads. Therefore,

$$\frac{F_{ext}}{32} = \frac{F_{int}}{12} \quad \text{and} \quad F_{int} = \tfrac{3}{8}F_{ext}$$

Since the columns for the right portion of the structure are in compression with a similar distribution, the resisting moment of all the columns at a given midheight of the columns is

$$M = 64F_{ext} + 24F_{int} = 73F_{ext}$$

Considering a free-body diagram consisting of the structure above the points of contra-flexure in the top columns and summing moments about the centerline of the structure gives

$$73F_{cd} = 3(6) = 18 \quad \text{and} \quad F_{cd} = 0.25 \text{ kip}$$

Thus,

$$F_{gh} = \tfrac{3}{8}F_{cd} = 0.09 \text{ kip}$$

The left free-body diagram of Fig. E13.2b can be used to calculate the unknown girder forces and the column shear. Subsequently, the right free-body diagram gives the other unknown forces at the inflection points. This same process is performed for all three levels. Note that for the column axial forces at the other levels, the force distribution in Fig. E13.2a is the same, but the bending moment changes as we progress down the building; thus,

$$73F_{bc} = 3(18) + 6(6) = 90$$

$$F_{bc} = 1.23 \text{ kips} \quad \text{and} \quad F_{fg} = 0.46 \text{ kip}$$

and

$$73F_{ab} = 3(30) + 6(18) + 6(6) = 234$$

$$F_{ab} = 3.21 \text{ kips} \quad \text{and} \quad F_{ef} = 1.20 \text{ kips}$$

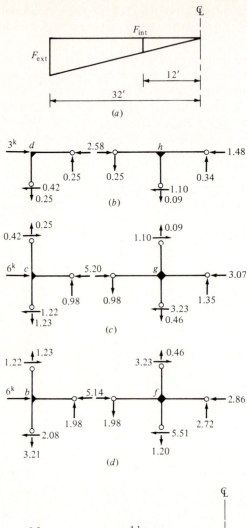

Figure E13.2 (*a*) Distribution of column axial forces (and stresses); (*b*) roof level with inflection points at midlengths of girders and columns (forces in kips); (*c*) second-floor level with inflection points at midlengths of girders and columns (forces in kips); (*d*) first-floor level with inflection points at midlengths of girders and columns (forces in kips); (*e*) final moment diagram (moments in kip•feet).

The final moment diagram in Fig. E13.2e is obtained using the results from Fig. E13.2b to d, together with the equations of equilibrium. Note that even for a structure with a regular framing plan, like this one, the results from the portal and cantilever methods differ (see Example 13.1).

The cantilever method gives accurate results for structures of 25 to 35 stories with moderate height-to-width ratios and girders of similar span and elastic properties. Generally, this method is more suitable than the portal method for high narrow buildings. For structures with height-to-width ratios greater than 4 it is advisable to use a method that incorporates the effects of the elastic properties of the columns. The cantilever method does not account for the effect of column bending on the distortion of the building; nor does it reveal the effect of unequal drift (differential horizontal deflection of adjacent floor levels) induced by girder bending. This latter effect stems from the assumption that drift is uniform if the building deforms as a beam; however, additional rotations at the floor levels could be introduced by girder bending.

PROBLEMS

13.1 to 13.3 Calculate the forces in all the members of these trusses:
   (a) Assuming that the diagonals can resist tension only.
   (b) Assuming that the diagonals in a panel each resist half the shear.
   (c) If all the intermediate vertical members are removed and assuming that the shear at a cross section is resisted as though the truss functioned as two independent component statically determinate trusses.

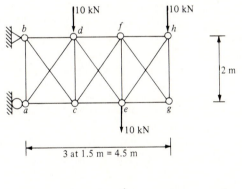

Figure P13.1

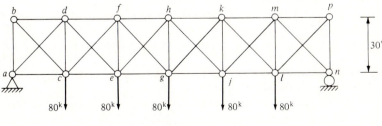

Figure P13.2

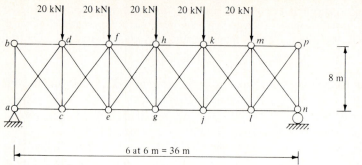

Figure P13.3

13.4 This truss, patented by Squire Whipple in 1847, can be analyzed approximately by assuming that it consists of two individual statically determinate trusses as shown by the solid and dashed lines. Calculate all the member forces.

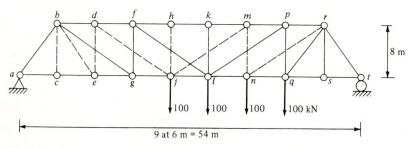

Figure P13.4

13.5 Calculate the forces in all the members of this tower:

    (*a*) Assuming that the diagonals can resist tension only.

    (*b*) Assuming that the external horizontal shear at a cross section is resisted equally by the two diagonals.

    (*c*) If horizontal members *cd*, *ef*, and *gh* are removed and assuming that the unbalanced horizontal shear at a cross section is resisted as though the tower functioned as two independent component statically determinate towers.

    In Probs. 13.6 to 13.10 use the assumptions developed in Sec. 13.3 for portals and mill bents.

13.6 Draw the moment diagram and the deflected shape of this portal.

13.7 Calculate the forces in the knee braces and draw the moment diagram and the deflected shape of this portal.

13.8 Calculate the forces in all the truss members and draw the moment diagram for the columns of this trussed portal.

13.9 Calculate the forces in the members framing into the columns and draw the moment diagram for the columns of this mill bent. Roof loads are applied normal to the surface.

13.10 Calculate the forces in the members framing into the columns and draw the moment diagram for the columns of this trussed portal.

13.11–13.12 Using the assumptions of the portal method, calculate the forces in the members framing into the columns and draw the moment diagrams for the columns.

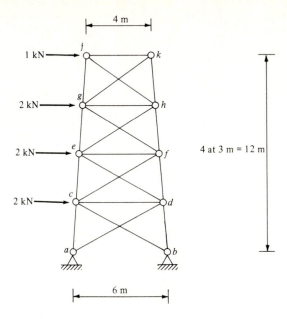

Figure P13.5

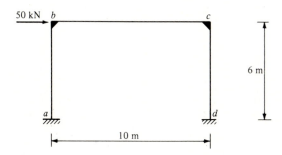

Figure P13.6

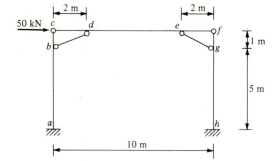

Figure P13.7

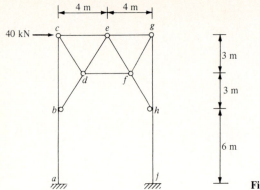

Figure P13.8

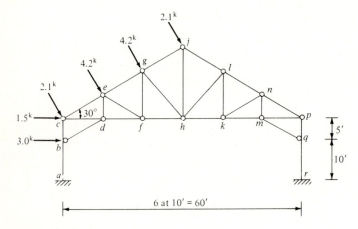

Figure P13.9

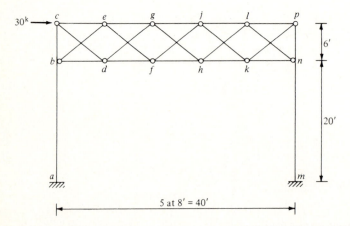

Figure P13.10

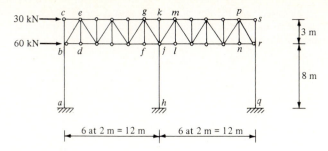

Figure P13.11

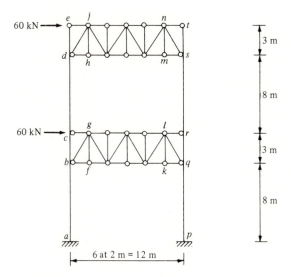

Figure P13.12

13.13 The building in Fig. 13.9 is to be analyzed for the effect of vertical loads only, assuming that the uniformly distributed live and dead loads on the floors and roof are 150 and 80 lb/ft^2, respectively. The rigid frames have a column spacing of 30 ft, the frames are spaced 25 ft on center, and the story height is 12 ft (center-to-center girder spacing). Using the assumptions of Sec. 13.4, calculate:

(a) The maximum positive girder moment.
(b) The negative girder moment of largest magnitude.
(c) The maximum girder shear.
(d) The maximum interior column axial load.
(e) The maximum exterior column axial load.
(f) The maximum interior column bending moment.
(g) The maximum exterior column bending moment.

13.14 The structure of Prob. 13.13 is to be analyzed for the effect of lateral loads only. The lateral forces applied to the roof and second- and first-floor levels, respectively, are 6, 12, and 12 kips. Obtain the moment diagram using (a) the portal method and (b) the cantilever method if all the columns have equal cross-sectional areas.

13.15 Calculate the moments in the rigid frame of Prob. 12.51 and draw the moment diagram using (a) the portal method and (b) the cantilever method if all the columns have equal cross-sectional areas.

13.16 These frames are spaced 6 m on centers, and the structure has a horizontal uniform pressure of 2500 Pa applied to the left side. Calculate the moments in the columns and girders at the eighth floor level using (a) the portal method and (b) the cantilever method if all the columns have equal cross-sectional areas.

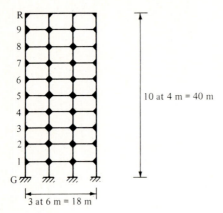

10 at 4 m = 40 m

3 at 6 m = 18 m

Figure P13.16

13.17 These frames are spaced 8 m on centers, and the structure has a horizontal uniform pressure of 3000 Pa applied to the left side.

(a) Use the portal method to calculate the moments in the columns between the tenth and eleventh floors and the third and fourth floors.

(b) Use the cantilever method to calculate the moments in the girders for the tenth floor and the third floor (assume that all column cross-sectional areas are equal).

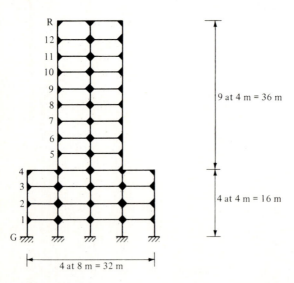

9 at 4 m = 36 m

4 at 4 m = 16 m

4 at 8 m = 32 m

Figure P13.17

13.18 Calculate the bending moments in this Vierendeel truss and draw the moment diagram (a) using the portal method and (b) using the cantilever method if the upper and lower chords have the same cross-sectional areas. *Hint:* It may be useful to rotate the structure 90° counterclockwise in analogy with a multistory building.

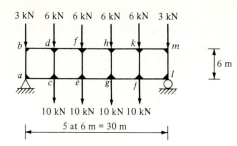

Figure P13.18

REFERENCES

1. Smith, A.: Wind Stresses in the Steel Frames of Office Buildings, *J. West Soc. Eng.,* vol. XX, no. 4, pp. 341–364, April 1915.
2. Wilson, A. C.: Wind Bracing with Knee Braces or Gusset Plates, *Eng. Rec.,* vol. 58, no. 10, pp. 272–274, Sept. 5, 1908.
3. American Society of Civil Engineers, Wind Bracing in Steel Buildings, *Trans. ASCE,* vol. 105, pp. 1713–1727, 1940.

ADDITIONAL READING

Andersen, P., and G. M. Nordby: *Introduction to Structural Mechanics,* chap. 13, Ronald, New York, 1960.
Benjamin, J. R.: *Statically Indeterminate Structures,* chaps. 3–6, McGraw-Hill, New York, 1959.
Hsieh, Yuan-Yu: *Elementary Theory of Structures,* 2d ed., chap. 5, Prentice-Hall, Englewood Cliffs, N.J., 1982.
Laursen, H. I.: *Structural Analysis,* chap. 6, McGraw-Hill, New York, 1978.
McCormac, J. C.: *Structural Analysis,* 3d ed., chap. 16, Intext Educational, New York, 1975.
Norris, C. H., J. B. Wilbur, and S. Utku: *Elementary Structural Analysis,* 3d ed., chap. 7, McGraw-Hill, New York, 1976.

FOURTEEN

INFLUENCE LINES FOR
INDETERMINATE STRUCTURES

14.1 INTRODUCTION

Statically indeterminate structures sustaining moving loads exhibit behavior which requires unique methods of analysis similar to those described for determinate beams, trusses, etc. The discussion in Chap. 5, which was limited to determinate structures, is extended to statically indeterminate structures in this chapter. An example of a structure requiring this type of analysis is the railroad bridge in Fig. 14.1. Recall that an influence line is a diagram showing the change in the value of a particular function (reaction, axial load, bending moment, etc.) as a unit concentrated load traverses the structure. For statically determinate structures these diagrams are composed of straight-line segments, but the response of indeterminate beams yields influence lines which must be described using higher-order curved shapes while those for indeterminate trusses are formed by straight lines.

To obtain the influence lines for indeterminate structures the methods described in Chaps. 11 to 13 must be combined with the fundamental ideas of influence lines in Chap. 5. The Müller-Breslau principle, which stems from energy considerations, is also applied to the construction of influence lines in this chapter.

Various design applications of influence lines described in this chapter include (1) the relationships between influence lines and the actual structural loading, discussed in Chap. 5, (2) the extension of influence lines to the behavior of buildings subjected to fixed loading to determine the critical arrangements of the live load, and (3) the construction of design envelopes.

14.2 SUCCESSIVE POSITIONS OF THE UNIT LOAD

In Chap. 5 the influence lines for statically determinate structures were obtained by envisioning a unit load moving across the structure and calculating the values of the function (reaction, axial load, etc.) for all positions of the load. When the load is

Figure 14.1 Latah Creek Bridge, Spokane, Washington. (*Burlington Northern.*)

applied to the structure directly, an expression for the influence ordinate as a function of the location of the unit load can be obtained. When the load is introduced to the primary structure at discrete points via floor beams, etc., the unit load must be placed at each point and the corresponding influence ordinate computed. The entire analysis for statically determinate structures is carried out using only the equations of equilibrium.

In contrast, we observed in Chaps. 11 and 12 that the equations of static equilibrium must be augmented with deflection equations to yield a tractable solution method for indeterminate structures. It is useful to introduce the study of the behavior of indeterminate structures under moving loads in a way similar to that used for determinate structures, but since each position of the unit load can require a separate investigation, the analysis of a structure with multiple redundancies may require a significant number of calculations. Since the approach employing successive positions of the unit load can be extremely useful and economical if the calculations are done on a computer, the reader is urged to reread the material in this chapter after studying Part Five.

Consider the uniform indeterminate beam in Fig. 14.2a and assume that the loads can move between points a and b. If this structure is to be investigated using the method of consistent displacements, only one reaction must be removed to yield a determinate primary structure. Considering the moment at point a as the redundancy, and placing the unit load at a distance x from the left support gives the rotation at a due to this load as

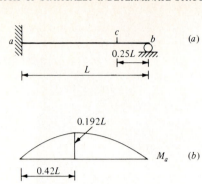

Figure 14.2 An indeterminate beam with various influence lines.

$$\Theta_{a0} = \frac{x}{6EIL}(L - x)(2L - x) \; (\frown) \tag{a}$$

The rotation at a due to a counterclockwise unit moment at a is

$$\theta_{aa} = \frac{-L}{3EI} \; (\frown) \tag{b}$$

Since the final rotation at a must be zero,

$$M_a\theta_{aa} + \Theta_{a0} = 0$$

giving
$$M_a = \frac{x}{2L^2}(L - x)(2L - x) \tag{14.1}$$

Thus, this reaction is expressed as a function of the position of the unit load, and plotting Eq. (14.1) gives the influence line in Fig. 14.2b. Alternatively, the moment influence line could have been obtained using other methods of indeterminate analysis described in Chaps. 11 and 12 (see Prob. 14.1). The reactions R_a and R_b are obtained using considerations of equilibrium, and they also can be expressed as functions of the location of the load (see Fig. 14.2c and d). Thus, the reaction influence lines can all be obtained by performing only one indeterminate analysis with the unit load placed in any general position, but this is not generally true for indeterminate structures. Each of these three influence lines, described with a cubic polynomial, should be contrasted with the simpler shapes of the influence lines for determinate structures. The shear and moment influence lines for any point in the structure can now be calculated using equilibrium (see, for example, Fig. 14.2e and f). The influence lines for a two-span beam are calculated in Example 14.1.

Example 14.1 Construct the influence lines for R_a, R_b, R_c, M_b, M_d, and V_d; $EI = $ const.

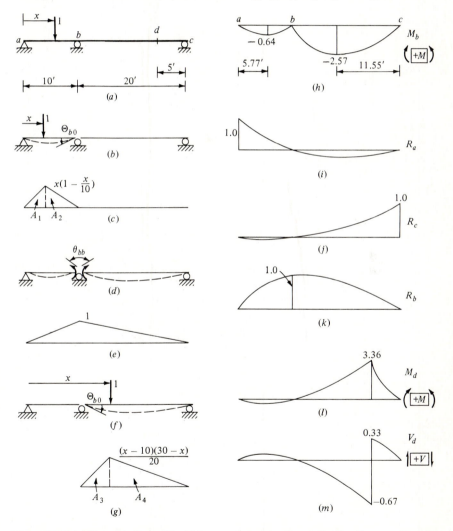

Figure E14.1 (a) Two-span beam; (b), (c) beam and moment diagram for Θ_{b0} ($0 \le x \le 10$); (d), (e) beam and moment diagram for θ_{bb}; (f), (g) beam and moment diagram for Θ_{b0} ($10 \le x \le 30$); (h) influence line for M_b; (i) influence line for R_a; (j) influence line for R_c; (k) influence line for R_b; (l) influence line for M_d; (m) influence line for V_d.

SOLUTION From Fig. E14.1b and c for $0 \leq x \leq 10$

$$EI\Theta_{b0} = \frac{A_1(2x/3)}{10} + \frac{A_2[x + \frac{1}{3}(10 - x)]}{10} = \frac{x}{60}(10 - x)(10 + x)$$

From Fig. E14.1d and e

$$EI\theta_{bb} = \tfrac{2}{3}(5 + 10) = 10$$

Since

$$M_b\theta_{bb} + \Theta_{b0} = 0$$

thus

$$M_b = \frac{x}{600}(x - 10)(x + 10) \qquad 0 \leq x \leq 10$$

From Fig. E14.1f and g for $10 \leq x \leq 30$

$$EI\Theta_{b0} = \frac{A_4[\frac{2}{3}(30 - x)]}{20} + \frac{A_3[30 - x + \frac{1}{3}(x - 10)]}{20}$$

$$= \frac{1}{120}(x^3 - 90x^2 + 2300x - 15{,}000)$$

Enforcing compatibility of slope at b gives

$$M_b = \frac{1}{1200}(15{,}000 - 2300x + 90x^2 - x^3) \qquad 10 \leq x \leq 30$$

(see Fig. E14.1h). Expressing R_a as a function of M_b yields

$$R_a = \begin{cases} \dfrac{1}{10}(10 - x + M_b) = \dfrac{1}{6000}(x^3 - 700x + 6000) & 0 \leq x \leq 10 \\[2ex] \dfrac{1}{10}M_b = \dfrac{1}{12{,}000}(-x^3 + 90x^2 - 2300x + 15{,}000) & 10 \leq x \leq 30 \end{cases}$$

(see Fig. E14.1i). Similarly, R_c can be expressed in terms of M_b as follows:

$$R_c = \begin{cases} \dfrac{1}{20}M_b = \dfrac{x}{12{,}000}(x^2 - 100) & 0 \leq x \leq 10 \\[2ex] \dfrac{1}{20}(x - 10 + M_b) = \dfrac{1}{24{,}000}(-x^3 + 90x^2 - 1100x + 3000) & 10 \leq x \leq 30 \end{cases}$$

(see Fig. E14.1j). Equilibrium of forces gives

$$R_b = 1 - R_a - R_c$$

Thus

$$R_b = \begin{cases} \dfrac{x}{4000}(-x^2 + 500) & 0 \leq x \leq 10 \\[2ex] \dfrac{1}{8000}(x^3 - 90x^2 + 1900x - 3000) & 10 \leq x \leq 30 \end{cases}$$

(see Fig. E14.1k).

DISCUSSION Figure E14.1*b* and *c* are associated with the rotation at *b* when the load is positioned in span *ab*, and Fig. E14.1*f* and *g* relate to the load on span *bc*. Both these situations must be used to obtain the expressions for M_b. The influence line for M_b in Fig. E14.1*h*, consisting of two cubic curves, is constructed using the strength-of-materials sign convention from Chap. 4. The reaction influence lines are obtained using M_b and the equations of static equilibrium (see Fig. E14.1*i* to *k*).

The moment and shear influence lines at point *d* are constructed using the reaction influence lines and the equations of equilibrium. Separate equations for *ab*, *bd*, and *dc* must be written for both these cases:

Segment *ab* $(0 \leq x \leq 10)$:

$$M_d = 5R_c \qquad M_d = \frac{x}{2400}(x^2 - 100)$$

$$V_d = -R_c \qquad V_d = \frac{x}{12,000}(-x^2 + 100)$$

Segment *bd* $(10 \leq x \leq 25$ for M_d; $10 \leq x < 25$ for $V_d)$:

$$M_d = 5R_c \qquad M_d = \frac{1}{4800}(-x^3 + 90x^2 - 1100x + 3000)$$

$$V_d = -R_c \qquad V_d = \frac{1}{24,000}(x^3 - 90x^2 + 1100x - 3000)$$

Segment *dc* $(25 \leq x \leq 30$ for M_d; $25 < x \leq 30$ for $V_d)$:

$$M_d = 5R_c - 1(x - 25) \qquad M_d = \frac{1}{4800}(-x^3 + 90x^2 - 5900x + 123,000)$$

$$V_d = 1 - R_c \qquad V_d = \frac{1}{24,000}(x^3 - 90x^2 + 1100x + 21,000)$$

See Fig. E14.1*l* and *m*.

If the structure has more than one redundant, the calculations can be considerably more involved. For example, for a beam which is indeterminate to the second degree, two simultaneous equations will result if the method of consistent displacements is being used. After obtaining expressions for the redundants and constructing their influence lines, the process continues as described for structures with single redundants. The three-moment equations and the methods of slope-deflection and moment distribution can be very useful for multiredundant structures.

Influence lines for statically indeterminate trusses are constructed in a fashion similar to that for beams by systematically placing the unit load at each panel point. The truss analysis required for each position of the unit load can be accomplished by various methods, e.g., consistent displacements together with the theorem of complementary virtual work. Example 14.2 illustrates the process for a truss with an external (reaction) redundant, and Example 14.3 presents the calculations for an internal (member) redundant.

Example 14.2 Construct the influence lines for the members. The load is applied to the bottom chord; AE = const for all members.

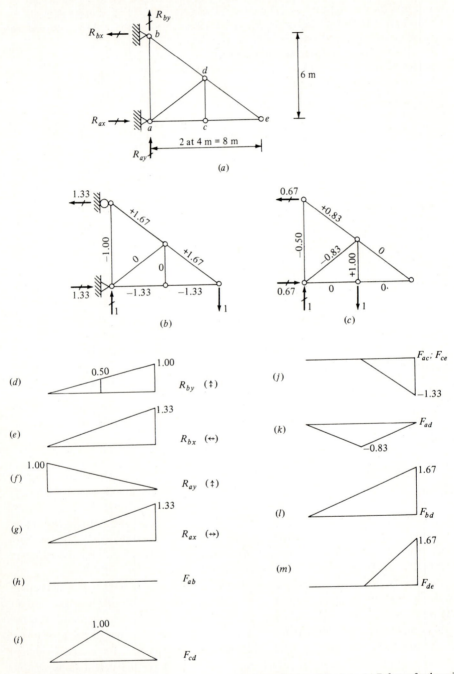

Figure E14.2 (a) Truss with one redundant reaction; (b) F_P forces for the unit load at e; (c) F_P forces for the unit load at c; (d) to (g) reaction influence lines; (h) to (m) member influence lines.

SOLUTION See Table E14.2.

Table E14.2

Member	L, m	F_Q	F_p unit load at: e	F_p unit load at: c	$F_p + R_{by}F_Q$ unit load at: e	$F_p + R_{by}F_Q$ unit load at: c
ab	6	1	−1.00	−0.50	0	0
cd	3	0	0	+1.00	0	1.00
ac	4	0	−1.33	0	−1.33	0
ce	4	0	−1.33	0	−1.33	0
ad	5	0	0	−0.83	0	−0.83
bd	5	0	+1.67	+0.83	+1.67	+0.83
de	5	0	+1.67	0	+1.67	0

$$\Delta_{bo} = \frac{1}{AE}\sum F_p F_Q L \qquad \delta_{bb} = \frac{1}{AE}\sum F_Q^2 L \qquad R_{by} = -\frac{\sum F_p F_Q L}{\sum F_Q^2 L}$$

$$R_{by} = \begin{cases} -\dfrac{-1.00}{1.00} = +1.00 & \text{unit load at } e \text{ (Fig. E14.2b)} \\[2ex] -\dfrac{-0.50}{1.00} = +0.50 & \text{unit load at } c \text{ (Fig. E14.2c)} \end{cases}$$

DISCUSSION The vertical reaction at b was selected as the redundant, and the member forces in the primary structure are shown in Table E14.2. Note that the equations of consistent displacement enforcing zero vertical displacement at point b involve exclusively the forces in member ab since only this member has a nonzero value for the applied virtual load F_Q. The last two columns of Table E14.2 show the final member forces for the two positions of the unit load and are used to construct the influence lines.

This is the truss analyzed in Example 11.2 with applied loads of 20 and 40 kN at joints c and e, respectively. The reader is urged to use the influence lines to check the values of the reactions obtained in that example.

Example 14.3 Draw influence lines for members cd, bd, ce, bc, de, and be; $AE = \text{const}$ for all members.

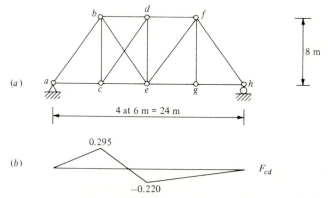

Figure E14.3 (a) Truss with one internal redundant; (b) to (g) influence lines for selected members.

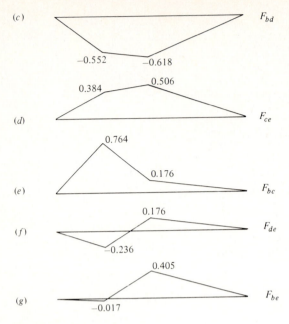

Figure E14.3 continued

SOLUTION See Table E14.3.

Table E14.3 †

				F_p unit load at:			F_pF_QL unit load at:			$F_p + XF_Q$ unit load at:		
	L, m	F_Q	F_Q^2L	c	e	g	c	e	g	c	e	g
bd	6	−0.6	2.16	−0.375	−0.750	−0.375	+1.350	+2.700	+1.350	−0.552	−0.618	−0.309
df	6	0	0	—	—	—	0	0	0	—	—	—
ac	6	0	0	—	—	—	0	0	0	—	—	—
ce	6	−0.6	2.16	+0.561	+0.374	+0.187	−2.020	−1.346	−0.675	+0.384	+0.506	+0.253
eg	6	0	0	—	—	—	0	0	0	—	—	—
gh	6	0	0	—	—	—	0	0	0	—	—	—
bc	8	--0.8	5.12	+1.000	0	0	−6.400	0	0	+0.764	+0.176	+0.088
de	8	−0.8	5.12	0	0	0	0	0	0	−0.236	+0.176	+0.088
fg	8	0	0	—	—	—	0	0	0	—	—	—
ab	10	0	0	—	—	—	0	0	0	—	—	—
be	10	+1.0	10.0	−0.312	+0.625	+0.312	−3.120	+6.250	+3.120	−0.017	+0.405	+0.202
ef	10	0	0	—	—	—	0	0	0	—	—	—
fh	10	0	0	—	—	—	0	0	0	—	—	—
cd	10	+1.0	10.0	0	0	0	0	0	0	+0.295	−0.220	−0.110
Total		Σ = 34.56					Σ = −10.190	+7.604	+3.795			

† Entries shown by a dash do not enter into computation of the redundancy and have therefore been disregarded.

$$\Delta_{x0} = \frac{1}{AE} \sum F_p F_Q L \qquad \delta_{xx} = \frac{1}{AE} \sum F_Q^2 L \qquad X = -\frac{\sum F_p F_Q L}{\sum F_Q^2 L}$$

$$X = \begin{cases} 0.295 & \text{for unit load at } c \\ -0.220 & \text{for unit load at } e \\ -0.110 & \text{for unit load at } g \end{cases}$$

DISCUSSION Regarding member cd as the redundant, the primary structure corresponds to a symmetrical truss; several of the influence lines for this particular structure are illustrated in Fig. 5.15.

14.3 THE MÜLLER-BRESLAU PRINCIPLE

Influence lines can also be constructed using a method which stems from the reciprocal theorem of Maxwell (Chap. 10) but is generally attributed to Müller-Breslau (1886).

Müller-Breslau principle A linear elastic structure will assume the shape of its own influence line if the restraint, corresponding to the generalized force of interest, is removed and the structure is given a corresponding generalized displacement. The resulting deflections are proportional to the influence ordinates.

The validity of this principle can be demonstrated by considering the statically indeterminate beam investigated in the previous section (see Fig. 14.3). Using the method of consistent displacements and considering R_b as the redundant, we first obtain the vertical displacement at the redundant due to a unit vertical load in a general position and denote it δ_{bx}. Next, the vertical displacement δ_{bb} at point b due to a unit load at that point is calculated. The requirement that the displacement at the redundant be zero gives

$$R_b \delta_{bb} - \delta_{bx} = 0$$

or

$$R_b = \frac{\delta_{bx}}{\delta_{bb}} \tag{a}$$

From the reciprocal theorem $\delta_{bx} = \delta_{xb}$; thus,

$$R_b = \frac{\delta_{xb}}{\delta_{bb}} \tag{14.2}$$

That is, the deflected configuration of the beam defines the shape of the influence line, and if point b were given a unit displacement, the deflections would correspond to the influence ordinates.

Application of this principle will yield other influence lines for the beam. For example, the shape of the influence line for R_a is obtained by imparting a displacement at a as shown in Fig. 14.3d, and the influence ordinates are

$$R_a = \frac{\delta_{xa}}{\delta_{aa}} \tag{b}$$

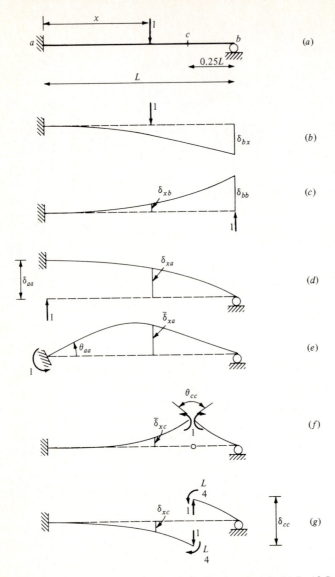

Figure 14.3 Influence lines using the Müller-Breslau principle: (c) R_b; (d) R_a; (e) M_a; (f) M_c; (g) V_c.

The influence line for the reaction M_a is formed by inducing a rotation at a (see Fig. 14.3e), and the condition that the slope at this point be zero results in

$$M_a\theta_{aa} - \theta_{ax} = 0 \tag{c}$$

where θ_{aa} is the rotation at a due to a unit moment at a and θ_{ax} is the rotation at a due to a unit vertical load positioned at some point along the beam [this will result in a clockwise rotation, hence the minus sign in Eq. (c)]. From the reciprocal theorem, the deflection $\bar{\delta}_{xa}$ at point x due to a unit moment at a is equal to θ_{ax}; thus,

$$M_a = \frac{\bar{\delta}_{xa}}{\theta_{aa}} \tag{d}$$

The influence line for the moment at an interior point such as c is obtained by introducing a discontinuity in the slope at that point. This yields a beam as shown in Fig. 14.3f with an internal hinge inserted at point c and unit moments as shown, which are applied to the left and right portions of the beam. This gives rise to the deflected shape depicted with the discontinuity in slope θ_{cc}. From the reciprocal theorem, the rotation θ_{cx} at c due to a unit vertical load at x equals the displacement $\bar{\delta}_{xc}$ at x due to a unit moment at c; therefore

$$M_c = \frac{\bar{\delta}_{xc}}{\theta_{cc}} \tag{e}$$

To formulate the influence line for the shear at c the slope continuity at this point is maintained, a slide mechanism is inserted, and a unit discontinuity in the displacement is introduced as shown in Fig. 14.3g. The influence ordinates are

$$V_c = \frac{\delta_{xc}}{\delta_{cc}} \tag{f}$$

where δ_{xc} is the displacement at x due to a unit shear at c and δ_{cc} is the relative displacement of the cut at c.

The influence lines in Fig. 14.3 are identical to those in Fig. 14.2, which are obtained by successive positioning of the unit load. The calculation of the ordinates for these influence lines by the Müller-Breslau principle is left for Prob. 14.12. The method is demonstrated in Example 14.4 for the two-span beam investigated in the previous section.

Example 14.4 Obtain the influence lines for R_a, R_b, R_c, M_d, and V_d using the Müller-Breslau principle; $EI = $ const.

SOLUTION From Fig. E14.4b and c

$$R_a = \frac{\delta_{xa}}{\delta_{aa}} \qquad \delta_{aa} = \frac{1000}{EI}$$

$$\delta_{xa} = \begin{cases} \dfrac{1}{6EI}(x^3 - 700x + 6000) & 0 \le x \le 10 \\[2ex] \dfrac{1}{12EI}(-x^3 + 90x^2 - 2300x + 15{,}000) & 10 \le x \le 30 \end{cases}$$

From Fig. E14.4d and e

$$R_b = \frac{\delta_{xb}}{\delta_{bb}} \qquad \delta_{bb} = \frac{4000}{9EI}$$

$$\delta_{xb} = \begin{cases} \dfrac{x}{9EI}(-x^2 + 500) & 0 \le x \le 10 \\[2ex] \dfrac{1}{18EI}(x^3 - 90x^2 + 1900x - 3000) & 10 \le x \le 30 \end{cases}$$

From Fig. E14.4f and g

$$R_c = \frac{\delta_{xc}}{\delta_{cc}} \qquad \delta_{cc} = \frac{4000}{EI}$$

$$\delta_{xc} = \begin{cases} \dfrac{x}{3EI}(x^2 - 100) & 0 \le x \le 10 \\[2ex] \dfrac{1}{6EI}(-x^3 + 90x^2 - 1100x + 3000) & 10 \le x \le 30 \end{cases}$$

From Fig. E14.4h and i

$$M_d = \frac{\bar{\delta}_{xd}}{\theta_{dd}} \qquad \theta_{dd} = \frac{160}{EI}$$

$$\bar{\delta}_{xd} = \begin{cases} \dfrac{x}{15EI}(x^2 - 100) & 0 \le x \le 10 \\[2ex] \dfrac{1}{30EI}(-x^3 + 90x^2 - 1100x + 3000) & 10 \le x \le 25 \\[2ex] \dfrac{1}{30EI}(-x^3 + 90x^2 - 5900x + 123{,}000) & 25 \le x \le 30 \end{cases}$$

From Fig. 14.4j and k

$$V_d = \frac{\delta_{xd}}{\delta_{dd}} \qquad \delta_{dd} = \frac{4000}{EI}$$

$$\delta_{xd} = \begin{cases} \dfrac{x}{3EI}(-x^2 + 100) & 0 \le x \le 10 \\[2ex] \dfrac{1}{6EI}(x^3 - 90x^2 + 1100x - 3000) & 10 \le x < 25 \\[2ex] \dfrac{1}{6EI}(x^3 - 90x^2 + 1100x + 21{,}000) & 25 < x \le 30 \end{cases}$$

DISCUSSION To obtain the influence line for R_a the beam is deformed as shown in Fig. E14.4b, resulting in the conjugate beam in Fig. E14.4c. Note that other methods described in Chaps. 9 and 10 could also have been used to calculate the displacements. The final computations giving the expressions for R_a have been indicated but not performed, but the results will be identical to those in Fig. E14.1i. The other two reaction influence lines are constructed in a similar manner; Fig. E14.4d and f show the respective deformed shapes and Fig. E14.4e and g the corresponding conjugate beams.

The influence line for the moment at d (Fig. E14.4h) is formed by giving a discontinuity in slope at this point, accomplished by inserting an internal hinge and applying a unit

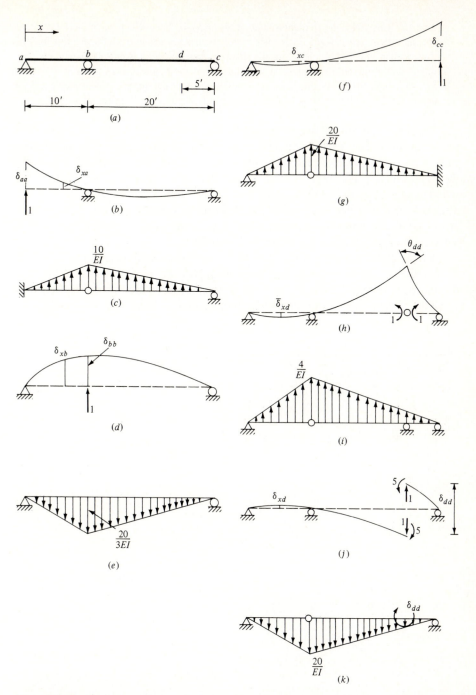

Figure E14.4 (*a*) Two-span beam; (*b*), (*c*) influence line and conjugate beam for R_a; (*d*), (*e*) influence line and conjugate beam for R_b; (*f*), (*g*) influence line and conjugate beam for R_c; (*h*), (*i*) influence line and conjugate beam for M_d; (*j*), (*k*) influence line and conjugate beam for V_d.

moment. This yields the conjugate beam of Fig. E14.4i, from which the influence ordinates are calculated. The deformation configuration required to obtain the influence line for V_d is illustrated in Fig. E14.4j. Note that the slope continuity at point d must be maintained. The required unit internal shear must be accompanied by the moments shown in Fig. E14.4j to maintain equilibrium. In Fig. E14.4k note that the external applied elastic moment δ_{dd} at d renders the conjugate beam stable, allowing the displacements to be calculated.

The use of the Müller-Breslau principle is illustrated in Fig. 14.4 for a beam with two redundants. The reaction influence lines in Fig. 14.4b to e are formed by giving each of the points a unit displacement and sketching the resulting deflected shape. The moment influence line (Fig. 14.4f) is constructed by inserting a hinge at point e and introducing a discontinuity in the slope at this point, giving the deflected shape shown. The shear influence line in Fig. 14.4g is obtained by incorporating a slide mechanism at this point and imposing a unit relative displacement to the ends of the cut. The calculation of the influence ordinates is left for Prob. 14.13.

The Müller-Breslau principle can also be applied to determinate structures. It is suggested that the reader review selected examples and problems from Chap. 5 to demonstrate this fact. Note that a statically determinate structure becomes a mechanism when the capacity for transmitting one generalized force is removed. It is implied that

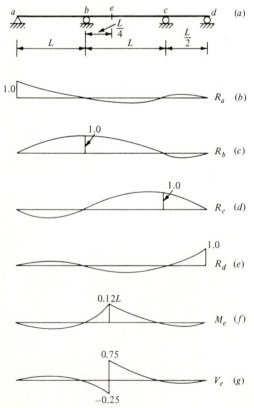

Figure 14.4 Influence lines for a beam with two redundant reactions.

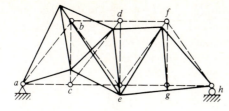

Figure 14.5 Influence line for F_{cd} using the Müller-Breslau principle.

the displacement imparted to the structure to give the influence line will be accompanied by appropriate forces so that equilibrium is satisfied.

Influence lines for trusses can also be obtained using the Müller-Breslau principle. For example, if the influence line for F_{cd} of the truss in Example 14.3 is required for loads applied to the bottom chord it can be envisioned that a unit displacement must be introduced between points c and d. This can be accomplished by inserting an imaginary turnbuckle in the member and imposing a deformation which results in a displacement of the truss as sketched in Fig. 14.5. The shape of the deflected lower chord forms the influence line for this member, and the ordinates are equal to the respective displacements divided by the imposed deformation of member cd.

14.4 APPLICATIONS

The relationships between influence lines and the structural loading developed in Chap. 5 are also applicable to influence lines for indeterminate structures. Thus, for example, if the beam of Example 14.1 were used for a bridge, the various design live loads would be placed on the structure in the critical positions as indicated by the influence lines and the corresponding functions calculated. Assume that this beam is subjected to a variable-length live load of 2 kips/ft and we want to calculate the maximum upward reaction at c and the maximum positive moment at point d. From the influence line for R_c we deduce that the load must be positioned on span bc; thus, using the expression for R_c for loading in this span, we have

$$R_c = 2 \int_{10}^{30} \frac{1}{24{,}000} (-x^3 + 90x^2 - 1100x + 3000) \, dx$$

which upon integration and evaluation at the limits gives $R_c = 16.7$ kips. Similarly, the maximum positive moment at d is obtained by loading the structure between points b and c, and the value of the moment is obtained by integrating the expressions from Example 14.1

$$M_d = 2 \int_{10}^{25} \frac{1}{4800} (-x^3 + 90x^2 - 1100x + 3000) \, dx$$

$$+ 2 \int_{25}^{30} \frac{1}{4800} (-x^3 + 90x^2 - 5900x + 123{,}000) \, dx$$

This gives $M_d = 58.3$ kip · ft. The values are identical to those obtained for this structure in Example 11.6, which was solved by the method of consistent displacements. Unfortunately, this approach is not as useful as this example implies since the functions for the influence lines are frequently too complex for convenient integration and some other approach, e.g., numerical integration, must be used.

The properties of influence lines can be used qualitatively in still another manner. Assume that the beam of Fig. 14.6a is a girder in a building such as a warehouse, where the live load can be arranged in various patterns. In order to obtain the critical design forces for this structure, influence lines can be used to determine where the uniform live load should be placed to give the largest magnitudes of the various functions. The influence lines for the reaction at c, the maximum positive moment in span cd, and the negative moments at points c and d have been sketched using the Müller-Breslau principle (Fig. 14.6b to e). Thus, for example, the maximum upward reaction at c is obtained by placing the live load from b to d and also on span ef, as shown in Fig. 14.6b. The critical loading patterns for the other three functions are depicted in Fig. 14.6c to e. Note that the final design values are obtained by combining these live-load values with those for dead load (which is applied over the entire length of the girder).

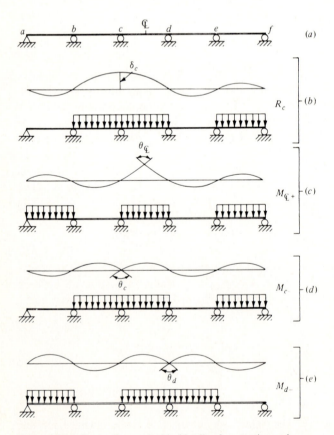

Figure 14.6 Influence lines and critical loading for a continuous beam.

Influence lines can also be used in this way for establishing the critical load patterns for the members in a rigid frame. For example, to determine the critical loading arrangement that yields the maximum positive moment in span cd of the three-story five-bay rigid frame in Fig. 14.7a the influence line is first constructed using the Müller-Breslau principle, which yields the deflected shape shown. Applying a uniformly distributed live load to only those spans with positive ordinates results in the checkerboard pattern of load which produces the maximum positive moment in this span. The shape of the influence line for the negative moment to the left of point c is presented in Fig. 14.7b. Constructing it by the Müller-Breslau principle requires a hinge to be placed in the girder to the left of joint c; thus, the columns are constrained to rotate with the girder to the right of c. The deflection pattern of the girders in the stories above and below this floor will influence the deformed shape of the columns immediately above and below joint c; therefore, the shape illustrated is the correct configuration if all girders have approximately the same stiffness. Since the deflections are small for the first- and third-floor girders in bay bc, they will be ignored. The critical loading pattern giving the largest magnitude of the negative moment at c is shown in Fig. 14.7b.

In the design of buildings it is often necessary to construct a design envelope for the members to indicate the range of forces to be sustained. Such an envelope can be

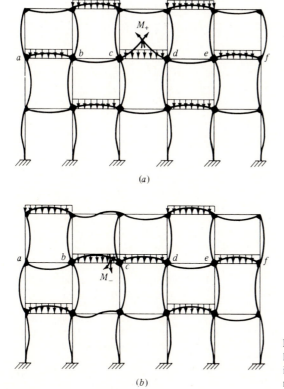

(a)

(b)

Figure 14.7 Influence line and critical loading for (a) maximum positive moment in span cd; and (b) largest magnitude of negative moment to the left of c.

prepared using conventional analysis methods in conjunction with the qualitative properties of influence lines. As an example, the shape of the design envelope for the bending moments in span cd for the continuous beam in Fig. 14.6a will be constructed. For the sake of clarity and simplicity the dead load of the structure is ignored, but *this must not be done in an actual design situation.* The influence lines in Fig. 14.6 are used to establish the critical locations of the live load. The loading and the approximate shapes of the bending-moment diagrams for the critical situations are sketched in Fig. 14.8. For example, the bending-moment diagram in Fig. 14.8c gives the maximum positive bending moment in span cd. In addition, we can observe from the influence line of Fig. 14.8b that if the live load is placed in spans bc and de, a negative moment is introduced into span cd. This loading and the corresponding bending moment are sketched in Fig. 14.8d. The bending-moment diagram giving the largest magnitude of the negative bending moment at point c is shown in Fig. 14.8f; furthermore, a positive moment would occur at point c if spans ab and de were loaded. This can be observed from the influence line in Fig. 14.8e, or a sketch of the deflected beam under this loading would reveal this to be the case. The design envelope for

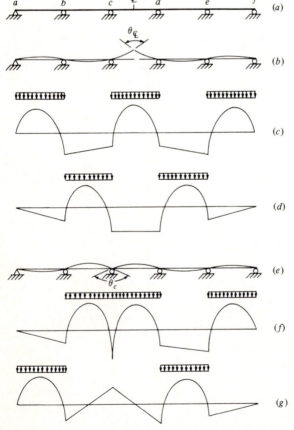

Figure 14.8 (a) Beam subjected to live load only; (b) influence line for $M_{\mathbb{C}+}$; (c) loading and moment diagram for maximum $M_{\mathbb{C}+}$; (d) loading and moment diagram for largest magnitude of $M_{\mathbb{C}-}$; (e) influence line for M_{c-}; (f) loading and moment diagram for largest magnitude of M_{c-}; (g) loading and moment diagram for maximum M_{c+}.

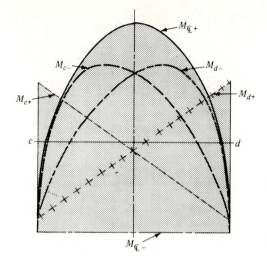

Figure 14.9 Design envelope for span cd for the beam in Fig. 14.8.

member cd can now be constructed using the four bending-moment diagrams in Fig. 14.8, together with the observation that the negative and positive moment diagrams for point d are mirror images of those for point c. In Fig. 14.9 the six bending-moment diagrams are plotted to portray the behavior of the member, and the shaded area between the extrema constitutes the design envelope for the member for live load only. From this the structural engineer can observe that both the center and the ends of the member will undergo bending moments that range from negative to positive, and the design values are the extrema.

PROBLEMS

14.1 Obtain the influence line for M_a in Fig. 14.2 using (*a*) the three-moment equation, (*b*) the slope-deflection equations, and (*c*) the moment-distribution method.

In Probs. 14.2 to 14.11 construct the designated influence lines using the method of successively positioning the unit load.

14.2 Refer to Prob. 11.5; the influence lines for R_b, R_d, M_c, M_d, and V_c.

14.3 Refer to Prob. 11.7; the influence lines for the reactions.

14.4 Refer to Prob. 11.8; the influence lines for the reactions plus M_c and V_c.

14.5 Refer to Prob. 11.10; the influence lines for the reactions and all members if: (*a*) vertical loads are applied to the bottom chord; and (*b*) horizontal loads are applied to the bottom chord.

14.6 Refer to Prob. 11.14; the influence lines for all members if vertical loads are applied to the bottom chord.

14.7 Refer to Prob. 11.15; the influence lines for cf, de, and df if vertical loads are applied to the bottom chord.

14.8 Refer to Prob. 11.16; the influence lines for ef, dg, and df if vertical loads are applied to the bottom chord.

14.9 The influence lines for members *be*, *dg*, *df*, and *bd* in the figure for vertical loads applied to the bottom chord if: (*a*) the diagonal members can resist tension only; and (*b*) the diagonal members resist equally the shear in a given panel.

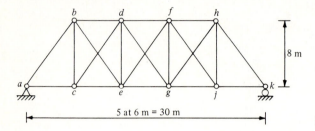

Figure P14.9

14.10 Refer to Prob. 11.17; the influence lines for the reactions at *a* plus M_b and V_b if vertical loads are applied to beam *ac*.

14.11 Refer to Prob. 11.19; the influence lines for the reactions at *a* plus the force in member *cd* if vertical loads are applied to beam *ac*.

14.12 Obtain the ordinates of the influence lines in Fig. 14.3 using the Müller-Breslau principle.

14.13 Obtain the ordinates of the influence lines in Fig. 14.4 using the Müller-Breslau principle. EI = const.

   In Probs. 14.14 to 14.20 construct the designated influence lines for the original structure using the Müller-Breslau principle.

14.14 Problem 14.2

14.16 Problem 14.4

14.15 Problem 14.3

14.17 Problem 14.10

14.18 Refer to Prob. 12.9. Sketch the shape of the influence lines for the reactions at *c* if loads are applied normal to all members.

14.19 Refer to Prob. 12.17. Sketch the shape of the influence lines for the reactions at *a* if loads are applied normal to all members.

14.20 Refer to Prob. 12.18. Sketch the shape of the influence lines for the reactions at *a* if loads are applied normal to all members.

14.21 Refer to Fig. 14.7. Determine the pattern of uniform vertical loading using the shapes of the influence lines to obtain the designated function:

   (*a*) The largest magnitude of negative moment in the girder to the right of *c*.

   (*b*) The maximum positive moment in span *ab*.

   (*c*) The largest magnitude of negative moment in the girder at *a*.

14.22 Determine the pattern of uniform vertical loading using the shapes of the influence lines to obtain the designated function:

   (*a*) The largest magnitude of negative moment in the girder at *a*.

   (*b*) The maximum positive moment in span *ab*.

   (*c*) The largest magnitude of negative moment in the girder to the left of *b*.

   (*d*) The largest magnitude of negative moment in the girder to the right of *b*.

   (*e*) The maximum positive moment in span *bc*.

14.23 Construct the design envelope for the bending moment in span *bc* of this uniform beam if it is subjected to a uniformly distributed live load of 3 kips/ft which can be positioned on the structure to give the critical design situation. Ignore the dead load of the beam.

14.24 Construct the design envelope for the bending moment in span *bc* of this uniform beam if it is subjected to a uniformly distributed live load of 12 kN/m which can be positioned on the structure to give the critical design situation. Ignore the dead load of the beam.

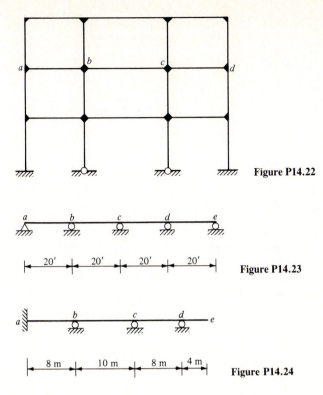

Figure P14.22

Figure P14.23

Figure P14.24

ADDITIONAL READING

Hsieh, Yuan-Yu: *Elementary Theory of Structures,* 2d ed., chap. 11, Prentice-Hall, Englewood Cliffs, N. J., 1982.

McCormac, J. C.: *Structural Analysis,* 3d ed., chaps. 19 and 20, Intext Educational, New York, 1975.

Norris, C. H., J. B. Wilbur, and S. Utku: *Elementary Structural Analysis,* 3d ed., chap. 10, McGraw-Hill, New York, 1976.

West, H. H.: *Analysis of Structures,* chap. 17, Wiley, New York, 1980.

Willems, N., and W. M. Lucas, Jr.: *Structural Analysis for Engineers,* chap. 5, McGraw-Hill, New York, 1978.

MATRIX STRUCTURAL ANALYSIS

FIFTEEN

THE STIFFNESS METHOD

15.1 INTRODUCTION

The digital computer, which has changed many aspects of our culture, has had a significant effect on structural analysis. Structural systems impossible to investigate by traditional methods can be analyzed and designed using the computer. For example, large problems with thousands of unknowns, nonlinear material behavior, and large-deformation phenomena are all tractable using current computer technology. The structures shown in Fig. 15.1 are typical of those which could not have been built before the computer age.

These advances in structural analysis have been made possible not only by the development of the digital computer, but also by a reformulation of the governing equations. In the 1940s and 1950s structural engineers in the airframe industry were faced with the necessity of analyzing and designing both the swept-wing and delta-wing aircraft. Since the structural systems for these wing configurations are statically indeterminate to a high degree, it was necessary to reformulate the governing equations to eliminate cumbersome hand calculations (the initial contributions on the swept- and delta-wing aircraft, both by S. Levy,[1,2] are the forerunners of the flexibility and stiffness methods, respectively).

The matrix methods of structural analysis were developed and initially formulated for truss and beam structures. The structural engineers responsible for these early developments enforced compatibility at discrete points throughout the structure to derive the basic relationships. This approach yields a set of equations in which the forces are the unknowns and the coefficients in the equations are the flexibility (influence) quantities. Hence, the method is termed either the *force method* or the *flexibility method*. Almost simultaneously, work was being carried out on a companion method, in which equilibrium is enforced at discrete points, giving a set of equations with displacements as the unknowns and the coefficients as the stiffness quantities. Hence, the method is referred to as either the *displacement method* or the *stiffness method*. These two methods were used and developed in parallel until near the middle of the 1960s (an interesting dual formulation of the force and displacement methods is

(a)

(b)

Figure 15.1 (a) Sears Tower, Chicago (*American Institute of Steel Construction*); (b) assembly of the 757 airplane (*The Boeing Company*).

presented in Ref. 3). A key contribution by Turner et al.,[4] together with subsequent significant advances, greatly strengthened the displacement method for handling complex problems, and as a result it has almost totally supplanted the force method.

In both methods the governing equations are constituted in matrix form. Matrix notation is a compact and convenient method for writing systems of equations, and the arrays represent an efficient format for manipulating and storing information in the digital computer. During 1947 to 1965, when the force and displacement methods were being formulated, the digital computer was also experiencing significant advances in both hardware and software design. The resulting combined developments of the computer and structural analysis technologies revolutionized the entire field of structural analysis. Computing efficiency has since increased greatly, and the stiffness method has been expanded and improved so that it can be used not only for trusses, beams, and rigid frames but also for such complex continuous structures as plates, shells, and solids.[†] Most structural engineering offices now have access to computers that permit use of the stiffness method for analyzing traditional civil engineering structures, airframes, or more exotic structures such as offshore oil drilling towers and space structures.

Although the method of virtual work (Chap. 10) is the basis for the stiffness method, for trusses, beams, and rigid frames the equations can also be derived from the fundamental equations of force equilibrium, compatibility, stress-strain, strain-displacement, and the boundary conditions. Since the stiffness method is a general form of the *equilibrium method*, it is categorized with the classical slope-deflection and moment-distribution methods described in Chap. 12. Unlike the classical methods, the stiffness method is designed for use with the digital computer; consequently, it is probably the best alternative for solving complex structural problems. Thus, the stiffness method augments the various methods previously studied and represents another approach for analyzing structures.

The purpose of this chapter is to present the concepts of the stiffness method. Energy principles from Chap. 10 are included in the initial discussion of flexibility and stiffness coefficients. Two springs connected in series are then used to demonstrate the stiffness method. This physical model allows the concept to be visualized without encumbrances. The springs are analyzed in two different ways: (1) the solution is formulated using basic principles from strength of materials, and (2) the equations are derived using the stiffness method. This sequence allows the reader to progress logically from familiar concepts. Finally, the stiffness method is heuristically generalized from the results of the two-spring problem, and the fundamental procedure of the displacement method is revealed. The basic steps require computations to obtain (1) member stiffness properties, (2) structural stiffness properties, (3) nodal displacements, and (4) member internal forces.

Since matrix notation and its associated mathematics will be used extensively throughout the rest of the book, readers should have at least a rudimentary knowledge of basic matrix operations before proceeding. It is a great asset to be able to handle structural concepts in the succinct notation of matrices. Development of this skill is merely a matter of practice once the basic operations have been mastered. Studying the

---

[†]When the stiffness method is applied to continua, it is generally called the *finite element method*. This designation will not be used here because the analysis in this book is restricted to discrete structural systems.

displacement method offers a bonus since facility in thinking in terms of matrices develops simultaneously with one's grasp of the primary structural theory.

15.2 FLEXIBILITY AND STIFFNESS COEFFICIENTS

The response of any complex system can be cast as a set of algebraic equations that relate the forces and displacements functionally. This dependency can be described by either the flexibility or stiffness coefficients. All energy theorems stem from the principle of virtual work or from the principle of complementary virtual work (Chap. 10). In Chaps. 11 and 12 we noted that the three-moment equations arise from the latter whereas both the slope-deflection method and the iterative moment-distribution method originate from the former.

The principle of complementary virtual work gives the flexibility coefficients, F_{ij}, which describe the deflection at point i due to a unit force applied at point j. Similarly the principle of virtual work can be used to obtain the stiffness coefficients K_{ij}, each of which defines the force at point i necessary to impose a unit deflection at point j. From a casual inspection the algebraic equations generated by using the flexibility or stiffness coefficients appear to be simple inverse relations of each other, but it is important to note that the underlying theories from which the equations are obtained are quite different.

For a linear elastic structure supported against rigid-body displacements and subjected to a set of forces X_j the deflection U_i at a point i is calculated as

$$U_i = F_{i1}X_1 + F_{i2}X_2 + \cdots + F_{ij}X_j + \cdots + F_{in}X_n \tag{15.1}$$

where there are n applied forces on the structural system. It is possible to obtain such an algebraic equation for each displacement, giving the set

$$
\begin{bmatrix} U_1 \\ U_2 \\ \cdot \\ U_i \\ \cdot \\ U_n \end{bmatrix}
=
\begin{bmatrix}
F_{11} & F_{12} & \cdots & F_{1i} & \cdots & F_{1n} \\
F_{21} & F_{22} & \cdots & F_{2i} & \cdots & F_{2n} \\
\multicolumn{6}{c}{\dotfill} \\
F_{i1} & F_{i2} & \cdots & F_{ii} & \cdots & F_{in} \\
\multicolumn{6}{c}{\dotfill} \\
F_{n1} & F_{n2} & \cdots & F_{ni} & \cdots & F_{nn}
\end{bmatrix}
\begin{bmatrix} X_1 \\ X_2 \\ \cdot \\ X_i \\ \cdot \\ X_n \end{bmatrix}
\tag{15.2}
$$

$$\qquad\qquad \mathbf{U} \qquad\qquad\qquad\qquad \mathbf{F} \qquad\qquad\qquad \mathbf{X}$$

or
$$\mathbf{U} = \mathbf{FX} \tag{15.3}$$

where the column matrices $\mathbf{U}$ and $\mathbf{X}$ contain all the components of displacement and force, respectively, and the square matrix $\mathbf{F}$ is the structural flexibility matrix. Note that these one-dimensional arrays (column matrices or vectors) do not contain the components of a physical vector, e.g., a force vector. If $\mathbf{F}$ is known, the displacements resulting from a known force matrix can be obtained by simple matrix multiplication, as indicated by Eq. (15.3). By matrix manipulation it is possible to solve the equations, i.e.,

$$\mathbf{X} = \mathbf{F}^{-1}\mathbf{U} \tag{15.4}$$

where $\mathbf{F}^{-1}$ is defined as the inverse of the matrix $\mathbf{F}$. In this form matrix $\mathbf{F}^{-1}$ will contain the stiffness coefficients K_{ij}, which represent the force at point i due to a unit deflection imposed at point j. Hence

$$\mathbf{K} = \mathbf{F}^{-1} \tag{15.5}$$

The limitations in going from Eq. (15.3) to Eq. (15.4) will be discussed later. Typically, Eq. (15.5) is valid, but the physical meaning of F_{ij} is quite different from that of K_{ij}, and the two matrices originate from distinctly different bases.

In expanded form the matrix equation (15.4) can be written

$$
\begin{bmatrix} X_1 \\ X_2 \\ \cdot \\ X_i \\ \cdot \\ X_n \end{bmatrix} = \begin{bmatrix} K_{11} & K_{12} & \cdots & K_{1i} & \cdots & K_{1n} \\ K_{21} & K_{22} & \cdots & K_{2i} & \cdots & K_{2n} \\ \multicolumn{6}{c}{\dotfill} \\ K_{i1} & K_{i2} & \cdots & K_{ii} & \cdots & K_{in} \\ \multicolumn{6}{c}{\dotfill} \\ K_{n1} & K_{n2} & \cdots & K_{ni} & \cdots & K_{nn} \end{bmatrix} \begin{bmatrix} U_1 \\ U_2 \\ \cdot \\ U_i \\ \cdot \\ U_n \end{bmatrix} \tag{15.6}
$$

Deflecting the structure into a configuration such that $U_1 = 1$ and $U_2 = U_3 = \cdots = U_n = 0$ gives

$$X_1 = K_{11}, \ X_2 = K_{21}, \cdots, \ X_n = K_{n1} \tag{15.7}$$

Equations (15.7) contain all the stiffness elements in the first column of $\mathbf{K}$. These stiffness coefficients represent the full set of forces necessary to maintain the deformed shape with $U_1 = 1$ and the other displacements all zero. In general, the ith column of the stiffness matrix contains the complete set of forces necessary to deform the structure into the shape $U_i = 1$ and all other deflections equal to zero. Or, as stated earlier, the element K_{ij} represents the force at node i to maintain equilibrium with a unit displacement imposed at node j.

The reciprocal theorem of Maxwell-Betti states that a linear elastic structure with unyielding supports subjected only to applied loads will behave in such a way that the deflection at point i due to a load P at point j will be equal to the deflection at point j due to a load P applied at point i. As a direct consequence $F_{ij} = F_{ji}$; that is, the matrix $\mathbf{F}$ is symmetric [see Eq. (15.2)]. Since $\mathbf{K}$ is the inverse of $\mathbf{F}$, as shown in Eq. (15.5), and since the inverse of a symmetric matrix is also a symmetric matrix, the matrix $\mathbf{K}$ is symmetric; that is, $K_{ij} = K_{ji}$. This fact is extremely useful in dealing with the equations of matrix structural analysis.

For most structural systems many manipulations are required to obtain the governing equations in the form shown in either Eq. (15.2) or (15.6). It is desirable to have these equations generated systematically so they can be programmed for the computer. The logic associated with the form shown in Eq. (15.2) is referred to as the *flexibility (force) method*; if the resulting equations are in the form of Eq. (15.6), the method used to generate the equations is called the *stiffness (displacement)* method. Since the former embodies the concepts of the principle of complementary virtual work and the latter incorporates the principle of virtual work, the methods should not be envisioned as simple inverse relations of each other. The stiffness method is used almost exclusively and the equations can be easily generated with the computer; consequently, the flexibility method will be abandoned at this point in the text.

15.3 THE ELASTIC SPRING ASSEMBLAGE

The solution of an assemblage consisting of two linear elastic springs in series can be used to illustrate many of the concepts and problems associated with the stiffness method. This elementary problem will initially be solved using strength-of-materials principles to calculate displacements at various connection points and to obtain forces in the springs. The governing algebraic equations will be cast in the proper form and systematized to illustrate the basic steps in the method.

First, the relationships between forces and displacements at points a, b, and c must be developed for the assemblage shown in Fig. 15.2. Note that the connection (nodal) points are designated by letters and the spring members are denoted by their two endpoints e.g., spring ab. The X_i and U_i denote the nodal point forces and displacements, respectively. The symbols k_1 and k_2 represent the individual stiffness of the two linear elastic springs, which can be obtained experimentally in the laboratory. If the ends of a linear elastic spring are simultaneously displaced as shown in Fig. 15.3a, it will translate and deform. The spring force depends upon the relative displacement of the ends $u_j - u_i$. A graph of spring force versus deformation can be constructed (see Fig. 15.3b), and the slope of the line k is defined as the stiffness of the spring.

There are two basic methods for constructing the desired relationships for this two-spring assemblage (Fig. 15.2). Individual forces can be alternately applied at the three points and the resulting deflections established, or deflections can be systematically imposed at the points and the forces found so that the system is maintained in equilibrium under the prescribed displacement configuration. The first approach will give the flexibility matrix in the form shown in Eq. (15.2). This method probably can be conveniently carried out in the laboratory. The second formulation, which adopts the viewpoint of the stiffness method, will yield the stiffness matrix in the form shown in Eq. (15.6).

In order to form **K** for the two-spring assemblage each of the three points will be individually deflected and all nine of the K_{ij} in the 3×3 stiffness matrix calculated. Figure 15.2 gives the positive directions of force and displacement.

Deflection Case 1: $U_a \neq 0$; $U_b = U_c = 0$

Only spring ab will be deformed; therefore using the force-deformation relations and force equilibrium, we calculate the forces on the points and springs as shown in Fig 15.4a.

Deflection Case 2: $U_b \neq 0$; $U_a = U_c = 0$

Since both springs are attached to node b, continuity requires that deformations and hence forces occur in springs ab and bc. Again, invoking the force-deformation relations

Figure 15.2 The two-spring assemblage.

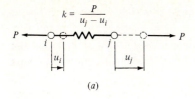

$$k = \frac{P}{u_j - u_i}$$

(a)

Force (P)

k

1

Deformation $(u_j - u_i)$

(b)

Figure 15.3 An individual spring; *(a)* the loaded spring; *(b)* the force-deformation graph.

and the equilibrium relations for both springs individually, we obtain the forces shown in Fig. 15.4*b*.

Deflection Case 3: $U_c \neq 0$; $U_a = U_b = 0$

This deformed state is similar to that in case 1 in that only one spring has been deformed. Hence, applying the appropriate governing equations gives the forces shown in Fig. 15.4*c*.

If nonzero values of U_a, U_b, and U_c are imposed simultaneously, the relationships between these deflections and the corresponding forces are obtained by superposing the

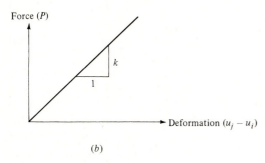

(a)

(b)

(c)

Figure 15.4 Two-spring assemblage showing equilibrating forces for points and members for three deflection cases: *(a)* case 1 deflection $(U_a \neq 0)$; *(b)* case 2 deflection $(U_b \neq 0)$; *(c)* case 3 deflection $(U_c \neq 0)$.

results from cases 1 to 3. The positive coordinate directions shown in Fig. 15.2 must be observed, and since the springs are connected at node b, the total force at that node is obtained by adding the forces from each individual spring. Hence, combining the results from deflection cases 1 to 3 gives the equilibrium equations for the three points

$$X_a^{(1)} + X_a^{(2)} + X_a^{(3)} = X_a = \quad k_1 U_a - k_1 U_b + \quad 0 U_c$$

$$X_b^{(1)} + X_b^{(2)} + X_b^{(3)} = X_b = -k_1 U_a + (k_1 + k_2) U_b - k_2 U_c \qquad (15.8)$$

$$X_c^{(1)} + X_c^{(2)} + X_c^{(3)} = X_c = \quad 0 U_a - k_2 U_b + k_2 U_c$$

In matrix form these force-displacement equations become

$$
\begin{bmatrix} X_a \\ X_b \\ X_c \end{bmatrix} =
\begin{bmatrix} k_1 & -k_1 & 0 \\ -k_1 & (k_1 + k_2) & -k_2 \\ 0 & -k_2 & k_2 \end{bmatrix}
\begin{bmatrix} U_a \\ U_b \\ U_c \end{bmatrix} \qquad (15.9)
$$

$$\qquad\qquad \mathbf{X} \qquad\qquad\qquad \mathbf{K} \qquad\qquad\quad \mathbf{U}$$

or
$$\mathbf{X} = \mathbf{KU} \qquad (15.10)$$

where $\mathbf{X}$ = column matrix of nodal forces
$\mathbf{U}$ = column matrix of nodal displacements
$\mathbf{K}$ = stiffness matrix

A perusal of Eq. (15.9) reveals that a given column of the stiffness matrix represents the equilibrating forces for a prescribed deformed shape of the assemblage. Thus, K_{11} is the force X_a due to a unit displacement U_a, K_{21} is the force X_b due to a unit displacement U_a, etc. Since all forces act in the X direction, a given column must sum to zero for assemblage equilibrium. This suggests that the stiffness matrix can be generated directly from the forces required to maintain equilibrium by alternately displacing each point (this approach is sometimes referred to as the *direct stiffness method*, but the term is not universal).

It appears that if the forces were known and the deflections unknown, the stiffness matrix $\mathbf{K}$ could be inverted and the deflections obtained by premultiplying the load matrix $\mathbf{X}$ by the inverse of $\mathbf{K}$. Note from Fig. 15.2, however, that the two springs have not been restrained against motion. Therefore, if an arbitrary set of loads X_a, X_b, and X_c is applied so that equilibrium in the horizontal direction is not attained, i.e.,

$$X_a + X_b + X_c \neq 0$$

the springs will exhibit elastic deformations plus rigid-body translation in the x direction. Since the assemblage does not have at least one point restrained, the $\mathbf{K}$ matrix is singular (has no inverse); consequently, there are no unique displacements for an arbitrary force matrix. The reader can verify this fact by attempting to invert the coefficient matrix $\mathbf{K}$ as it appears in Eq. (15.9).

The characteristics embodied in Eq. (15.9) can be examined in another way. The deformations (relative displacements) of the two springs (e_{ab} and e_{bc}) are

$$e_{ab} = U_b - U_a \qquad e_{bc} = U_c - U_b \qquad (a)$$

or
$$U_a = U_b - e_{ab} \qquad U_b = U_c - e_{bc} \qquad (b)$$

Combining Eq. (*b*) gives

$$U_a = U_c - e_{ab} - e_{bc} \tag{c}$$

When the new variables e_{ab}, e_{bc}, and U_c are introduced into Eq. (15.9), we have

$$\begin{bmatrix} X_a \\ X_b \\ X_c \end{bmatrix} = \begin{bmatrix} -k_1 & 0 & 0 \\ k_1 & -k_2 & 0 \\ 0 & k_2 & 0 \end{bmatrix} \begin{bmatrix} e_{ab} \\ e_{bc} \\ U_c \end{bmatrix} \tag{d}$$

It can be observed that the coefficient matrix is singular with a rank of 2. A solution for the forces will exist only if the three equations are linearly dependent; i.e., a solution exists only for certain values of the forces. Expansion of the matrix equation gives

$$X_a = -k_1 e_{ab} \tag{e}$$

$$X_b = k_1 e_{ab} - k_2 e_{bc} \tag{f}$$

$$X_c = k_2 e_{bc} \tag{g}$$

The solution of Eqs. (*e*) and (*g*) yields

$$e_{ab} = \frac{-X_a}{k_1} \qquad e_{bc} = \frac{X_c}{k_2}$$

Substituting these two deformations into Eq. (*f*), we have

$$X_b = -X_a - X_c$$

which is the equation of equilibrium for the assemblage. Thus, the system of equations can have a solution only if the three forces are in equilibrium.

Note that the displacement U_c cannot be evaluated whereas the relative displacements e_{ab} and e_{bc} can be determined. If U_c is specified, the three displacements can be calculated from Eqs. (*b*).

From this we have established that (1) the coefficient matrix is singular and of rank 2, (2) the forces must be in equilibrium, (3) only the deformations of the springs can be determined, and (4) calculation of the displacements of the points requires that at least one of the displacements be specified.

If the two springs are constrained at some point, the problem of calculating deflections is tractable. If node c is constrained as shown in Fig. 15.5 and if computation of the reaction force X_c is not required, the third equation in Eq. (15.8) can be ignored. Therefore, the remaining equations for this configuration are

$$X_a = k_1 U_a - k_1 U_b \qquad X_b = -k_1 U_a + (k_1 + k_2) U_b \tag{15.11}$$

or in matrix form

$$\begin{bmatrix} X_a \\ X_b \end{bmatrix} = \begin{bmatrix} k_1 & -k_1 \\ -k_1 & (k_1 + k_2) \end{bmatrix} \begin{bmatrix} U_a \\ U_b \end{bmatrix} \tag{15.12}$$

We write the load matrix $\{X_a \ X_b\}$, where the braces indicate a column matrix written as a row matrix to conserve space. Premultiplying $\{X_a \ X_b\}$ by the inverse of the stiffness matrix gives the deflections

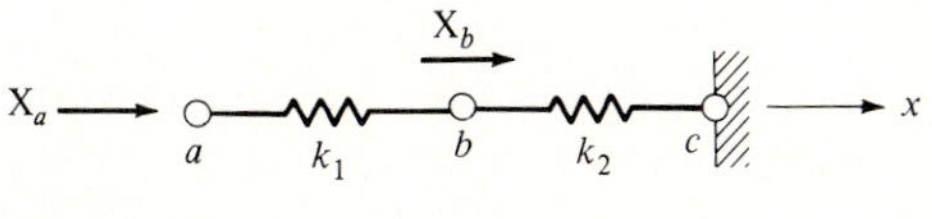

Figure 15.5 Constrained two-spring assemblage showing applied point forces.

$$\begin{bmatrix} U_a \\ U_b \end{bmatrix} = \frac{1}{k_1 k_2} \begin{bmatrix} (k_1 + k_2) & k_1 \\ k_1 & k_1 \end{bmatrix} \begin{bmatrix} X_a \\ X_b \end{bmatrix} \tag{15.13}$$

that is,
$$U_a = \left(\frac{1}{k_1} + \frac{1}{k_2} \right) X_a + \frac{1}{k_2} X_b \qquad U_b = \frac{1}{k_2} X_a + \frac{1}{k_2} X_b \tag{15.14}$$

Now the internal forces s_{ab} and s_{bc} in each of the two springs can be calculated by multiplying the net deflection of the endpoints by the respective spring stiffnesses, giving

$$s_{ab} = k_1(U_b - U_a) = k_1 \left[\frac{1}{k_2}(X_a + X_b) - \frac{1}{k_1} X_a - \frac{1}{k_2}(X_a + X_b) \right] = -X_a$$
$$\tag{15.15}$$
$$s_{bc} = k_2(U_c - U_b) = k_2 \left[0 - \frac{1}{k_2}(X_a + X_b) \right] = -X_a - X_b$$

This problem can be solved more easily by applying elementary force equilibrium to free-body diagrams of each spring and obtaining the deflections U_a and U_b in Eq. (15.14) using the force-deformation relations. However, it should be pointed out that the steps used to solve this problem embody the logic of the stiffness method, which in its general form can be used to investigate truss, beam, and rigid-frame systems. The schematic shown in Fig. 15.6 shows the basic steps necessary to solve a problem similar to the two-spring assemblage. In our situation the individual member (spring) properties were k_1 and k_2. The force-deflection relations of the combined members (springs) are shown in Eq. (15.9) and again in Eq. (15.12) after the points of restraint (reactions) have been defined. The nodal deflections are contained in Eq. (15.13), and the member (spring) internal forces are calculated in Eq. (15.15). Numerical results for a two-spring assemblage are shown in Example 15.1.

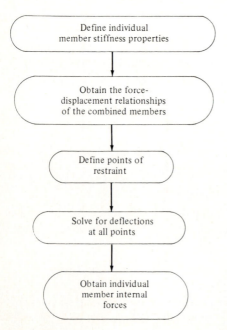

Define individual
member stiffness properties

Obtain the force-
displacement relationships
of the combined members

Define points of
restraint

Solve for deflections
at all points

Obtain individual
member internal
forces

Figure 15.6 Schematic of basic steps for solving the two-spring assemblage.

Example 15.1 Obtain the stiffness matrix for this two-spring assemblage and solve the resulting equations for the displacements at points b and c. Also, calculate the forces in springs ab and bc; $X_b = X_c = 90$ kN.

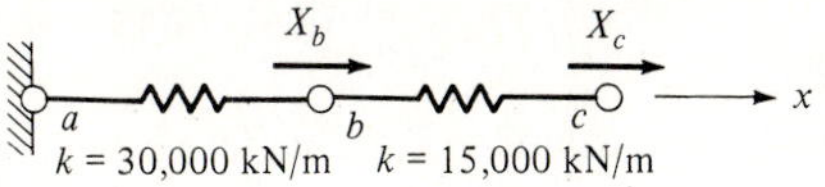

Figure E15.1

SOLUTION For the unrestrained assemblage first let $U_a \neq 0$ and $U_b = U_c = 0$; then

$$X_a = 30 \times 10^3 U_a \qquad X_b = -X_a \qquad X_c = 0$$

Next let $U_b \neq 0$ and $U_a = U_c = 0$; then

$$X_a = -30 \times 10^3 U_b \qquad X_b = 45 \times 10^3 U_b \qquad X_c = -15 \times 10^3 U_b$$

Finally let $U_c \neq 0$ and $U_a = U_b = 0$; then

$$X_a = 0 \qquad X_b = -15 \times 10^3 U_c \qquad U_c = 15 \times 10^3 U_c$$

If $U_a \neq 0$, $U_b \neq 0$, and $U_c \neq 0$ simultaneously,

$$X_a = \quad 30 \times 10^3 U_a - 30 \times 10^3 U_b \qquad + \ 0 U_c$$

$$X_b = -30 \times 10^3 U_a + 45 \times 10^3 U_b - 15 \times 10^3 U_c$$

$$X_c = \qquad 0 U_a - 15 \times 10^3 U_b + 15 \times 10^3 U_c$$

or

$$\begin{bmatrix} X_a \\ X_b \\ X_c \end{bmatrix} = 15 \times 10^3 \begin{bmatrix} 2 & -2 & 0 \\ -2 & 3 & -1 \\ 0 & -1 & 1 \end{bmatrix} \begin{bmatrix} U_a \\ U_b \\ U_c \end{bmatrix}$$

Since point a is constrained, $U_a = 0$ and the resulting equations are

$$\begin{bmatrix} X_b \\ X_c \end{bmatrix} = (15 \times 10^3) \begin{bmatrix} 3 & -1 \\ -1 & 1 \end{bmatrix} \begin{bmatrix} U_b \\ U_c \end{bmatrix}$$

When solved, this gives

$$\begin{bmatrix} U_b \\ U_c \end{bmatrix} = \frac{1}{30 \times 10^3} \begin{bmatrix} 1 & 1 \\ 1 & 3 \end{bmatrix} 90 \begin{bmatrix} 1 \\ 1 \end{bmatrix} = (3 \times 10^{-3}) \begin{bmatrix} 2 \\ 4 \end{bmatrix} \text{m}$$

Thus, the forces in the two springs are

$$S_{ab} = 30 \times 10^3 (U_b - U_a) = 180 \text{ kN} \qquad S_{bc} = 15 \times 10^3 (U_c - U_b) = 90 \text{ kN}$$

In the remainder of this chapter the concepts alluded to in this section for the two-spring problem will be used to develop a general algorithm for solving structures with discrete points of member connectivity. The spring members and assemblage will again be used to prevent the symbols and mathematics from masking the basic principles.

15.4 INDIVIDUAL MEMBER STIFFNESS COEFFICIENTS

The purpose of this section is to obtain the relationships between the forces and displacements at the endpoints of an individual linear elastic member, as shown in

Fig. 15.7. The spring has a stiffness constant k with units of force per length, and the ends are connected to points i and j. The positive directions for the forces x_i and x_j and the corresponding displacements u_i and u_j are toward the right, as illustrated in Fig. 15.7. Lowercase letters refer to quantities associated with individual members, and uppercase symbols denote assemblage values.

The relations between the forces and displacements are derived by deflecting points i and j independently and calculating the forces required to obtain equilibrium using the force-deformation properties of the spring.

First Deflection Case: $u_i \neq 0;\ u_j = 0$

From the force-deformation relationship of the member, the force necessary to sustain this deflection is

$$x_i = ku_i$$

In order to satisfy force equilibrium

$$x_j = -x_i = -ku_i$$

Second Deflection Case: $u_j \neq 0;\ u_i = 0$

Using logic similar to that in the first deflection case gives

$$x_j = ku_j \qquad \text{and} \qquad x_i = -x_j = -ku_j$$

If both nodes i and j of the element are displaced simultaneously so that $u_i \neq 0$ and $u_j \neq 0$, the force-deflection equations are obtained by combining the results from the two individual deflection cases

$$x_i = ku_i - ku_j \qquad x_j = -ku_i + ku_j \tag{15.16}$$

In matrix form, the equations become

$$\begin{bmatrix} x_i \\ x_j \end{bmatrix} = k \begin{bmatrix} 1 & -1 \\ -1 & 1 \end{bmatrix} \begin{bmatrix} u_i \\ u_j \end{bmatrix} \tag{15.17}$$

or simply
$$\mathbf{x} = \mathbf{ku} \tag{15.18}$$

where the force matrix is $\mathbf{x} = \{x_i \quad x_j\}$, the displacement matrix is $\mathbf{u} = \{u_i \quad u_j\}$, and the coefficient matrix $\mathbf{k}$, called the *member (element) stiffness matrix,* is represented by

$$\mathbf{k} = k \begin{bmatrix} 1 & -1 \\ -1 & 1 \end{bmatrix} \tag{15.19}$$

A typical entry k_{ij} in the element stiffness matrix represents the force at point i due to a unit displacement at point j. In general, the fact that $\mathbf{k}$ is symmetric is attributable

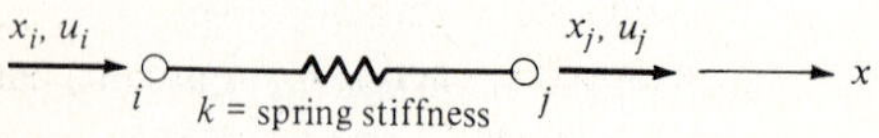

Figure 15.7 Individual spring member.

to the reciprocal theorem, discussed earlier in the chapter. In addition, **k** is singular (has no inverse) because neither point i nor point j of the element is fixed in position. Hence, the entire element can translate as a rigid body as well as deform elastically. This problem is alleviated when elements are combined to form a total system, which is then restrained against rigid-body motion.

The stiffness for a spring oriented in a general position is calculated in Example 15.2.

Example 15.2 Obtain the stiffness matrix relating the forces x_i, y_i, x_j, and y_j to the displacements u_i, v_i, u_j, and v_j for this individual spring member with stiffness k.

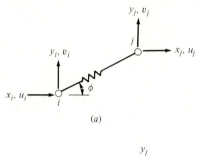

(a)

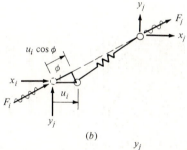

(b)

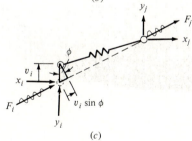

(c)

Figure E15.2 (a) Spring with general orientation; (b) first deformation case with $u_i \neq 0$ and $v_i = u_j = v_j = 0$; (c) second deformation case with $v_i \neq 0$ and $u_i = u_j = v_j = 0$.

SOLUTION From Fig. E15.2b

$$F_i = ku_i \cos \phi \qquad F_j = -F_i$$

$$x_i = F_i \cos \phi = (k \cos^2 \phi)u_i \qquad y_i = F_i \sin \phi = (k \sin \phi \cos \phi)u_i$$

$$x_j = -x_i \qquad y_j = -y_i$$

From Fig. E15.2c

$$F_i = kv_i \sin \phi \qquad F_j = -F_i$$

$$x_i = F_i \cos \phi = (k \cos \phi \sin \phi)v_i \qquad y_i = F_i \sin \phi = (k \sin^2 \phi)v_i$$

$$x_j = -x_i \qquad y_j = -y_i$$

Similarly, the third deformation case ($u_j \neq 0$, $u_i = v_i = v_j = 0$) gives

$$F_j = ku_j \cos \phi \qquad F_i = -F_j$$

$$x_j = -x_i = (k \cos^2 \phi)u_j \qquad y_j = -y_i = (k \sin \phi \cos \phi)u_j$$

And the fourth deformation case ($v_j \neq 0$, $u_i = v_i = u_j = 0$) gives

$$F_j = kv_j \sin \phi \qquad F_i = -F_j$$

$$x_j = -x_i = (k \sin \phi \cos \phi)v_j \qquad y_j = -y_i = (k \sin^2 \phi)v_j$$

Combining these four deformation cases yields the force-displacement equations for the member

$$\begin{bmatrix} x_i \\ y_i \\ x_j \\ y_j \end{bmatrix} = k \begin{bmatrix} \cos^2 \phi & \sin \phi \cos \phi & -\cos^2 \phi & -\sin \phi \cos \phi \\ \sin \phi \cos \phi & \sin^2 \phi & -\sin \phi \cos \phi & -\sin^2 \phi \\ -\cos^2 \phi & -\sin \phi \cos \phi & \cos^2 \phi & \sin \phi \cos \phi \\ -\sin \phi \cos \phi & -\sin^2 \phi & \sin \phi \cos \phi & \sin^2 \phi \end{bmatrix} \begin{bmatrix} u_i \\ v_i \\ u_j \\ v_j \end{bmatrix}$$

or

$$\mathbf{x} = \mathbf{ku}$$

15.5 EQUILIBRIUM OF SPRING ASSEMBLAGE

For the two-spring assemblage shown in Fig. 15.2 there are two member stiffness matrices with the general form exhibited in Eq. (15.19). For springs ab and bc they are

$$\mathbf{k}^{(ab)} = k_1 \begin{bmatrix} 1 & -1 \\ -1 & 1 \end{bmatrix} \qquad \text{and} \qquad \mathbf{k}^{(bc)} = k_2 \begin{bmatrix} 1 & -1 \\ -1 & 1 \end{bmatrix} \tag{15.20}$$

where the superscripts in parentheses identify the spring to which the member stiffness matrix refers. Alternatively, the member stiffness matrices in Eq. (15.20) can be expanded to reflect all forces and displacements involved in the assemblage. That is, if the element end forces are expressed in terms of all the system displacements, we have

$$\begin{bmatrix} X_a \\ X_b \\ X_c \end{bmatrix}^{(ab)} = k_1 \begin{bmatrix} 1 & -1 & 0 \\ -1 & 1 & 0 \\ 0 & 0 & 0 \end{bmatrix} \begin{bmatrix} U_a \\ U_b \\ U_c \end{bmatrix}^{(ab)} \qquad \begin{bmatrix} X_a \\ X_b \\ X_c \end{bmatrix}^{(bc)} = k_2 \begin{bmatrix} 0 & 0 & 0 \\ 0 & 1 & -1 \\ 0 & -1 & 1 \end{bmatrix} \begin{bmatrix} U_a \\ U_b \\ U_c \end{bmatrix}^{(bc)} \tag{15.21}$$

Here the superscripts on the force and displacement matrices identify the member with which they are associated. Although no information has been introduced into Eq. (15.21) beyond that contained in Eq. (15.20) (merely a row and column of zeros was introduced in the member stiffness matrices), in this form the two member stiffness matrices are conformable with respect to addition. Nodal equilibrium of all the nodes is invoked by summing the member end forces applied to the nodes [Eqs. (15.21)] and equating them to their respective applied nodal forces. This gives

$$
\begin{bmatrix} X_a \\ X_b \\ X_c \end{bmatrix} = \begin{bmatrix} X_a \\ X_b \\ X_c \end{bmatrix}^{(ab)} + \begin{bmatrix} X_a \\ X_b \\ X_c \end{bmatrix}^{(bc)} = \begin{bmatrix} k_1 & -k_1 & 0 \\ -k_1 & (k_1 + k_2) & -k_2 \\ 0 & -k_2 & k_2 \end{bmatrix} \begin{bmatrix} U_a \\ U_b \\ U_c \end{bmatrix} \tag{15.22}
$$

or
$$
\mathbf{X} = \mathbf{KU} \tag{15.23}
$$

The matrix $\mathbf{K}$ is the structural stiffness matrix and is identical to the result in Eq. (15.9) obtained previously for this assemblage. Note that in obtaining $\mathbf{K}$, not only has equilibrium been ensured by equating the member end forces to the applied nodal forces, but nodal compatibility is also invoked. For example, it is implied that the deflection at node b associated with member ab is the same as the deflection at node b identified with the equations describing member bc [Eq. (15.21)]. Hence, adding the two matrix equations of Eq. (15.21) ensures that the two springs will be attached at point b.

15.6 SOLVING FOR NODAL DISPLACEMENTS

The nodal equilibrium conditions shown in Eq. (15.23) contain a singular matrix $\mathbf{K}$; consequently, it is impossible to obtain the deflection matrix $\mathbf{U}$ for a given force matrix $\mathbf{X}$. As the problem now exists, no point on the system has been fixed in position, and the fact that $\mathbf{K}$ is singular can be interpreted physically as meaning that there are no unique displacements for a given set of forces. This occurs because the elastic deformations of the springs and the rigid-body motion of the system are both embodied in the displacements at this point in the development.

To overcome this problem we first partition the various matrices as follows:

$$
\begin{bmatrix} \mathbf{X}_f \\ \hline \mathbf{X}_s \end{bmatrix} = \begin{bmatrix} \mathbf{K}_{ff} & \vdots & \mathbf{K}_{fs} \\ \hline \mathbf{K}_{sf} & \vdots & \mathbf{K}_{ss} \end{bmatrix} \begin{bmatrix} \mathbf{U}_f \\ \hline \mathbf{U}_s \end{bmatrix} \tag{15.24}
$$

where, in general, the subscript f refers to the free or unrestrained nodes and subscript s refers to a supported or constrained node. Hence, $\mathbf{X}_f$ represents applied forces and $\mathbf{U}_f$ is the column matrix of corresponding displacements, while $\mathbf{U}_s$ represents prescribed displacements and $\mathbf{X}_s$ contains the associated forces. When $\mathbf{U}_s$ is null (unyielding supports), $\mathbf{X}_s$ will contain the reaction forces. Multiplying the partitioned matrices in Eq. (15.24) gives the two matrix equations

$$
\mathbf{X}_f = \mathbf{K}_{ff}\mathbf{U}_f + \mathbf{K}_{fs}\mathbf{U}_s \qquad \mathbf{X}_s = \mathbf{K}_{sf}\mathbf{U}_f + \mathbf{K}_{ss}\mathbf{U}_s \tag{15.25}
$$

If $\mathbf{U}_s = \mathbf{0}$, then

$$
\mathbf{X}_f = \mathbf{K}_{ff}\mathbf{U}_f \tag{15.26}
$$

and
$$
\mathbf{X}_s = \mathbf{K}_{sf}\mathbf{U}_f \tag{15.27}
$$

Usually it is neither necessary nor useful at this point in the solution to find the reaction forces $\mathbf{X}_s$; therefore, it is sufficient to work only with Eq. (15.26). A comparison of

Eqs. (15.26) and (15.24) reveals that three out of four partitions in the matrix **K** need not be created or if the complete **K** has been assembled in view of Eq. (15.26), it is possible to discard all but the $\mathbf{K}_{ff}$ partition in order to calculate the unknown displacements at the unconstrained nodes where loads $\mathbf{X}_f$ are specified. This solution process using Eq. (15.26) is

$$\mathbf{U}_f = \mathbf{K}_{ff}^{-1}\mathbf{X}_f \tag{15.28}$$

For the two-spring assemblage (Fig. 15.2) if the system is constrained at point c $(U_c = 0)$, then

$$\mathbf{K}_{ff} = \begin{bmatrix} k_1 & -k_1 \\ -k_1 & (k_1 + k_2) \end{bmatrix} \tag{15.29}$$

and consequently

$$\mathbf{K}_{ff}^{-1} = \begin{bmatrix} \left(\dfrac{1}{k_2}+\dfrac{1}{k_1}\right) & \dfrac{1}{k_2} \\ \dfrac{1}{k_2} & \dfrac{1}{k_2} \end{bmatrix} \tag{15.30}$$

Therefore, if $\mathbf{X}_f$ is specified, this matrix can be premultiplied by $\mathbf{K}_{ff}^{-1}$, as suggested by Eq. (15.28), and the corresponding deflections calculated.

It is important to observe that in this approach the reaction forces $\mathbf{X}_s$ are not explicitly calculated at this point in the solution. Usually only the partition $\mathbf{K}_{ff}$ is retained, $\mathbf{K}_{sf}$ being discarded. This makes it difficult to return to Eq. (15.27) to obtain these unknown forces. It should be noted, however, that the reaction forces can usually be obtained when the forces in the members are known. These calculations are discussed in the next section.

As a matter of interest, it is possible to return to Eq. (15.27) and compute the reaction force X_c for the two-spring assemblage in Fig. 15.5. Thus,

$$X_c = \begin{bmatrix} 0 & -k_2 \end{bmatrix} \begin{bmatrix} \left(\dfrac{1}{k_1} + \dfrac{1}{k_2}\right)X_a + \dfrac{1}{k_2}X_b \\ \dfrac{1}{k_2}X_a + \dfrac{1}{k_2}X_b \end{bmatrix} = -X_a - X_b$$

That is, the calculation reveals that the assemblage is in equilibrium.

The more general partitioning scheme in Eq. (15.25) must be used for a problem in which some of the forces and some of the displacements are specified. For example, assume that in Fig. 15.2 X_a, U_b, and U_c are given as nonzero and it is required to calculate U_a, X_b, and X_c. In this case, the various partitions are

$$\mathbf{X}_f = X_a \qquad \mathbf{U}_f = U_a \qquad \text{and} \qquad \mathbf{X}_s = \begin{bmatrix} X_b \\ X_c \end{bmatrix} \qquad \mathbf{U}_s = \begin{bmatrix} U_b \\ U_c \end{bmatrix}$$

$$\mathbf{K}_{ff} = k_1 \qquad \mathbf{K}_{fs} = \begin{bmatrix} -k_1 & 0 \end{bmatrix} \qquad \mathbf{K}_{sf} = \begin{bmatrix} -k_1 \\ 0 \end{bmatrix} \qquad \mathbf{K}_{ss} = \begin{bmatrix} (k_1 + k_2) & -k_2 \\ -k_2 & k_2 \end{bmatrix}$$

Equation (15.25) gives

$$X_a = k_1 U_a + [-k_1 \quad 0]\begin{bmatrix} U_b \\ U_c \end{bmatrix} \quad \text{or} \quad U_a = \frac{1}{k_1} X_a + U_b$$

and
$$\begin{bmatrix} X_b \\ X_c \end{bmatrix} = \begin{bmatrix} -k_1 \\ 0 \end{bmatrix}\left(\frac{1}{k_1} X_a + U_b\right) + \begin{bmatrix} (k_1 + k_2) & -k_2 \\ -k_2 & k_2 \end{bmatrix}\begin{bmatrix} U_b \\ U_c \end{bmatrix} = \begin{bmatrix} -X_a + k_2(U_b - U_c) \\ k_2(-U_b + U_c) \end{bmatrix}$$

indicating that the system is in equilibrium.

The assemblages analyzed in Examples 15.3 and 15.4 illustrate the concepts outlined in Secs. 15.5 and 15.6.

Example 15.3 The assemblage of Example 15.1 is unrestrained at joint a and has the following prescribed loads and displacements: $X_a = 120$ kN, $X_b = -75$ kN, and $U_c = 18$ mm. Calculate the unknown nodal displacements and forces.

SOLUTION Using the partitioning scheme shown in Eq. (15.24), gives

$$\begin{bmatrix} 120 \\ -75 \\ \hline X_c \end{bmatrix} = (15 \times 10^3)\begin{bmatrix} 2 & -2 & 0 \\ -2 & 3 & -1 \\ \hline 0 & -1 & 1 \end{bmatrix}\begin{bmatrix} U_a \\ U_b \\ \hline 0.018 \end{bmatrix}$$

From Eq. (15.25)

$$\mathbf{U}_f = \mathbf{K}_{ff}^{-1}(\mathbf{X}_f - \mathbf{K}_{fs}\mathbf{U}_s)$$

$$\begin{bmatrix} U_a \\ U_b \end{bmatrix} = \frac{1}{30 \times 10^3}\begin{bmatrix} 3 & 2 \\ 2 & 2 \end{bmatrix}\left\{\begin{bmatrix} 120 \\ -75 \end{bmatrix} - (15 \times 10^3)\begin{bmatrix} 0 \\ -1 \end{bmatrix}[0.018]\right\}$$

$$= \frac{1}{30 \times 10^3}\begin{bmatrix} 3 & 2 \\ 2 & 2 \end{bmatrix}\left\{\begin{bmatrix} 120 \\ -75 \end{bmatrix} - \begin{bmatrix} 0 \\ -270 \end{bmatrix}\right\}$$

$$= \frac{1}{30 \times 10^3}\begin{bmatrix} 3 & 2 \\ 2 & 2 \end{bmatrix}\begin{bmatrix} 120 \\ 195 \end{bmatrix} = 10^{-3}\begin{bmatrix} 25 \\ 21 \end{bmatrix} \text{ m}$$

Also from Eq. (15.25)

$$\mathbf{X}_s = \mathbf{K}_{sf}\mathbf{U}_f + \mathbf{K}_{ss}\mathbf{U}_s$$

$$X_c = (15 \times 10^3)[0 \quad -1] \times (10^{-3})\begin{bmatrix} 25 \\ 21 \end{bmatrix} + (15 \times 10^3)[1][0.018]$$

$$= -315 + 270 = -45 \text{ kN}$$

Example 15.4 Calculate the horizontal and vertical displacements at joint b due to the loads shown; $k_{ab} = 30{,}000$ kN/m; $k_{bc} = 15{,}000$ kN/m.

SOLUTION Use the member stiffness matrix obtained in Example 15.2. For member ab ($\phi = 0°$)

$$\mathbf{k}_{ab} = 10^3\begin{bmatrix} 30 & 0 & -30 & 0 \\ 0 & 0 & 0 & 0 \\ -30 & 0 & 30 & 0 \\ 0 & 0 & 0 & 0 \end{bmatrix}$$

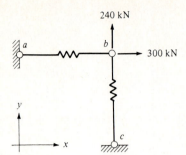

Figure E15.4

Expanding this to include all displacements for the assemblage gives

$$\begin{bmatrix} X_a \\ Y_a \\ X_b \\ Y_b \\ X_c \\ Y_c \end{bmatrix}^{(ab)} = 10^3 \begin{bmatrix} 30 & 0 & -30 & 0 & 0 & 0 \\ 0 & 0 & 0 & 0 & 0 & 0 \\ -30 & 0 & 30 & 0 & 0 & 0 \\ 0 & 0 & 0 & 0 & 0 & 0 \\ 0 & 0 & 0 & 0 & 0 & 0 \\ 0 & 0 & 0 & 0 & 0 & 0 \end{bmatrix} \begin{bmatrix} U_a \\ V_a \\ U_b \\ V_b \\ U_c \\ V_c \end{bmatrix}$$

For member bc ($\phi = -90°$)

$$\mathbf{k}_{bc} = 10^3 \begin{bmatrix} 0 & 0 & 0 & 0 \\ 0 & 15 & 0 & -15 \\ 0 & 0 & 0 & 0 \\ 0 & -15 & 0 & 15 \end{bmatrix}$$

Expanding this to include all displacements for the assemblage gives

$$\begin{bmatrix} X_a \\ Y_a \\ X_b \\ Y_b \\ X_c \\ Y_c \end{bmatrix}^{(bc)} = 10^3 \begin{bmatrix} 0 & 0 & 0 & 0 & 0 & 0 \\ 0 & 0 & 0 & 0 & 0 & 0 \\ 0 & 0 & 0 & 0 & 0 & 0 \\ 0 & 0 & 0 & 15 & 0 & -15 \\ 0 & 0 & 0 & 0 & 0 & 0 \\ 0 & 0 & 0 & -15 & 0 & 15 \end{bmatrix} \begin{bmatrix} U_a \\ V_a \\ U_b \\ V_b \\ U_c \\ V_c \end{bmatrix}$$

Adding the two force matrices gives

$$\begin{bmatrix} X_a \\ Y_a \\ X_b \\ Y_b \\ X_c \\ Y_c \end{bmatrix} = \begin{bmatrix} X_a \\ Y_a \\ X_b \\ Y_b \\ X_c \\ Y_c \end{bmatrix}^{(ab)} + \begin{bmatrix} X_a \\ Y_a \\ X_b \\ Y_b \\ X_c \\ Y_c \end{bmatrix}^{(bc)} = 10^3 \begin{bmatrix} 30 & 0 & -30 & 0 & 0 & 0 \\ 0 & 0 & 0 & 0 & 0 & 0 \\ -30 & 0 & 30 & 0 & 0 & 0 \\ 0 & 0 & 0 & 15 & 0 & -15 \\ 0 & 0 & 0 & 0 & 0 & 0 \\ 0 & 0 & 0 & -15 & 0 & 15 \end{bmatrix} \begin{bmatrix} U_a \\ V_a \\ U_b \\ V_b \\ U_c \\ V_c \end{bmatrix}$$

Since joints a and c are constrained, the equation for the unconstrained degrees of freedom $\mathbf{X}_f = \mathbf{K}_{ff}\mathbf{U}_f$ is

$$\begin{bmatrix} 300 \\ 240 \end{bmatrix} = 10^3 \begin{bmatrix} 30 & 0 \\ 0 & 15 \end{bmatrix} \begin{bmatrix} U_b \\ V_b \end{bmatrix}$$

The solution for the displacement gives

$$\begin{bmatrix} U_b \\ V_b \end{bmatrix} = 10^{-3} \begin{bmatrix} \frac{1}{30} & 0 \\ 0 & \frac{1}{15} \end{bmatrix} \begin{bmatrix} 300 \\ 240 \end{bmatrix} = 10^{-3} \begin{bmatrix} 10 \\ 16 \end{bmatrix} \text{ m}$$

15.7 CALCULATION OF MEMBER FORCES

The final step in the total problem solution is calculating the individual member forces as indicated in Fig. 15.6. After the nodal displacements have been calculated, the internal forces in the springs can be obtained by using the force-displacement relationships for each individual member. These relationships are expressed in matrix form in Eqs. (15.17) and (15.18); therefore, since the matrix **u** for each element is known, **u** can be premultiplied by each individual element stiffness matrix to obtain the element forces.

Although using Eq. (15.18) to obtain the element forces is correct, it is not the most efficient way of finding these unknowns. The formulation of the element properties incorporated force equilibrium, which stated that $x_i = -x_j$. Therefore, since the spring force is constant throughout the length of the member, it is apparent that when only one of the nodal forces is specified the state of force in the entire member is known. Note from Fig. 15.7 that if x_j is in the positive coordinate direction, the member will be in tension; however, a positive value for x_i implies that the member is in compression. Hence, if only the force x_j is calculated, the results can be conveniently interpreted in the usual way, i.e., a positive sign indicating a member in tension and a negative sign indicating a member in compression. Since the force throughout the member is being computed, the second equation of Eq. (15.16) will be formulated as

$$s_{ij} = k[-1 \quad 1] \begin{bmatrix} u_i \\ u_j \end{bmatrix} \tag{15.31}$$

or

$$s_{ij} = \mathbf{s} \mathbf{u}_{ij} \tag{15.32}$$

where s_{ij} = force in member ij which is connected to nodes i and j
  $\mathbf{s}$ = stress matrix for individual spring member
  $\mathbf{u}_{ij}$ = displacement matrix for member ij

The displacement matrix is formed by selecting entries from **U**, which contains all the displacements for the structure. For example, for the two-spring assemblage (Fig. 15.5)

$$s_{bc} = k_2[-1 \quad 1] \frac{1}{k_2} \begin{bmatrix} X_a + X_b \\ 0 \end{bmatrix} = -X_a - X_b$$

which agrees with the result in Eq. (15.15).

From a free-body diagram it can be observed that the force in member bc is the reaction force at node c for this assemblage (see Fig. 15.5). Hence, it is not necessary to calculate reaction forces formally by retaining $\mathbf{K}_{sf}$ and $\mathbf{K}_{ss}$, as suggested in

Eq. (15.25). Example 15.5 illustrates the use of the stress matrix **s** for finding individual member internal forces.

Example 15.5 Calculate the forces in the two individual spring members of Example 15.3 using the stress matrix from Eq. (15.32).

SOLUTION For member ab

$$\mathbf{s} = (30 \times 10^3)[-1 \quad 1] \qquad \mathbf{u}_{ab} = \begin{bmatrix} U_a \\ U_b \end{bmatrix} = \begin{bmatrix} 0.025 \\ 0.021 \end{bmatrix} \text{m}$$

$$S_{ab} = \mathbf{su} = (30 \times 10^3)[-1 \quad 1]\begin{bmatrix} 0.025 \\ 0.021 \end{bmatrix} = -120 \text{ kN (c)}$$

For member bc

$$\mathbf{s} = (15 \times 10^3)[-1 \quad 1] \qquad \mathbf{u}_{bc} = \begin{bmatrix} U_b \\ U_c \end{bmatrix} = \begin{bmatrix} 0.021 \\ 0.018 \end{bmatrix} \text{m}$$

$$S_{bc} = (15 \times 10^3)[-1 \quad 1]\begin{bmatrix} 0.021 \\ 0.018 \end{bmatrix} = -45 \text{ kN (c)}$$

15.8 GENERALIZATION OF THE METHOD

The logic in Fig. 15.6 for solving the two-spring problem can be applied to a broad category of problems in structural analysis. The solution for a structure is valid if it satisfies the fundamental strain-displacement, stress-strain, force-equilibrium, and compatibility equations, in addition to the boundary conditions. In the discussion of the two-spring problem, the usual strain-displacement and stress-strain relations applicable to continuous bodies were supplanted by the elementary force-deformation relationship for each individual member. Force equilibrium was satisfied within each element and was enforced at the nodal points for the total assemblage by merging the individual stiffness matrices. Compatibility was satisfied at the nodal points by attaching contiguous members. Since the merged stiffness matrix contains rigid-body modes, the nodes that are supported (constrained against deflection) are removed from the merged stiffness matrix by a partitioning scheme. Hence, this matrix manipulation satisfies the prescribed boundary conditions.

Since the basic equations have been satisfied, the solution is valid; furthermore, the logical flow of information implies that it should be possible to solve structural assemblages composed of members connected at discrete points. In later chapters it will be demonstrated that the stiffness method is applicable to truss, beam, and rigid-frame structures. Although the logical flow of information will resemble that in Fig. 15.6 for the spring assemblage, the detailed information implied by each box in the schematic will be different for various types of structures. Based upon the arguments given previously in this section, the schematic for the generalization of the stiffness method is presented in Fig. 15.8, together with explanatory notation.

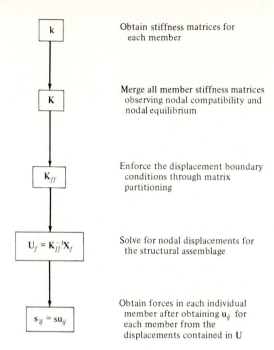

k — Obtain stiffness matrices for each member

K — Merge all member stiffness matrices observing nodal compatibility and nodal equilibrium

$\mathbf{K}_{ff}$ — Enforce the displacement boundary conditions through matrix partitioning

$\mathbf{U}_f = \mathbf{K}_{ff}^{-1}\mathbf{X}_f$ — Solve for nodal displacements for the structural assemblage

$\mathbf{s}_{ij} = \mathbf{s}\mathbf{u}_{ij}$ — Obtain forces in each individual member after obtaining $\mathbf{u}_{ij}$ for each member from the displacements contained in U

Figure 15.8 Schematic showing the basic steps for the stiffness method.

15.9 CONCLUDING REMARKS

The stiffness method as described for the spring assemblage is in a form that can conveniently be programmed for a computer. The various arrays can be routinely stored and manipulated using standard FORTRAN statements, but using assembly language for some portions improves efficiency. Logic such as that of the merge process, wherein the structural stiffness is assembled using the member stiffness matrices, can be efficiently programmed in a straightforward manner. The most time-consuming part of the analysis generally involves the solution of the force-displacement equations. With current computer technology, however, it is possible to solve large sets of equations efficiently and accurately provided the equations are well-conditioned with respect to solution.

The emphasis in this book is on the concepts; computer implementation of the theory will not be discussed. The reader is urged to gain a thorough understanding of the stiffness method before trying to automate it on the computer. Nevertheless an instructive way of learning the theory in conjunction with the computer is to use a symbolic matrix interpretive system in which the matrices can be manipulated by the input commands. In this way the student can follow the theory and exploit the power of the computer to manipulate matrices and solve the assembled equations.

Observe that no consideration has been given to the statical redundancy of the structural system being investigated. This is characteristic of the stiffness method since the displacements, and not the forces, are considered as the independent variables. As a result, all structures can be treated similarly; i. e., the solution process can proceed without any distinction between statically determinate and indeterminate structures.

PROBLEMS

Use the stiffness method.

15.1 Obtain the stiffness matrix for the assemblage by imposing the four possible distinct displacement cases. Note in the calculations where equilibrium, continuity, and the member force-deformation relations are imposed; $k_{ab} = 48,000$ kN/m; $k_{bc} = 24,000$ kN/m; $k_{cd} = 63,000$ kN/m.

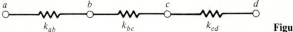

Figure P15.1

15.2 Form the stiffness matrix for the assemblage of Prob. 15.1 by superimposing (merging) the stiffness matrices for the individual spring members.

15.3 The assemblage shown in Prob. 15.1 has point a fixed in position, and it is subjected to the applied forces $X_b = X_c = X_d = 192$ kN. Calculate the unknown displacements at points b, c, and d and the reaction force at point a.

15.4 The assemblage shown in Prob. 15.1 has points a and d fixed in position, and it is subjected to the two forces $X_b = X_c = 192$ kN. Calculate the displacements at points b and c and the reactions at points a and d.

15.5 The assemblage shown in Prob. 15.1 has points a and d fixed in position. If node b is given a displacement of 24 mm and point c has an applied load of 192 kN, calculate U_c, X_b, and the unknown reactions.

15.6 Calculate the stress matrix for the generally oriented spring member of Example 15.2.

15.7 Using the result of Prob. 15.6, calculate the member forces for the assemblage of Example 15.4.

15.8 Calculate the displacements at point e, the reactions, and the member forces if $X_e = 600$ kN and $Y_e = 1200$ kN; $k_{ae} = 30,000$ kN/m, $k_{be} = 45,000$ kN/m, $k_{ce} = 15,000$ kN/m, and $k_{de} = 60,000$ kN/m.

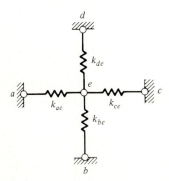

Figure P15.8

15.9 The assemblage of Prob. 15.8 has a prescribed displacement $V_e = 15$ mm and an applied load $X_e = 750$ kN. Calculate Y_e, U_e, the reactions, and the member forces.

15.10 Calculate the displacements at point c, the reactions, and the member forces if $X_c = 100$ kN and $Y_c = -96$ kN; $k_{ac} = k_{bc} = 5000$ kN/m.

15.11 The assemblage of Prob. 15.10 has a prescribed displacement $V_c = 24$ mm and an applied load $X_c = 80$ kN. Calculate Y_c, U_c, the reactions, and the member forces.

15.12 Points a, c, and d are fixed in position, and the applied loads are $X_b = 175$ kips and $Y_b = -200$ kips. Calculate the unknown displacements, the reactions, and the member forces; $k_{ab} = 100$ kips/in, $k_{bc} = 141$ kips/in, and $k_{bd} = 200$ kips/in.

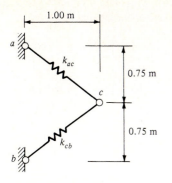

Figure P15.10

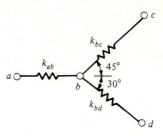

Figure P15.12

15.13 The assemblage in Fig. P15.12 has points c and d fixed in position. If $X_a = 100$ kips, $X_b = 75$ kips, and $Y_b = -200$ kips, calculate the unknown displacements, the reactions, and the member forces. If Y_a were nonzero, would the stiffness method give unique solutions for the displacements?

15.14 Calculate the unknown displacements, the reactions, and the member forces; $k_{ae} = 24{,}000$ kN/m, $k_{be} = 12{,}000$ kN/m, $k_{ce} = 16{,}000$ kN/m, and $k_{de} = 14{,}141$ kN/m.

15.15 Calculate U_b, U_d, and the reactions and forces in the two springs; $k_{ab} = 1500$ kN/m, and $k_{cd} = 2000$ kN/m.

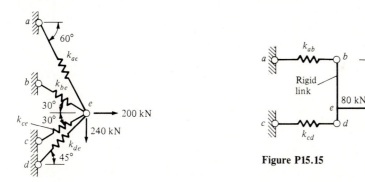

Figure P15.14

Figure P15.15

15.16 Calculate the unknown displacements, the reactions, and the member forces if $U_a = V_a = U_c = 0$, $X_b = 50$ kips, and $Y_b = -35$ kips; $k_{ab} = 50$ kips/in, $k_{ac} = 25$ kips/in, and $k_{cb} = 75$ kips/in.

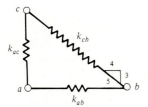

Figure P15.16

15.17 Calculate the unknown displacements, the reactions, and the member forces for the assemblage of Fig. P15.16 if $U_a = V_a = U_c = V_c = 0$, $X_b = 50$ kips, and $Y_b = -35$ kips.

15.18 Calculate the unknown displacements, the reactions, and the member forces for the assemblage of Fig. P15.16 if $U_a = V_a = V_b = 0$, $X_c = 35$ kips, and $Y_c = -80$ kips.

15.19 Calculate U_b, U_c, Y_c, the reactions, and the member forces for the assemblage of Fig. P15.16 if $U_a = V_a = V_b = 0$, $X_c = 40$ kips, and $V_c = 0.50$ in.

15.20 Calculate the unknown displacements, the reactions, and the member forces if $U_a = V_a = U_b = V_b = 0$, $X_c = Y_c = 0$, $X_d = 25$ kips, and $Y_d = -10$ kips; $k_{ab} = k_{cd} = 50$ kips/in, $k_{ac} = k_{bd} = 25$ kips/in, and $k_{cb} = 75$ kips/in.

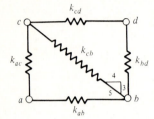

Figure P15.20

15.21 Calculate the unknown displacements, the reactions, and the member forces for the assemblage of Fig. P15.20 if it is constrained in the same manner but loaded as follows: $X_c = 25$ kips, $Y_c = -25$ kips, $X_d = 25$ kips, and $Y_d = -10$ kips.

15.22 Calculate the unknown displacements, the reactions, and the member forces for the assemblage of Fig. P15.20 if it is identically loaded but constrained as follows: $U_a = V_a = U_c = V_c = 0$.

15.23 The assemblage of Fig. P15.20 has a member added from a to d with $k_{ad} = k_{cb}$. Calculate the unknown displacements, the reactions, and the member forces if it is constrained and loaded like Prob. 15.20.

REFERENCES

1. Levy, S.: Computation of Influence Coefficients for Aircraft Structures with Discontinuities and Sweepback, *J. Aeron. Sci.*, vol. 14, no. 10, pp. 547–560, October 1947.
2. Levy, S.: Structural Analysis and Influence Coefficients for Delta Wings, *J. Aeron. Sci.*, vol. 20, no. 7, pp. 449–454, July 1953.
3. Argyris, J. H., and S. Kelsey: *Energy Theorems and Structural Analysis*, Butterworth, London, 1960.
4. Turner, M. J., R. W. Clough, H. C. Martin, and L. J. Topp: Stiffness and Deflection Analysis of Complex Structures, *J. Aeron. Sci.*, vol. 23, no. 9, pp. 805–824, September 1956.

ADDITIONAL READING

Beaufait, F. W.: *Basic Concepts of Structural Analysis*, chap. 9, Prentice-Hall, Englewood Cliffs, N.J., 1977.
Gutkowski, R. M.: *Structures: Fundamental Theory and Behavior*, chap. 7, Van Nostrand Reinhold, New York, 1981.
Kardestuncer, H.: *Elementary Matrix Analysis of Structures*, chap. 5, McGraw-Hill, New York, 1974.
Laursen, H. I.: *Structural Analysis*, chap. 14, McGraw-Hill, New York, 1978.
McGuire, W., and R. H. Gallagher: *Matrix Structural Analysis*, chaps. 2 and 3, Wiley, New York, 1979.
Martin, H. C.: *Introduction to Matrix Methods of Structural Analysis*, chap. 2, McGraw-Hill, New York, 1966.
Weaver, W., Jr., and J. M. Gere: *Matrix Analysis of Framed Structures*, 2d ed., chaps. 3 and 4, Van Nostrand, New York, 1980.

SIXTEEN

ANALYSIS OF TRUSSES USING THE STIFFNESS METHOD

16.1 INTRODUCTION

The stiffness method was outlined and illustrated using simple spring assemblages in Chap. 15. This chapter broadens these ideas so that the method can be used to analyze two-dimensional truss structures. In Chap. 15 it was pointed out that the basic equilibrium, stress-strain (force-deformation), and compatibility equations are satisfied within a given member. Also, equilibrium, compatibility, and boundary conditions are satisfied for the assembled structure at the points of connection. To generalize the method to truss structures we must derive two basic force-displacement relationships for an individual truss member, which will yield the stiffness matrix and the stress matrix. Since springs and truss members both act as two-force members, their basic matrices are also similar.

Recall that the elements in a truss function as two-force members if (1) the members are connected with smooth frictionless pins, (2) the lines of action of all members at the pin connections meet at a point, and (3) all loads are applied at the pin connections. Figure 16.1 shows a truss structure amenable to analysis using matrix structural analysis.

Usually the horizontal and vertical components of the joint displacements for a truss are required and can all be referenced to a single rectangular cartesian coordinate system. In contrast, the force in each individual member is described with respect to a coordinate system that is oriented along the member axis. Thus, it is necessary to have one coordinate system for the entire truss and another for each individual member. This requires the displacements, forces, and member stiffness relationships to undergo coordinate transformations during the analysis, an aspect of truss analysis that is also discussed in this chapter.

16.2 THE TRUSS MEMBER

The first step in formulating the analysis method for trusses is to obtain the relationships between force and displacement for an individual member. The procedure is similar to

Figure 16.1 Sidewall trusses, Bendan Byrne Arena, East Rutherford Borough, New Jersey. (*Bethlehem Steel Corp.*)

that used to obtain the stiffness matrix for a spring element [see Eq. (15.19)]. That is, each displacement at the ends of the member will be given a nonzero value while the other displacements are constrained to zero. The forces required to constrain the member in this deformed configuration are obtained using the equilibrium equations and force-deformation (stress-strain) relationships. When this has been accomplished for all individual displacements, the resulting equations are combined and organized into a succinct form using matrix equations.

The uniform (prismatic) truss member in Fig. 16.2 is shown in a general orientation relative to the coordinate system. Assume that this member has a uniform

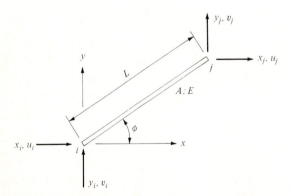

Figure 16.2 The truss member oriented in a general position.

cross-sectional area A over its length L and that it is made from a homogeneous, isotropic, linear elastic material with a modulus of elasticity E. It is required to obtain the force-deflection relationships at the two ends i and j of the member in the xy coordinate system. The displacements at joint i in the two positive coordinate directions are denoted u_i and v_i, respectively, and x_i and y_i are the corresponding forces. The subscript on the displacements and forces indicates the joint; thus there are similar quantities at joint j.

First Displacement Case: $u_i \neq 0$; $v_i = u_j = v_j = 0$

The member in this deflected configuration is shown in Fig. 16.3a. Note that it has shortened an amount equal to $u_i \cos \phi$; therefore, the compressive axial force P_i in the member at point i is obtained using Hooke's law

$$P_i = \frac{AE}{L} u_i \cos \phi \qquad (16.1)$$

The rectangular cartesian components of the forces at the end of the member are

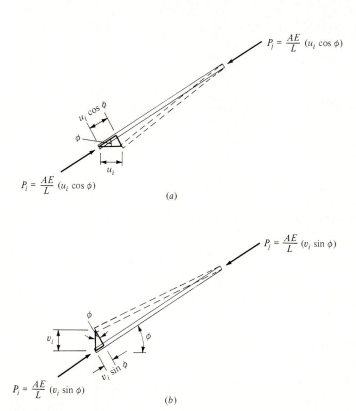

$$P_j = \frac{AE}{L} (u_i \cos \phi)$$

$$P_i = \frac{AE}{L} (u_i \cos \phi)$$

(a)

$$P_j = \frac{AE}{L} (v_i \sin \phi)$$

$$P_i = \frac{AE}{L} (v_i \sin \phi)$$

(b)

Figure 16.3 Forces acting on the truss member with various displacements: (a) $u_i \neq 0$ and $v_i = u_j = v_j = 0$; (b) $v_i \neq 0$ and $u_i = u_j = v_j = 0$.

$$x_i = P_i \cos \phi = \frac{AE}{L}(u_i \cos \phi)(\cos \phi) = \frac{AE}{L}u_i \cos^2 \phi \qquad (16.2)$$

and
$$y_i = P_i \sin \phi = \frac{AE}{L}(u_i \cos \phi)(\sin \phi) = \frac{AE}{L}u_i \sin \phi \cos \phi \qquad (16.3)$$

In order to maintain equilibrium of the member, the magnitude of the compressive force P_j must be equal to that of P_i and the force must act in the direction shown in Fig. 16.3a. Therefore, the forces at joint j in the x and y directions are respectively

$$x_j = -x_i = -\frac{AE}{L}u_i \cos^2 \phi \qquad (16.4)$$

and
$$y_j = -y_i = -\frac{AE}{L}u_i \sin \phi \cos \phi \qquad (16.5)$$

Second Displacement Case: $v_i \neq 0$; $u_i = u_j = v_j = 0$

End i of the truss member is given a displacement in the y direction while the other three displacements at the two reference points are not allowed to displace. This configuration is shown in Fig. 16.3b. Since the deformation along the member axis is equal to $v_i \sin \phi$, from Hooke's law the axial compressive force at joint i is

$$P_i = \frac{AE}{L}v_i \sin \phi \qquad (16.6)$$

Consequently, the x and y components of this force are expressed as

$$x_i = P_i \cos \phi = \frac{AE}{L}v_i \sin \phi \cos \phi \qquad (16.7)$$

and
$$y_i = P_i \sin \phi = \frac{AE}{L}v_i \sin^2 \phi \qquad (16.8)$$

Again, in order to maintain equilibrium, the forces at joint j are

$$x_j = -x_i = -\frac{AE}{L}v_i \sin \phi \cos \phi \qquad (16.9)$$

and
$$y_j = -y_i = -\frac{AE}{L}v_i \sin^2 \phi \qquad (16.10)$$

Third Displacement Case: $u_j \neq 0$; $u_i = v_i = v_j = 0$

This deflected configuration of the member will result in an elongation equal to $u_j \cos \phi$, and in a manner similar to that used above, the forces at the two joints can be described as

$$x_j = -x_i = \frac{AE}{L}u_j \cos^2 \phi \qquad (16.11)$$

and
$$y_j = -y_i = \frac{AE}{L} u_j \sin \phi \cos \phi \qquad (16.12)$$

Fourth Displacement Case: $v_j \neq 0$; $u_i = v_i = u_j = 0$

Finally, joint j is given a deflection in the y direction while all other deflection components are constrained. Using the equations of equilibrium, along with Hooke's law, gives the constraining forces

$$x_j = -x_i = \frac{AE}{L} v_j \sin \phi \cos \phi \qquad (16.13)$$

and
$$y_j = -y_i = \frac{AE}{L} v_j \sin^2 \phi \qquad (16.14)$$

General Displacement Case: $u_i \neq 0$; $v_i \neq 0$; $u_j \neq 0$; $v_j \neq 0$

If all four of the displacement components are simultaneously given nonzero values, the forces conforming to Hooke's law and maintaining overall equilibrium of the member can be obtained by combining Eqs. (16.4), (16.5), and (16.9) through (16.14). Thus, for this general displacement case, the following force-displacement relationships are obtained:

$$x_i = \frac{AE}{L} [(\cos^2 \phi)u_i + (\sin \phi \cos \phi)v_i - (\cos^2 \phi)u_j - (\sin \phi \cos \phi)v_j]$$

$$y_i = \frac{AE}{L} [(\sin \phi \cos \phi)u_i + (\sin^2 \phi)v_i - (\sin \phi \cos \phi)u_j - (\sin^2 \phi)v_j]$$
$$\qquad (16.15)$$
$$x_j = \frac{AE}{L} [-(\cos^2 \phi)u_i - (\sin \phi \cos \phi)v_i + (\cos^2 \phi)u_j + (\sin \phi \cos \phi)v_j]$$

$$y_j = \frac{AE}{L} [-(\sin \phi \cos \phi)u_i - (\sin^2 \phi)v_i + (\sin \phi \cos \phi)u_j + (\sin^2 \phi)v_j]$$

expressed in matrix form as

$$\begin{bmatrix} x_i \\ y_i \\ x_j \\ y_j \end{bmatrix} = \frac{AE}{L} \begin{bmatrix} \cos^2 \phi & \sin \phi \cos \phi & -\cos^2 \phi & -\sin \phi \cos \phi \\ \sin \phi \cos \phi & \sin^2 \phi & -\sin \phi \cos \phi & -\sin^2 \phi \\ -\cos^2 \phi & -\sin \phi \cos \phi & \cos^2 \phi & \sin \phi \cos \phi \\ -\sin \phi \cos \phi & -\sin^2 \phi & \sin \phi \cos \phi & \sin^2 \phi \end{bmatrix} \begin{bmatrix} u_i \\ v_i \\ u_j \\ v_j \end{bmatrix}$$

$$\mathbf{x} \qquad\qquad\qquad \mathbf{k} \qquad\qquad\qquad\qquad \mathbf{u}$$
$$\qquad (16.16)$$

or
$$\mathbf{x = ku} \qquad (16.16a)$$

Equation (16.16) describes the relationship between the forces and the displacements for an individual truss member in a general orientation in two-dimensional space.

The forces are contained in the column matrix **x**, and the displacements are in the column matrix **u**. The entries in the displacement matrix are placed in the same location as the corresponding component in the force matrix. These two matrices are related through the square matrix **k**, which is the member stiffness matrix. Note that **k** is symmetric. In Chap. 15 it was pointed out that this is a consequence of the Maxwell-Betti reciprocal theorem. Also, **k** has no inverse since the member is not constrained to prevent rigid-body displacements. This so-called *singularity problem* is alleviated when a number of members are combined into a structure that is supported in such a way that the general rigid-body displacements (two perpendicular translations and a rotation in two dimensions) are constrained. Furthermore, it can be observed that the columns of **k** add to zero. This is attributed to the fact that a given column contains the force components necessary to maintain the member in a prescribed deformed position. For example, the second column of **k** contains the forces in the x and y directions, with the displacements $v_i = 1$, and $u_i = u_j = v_j = 0$. Since equilibrium requires the forces in these orthogonal directions to sum to zero, the four quantities in this column will sum to zero.

Example 16.1 demonstrates the use of Eq. (16.16) to analyze an individual truss member that is properly constrained and loaded at one joint.

Example 16.1 Calculate the displacements and forces for this single strut using the stiffness matrix for the truss member [Eq. (16.16)]; $A = 2.07$ in^2 and $E = 29 \times 10^3$ kips/in^2.

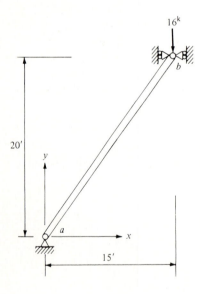

Figure E16.1

SOLUTION

$$\frac{AE}{L} = \frac{2.07(29 \times 10^3)}{25(12)} = 200 \text{ kips/in}$$

$$\cos \phi = \tfrac{3}{5} \qquad \sin \phi = \tfrac{4}{5}$$

From Eq. (16.16)

$$
\begin{bmatrix} x_a \\ y_a \\ x_b \\ y_b \end{bmatrix} = \frac{200}{25} \begin{bmatrix} 9 & 12 & -9 & -12 \\ 12 & 16 & -12 & -16 \\ -9 & -12 & 9 & 12 \\ -12 & -16 & 12 & 16 \end{bmatrix} \begin{bmatrix} u_a \\ v_a \\ u_b \\ v_b \end{bmatrix}
$$

But $u_a = v_a = u_b = 0$ and $y_b = -16$. From the fourth equation

$$-16 = 8(16v_b)$$

$$v_b = -0.125 \text{ in}$$

Hence, the first three equations give

$$x_a = 8(-12v_b) = +12 \text{ kips}$$

$$y_a = 8(-16v_b) = +16 \text{ kips}$$

$$x_b = 8(12v_b) = -12 \text{ kips}$$

16.3 NODAL EQUILIBRIUM OF THE ASSEMBLAGE

The force-displacement relationships for a truss composed of uniform prismatic members is formulated by writing the equilibrium equations for each individual point of connectivity, referred to as *nodes* or *nodal points*. The truss in Fig. 16.4a has 12 nodes; thus, with two displacements per node, both the force and displacement matrices are 24 × 1 and the stiffness matrix is 24 × 24. Consider node i and the connected members in Fig. 16.4b. The force-displacement equations for each of the three members attached to this node can be expressed using Eq. (16.16) to give

$$
\begin{bmatrix} x_l^{(li)} \\ y_l^{(li)} \\ x_i^{(li)} \\ y_i^{(li)} \end{bmatrix} = \begin{bmatrix} k_{11}^{(li)} & k_{12}^{(li)} & k_{13}^{(li)} & k_{14}^{(li)} \\ k_{21}^{(li)} & k_{22}^{(li)} & k_{23}^{(li)} & k_{24}^{(li)} \\ k_{31}^{(li)} & k_{32}^{(li)} & k_{33}^{(li)} & k_{34}^{(li)} \\ k_{41}^{(li)} & k_{42}^{(li)} & k_{43}^{(li)} & k_{44}^{(li)} \end{bmatrix} \begin{bmatrix} u_l \\ v_l \\ u_i \\ v_i \end{bmatrix}
\tag{16.17a}
$$

$$
\begin{bmatrix} x_m^{(mi)} \\ y_m^{(mi)} \\ x_i^{(mi)} \\ y_i^{(mi)} \end{bmatrix} = \begin{bmatrix} k_{11}^{(mi)} & k_{12}^{(mi)} & k_{13}^{(mi)} & k_{14}^{(mi)} \\ k_{21}^{(mi)} & k_{22}^{(mi)} & k_{23}^{(mi)} & k_{24}^{(mi)} \\ k_{31}^{(mi)} & k_{32}^{(mi)} & k_{33}^{(mi)} & k_{34}^{(mi)} \\ k_{41}^{(mi)} & k_{42}^{(mi)} & k_{43}^{(mi)} & k_{44}^{(mi)} \end{bmatrix} \begin{bmatrix} u_m \\ v_m \\ u_i \\ v_i \end{bmatrix}
\tag{16.17b}
$$

$$
\begin{bmatrix} x_i^{(in)} \\ y_i^{(in)} \\ x_n^{(in)} \\ y_n^{(in)} \end{bmatrix} = \begin{bmatrix} k_{11}^{(in)} & k_{12}^{(in)} & k_{13}^{(in)} & k_{14}^{(in)} \\ k_{21}^{(in)} & k_{22}^{(in)} & k_{23}^{(in)} & k_{24}^{(in)} \\ k_{31}^{(in)} & k_{32}^{(in)} & k_{33}^{(in)} & k_{34}^{(in)} \\ k_{41}^{(in)} & k_{42}^{(in)} & k_{43}^{(in)} & k_{44}^{(in)} \end{bmatrix} \begin{bmatrix} u_i \\ v_i \\ u_n \\ v_n \end{bmatrix}
\tag{16.17c}
$$

where the superscripts denote the end nodes of the member associated with the quantity. No superscripts are necessary for the displacements since for compatibility all mem-

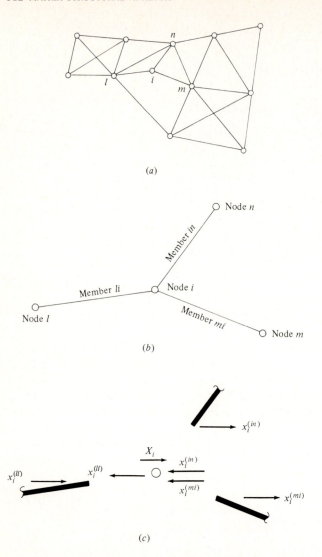

(a)

(b)

(c)

Figure 16.4 (a) Truss showing the locations of members; (b) member connectivity at node i; (c) free-body diagram of node i; for clarity the y force components are not shown.

bers must be connected to a common node of attachment, for example $u_i^{(li)} = u_i^{(mi)} = u_i^{(in)} = u_i = U_i$.

The ensuing discussion will treat only the forces in the x direction; a similar description for the y direction can be envisioned. Figure 16.4c shows the free-body diagrams for node i and the associated members, with only the x forces shown for clarity. The force applied externally to the node in the x direction is denoted by X_i. Enforcing equilibrium in the x direction gives

$$X_i = x_i^{(li)} + x_i^{(mi)} + x_i^{(in)}$$

Upon substituting from Eqs. (16.17a) to (16.17c), the equation becomes

$$X_i = (k_{33}^{(li)} + k_{33}^{(mi)} + k_{11}^{(in)})U_i + (k_{34}^{(li)} + k_{34}^{(mi)} + k_{12}^{(in)})V_i + k_{31}^{(li)}U_l + k_{32}^{(li)}V_l$$

$$+ k_{31}^{(mi)}U_m + k_{32}^{(mi)}V_m + k_{13}^{(in)}U_n + k_{14}^{(in)}V_n \tag{16.18}$$

or $\quad X_i = K_{gg}U_i + K_{gh}V_i + K_{go}U_l + K_{gp}V_l + K_{gq}U_m + K_{gr}V_m + K_{gs}U_n + K_{gt}V_n$

where $\qquad K_{gg} = k_{33}^{(li)} + k_{33}^{(mi)} + k_{11}^{(in)} \qquad K_{gh} = k_{34}^{(li)} + k_{34}^{(mi)} + k_{12}^{(in)} \tag{16.19}$

and so forth. Thus, equilibrium at a node is attained by combining the stiffness coefficients from the individual members connected to it, as shown in Fig. 16.5. Equilibrium is ensured by summing the element end forces at the node, and compatibility is invoked by associating member end displacements with nodal displacements. Equilibrium must be enforced for all independent force components that exist at each nodal point. The combination of the individual member stiffness matrices into the structural stiffness matrix for the entire truss is referred to as the *merge process*.

After performing the merge process for each node of the structure, a matrix equation will be obtained that relates all the nodal displacements to the nodal forces

$$\mathbf{X} = \mathbf{KU} \tag{16.20}$$

where $\mathbf{X}$ = column matrix containing nodal forces
$\quad \mathbf{U}$ = column matrix of nodal displacements (ordered like corresponding forces in $\mathbf{X}$)
$\quad \mathbf{K}$ = square structural stiffness matrix

Uppercase letters distinguish these quantities from the lowercase symbols used to describe the force-displacement relationship for individual members [see Eq. (16.16a)].

The process of obtaining the structural stiffness matrix for a three-member truss is illustrated in Example 16.2.

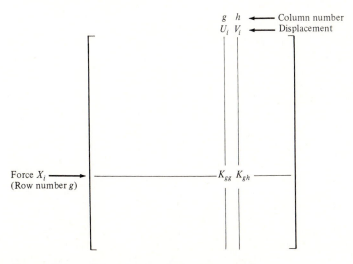

Figure 16.5 Arrangement of the structural stiffness matrix.

Example 16.2 Obtain the structural stiffness matrix for this three-member truss; $A_{ab} = 3 \times 10^{-4}$ m^2, $A_{bc} = 5 \times 10^{-4}$ m^2, $A_{ac} = 4 \times 10^{-4}$ m^2, and $E = 200$ GPa.

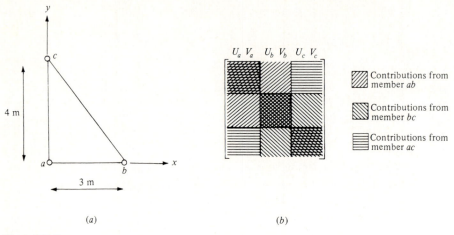

(a)   (b)

Figure E16.2

SOLUTION For all members $AE/L = 20,000$ kN/m. For member ab

$$\phi = 0° \qquad \cos \phi = 1.0 \qquad \sin \phi = 0.0$$

$$k_{ab} = 20,000 \begin{array}{c} \begin{array}{cccc} U_a & V_a & U_b & V_b \end{array} \\ \begin{bmatrix} 1.00 & 0.00 & -1.00 & 0.00 \\ 0.00 & 0.00 & 0.00 & 0.00 \\ -1.00 & 0.00 & 1.00 & 0.00 \\ 0.00 & 0.00 & 0.00 & 0.00 \end{bmatrix} \end{array}$$

For member bc

$$\phi = 126.9° \qquad \cos \phi = -0.6 \qquad \sin \phi = 0.8$$

$$k_{bc} = 20,000 \begin{array}{c} \begin{array}{cccc} U_b & V_b & U_c & V_c \end{array} \\ \begin{bmatrix} 0.36 & -0.48 & -0.36 & 0.48 \\ -0.48 & 0.64 & 0.48 & -0.64 \\ -0.36 & 0.48 & 0.36 & -0.48 \\ 0.48 & -0.64 & -0.48 & 0.64 \end{bmatrix} \end{array}$$

For member ac

$$\phi = 90° \qquad \cos \phi = 0.0 \qquad \sin \phi = 1.0$$

$$k_{ac} = 20,000 \begin{array}{c} \begin{array}{cccc} U_a & V_a & U_c & V_c \end{array} \\ \begin{bmatrix} 0.00 & 0.00 & 0.00 & 0.00 \\ 0.00 & 1.00 & 0.00 & -1.00 \\ 0.00 & 0.00 & 0.00 & 0.00 \\ 0.00 & -1.00 & 0.00 & 1.00 \end{bmatrix} \end{array}$$

The element stiffness matrices must be combined as shown in Fig. E16.2b to obtain **K**.

$$
\begin{bmatrix} X_a \\ Y_a \\ X_b \\ Y_b \\ X_c \\ Y_c \end{bmatrix} = 20{,}000 \begin{bmatrix} 1.00 & 0.00 & -1.00 & 0.00 & 0.00 & 0.00 \\ 0.00 & 1.00 & 0.00 & 0.00 & 0.00 & -1.00 \\ -1.00 & 0.00 & 1.36 & -0.48 & -0.36 & 0.48 \\ 0.00 & 0.00 & -0.48 & 0.64 & 0.48 & -0.64 \\ 0.00 & 0.00 & -0.36 & 0.48 & 0.36 & -0.48 \\ 0.00 & -1.00 & 0.48 & -0.64 & -0.48 & 1.64 \end{bmatrix} \begin{bmatrix} U_a \\ V_a \\ U_b \\ V_b \\ U_c \\ V_c \end{bmatrix}
$$

DISCUSSION Each individual stiffness matrix is obtained using Eq. (16.16), and the structural stiffness matrix is constructed by merging these three matrices using the logic developed in the foregoing discussion. For several reasons the merge cannot be accomplished by simple matrix addition. The graphic merge representation in Fig. E16.2b assumes that **K** is initially null and the member stiffness matrices are added to the corresponding matrix elements as they are placed in the appropriate locations.

Merge Process

No attempt will be made to represent the merge process in a more succinct notation amenable to computer programming or mathematical representation. It is worth noting, however, that this can be accomplished quite easily by numbering the generalized displacements (degrees of freedom) sequentially. For example, assume the following correspondence:

Displacement	u_i	v_i	u_l	v_l	u_m	v_m	u_n	v_n
Degree of freedom	21	22	25	26	27	28	29	30

Thus, Eq. (16.19) is the twenty-first equation of the force-displacement equations, and the structural stiffness coefficients would correspond to the following locations in the matrix **K**:

Stiffness coefficient	K_{gg}	K_{gh}	K_{go}	K_{gp}	K_{gq}	K_{gr}	K_{gs}	K_{gt}
Location (row, column)	(21, 21)	(21, 22)	(21, 25)	(21, 26)	(21, 27)	(21, 28)	(21, 29)	(21, 30)

16.4 NODAL DISPLACEMENTS

The next step in the process after the structural stiffness matrix **K** has been assembled is to enforce the boundary conditions; i.e., to observe how the structure is supported and take advantage of the fact that certain displacements are known. This usually requires that **K** be arranged according to the partitioning scheme

$$
\begin{bmatrix} \mathbf{X}_f \\ \cdots \\ \mathbf{X}_s \end{bmatrix} = \begin{bmatrix} \mathbf{K}_{ff} & \vdots & \mathbf{K}_{fs} \\ \cdots & \vdots & \cdots \\ \mathbf{K}_{sf} & \vdots & \mathbf{K}_{ss} \end{bmatrix} \begin{bmatrix} \mathbf{U}_f \\ \cdots \\ \mathbf{U}_s \end{bmatrix} \tag{15.24}
$$

Recall that subscript f refers to the free or unrestrained nodal quantities and that subscript s denotes supported or nodal degrees of freedom where the displacements are known; e.g., if the supports settle, $\mathbf{U}_s$ has nonzero entries. It was shown in Chap. 15 that expansion of these equations gives

$$\mathbf{X}_f = \mathbf{K}_{ff}\mathbf{U}_f + \mathbf{K}_{fs}\mathbf{U}_s \qquad \mathbf{X}_s = \mathbf{K}_{sf}\mathbf{U}_f + \mathbf{K}_{ss}\mathbf{U}_s \qquad (15.25)$$

If $\mathbf{U}_s = \mathbf{0}$, that is, all the degrees of freedom are zero where displacements are specified, the matrix partition of unknown displacements $\mathbf{U}_f$ can be obtained as

$$\mathbf{U}_f = \mathbf{K}_{ff}^{-1}\mathbf{X}_f \qquad (15.28)$$

In this case, the reactions are contained in the partition $\mathbf{X}_s$ and are calculated from the second of Eqs. (15.25) to give

$$\mathbf{X}_s = \mathbf{K}_{sf}\mathbf{U}_f \qquad (15.27)$$

Examples 16.3 through 16.5 illustrate the solution for the nodal displacements and also for the reaction forces using the above equations. The statically indeterminate structure in Example 16.5 is handled exactly the same as the statically determinate trusses analyzed in Examples 16.3 and 16.4; that is, statical determinancy is not a relevant consideration for this method of analysis.

Example 16.3 Calculate the displacements and reactions for the truss of Example 16.2 if it is supported with $V_b = U_c = V_c = 0$; $X_a = 60$ kN, $Y_a = 90$ kN.

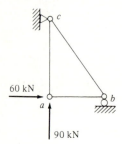

Figure E16.3

SOLUTION

$$\begin{bmatrix} X_a = 60 \\ Y_a = 90 \\ X_b = 0 \\ \hline Y_b \\ X_c \\ Y_c \end{bmatrix} = (2 \times 10^4) \begin{bmatrix} 1.00 & 0.00 & -1.00 & \vdots & 0.00 & 0.00 & 0.00 \\ 0.00 & 1.00 & 0.00 & \vdots & 0.00 & 0.00 & -1.00 \\ -1.00 & 0.00 & 1.36 & \vdots & -0.48 & -0.36 & 0.48 \\ \hline 0.00 & 0.00 & -0.48 & \vdots & 0.64 & 0.48 & -0.64 \\ 0.00 & 0.00 & -0.36 & \vdots & 0.48 & 0.36 & -0.48 \\ 0.00 & -1.00 & 0.48 & \vdots & -0.64 & -0.48 & 1.64 \end{bmatrix} \begin{bmatrix} U_a \\ V_a \\ U_b \\ \hline V_b = 0 \\ U_c = 0 \\ V_c = 0 \end{bmatrix}$$

From $\mathbf{U}_f = \mathbf{K}_{ff}^{-1}\mathbf{X}_f$ we get

$$\begin{bmatrix} U_a \\ V_a \\ U_b \end{bmatrix} = \frac{1}{7200} \begin{bmatrix} 1.36 & 0.00 & 1.00 \\ 0.00 & 0.36 & 0.00 \\ 1.00 & 0.00 & 1.00 \end{bmatrix} \begin{bmatrix} 60 \\ 90 \\ 0 \end{bmatrix} = 10^{-4} \begin{bmatrix} 113.3 \\ 45.0 \\ 83.3 \end{bmatrix} \text{ m}$$

From $\mathbf{X}_s = \mathbf{K}_{sf}\mathbf{U}_f$ we get

$$\begin{bmatrix} Y_b \\ X_c \\ Y_c \end{bmatrix} = (2 \times 10^4) \begin{bmatrix} 0.00 & 0.00 & -0.48 \\ 0.00 & 0.00 & -0.36 \\ 0.00 & -1.00 & 0.48 \end{bmatrix} \left(10^{-4} \begin{bmatrix} 113.3 \\ 45.0 \\ 83.3 \end{bmatrix} \right) = \begin{bmatrix} -80 \\ -60 \\ -10 \end{bmatrix} \text{ kN}$$

Example 16.4 Calculate the displacements and reactions for the truss of Example 16.2 if it is supported with $U_a = V_a = U_c = 0$; $X_b = 72$ kN, $Y_b = 90$ kN.

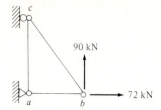

90 kN

72 kN **Figure E16.4**

SOLUTION Rearranging the displacements and forces in the matrices of Example 16.3 gives

$$
\begin{bmatrix} X_b = 72 \\ Y_b = 90 \\ Y_c = 0 \\ \hline X_a \\ Y_a \\ X_c \end{bmatrix} = (2 \times 10^4)
\begin{bmatrix}
1.36 & -0.48 & 0.48 & | & -1.00 & 0.00 & -0.36 \\
-0.48 & 0.64 & -0.64 & | & 0.00 & 0.00 & 0.48 \\
0.48 & -0.64 & 1.64 & | & 0.00 & -1.00 & -0.48 \\
\hline
-1.00 & 0.00 & 0.00 & | & 1.00 & 0.00 & 0.00 \\
0.00 & 0.00 & -1.00 & | & 0.00 & 1.00 & 0.00 \\
-0.36 & 0.48 & -0.48 & | & 0.00 & 0.00 & 0.36
\end{bmatrix}
\begin{bmatrix} U_b \\ V_b \\ V_c \\ \hline U_a = 0 \\ V_a = 0 \\ U_c = 0 \end{bmatrix}
$$

From $\mathbf{U}_f = \mathbf{K}_{ff}^{-1}\mathbf{X}_f$ we get

$$
\begin{bmatrix} U_b \\ V_b \\ V_c \end{bmatrix} = \frac{1}{12,800}
\begin{bmatrix}
0.64 & 0.48 & 0.00 \\
0.48 & 2.00 & 0.64 \\
0.00 & 0.64 & 0.64
\end{bmatrix}
\begin{bmatrix} 72 \\ 90 \\ 0 \end{bmatrix}
= 10^{-4}
\begin{bmatrix} 69.75 \\ 167.62 \\ 45.00 \end{bmatrix} \text{ m}
$$

From $\mathbf{X}_s = \mathbf{K}_{sf}\mathbf{U}_f$

$$
\begin{bmatrix} X_a \\ Y_a \\ X_c \end{bmatrix} = (2 \times 10^4)
\begin{bmatrix}
-1.00 & 0.00 & 0.00 \\
0.00 & 0.00 & -1.00 \\
-0.36 & 0.48 & -0.48
\end{bmatrix}
\left(10^{-4} \begin{bmatrix} 69.75 \\ 167.62 \\ 45.00 \end{bmatrix} \right)
= \begin{bmatrix} -139.5 \\ -90.0 \\ 67.5 \end{bmatrix} \text{ kN}
$$

DISCUSSION Partitioning the matrices according to Eq. (15.24) requires a reordering of the equations and displacements, as shown. The rest of the calculations proceed as in Example 16.3.

Example 16.5 Obtain the displacements and reactions for this truss. For all members $E = 29 \times 10^3$ kips/in^2 and $A = 0.993$ in^2.

SOLUTION

$$
\frac{AE}{L} = \begin{cases} 200 \text{ kips/in} & \text{for members } ab, cd, ac, bd \\ 141.4 \text{ kips/in} & \text{for members } ad, bc \end{cases}
$$

$$
\mathbf{k}_{ab} = \mathbf{k}_{cd} = 200
\begin{bmatrix}
1 & 0 & -1 & 0 \\
0 & 0 & 0 & 0 \\
-1 & 0 & 1 & 0 \\
0 & 0 & 0 & 0
\end{bmatrix}
\qquad
\mathbf{k}_{ac} = \mathbf{k}_{bd} = 200
\begin{bmatrix}
0 & 0 & 0 & 0 \\
0 & 1 & 0 & -1 \\
0 & 0 & 0 & 0 \\
0 & -1 & 0 & 1
\end{bmatrix}
$$

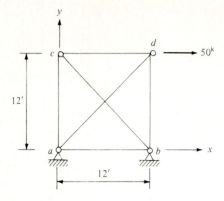

Figure E16.5

Table E16.5

Member	ϕ	$\cos \phi$	$\sin \phi$
ab, cd	0°	1	0
ac, bd	90°	0	1
ad	45°	$1/\sqrt{2}$	$1/\sqrt{2}$
bc	135°	$-1/\sqrt{2}$	$1/\sqrt{2}$

$$\mathbf{k}_{ad} = \frac{141.4}{2} \begin{bmatrix} 1 & 1 & -1 & -1 \\ 1 & 1 & -1 & -1 \\ -1 & -1 & 1 & 1 \\ -1 & -1 & 1 & 1 \end{bmatrix} \qquad \mathbf{k}_{bc} = \frac{141.4}{2} \begin{bmatrix} 1 & -1 & -1 & 1 \\ -1 & 1 & 1 & -1 \\ -1 & 1 & 1 & -1 \\ 1 & -1 & -1 & 1 \end{bmatrix}$$

$$\mathbf{K} = 200 \begin{bmatrix}
U_a & V_a & U_b & V_b & U_c & V_c & U_d & V_d \\
1.354 & 0.354 & -1.000 & 0.000 & 0.000 & 0.000 & -0.354 & -0.354 \\
0.354 & 1.354 & 0.000 & 0.000 & 0.000 & -1.000 & -0.354 & -0.354 \\
-1.000 & 0.000 & 1.354 & -0.354 & -0.354 & 0.354 & 0.000 & 0.000 \\
0.000 & 0.000 & -0.354 & 1.354 & 0.354 & -0.354 & 0.000 & -1.000 \\
0.000 & 0.000 & -0.354 & 0.354 & 1.354 & -0.354 & -1.000 & 0.000 \\
0.000 & -1.000 & 0.354 & -0.354 & -0.354 & 1.354 & 0.000 & 0.000 \\
-0.354 & -0.354 & 0.000 & 0.000 & -1.000 & 0.000 & 1.354 & 0.354 \\
-0.354 & -0.354 & 0.000 & -1.000 & 0.000 & 0.000 & 0.354 & 1.354
\end{bmatrix}$$

Then from $\mathbf{U}_f = \mathbf{K}_{ff}^{-1} \mathbf{X}_f$

$$\begin{bmatrix} U_c \\ V_c \\ U_d \\ V_d \end{bmatrix} = \frac{1}{200} \begin{bmatrix} 1.354 & -0.354 & -1.000 & 0.000 \\ -0.354 & 1.354 & 0.000 & 0.000 \\ -1.000 & 0.000 & 1.354 & 0.354 \\ 0.000 & 0.000 & 0.354 & 1.354 \end{bmatrix}^{-1} \begin{bmatrix} 0 \\ 0 \\ 50 \\ 0 \end{bmatrix}$$

$$\begin{bmatrix} U_c \\ V_c \\ U_d \\ V_d \end{bmatrix} = \frac{1}{200} \begin{bmatrix} 2.135 & 0.558 & 1.693 & -0.442 \\ 0.558 & 0.884 & 0.442 & -0.116 \\ 1.693 & 0.442 & 2.135 & -0.558 \\ -0.442 & -0.116 & -0.558 & 0.884 \end{bmatrix} \begin{bmatrix} 0 \\ 0 \\ 50 \\ 0 \end{bmatrix} = \begin{bmatrix} 0.423 \\ 0.110 \\ 0.534 \\ -0.140 \end{bmatrix} \text{ in}$$

and from $\mathbf{X}_s = \mathbf{K}_{sf} \mathbf{U}_f$

$$\begin{bmatrix} X_a \\ Y_a \\ X_b \\ Y_b \end{bmatrix} = 200 \begin{bmatrix} 0.000 & 0.000 & -0.354 & -0.354 \\ 0.000 & -1.000 & -0.354 & -0.354 \\ -0.354 & 0.354 & 0.000 & 0.000 \\ 0.354 & -0.354 & 0.000 & -1.000 \end{bmatrix} \begin{bmatrix} 0.423 \\ 0.110 \\ 0.534 \\ -0.140 \end{bmatrix} = \begin{bmatrix} -27.9 \\ -50.0 \\ -22.1 \\ 50.0 \end{bmatrix} \text{ kips}$$

DISCUSSION Although this truss is statically indeterminate to the second degree, the analysis proceeds exactly the same as for the statically determinate trusses of the previous two examples; in the stiffness method the statical indeterminacy of the structure is not a consideration. A partial explanation can be obtained by recalling that statical indeterminacy is significant when one of the so-called compatibility methods is used but not when equilibrium methods are used. Since the stiffness method belongs to this second group, this consideration is suppressed during the investigation. It is sufficient for the reader to note at this point that all the pertinent equations (equilibrium, force-deformation, etc.) are satisfied; thus the solution is valid.

After merging the individual member stiffness matrices, the structural stiffness matrix is obtained in the form shown; therefore, it must be reordered to obtain the form shown in Eq. (15.24). For this truss, however, it is a simple matter to extract the required partitions $\mathbf{K}_{ff}$ and $\mathbf{K}_{sf}$ without going through the reordering.

16.5 STRUCTURAL AND MEMBER COORDINATE SYSTEMS

Although the methods for obtaining the nodal displacements for trusses have now been developed, the forces in the individual members must be calculated to make the solution complete. By drawing an analogy between the analysis of trusses and the investigation of spring assemblages (from the previous chapter) it can be seen that the next logical step in the analysis requires the development of a general method for obtaining the member forces (see Fig. 15.6).

A perusal of Eq. (16.16) reveals that it is possible to obtain the column matrix of member forces $\mathbf{x}$ by a simple matrix multiplication once the displacements at the nodal points $\mathbf{u}$ have been obtained. Unfortunately, usually none of these force components is the axial force in the member; i.e., these forces are described in the xy coordinate system which is the basis for calculating the displacements. The force directed along the member axis is the quantity required to predict the behavior of the member and to design the cross section.

The axial force in the various truss members can be obtained using a straightforward coordinate transformation. Thus, a number of coordinate systems are necessary, one for the assembled structure (*structural coordinate system*) and one for each of the members of the truss (*member coordinate systems*). For example, the displacements at all the nodes for the truss shown in Fig. 16.6a will be described in terms of the xy structural coordinate system. However, the axial force in member ij must be calculated in terms of the $\bar{x}\bar{y}$ member coordinate system, in which the $\bar{x}$ axis is aligned along the length of the truss member. All quantities referred to the member coordinate system will henceforth be denoted using a bar over the associated quantity ($\bar{x}_i, \bar{y}_i$ for the forces and $\bar{u}_i, \bar{v}_i$ for the displacements, etc.).

16.6 COORDINATE TRANSFORMATIONS

From Fig. 16.6b it can be observed that the axial force in the member can be obtained from the force components expressed in the structural coordinate system at points i and j, respectively, as

$$\bar{x}_i = x_i \cos \phi + y_i \sin \phi \qquad (16.21)$$

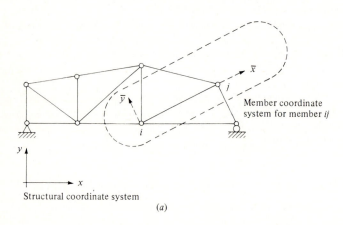

(a)

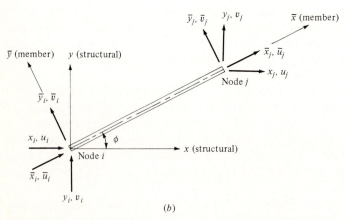

(b)

Figure 16.6 (a) The member and structural coordinate systems; (b) shown for one member of a truss with nodal forces and displacements.

and
$$\bar{x}_j = x_j \cos \phi + y_j \sin \phi \tag{16.22}$$

Since the member must be in equilibrium, $\bar{x}_i$ and $\bar{x}_j$ will be equal in magnitude but opposite in direction, but the dependency between these two quantities will not be used at this point in the discussion. The member can support only forces parallel to its longitudinal axis; thus the two forces $\bar{y}_i$ and $\bar{y}_j$ normal to the member axis must turn out to equal zero; however, they can be expressed in terms of the structural force components at points i and j, respectively, as

$$\bar{y}_i = -x_i \sin \phi + y_i \cos \phi \tag{16.23}$$

and
$$\bar{y}_j = -x_j \sin \phi + y_j \cos \phi \tag{16.24}$$

Equations (16.21) through (16.24) can be combined into a single matrix equation that transforms the forces from structural to member coordinates

$$\underbrace{\begin{bmatrix} \bar{x}_i \\ \bar{y}_i \\ \bar{x}_j \\ \bar{y}_j \end{bmatrix}}_{\bar{\mathbf{x}}} = \underbrace{\begin{bmatrix} \cos \phi & \sin \phi & 0 & 0 \\ -\sin \phi & \cos \phi & 0 & 0 \\ 0 & 0 & \cos \phi & \sin \phi \\ 0 & 0 & -\sin \phi & \cos \phi \end{bmatrix}}_{\mathbf{T}} \underbrace{\begin{bmatrix} x_i \\ y_i \\ x_j \\ y_j \end{bmatrix}}_{\mathbf{x}} \tag{16.25}$$

or
$$\bar{\mathbf{x}} = \mathbf{Tx} \tag{16.25a}$$

where $\bar{\mathbf{x}}$ = column matrix containing nodal force components in member coordinate system

$\mathbf{x}$ = column matrix containing nodal force components in structural coordinate system

$\mathbf{T}$ = coordinate transformation matrix

The matrix $\mathbf{T}$ is an *orthogonal matrix,* defined as a square matrix with its inverse equal to its its transpose

$$\mathbf{T}^{-1} = \mathbf{T}^T \tag{16.26}$$

It is left as an exercise for the reader to demonstrate this fact, i.e., to show that $\mathbf{T}^T\mathbf{T}$ (or $\mathbf{TT}^T$) is equal to the identity matrix $\mathbf{I}$. This property will be useful later when the transformation of the stiffness matrix is discussed.

It is also possible to obtain $\mathbf{x}$ in terms of $\bar{\mathbf{x}}$ (the inverse transform) by premultiplying both sides of Eq. (16.25a) by $\mathbf{T}^{-1}$

$$\mathbf{T}^{-1}\bar{\mathbf{x}} = \mathbf{T}^{-1}\mathbf{Tx} = \mathbf{Ix} = \mathbf{x}$$

Using Eq. (16.26) gives

$$\mathbf{x} = \mathbf{T}^T\bar{\mathbf{x}} \tag{16.27}$$

Equation (16.25) is a coordinate transformation between the two force column matrices $\bar{\mathbf{x}}$ and $\mathbf{x}$. This same transformation can be applied to the corresponding column matrices $\bar{\mathbf{u}}$ and $\mathbf{u}$, as well.

$$\begin{bmatrix} \bar{u}_i \\ \bar{v}_i \\ \bar{u}_j \\ \bar{v}_j \end{bmatrix} = \begin{bmatrix} \cos\phi & \sin\phi & 0 & 0 \\ -\sin\phi & \cos\phi & 0 & 0 \\ 0 & 0 & \cos\phi & \sin\phi \\ 0 & 0 & -\sin\phi & \cos\phi \end{bmatrix} \begin{bmatrix} u_i \\ v_i \\ u_j \\ v_j \end{bmatrix} \qquad (16.28)$$

$$\underset{\bar{\mathbf{u}}}{} \qquad \underset{\mathbf{T}}{} \qquad \underset{\mathbf{u}}{}$$

or

$$\bar{\mathbf{u}} = \mathbf{T}\mathbf{u} \qquad (16.28a)$$

where $\mathbf{u}$ and $\bar{\mathbf{u}}$ are the column matrices containing the displacements expressed in the structural and member coordinates, respectively (see Fig. 16.6b). The inverse transformation is obtained by premultiplying both sides of Eq. (16.28a) by $\mathbf{T}^{-1}$ and using Eq. (16.26), giving

$$\mathbf{u} = \mathbf{T}^T\bar{\mathbf{u}} \qquad (16.29)$$

Having investigated the transformations for the forces and displacements, we must see how a coordinate change alters the form of the stiffness matrix. Equation (16.16a) described the force-displacement relation for a single member in structural coordinates. Substituting the force and displacement transformations from Eqs. (16.27) and (16.29), respectively, into Eq. (16.16a) gives

$$\mathbf{T}^T\bar{\mathbf{x}} = \mathbf{k}\mathbf{T}^T\bar{\mathbf{u}} \qquad (16.30)$$

Premultiplying both sides by $\mathbf{T}$ and making use of the orthogonality property (16.26) yields

$$\bar{\mathbf{x}} = \mathbf{T}\mathbf{k}\mathbf{T}^T\bar{\mathbf{u}} \qquad (16.31)$$

or

$$\bar{\mathbf{x}} = \bar{\mathbf{k}}\bar{\mathbf{u}} \qquad (16.32)$$

That is, since the forces and displacements in member coordinates are related as shown in Eq. (16.31), we can conclude that the matrix triple product $\mathbf{T}\mathbf{k}\mathbf{T}^T$ is the stiffness matrix expressed in member coordinates

$$\bar{\mathbf{k}} = \mathbf{T}\mathbf{k}\mathbf{T}^T \qquad (16.33)$$

If the multiplication indicated in Eq. (16.33) is carried out, the element stiffness matrix in member coordinates is

$$\bar{\mathbf{k}} = \frac{AE}{L} \begin{bmatrix} 1 & 0 & -1 & 0 \\ 0 & 0 & 0 & 0 \\ -1 & 0 & 1 & 0 \\ 0 & 0 & 0 & 0 \end{bmatrix} \qquad (16.34)$$

Hence, the complete force-displacement equation for the member is

$$\begin{bmatrix} \bar{x}_i \\ \bar{y}_i \\ \bar{x}_j \\ \bar{y}_j \end{bmatrix} = \frac{AE}{L} \begin{bmatrix} 1 & 0 & -1 & 0 \\ 0 & 0 & 0 & 0 \\ -1 & 0 & 1 & 0 \\ 0 & 0 & 0 & 0 \end{bmatrix} \begin{bmatrix} \bar{u}_i \\ \bar{v}_i \\ \bar{u}_j \\ \bar{v}_j \end{bmatrix} \qquad (16.35)$$

The information contained in Eq. (16.35) is basic. The values of $\bar{y}_i$ and $\bar{y}_j$ are zero, as one would assume for a truss member. Also, $\bar{x}_j = -\bar{x}_i$, which is an expression of equilibrium; furthermore,

$$\bar{x}_i = \frac{AE}{L}(\bar{u}_i - \bar{u}_j)$$

which is merely a statement of Hooke's law for a two-force member.

The stiffness matrix can also be transformed from member to structural coordinates. Starting with Eq. (16.32) and using the force and displacement transformations (16.25a) and (16.28a) along with the orthogonality condition for **T** gives

$$\mathbf{x} = \mathbf{T}^T \bar{\mathbf{k}} \mathbf{T} \mathbf{u} \qquad (16.36)$$

Therefore,

$$\mathbf{k} = \mathbf{T}^T \bar{\mathbf{k}} \mathbf{T} \qquad (16.37)$$

The transformation shown in Eq. (16.33) was derived first since the **k** matrix had already been obtained, but the transformation in Eq. (16.37) is the one used for many computer applications. That is, the stiffness matrix for each individual member [in member coordinates shown in Eq. (16.34)] can be conveniently programmed and stored, along with the appropriate transformation matrix **T**. Then the transformation from member to structural coordinates can be executed using the matrix triple product in Eq. (16.37). This approach is easy to program if a matrix interpretive system is used, but it may be more efficient to generate **k** explicitly in the form shown in Eq. (16.16) using a carefully written subprogram.

16.7 MEMBER FORCES

A number of alternative procedures can be used to calculate the member forces once the nodal displacements are known. One option consists of using Eq. (16.16a) to obtain the nodal force matrix **x**. These force components must then be transformed into the member coordinates using Eq. (16.25a) to yield the axial force in the member. A second option involves transforming the displacements from structural to member coordinates using Eq. (16.28a) and then calculating the nodal forces in member coordinates using Eq. (16.32). These force components are aligned with the $\bar{x}\bar{y}$ axes, and the axial force is one of them. Unfortunately, in this form, $\bar{\mathbf{x}}$ contains redundant information; i.e., the forces $\bar{y}_i$ and $\bar{y}_j$ are both zero since this is a two-force member and it is known from equilibrium that $\bar{x}_i$ is equal in magnitude but opposite in direction to $\bar{x}_j$. Therefore, this second option can be simplified in the light of these facts, and only $\bar{x}_i$ or $\bar{x}_j$ need be calculated. By selecting the third row of the $\bar{\mathbf{k}}$ matrix [shown in Eq. (16.34)] the force in a truss member can be obtained using

$$\underset{s}{S_{ij}} = \frac{AE}{L}\begin{bmatrix} -1 & 0 & 1 & 0 \end{bmatrix}\underset{\bar{\mathbf{u}}_{ij}}{\begin{bmatrix} \bar{u}_i \\ \bar{v}_i \\ \bar{u}_j \\ \bar{v}_j \end{bmatrix}} \qquad (16.38)$$

or
$$s_{ij} = \mathbf{s}\bar{\mathbf{u}}_{ij} \qquad\qquad (16.38a)$$

where s_{ij} is the force in member ij and the row matrix $\mathbf{s}$ is the stress matrix. The subscripts are placed on the displacement matrix to emphasize that it must be constructed by removing from the structural displacement matrix $\mathbf{U}$ the displacements associated with the member ij.

Note that by picking the third row of $\bar{\mathbf{k}}$ for use as the matrix $\mathbf{s}$ the force $\bar{x}_j$ is calculated as in Eq. (16.38). This choice was made primarily to simplify interpretation of the sign of the force that results; a positive value of $\bar{x}_j$ implies that this force is directed in the positive coordinate direction, placing the truss member in tension. Thus, a positive value of s_{ij} denotes that the member is in tension; therefore, interpretation of the element forces is extremely convenient. This would not have been the case if the first row of $\bar{\mathbf{k}}$ had been used.

The element forces are calculated using Eq. (16.38) in Examples 16.6 and 16.7 for the trusses analyzed in Examples 16.4 and 16.5, respectively.

Example 16.6 Calculate the member forces for the structure of Example 16.4.

SOLUTION For member ab

$$\mathbf{T} = \mathbf{I} \qquad \bar{\mathbf{u}}_{ab} = \mathbf{u}_{ab} = \{U_a \quad V_a \quad U_b \quad V_b\} = 10^{-4}\{0.00 \quad 0.00 \quad 69.75 \quad 167.62\}$$

$$s_{ab} = \mathbf{s}\bar{\mathbf{u}}_{ab} = (2 \times 10^4)[-1 \quad 0 \quad 1 \quad 0]\left(10^{-4}\begin{bmatrix} 0.00 \\ 0.00 \\ 69.75 \\ 167.62 \end{bmatrix}\right) = 139.5 \text{ kN (t)}$$

For member bc

$$\mathbf{u}_{bc} = \{U_b \quad V_b \quad U_c \quad V_c\} = 10^{-4}\{69.75 \quad 167.62 \quad 0.00 \quad 45.00\}$$

$$\bar{\mathbf{u}}_{bc} = \mathbf{T}\mathbf{u}_{bc} = \begin{bmatrix} -0.6 & 0.8 & 0.0 & 0.0 \\ -0.8 & -0.6 & 0.0 & 0.0 \\ 0.0 & 0.0 & -0.6 & 0.8 \\ 0.0 & 0.0 & -0.8 & -0.6 \end{bmatrix}\left(10^{-4}\begin{bmatrix} 69.75 \\ 167.62 \\ 0.00 \\ 45.00 \end{bmatrix}\right) = 10^{-4}\begin{bmatrix} 92.25 \\ -156.37 \\ 36.00 \\ -27.00 \end{bmatrix}$$

$$s_{bc} = \mathbf{s}\bar{\mathbf{u}}_{bc} = (2 \times 10^4)[-1 \quad 0 \quad 1 \quad 0]\left(10^{-4}\begin{bmatrix} 92.25 \\ -156.37 \\ 36.00 \\ -27.00 \end{bmatrix}\right) = -112.5 \text{ kN (c)}$$

For member ac

$$\mathbf{u}_{ac} = \{U_a \quad V_a \quad U_c \quad V_c\} = 10^{-4}\{0.00 \quad 0.00 \quad 0.00 \quad 45.00\}$$

$$\bar{\mathbf{u}}_{ac} = \mathbf{T}\mathbf{u}_{ac} = \begin{bmatrix} 0.0 & 1.0 & 0.0 & 0.0 \\ -1.0 & 0.0 & 0.0 & 0.0 \\ 0.0 & 0.0 & 0.0 & 1.0 \\ 0.0 & 0.0 & -1.0 & 0.0 \end{bmatrix}\left(10^{-4}\begin{bmatrix} 0.00 \\ 0.00 \\ 0.00 \\ 45.00 \end{bmatrix}\right) = 10^{-4}\begin{bmatrix} 0.00 \\ 0.00 \\ 45.00 \\ 0.00 \end{bmatrix}$$

$$s_{ac} = \mathbf{s\bar{u}}_{ac} = (2 \times 10^4)[-1 \quad 0 \quad 1 \quad 0]\left(10^{-4}\begin{bmatrix} 0.00 \\ 0.00 \\ 45.00 \\ 0.00 \end{bmatrix}\right) = 90.0 \text{ kN (t)}$$

Example 16.7 Calculate the member forces for the structure of Example 16.5.

SOLUTION For members *ab* and *cd*

$$\mathbf{T} = \mathbf{I} \qquad \mathbf{\bar{u}} = \mathbf{u}$$

$$s_{ab} = 200[-1 \quad 0 \quad 1 \quad 0]\begin{bmatrix} 0.000 \\ 0.000 \\ 0.000 \\ 0.000 \end{bmatrix} = 0.0 \text{ kip}$$

$$s_{cd} = 200[-1 \quad 0 \quad 1 \quad 0]\begin{bmatrix} 0.423 \\ 0.110 \\ 0.534 \\ -0.140 \end{bmatrix} = 22.0 \text{ kips (t)}$$

For members *ac* and *bd*

$$\mathbf{T} = \begin{bmatrix} 0.0 & 1.0 & 0.0 & 0.0 \\ -1.0 & 0.0 & 0.0 & 0.0 \\ 0.0 & 0.0 & 0.0 & 1.0 \\ 0.0 & 0.0 & -1.0 & 0.0 \end{bmatrix}$$

$$s_{ac} = 200[-1 \quad 0 \quad 1 \quad 0]\begin{bmatrix} 0.000 \\ 0.000 \\ 0.110 \\ -0.423 \end{bmatrix} = 22.0 \text{ kips (t)}$$

$$s_{bd} = 200[-1 \quad 0 \quad 1 \quad 0]\begin{bmatrix} 0.000 \\ 0.000 \\ -0.140 \\ -0.534 \end{bmatrix} = -28.0 \text{ kips (c)}$$

For member *ad*

$$\mathbf{u}_{ad} = \{0.000 \quad 0.000 \quad 0.534 \quad -0.140\}$$

$$\mathbf{\bar{u}}_{ad} = \mathbf{Tu}_{ad} = \frac{1}{\sqrt{2}}\begin{bmatrix} 1 & 1 & 0 & 0 \\ -1 & 1 & 0 & 0 \\ 0 & 0 & 1 & 1 \\ 0 & 0 & -1 & 1 \end{bmatrix}\begin{bmatrix} 0.000 \\ 0.000 \\ 0.534 \\ -0.140 \end{bmatrix} = \begin{bmatrix} 0.000 \\ 0.000 \\ 0.279 \\ -0.477 \end{bmatrix}$$

$$s_{ad} = 141.4[-1 \quad 0 \quad 1 \quad 0]\begin{bmatrix} 0.000 \\ 0.000 \\ 0.279 \\ -0.477 \end{bmatrix} = 39.4 \text{ kips (t)}$$

For member bc

$$\mathbf{u}_{bc} = \{0.000 \quad 0.000 \quad 0.423 \quad 0.110\}$$

$$\bar{\mathbf{u}}_{bc} = \mathbf{T}\mathbf{u}_{bc} = \frac{1}{\sqrt{2}}\begin{bmatrix} -1 & 1 & 0 & 0 \\ -1 & -1 & 0 & 0 \\ 0 & 0 & -1 & 1 \\ 0 & 0 & -1 & -1 \end{bmatrix}\begin{bmatrix} 0.000 \\ 0.000 \\ 0.423 \\ 0.110 \end{bmatrix} = \begin{bmatrix} 0.000 \\ 0.000 \\ -0.221 \\ -0.377 \end{bmatrix}$$

$$s_{bc} = 141.4[-1 \quad 0 \quad 1 \quad 0]\begin{bmatrix} 0.000 \\ 0.000 \\ -0.221 \\ -0.377 \end{bmatrix} = -31.2 \text{ kips (c)}$$

16.8 PROBLEMS INVOLVING SETTLEMENT, TEMPERATURE, ETC.

A truss with differential settlement of the supports can be analyzed using the nodal equilibrium equations discussed previously. In this case, however, the partitioned form of these equations shown in Eq. (15.25) must be used in its complete form since $\mathbf{U}_s \neq \mathbf{0}$. Thus, the nodal displacements can be obtained by solving the first of these equations

$$\mathbf{U}_f = \mathbf{K}_{ff}^{-1}(\mathbf{X}_f - \mathbf{K}_{fs}\mathbf{U}_s)$$

After calculating the unknown nodal displacements, the second of the matrix equations in Eq. (15.25) can be used to obtain the unknown forces $\mathbf{X}_s$ associated with $\mathbf{U}_s$

$$\mathbf{X}_s = \mathbf{K}_{sf}\mathbf{U}_f + \mathbf{K}_{ss}\mathbf{U}_s$$

The forces in the members are found using Eq. (16.38a) in the same way as for structures without support settlement.

Examples 16.8 and 16.9 illustrate the analysis of support displacement for a statically determinate and a statically indeterminate truss, respectively.

Example 16.8 The loads at joint b are removed from the truss in Example 16.4, and the following support displacements occur: $U_a = 15$ mm; $V_a = 20$ mm; and $U_c = 27$ mm. Calculate (a) the resulting displacements $\mathbf{U}_f$, (b) the forces where the displacements are prescribed $\mathbf{X}_s$, and (c) the forces in the three members.

SOLUTION (a) Use the first matrix equation in Eq. (15.25) and the matrix partitions in Example 16.4

$$\begin{bmatrix} 0 \\ 0 \\ 0 \end{bmatrix} = \mathbf{K}_{ff}\mathbf{U}_f + 20{,}000 \begin{bmatrix} -1.00 & 0.00 & -0.36 \\ 0.00 & 0.00 & 0.48 \\ 0.00 & -1.00 & -0.48 \end{bmatrix} \begin{bmatrix} 0.015 \\ 0.020 \\ 0.027 \end{bmatrix}$$

$$\begin{bmatrix} U_b \\ V_b \\ V_c \end{bmatrix} = \frac{1}{12{,}800} \begin{bmatrix} 0.64 & 0.48 & 0.00 \\ 0.48 & 2.00 & 0.64 \\ 0.00 & 0.64 & 0.64 \end{bmatrix} \begin{bmatrix} 494.4 \\ -259.2 \\ 659.2 \end{bmatrix} = \begin{bmatrix} 0.015 \\ 0.011 \\ 0.020 \end{bmatrix} \text{m}$$

(b) The forces associated with the known displacements $\mathbf{X}_s$ are obtained using the second matrix equation (15.25) and the appropriate matrix partitions from Example 16.4

$$\mathbf{X}_s = \begin{bmatrix} X_a \\ Y_a \\ X_c \end{bmatrix}$$

$$= 20{,}000 \begin{bmatrix} -1.00 & 0.00 & 0.00 \\ 0.00 & 0.00 & -1.00 \\ -0.36 & 0.48 & -0.48 \end{bmatrix} \begin{bmatrix} 0.015 \\ 0.011 \\ 0.020 \end{bmatrix} + 20{,}000 \begin{bmatrix} 1.00 & 0.00 & 0.00 \\ 0.00 & 1.00 & 0.00 \\ 0.00 & 0.00 & 0.36 \end{bmatrix} \begin{bmatrix} 0.015 \\ 0.020 \\ 0.027 \end{bmatrix}$$

$$= \begin{bmatrix} -300 \\ -400 \\ -194 \end{bmatrix} + \begin{bmatrix} 300 \\ 400 \\ 194 \end{bmatrix} = \begin{bmatrix} 0 \\ 0 \\ 0 \end{bmatrix}$$

(c) See the basic matrices in Example 16.6. For member ab

$$\bar{\mathbf{u}}_{ab} = \mathbf{u}_{ab}$$

$$s_{ab} = 20{,}000[-1 \quad 0 \quad 1 \quad 0] \begin{bmatrix} 0.015 \\ 0.020 \\ 0.015 \\ 0.011 \end{bmatrix} = 0$$

For member bc, transforming the displacements and substituting into Eq. (16.38) gives

$$s_{bc} = 20{,}000[-1 \quad 0 \quad 1 \quad 0] \begin{bmatrix} -0.0002 \\ -0.0186 \\ -0.0002 \\ -0.0336 \end{bmatrix} = 0$$

For member ac, transforming the displacements and substituting into Eq. (16.38) gives

$$s_{ac} = 20{,}000[-1 \quad 0 \quad 1 \quad 0] \begin{bmatrix} 0.020 \\ -0.015 \\ 0.020 \\ -0.027 \end{bmatrix} = 0$$

DISCUSSION In general, such a problem does not warrant this general type of analysis, but this example is intended to demonstrate the use of the stiffness method when known displacements are involved. Since the truss is statically determinate, it should be anticipated from the outset that there will be neither support forces nor member forces induced by the support movements. The various nodal displacements can be checked using simple kinematics; i.e., the truss experiences rigid-body translations in the x and y directions plus a rigid-body rotation.

Example 16.9 The load at joint d is removed from the truss in Example 16.5, and the following support displacements occur: $U_a = -0.25$ in; $U_b = -0.50$ in; and $V_b = 0.25$ in. Calculate (a) the resulting displacements $\mathbf{U}_f$, (b) the forces where the displacements are prescribed $\mathbf{X}_s$, and (c) the forces in all the members.

SOLUTION (a) Using the first matrix equation in Eq. (15.25) and the matrix partitions in Example 16.5 gives

$$
\begin{bmatrix} 0 \\ 0 \\ 0 \\ 0 \end{bmatrix} = \mathbf{K}_{ff}\mathbf{U}_f + 200 \begin{bmatrix} 0.000 & 0.000 & -0.354 & 0.354 \\ 0.000 & -1.000 & 0.354 & -0.354 \\ -0.354 & -0.354 & 0.000 & 0.000 \\ -0.354 & -0.354 & 0.000 & -1.000 \end{bmatrix} \begin{bmatrix} -0.25 \\ 0.00 \\ -0.50 \\ 0.25 \end{bmatrix}
$$

$$
\begin{bmatrix} U_c \\ V_c \\ U_d \\ V_d \end{bmatrix} = \frac{1}{200} \begin{bmatrix} 2.135 & 0.558 & 1.693 & -0.442 \\ 0.558 & 0.884 & 0.442 & -0.116 \\ 1.693 & 0.442 & 2.135 & -0.558 \\ -0.442 & -0.116 & -0.558 & 0.884 \end{bmatrix} \begin{bmatrix} -53.10 \\ 53.10 \\ -17.70 \\ 32.30 \end{bmatrix} = \begin{bmatrix} -0.6399 \\ 0.0287 \\ -0.6112 \\ 0.2787 \end{bmatrix} \text{ in}
$$

(b) The forces associated with the known displacements $\mathbf{X}_s$ are obtained using the second matrix equation in Eq. (15.25) and the appropriate matrix partitions from Example 16.5

$$
\begin{bmatrix} X_a \\ Y_a \\ X_b \\ Y_b \end{bmatrix} = 200 \begin{bmatrix} 0.000 & 0.000 & -0.354 & -0.354 \\ 0.000 & -1.000 & -0.354 & -0.354 \\ -0.354 & 0.354 & 0.000 & 0.000 \\ 0.354 & -0.354 & 0.000 & -1.000 \end{bmatrix} \begin{bmatrix} -0.6399 \\ 0.0287 \\ -0.6112 \\ 0.2787 \end{bmatrix}
$$

$$
+ 200 \begin{bmatrix} 1.354 & 0.354 & -1.000 & 0.000 \\ 0.354 & 1.354 & 0.000 & 0.000 \\ -1.000 & 0.000 & 1.354 & -0.354 \\ 0.000 & 0.000 & -0.354 & 1.354 \end{bmatrix} \begin{bmatrix} -0.25 \\ 0.00 \\ -0.50 \\ 0.25 \end{bmatrix} = \begin{bmatrix} 55.74 \\ 0.00 \\ -55.74 \\ 0.00 \end{bmatrix} \text{ kips}
$$

(c) See the basic matrices and transformations in Example 16.7

$$
S_{ab} = \mathbf{s} \begin{bmatrix} -0.2500 \\ 0.0000 \\ -0.5000 \\ 0.2500 \end{bmatrix} = -50.00 \text{ kips (c)} \qquad S_{cd} = \mathbf{s} \begin{bmatrix} -0.6399 \\ 0.0287 \\ -0.6112 \\ 0.2787 \end{bmatrix} = 5.74 \text{ kips (t)}
$$

$$S_{ac} = \mathbf{s}\begin{bmatrix} 0.0000 \\ 0.2500 \\ 0.0287 \\ 0.6399 \end{bmatrix} = 5.74 \text{ kips (t)} \qquad S_{bd} = \mathbf{s}\begin{bmatrix} 0.2500 \\ 0.5000 \\ 0.2787 \\ 0.6112 \end{bmatrix} = 5.74 \text{ kips (t)}$$

$$S_{ad} = \mathbf{s}\begin{bmatrix} -0.1768 \\ 0.1768 \\ -0.2351 \\ 0.6293 \end{bmatrix} = -8.12 \text{ kips (c)} \qquad S_{bc} = \mathbf{s}\begin{bmatrix} 0.5303 \\ 0.1768 \\ 0.4728 \\ 0.4322 \end{bmatrix} = -8.12 \text{ kips (c)}$$

DISCUSSION This problem illustrates the effect of support movements on a statically indeterminate truss. Note that the resulting member forces and two of the support reactions are nonzero. The solution procedure is identical to that used in Example 16.8 for the analysis of a statically determinate truss, but here the difficulty warrants use of the stiffness method. Equilibrium is satisfied implicitly in the stiffness method, but the reader may wish to check independently that all the equations of equilibrium are satisfied. Many analysts use this procedure to review the numerical accuracy of the calculations.

When the truss is subjected to thermal changes, fabrication errors, etc., the solution is obtained using the principle of superposition. The procedure is summarized as follows:

1. Apply initial fixed-end forces to the members to prevent any nodal displacements introduced by thermal changes, fabrication errors, etc.
2. Remove the initial forces applied in step 1 and introduce forces at all nodal points that are equal in magnitude but opposite in direction to the initial fixed-end forces applied in step 1. Calculate the nodal displacements.
3. Superpose the results from steps 1 and 2 to obtain the actual solution for the truss.

The force-displacement relation for an individual truss member subjected to the initial forces described in step 1 is

$$\bar{\mathbf{x}} = \bar{\mathbf{k}}\bar{\mathbf{u}} + \bar{\mathbf{x}}^0 \tag{16.39}$$

where $\bar{\mathbf{x}}^0$ is the column matrix of forces (in member coordinates) introduced by the initial effects, i.e., the forces described in step 1 that are necessary to constrain the member with zero nodal displacements when it sustains thermal changes, etc. For example, if the member of Fig. 16.6b is subjected to a temperature rise of ΔT, the strain is $\alpha \, \Delta T$, the stress is $E\alpha \, \Delta T$, and

$$\bar{\mathbf{x}}^0 = AE\alpha \, \Delta T \begin{bmatrix} 1 \\ 0 \\ -1 \\ 0 \end{bmatrix} \tag{16.40a}$$

where A = member cross-sectional area
E = modulus of elasticity
α = coefficient of linear thermal expansion

Note that since the member has a tendency to increase in length, the initial forces must act in the directions indicated and will result in compression in the member. Or if the member is fabricated ΔL too long, the strain in the member is $\Delta L/L$; the stress is $E(\Delta L/L)$, and

$$\bar{\mathbf{x}}^0 = \frac{AE\ \Delta L}{L} \begin{bmatrix} 1 \\ 0 \\ -1 \\ 0 \end{bmatrix} \tag{16.40b}$$

Again, the axial forces are those necessary to constrain the member with zero nodal displacements; therefore, the force will be compressive with $\bar{x}_i$ directed in the positive coordinate direction and $\bar{x}_j$ in the negative direction.

The initial force matrices are transformed into structural coordinates using Eq. (16.27) and are then combined at the nodal points during the process of enforcing equilibrium for the assemblage. Therefore, the force-displacement relation for the total structure is

$$\mathbf{X} = \mathbf{KU} + \mathbf{X}^0 \tag{16.41}$$

where $\mathbf{X}^0$ is the column matrix of the initial forces obtained by adding the effects from the individual members to obtain nodal equilibrium. These equations can be partitioned as before in Eq. (15.24) to give

$$\begin{bmatrix} \mathbf{X}_f \\ \hline \mathbf{X}_s \end{bmatrix} = \begin{bmatrix} \mathbf{K}_{ff} & \mathbf{K}_{fs} \\ \hline \mathbf{K}_{sf} & \mathbf{K}_{ss} \end{bmatrix} \begin{bmatrix} \mathbf{U}_f \\ \hline \mathbf{U}_s \end{bmatrix} + \begin{bmatrix} \mathbf{X}_f^0 \\ \hline \mathbf{X}_s^0 \end{bmatrix} \tag{16.42}$$

where $\mathbf{X}_f^0$ and $\mathbf{X}_s^0$ are the column matrices of the initial forces at the nodes corresponding to $\mathbf{X}_f$ and $\mathbf{X}_s$, respectively. Multiplying the partitioned matrices in Eq. (16.42) gives

$$\mathbf{X}_f = \mathbf{K}_{ff}\mathbf{U}_f + \mathbf{K}_{fs}\mathbf{U}_s + \mathbf{X}_f^0$$
$$\mathbf{X}_s = \mathbf{K}_{sf}\mathbf{U}_f + \mathbf{K}_{ss}\mathbf{U}_s + \mathbf{X}_s^0 \tag{16.43}$$

Thus, if the supports are unyielding ($\mathbf{U}_s = \mathbf{0}$), the unknown displacements can be obtained by solving

$$\mathbf{X}_f = \mathbf{K}_{ff}\mathbf{U}_f + \mathbf{X}_f^0 \tag{16.44}$$

to give

$$\mathbf{U}_f = \mathbf{K}_{ff}^{-1}(\mathbf{X}_f - \mathbf{X}_f^0)$$

Subtracting $\mathbf{X}_f^0$ from both sides of Eq. (16.44) is essentially step 2 in the procedure. If these initial force effects are accompanied by support movement ($\mathbf{U}_s \neq \mathbf{0}$), the additional matrix $\mathbf{K}_{fs}\mathbf{U}_s$ must be retained in the solution process and Eq. (16.44) cannot be used.

After obtaining the nodal displacements, the member forces must be calculated (step 3); however, it is imperative to realize that according to Eq. (16.39) some of the members are initially subjected to end forces. The member forces can be obtained in a manner analogous to that used in Sec. 16.7, but the complete equation now becomes

$$s_{ij} = s\bar{u}_{ij} + \bar{s}^0 \tag{16.45}$$

where $\bar{s}^0$ contains the initial member force (in member coordinates). Recall that the stress matrix s was formed using the third row from the matrix $\bar{k}$ (the part that yields the local force $\bar{x}_j$); therefore, $\bar{s}^0$ is also obtained using the third row from $\bar{x}^0$. Hence, for a temperature change ΔT

$$\bar{s}^0 = -AE\alpha\,\Delta T \tag{16.46a}$$

and for an initial strain $\Delta L/L$

$$\bar{s}^0 = -\frac{AE\,\Delta L}{L} \tag{16.46b}$$

Examples 16.10 and 16.11 illustrate the calculations for a statically determinate and a statically indeterminate truss, respectively, subject to initial forces.

Example 16.10 The truss of Example 16.4 is fabricated with member bc 15 mm too long. Calculate (a) the resulting nodal displacements, (b) the reaction forces, and (c) the member forces. Assume that there are no applied loads at joint b as shown.

SOLUTION (a) The matrix of initial forces in member coordinates is obtained using Eq. (16.40b) and noting from Example 16.2 that $E = 200$ GPa and $A_{bc} = 5 \times 10^{-4}$ m^2

$$\bar{x}^0 = \begin{bmatrix} \bar{x}_b^0 \\ \bar{y}_b^0 \\ \bar{x}_c^0 \\ \bar{y}_c^0 \end{bmatrix} = \begin{bmatrix} 300 \\ 0 \\ -300 \\ 0 \end{bmatrix} \text{ kN}$$

Transforming this matrix into structural coordinates using Eq. (16.27) and the transformation matrix shown in Example 16.6 gives

$$x^0 = T^T\bar{x}^0 = \begin{bmatrix} x_b^0 \\ y_b^0 \\ x_c^0 \\ y_c^0 \end{bmatrix} = \begin{bmatrix} -0.6 & -0.8 & 0.0 & 0.0 \\ 0.8 & -0.6 & 0.0 & 0.0 \\ 0.0 & 0.0 & -0.6 & -0.8 \\ 0.0 & 0.0 & 0.8 & -0.6 \end{bmatrix} \begin{bmatrix} 300 \\ 0 \\ -300 \\ 0 \end{bmatrix} = \begin{bmatrix} -180 \\ 240 \\ 180 \\ -240 \end{bmatrix}$$

Using the first of the matrix equations in Eq. (16.43) and noting that X_f and U_s are both null column matrices, we have

$$U_f = K_{ff}^{-1}(-X_f^0)$$

where X_f^0 is obtained by combining the initial force matrices. Since in this case only one member is involved, this matrix is composed of entries taken from the initial force matrix for member bc only

$$\begin{bmatrix} U_b \\ V_b \\ V_c \end{bmatrix} = \frac{1}{12,800} \begin{bmatrix} 0.64 & 0.48 & 0.00 \\ 0.48 & 2.00 & 0.64 \\ 0.00 & 0.64 & 0.64 \end{bmatrix} \begin{bmatrix} 180 \\ -240 \\ 240 \end{bmatrix} = 10^{-3} \begin{bmatrix} 0.00 \\ -18.75 \\ 0.00 \end{bmatrix} \text{ m}$$

(b) The force matrix $\mathbf{X}_s$ is obtained using the second matrix equation in Eq. (16.43) and the appropriate matrix partitions from Example 16.4. Note that $\mathbf{X}_s^0 \neq \mathbf{0}$

$$\mathbf{X}_s = 20,000 \begin{bmatrix} -1.00 & 0.00 & 0.00 \\ 0.00 & 0.00 & -1.00 \\ -0.36 & 0.48 & -0.48 \end{bmatrix} \left(10^{-3} \begin{bmatrix} 0.00 \\ -18.75 \\ 0.00 \end{bmatrix} \right) + \begin{bmatrix} 0 \\ 0 \\ 180 \end{bmatrix} = \begin{bmatrix} 0 \\ 0 \\ 0 \end{bmatrix}$$

(c) See the basic matrices and transformations in Example 16.6

$$S_{ab} = (\mathbf{s} \times 10^{-3}) \begin{bmatrix} 0.00 \\ 0.00 \\ 0.00 \\ -18.75 \end{bmatrix} = 0 \qquad S_{bc} = (\mathbf{s} \times 10^{-3}) \begin{bmatrix} -15.00 \\ 11.25 \\ 0.00 \\ 0.00 \end{bmatrix} + [-300] = 0$$

$$S_{ac} = \mathbf{s} \begin{bmatrix} 0.00 \\ 0.00 \\ 0.00 \\ 0.00 \end{bmatrix} = 0$$

DISCUSSION Again, as in Example 16.8, this simple truss would not warrant such an elaborate method of analysis, but it is used to demonstrate the application of the stiffness method when initial forces are present. Note that the fabrication error results in neither support reactions nor forces in the members since the truss is statically determinate. After using Eq. (16.43) to calculate $\mathbf{U}_f$ and $\mathbf{X}_s$ it is important for any initial forces to be included in computing the member forces; i.e., Eq. (16.45) must be used. For member bc, $s^0 = -300$ kN, but it is zero for the other two members.

Example 16.11 Member cd of the truss in Example 16.5 is heated 100 Fahrenheit degrees above the temperature of the other members. Calculate (a) the resulting nodal displacements, (b) the reaction forces, and (c) the member forces. Assume that the load at node d is removed.

SOLUTION The matrix of initial forces in member coordinates is obtained using Eq. (16.40a). From Example 16.5, $E = 29 \times 10^6$ lb/in^2 and $A_{cd} = 0.993$ in^2; furthermore let $\alpha = 6.5 \times 10^{-6}$ in/in · °F. For member cd

$$\bar{\mathbf{x}}^0 = \mathbf{x}^0 = \begin{bmatrix} x_c^0 \\ y_c^0 \\ x_d^0 \\ y_d^0 \end{bmatrix} = \begin{bmatrix} 18.72 \\ 0.00 \\ -18.72 \\ 0.00 \end{bmatrix} \text{ kips}$$

Using the first of the matrix equations in Eq. (16.43) and noting that $\mathbf{X}_f$ and $\mathbf{U}_s$ are both null, we get

$$\mathbf{U}_f = \mathbf{K}_{ff}^{-1}(-\mathbf{X}_f^0)$$

where X_f^0 is obtained by combining the initial force matrices. Since in this case only one member is involved, this matrix is the matrix x^0 because the displacements in U_f are also the displacements of member cd

$$\begin{bmatrix} U_c \\ V_c \\ U_d \\ V_d \end{bmatrix} = \frac{1}{200} \begin{bmatrix} 2.135 & 0.558 & 1.693 & -0.442 \\ 0.558 & 0.884 & 0.442 & -0.116 \\ 1.693 & 0.442 & 2.135 & -0.558 \\ -0.442 & -0.116 & -0.558 & 0.884 \end{bmatrix} \begin{bmatrix} -18.72 \\ 0.00 \\ 18.72 \\ 0.00 \end{bmatrix} = \begin{bmatrix} -0.0414 \\ -0.0108 \\ 0.0414 \\ -0.0108 \end{bmatrix} \text{ in}$$

(b) The force matrix X_s is obtained using the second matrix equation in Eq. (16.43) and the appropriate matrix partitions from Example 16.5. Note that $X_s^0 = 0$

$$X_s = \begin{bmatrix} X_a \\ Y_a \\ X_b \\ Y_b \end{bmatrix} = 200 \begin{bmatrix} 0.000 & 0.000 & -0.354 & -0.354 \\ 0.000 & -1.000 & -0.354 & -0.354 \\ -0.354 & 0.354 & 0.000 & 0.000 \\ 0.354 & -0.354 & 0.000 & -1.000 \end{bmatrix} \begin{bmatrix} -0.0414 \\ -0.0108 \\ 0.0414 \\ -0.0108 \end{bmatrix} = \begin{bmatrix} -2.16 \\ 0.00 \\ 2.16 \\ 0.00 \end{bmatrix} \text{ kips}$$

(c) See the basic matrices and transformations in Example 16.7

$$S_{ab} = 0 \qquad S_{cd} = s \begin{bmatrix} -0.0414 \\ -0.0108 \\ 0.0414 \\ -0.0108 \end{bmatrix} + [-18.72] = 16.56 - 18.72 = -2.16 \text{ kips (c)}$$

$$S_{bd} = S_{ac} = s \begin{bmatrix} 0.0000 \\ 0.0000 \\ -0.0108 \\ 0.0414 \end{bmatrix} = -2.16 \text{ kips (c)} \qquad S_{ad} = s \begin{bmatrix} 0.0000 \\ 0.0000 \\ 0.0216 \\ -0.0369 \end{bmatrix} = 3.05 \text{ kips (t)} = S_{bc}$$

DISCUSSION The modest temperature increase in one member of this indeterminate truss results in nonzero forces in the members and two of the reactions. The solution differs from that for Example 16.10 in that X_s^0 is null and hence does not enter into the calculations for X_s. Since only member cd has initial forces s^0, Eq. (16.45) reduces to Eq. (16.38a) for all other members. Because the truss and the thermal expansion are both symmetric, it is to be expected that the deflections and member forces will also be symmetric.

16.9 DISCUSSION

The basic steps for solving assemblages composed of springs were developed in Chap. 15, and in this chapter the concepts of the stiffness method were used to solve truss structures. Therefore, by reviewing Fig. 15.8 the reader can observe that, indeed, only the matrices unique to a particular member type (spring or truss member) must be different in order to solve these two dissimilar types of structures. Thus, by using the stiffness matrix for each truss member [see Eq. (16.16)] the merge process for obtaining K is similar to that used for spring assemblages. The resulting merged stiffness matrix is partitioned with respect to the information known about each individual generalized displacement. That is, if the displacement is denoted as unsupported (free to displace in an unconstrained manner), it is put into the $\mathbf{K}_{ff}$ partition. The resulting simultaneous linear equations are solved using Eq. (15.28) to obtain $\mathbf{U}_f$. The final step of the analysis involves (1) obtaining the displacements associated with each individual member, (2) transforming these displacements into member coordinates using Eq. (16.28a), and (3) obtaining the element forces by Eq. (16.38). It is important to note in this final step that for each individual member, $\mathbf{u}_{ij}$ is a 4×1 matrix constructed by using the appropriate quantities selected from the matrix $\mathbf{U}$. Therefore, $\bar{\mathbf{u}}_{ij}$ for an individual truss member will be a 4×1 matrix with the generalized displacements specified in the coordinate system that is unique for that member.

The concept of coordinate transformation must be fully understood in order to analyze truss structures. Thus, the element stiffness matrix can be derived in its own coordinate system [Eq. (16.34)] and transformed into the structural coordinate system using Eq. (16.37); alternatively, the element stiffness matrix can be derived in a direct manner in the structural coordinate system, as shown in Eq. (16.16). The coordinate transformation becomes crucial when the member forces are to be calculated. The nodal displacements are obtained in the structural coordinate system, and the force in the member must be referenced to the member coordinate system.

Examples 16.12 and 16.13 illustrate the analysis of the nodal displacements and member forces using the stiffness method for two trusses with applied forces.

Example 16.12 Calculate the nodal displacements and member forces for this truss; $E = 200$ GPa, $A_{ab} = 5 \times 10^{-4}$ m^2, $A_{ac} = 4 \times 10^{-4}$ m^2, $A_{ad} = 10 \times 10^{-4}$ m^2, and $A_{ae} = 11.31 \times 10^{-4}$ m^2.

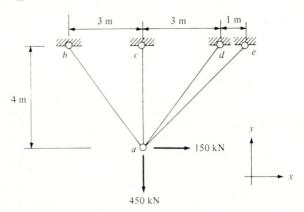

Figure E16.12

SOLUTION See Table E16.12.

Table E16.12

Member	ϕ, deg	$\cos \phi$	$\sin \phi$
ae	45	0.707	0.707
ad	53.13	0.600	0.800
ac	90	0.000	1.000
ab	126.87	−0.600	0.800

Since only node a is unconstrained for all members, the 2×2 partitions of the member stiffness matrices that will be merged into $\mathbf{K}_{ff}$ are

$$\mathbf{k}_{ae} = (2 \times 10^4)\begin{bmatrix} 1 & 1 \\ 1 & 1 \end{bmatrix} \qquad \mathbf{k}_{ad} = (4 \times 10^4)\begin{bmatrix} 0.36 & 0.48 \\ 0.48 & 0.64 \end{bmatrix}$$

$$\mathbf{k}_{ac} = (2 \times 10^4)\begin{bmatrix} 0 & 0 \\ 0 & 1 \end{bmatrix} \qquad \mathbf{k}_{ab} = (2 \times 10^4)\begin{bmatrix} 0.36 & -0.48 \\ -0.48 & 0.64 \end{bmatrix}$$

Therefore,

$$\mathbf{K}_{ff} = (2 \times 10^4)\begin{bmatrix} 2.08 & 1.48 \\ 1.48 & 3.92 \end{bmatrix} \qquad \begin{bmatrix} U_a \\ V_a \end{bmatrix} = 10^{-6}\begin{bmatrix} 32.87 & -12.41 \\ -12.41 & 17.44 \end{bmatrix}\begin{bmatrix} 150 \\ -450 \end{bmatrix}$$

$$= 10^{-3}\begin{bmatrix} 10.52 \\ -9.71 \end{bmatrix} \text{ m}$$

(b) From $\mathbf{s}_{ij} = \mathbf{s}\bar{\mathbf{u}}_{ij}$

$$\mathbf{s}_{ab} = (2 \times 10^4)\begin{bmatrix} -1 & 0 & 1 & 0 \end{bmatrix}\begin{bmatrix} -0.6 & 0.8 & 0.0 & 0.0 \\ -0.8 & -0.6 & 0.0 & 0.0 \\ 0.0 & 0.0 & -0.6 & 0.8 \\ 0.0 & 0.0 & -0.8 & -0.6 \end{bmatrix}\left(10^{-3}\begin{bmatrix} 10.52 \\ -9.71 \\ 0.00 \\ 0.00 \end{bmatrix}\right)$$

$$= +282 \text{ kN (t)}$$

$$\mathbf{s}_{ac} = (2 \times 10^4)\begin{bmatrix} -1 & 0 & 1 & 0 \end{bmatrix}10^{-3}\left(\begin{bmatrix} -9.71 \\ -10.52 \\ 0.00 \\ 0.00 \end{bmatrix}\right) = +194 \text{ kN (t)}$$

Similarly, $\qquad s_{ad} = +58 \text{ kN (t)} \qquad s_{ae} = -23 \text{ kN (c)}$

Example 16.13 Calculate (a) the nodal displacements and (b) member forces for this symmetrical truss. For all members $AE = 200{,}000$ kN.

SOLUTION (a)

$$\mathbf{k}_{ad} = \frac{2 \times 10^5}{6(1.414)(2)}\begin{bmatrix} 1 & 1 & -1 & -1 \\ 1 & 1 & -1 & -1 \\ -1 & -1 & 1 & 1 \\ -1 & -1 & 1 & 1 \end{bmatrix} \qquad \mathbf{k}_{bc} = \frac{2 \times 10^5}{6(1.414)(2)}\begin{bmatrix} 1 & -1 & -1 & 1 \\ -1 & 1 & 1 & -1 \\ -1 & 1 & 1 & -1 \\ 1 & -1 & -1 & 1 \end{bmatrix}$$

$$\mathbf{k}_{cd} = \frac{2 \times 10^5}{2.536}\begin{bmatrix} 1 & 0 & -1 & 0 \\ 0 & 0 & 0 & 0 \\ -1 & 0 & 1 & 0 \\ 0 & 0 & 0 & 0 \end{bmatrix}$$

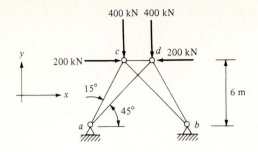

Figure E16.13

$$\mathbf{k}_{ac} = \frac{2 \times 10^5}{27.713} \begin{bmatrix} 1.0000 & 1.7321 & -1.0000 & -1.7321 \\ 1.7321 & 3.0000 & -1.7321 & -3.0000 \\ -1.0000 & -1.7321 & 1.0000 & 1.7321 \\ -1.7321 & -3.0000 & 1.7321 & 3.0000 \end{bmatrix}$$

$$\mathbf{k}_{bd} = \frac{2 \times 10^5}{27.713} \begin{bmatrix} 1.0000 & -1.7321 & -1.0000 & 1.7321 \\ -1.7321 & 3.0000 & 1.7321 & -3.0000 \\ -1.0000 & 1.7321 & 1.0000 & -1.7321 \\ 1.7321 & -3.0000 & -1.7321 & 3.0000 \end{bmatrix}$$

$$\begin{array}{cccc} U_c & V_c & U_d & V_d \end{array}$$
$$\mathbf{K}_{ff} = 10^5 \begin{bmatrix} 0.9788 & 0.0071 & -0.7887 & 0.0000 \\ 0.0071 & 0.3344 & 0.0000 & 0.0000 \\ -0.7887 & 0.0000 & 0.9788 & -0.0071 \\ 0.0000 & 0.0000 & -0.0071 & 0.3344 \end{bmatrix}$$

Note that for this loading, $U_c = -U_d$ and $V_c = V_d$; thus

$$\begin{bmatrix} U_c \\ V_c \end{bmatrix} = 10^{-5} \begin{bmatrix} 1.7675 & 0.0071 \\ 0.0071 & 0.3344 \end{bmatrix}^{-1} \begin{bmatrix} +200 \\ -400 \end{bmatrix} = 10^{-5} \begin{bmatrix} 0.5658 & -0.0120 \\ -0.0120 & 2.9907 \end{bmatrix} \begin{bmatrix} +200 \\ -400 \end{bmatrix}$$

$$= 10^{-3} \begin{bmatrix} 1.18 \\ -11.98 \end{bmatrix} \text{ m}$$

(b)

$$S_{cd} = \frac{2 \times 10^5}{2.54} \begin{bmatrix} -1 & 0 & 1 & 0 \end{bmatrix} \left(10^{-3} \begin{bmatrix} 1.18 \\ -11.98 \\ -1.18 \\ -11.98 \end{bmatrix} \right) = -186 \text{ kN (c)}$$

$$S_{ac} = \frac{2 \times 10^5}{6.93} \begin{bmatrix} -1 & 0 & 1 & 0 \end{bmatrix} \begin{bmatrix} 0.50 & 0.87 & 0.00 & 0.00 \\ -0.87 & 0.50 & 0.00 & 0.00 \\ 0.00 & 0.00 & 0.50 & 0.87 \\ 0.00 & 0.00 & -0.87 & 0.50 \end{bmatrix} \left(10^{-3} \begin{bmatrix} 0.00 \\ 0.00 \\ 1.18 \\ -11.98 \end{bmatrix} \right)$$

$$= -282 \text{ kN (c)}$$

$$S_{ad} = \frac{2 \times 10^5}{8.48}[-1 \quad 0 \quad 1 \quad 0]\begin{bmatrix} 0.71 & 0.71 & 0.00 & 0.00 \\ -0.71 & 0.71 & 0.00 & 0.00 \\ 0.00 & 0.00 & 0.71 & 0.71 \\ 0.00 & 0.00 & -0.71 & 0.71 \end{bmatrix}\left(10^{-3}\begin{bmatrix} 0.00 \\ 0.00 \\ -1.18 \\ -11.98 \end{bmatrix}\right)$$

$$= -219 \text{ kN (c)}$$

$$S_{bd} = S_{ac} \qquad S_{bc} = S_{ad}$$

DISCUSSION Because of the symmetry of the structure and the loading, the calculations can be minimized. Thus, the member forces in bd and bc can be inferred from their symmetric counterparts. The stiffness matrix $\mathbf{K}_{ff}$ was also reduced in size by using the symmetric behavior. Noting that $U_c = -U_d$ implies that the first column can be replaced by the difference between the first and the third columns. Similarly, since $V_c = V_d$, the second column can be replaced by the sum of the second and fourth columns. Rows (equations) 3 and 4 were accordingly omitted. There is an option of combining the rows in the same manner as the columns (this amounts to adding or subtracting the corresponding equations).

PROBLEMS

Use the stiffness method for all problems.

16.1 Describe how and at what point in the analysis the following concepts are incorporated into the theory of matrix structural analysis: (a) equilibrium, (b) compatibility, (c) stress-strain relations, and (d) rigid-body translations and rotations.

16.2 Describe four properties of the member stiffness matrix for a truss member and explain the basis for each property.

16.3 Write the stiffness matrix for a truss member if (a) $\phi = 0°$, (b) $\phi = 45°$, (c) $\phi = 90°$, (d) $\phi = 180°$ (see Fig. 16.2).

16.4 This strut with $A = 10^{-3}$ m^2 and $E = 200$ GPa is supported with $U_a = U_b = V_b = 0$. Calculate V_a if it is loaded with $Y_a = -120$ kN.

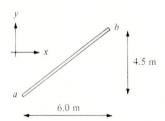

4.5 m

6.0 m

Figure P16.4

16.5 The strut of Prob. 16.4 has $V_a = 12$ mm; calculate the force Y_a.

16.6 The strut of Prob. 16.4 is constrained with $U_a = V_a = V_b = 0$. Calculate U_b if it is loaded with $X_b = 160$ kN.

16.7 The strut of Prob. 16.4 is constrained so that $U_a = V_a = 0$; $U_b = -20$ mm and $V_b = -16$ mm. Calculate the matrix $\mathbf{X}_s = \{X_b \quad Y_b\}$.

16.8 This strut with $A = 3.00$ in^2 and $E = 29,000$ kips/in^2 is supported with $U_a = V_a = V_b = 0$. Calculate U_b if it is loaded with $X_b = 75$ kips.

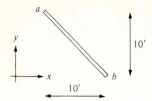

Figure P16.8

16.9 The strut of Prob. 16.8 has the imposed displacement $U_b = 0.30$ in. Calculate the force X_b.

16.10 The strut of Prob. 16.8 is constrained so that $U_a = V_a = 0$; $U_b = -V_b = 0.2828$ in. Calculate the matrix $\mathbf{X}_s = \{X_b \quad Y_b\}$.

16.11 The strut of Prob. 16.8 has the imposed displacements $U_a = -V_a = 0.50$ in and $U_b = -V_b = +0.25$ in. Calculate the matrix $\mathbf{X}_s = \{X_a \quad Y_a \quad X_b \quad Y_b\}$.

16.12 What is the size of the merged stiffness matrix for these three unconstrained trusses?

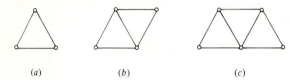

(a)   (b)   (c)

Figure P16.12

16.13 What are the sizes of the following matrices: (a) $\mathbf{K}$; (b) the partitions $\mathbf{K}_{ff}$, $\mathbf{K}_{fs}$, $\mathbf{K}_{sf}$, and $\mathbf{K}_{ss}$; (c) the partitions $\mathbf{X}_f$ and $\mathbf{X}_s$; and (d) the partitions $\mathbf{U}_f$ and $\mathbf{U}_s$?

 **Figure P16.13**

16.14 What are the sizes of the following matrices: $\mathbf{K}$; the partitions $\mathbf{K}_{ff}$, $\mathbf{K}_{fs}$, $\mathbf{K}_{sf}$, and $\mathbf{K}_{ss}$; the partitions $\mathbf{X}_f$ and $\mathbf{X}_s$; and the partitions $\mathbf{U}_f$ and $\mathbf{U}_s$?

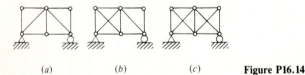

(a)   (b)   (c)   **Figure P16.14**

16.15 Label the nodes for this truss, and indicate the sizes of the following matrices: (a) $\mathbf{K}$; (b) the partitions $\mathbf{K}_{ff}$, $\mathbf{K}_{fs}$, $\mathbf{K}_{sf}$, and $\mathbf{K}_{ss}$; (c) the partitions of $\mathbf{X}_f$ and $\mathbf{X}_s$; and (d) the partitions $\mathbf{U}_f$ and $\mathbf{U}_s$. Write out the matrix $\mathbf{X}_f$.

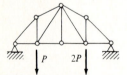

Figure P16.15

16.16 Indicate the sizes of the following matrices: (a) $\mathbf{K}$; (b) the partitions $\mathbf{K}_{ff}$, $\mathbf{K}_{fs}$, $\mathbf{K}_{sf}$, and $\mathbf{K}_{ss}$; (c) the partitions $\mathbf{X}_f$ and $\mathbf{X}_s$; and (d) the partitions $\mathbf{U}_f$ and $\mathbf{U}_s$. Label the nodes and indicate their order in $\mathbf{K}_{ff}$ that will result in the most efficient solution for $\mathbf{U}_f$. Discuss the criteria for an optimal nodal labeling and ordering of degrees of freedom. Write the matrix $\mathbf{X}_f$.

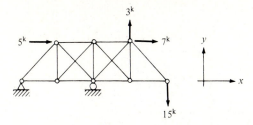

Figure P16.16

16.17 The stiffness matrix for member bd is merged into what locations of $\mathbf{K}$? When merging is completed, what location(s) of $\mathbf{K}$ will contain all zeros? What location(s) of $\mathbf{K}$ constitute $\mathbf{K}_{ff}$?

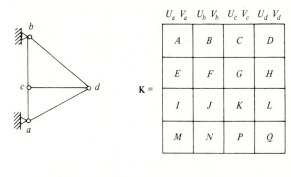

(a)   (b)   **Figure P16.17**

16.18 The stiffness matrix for member ad is merged into what locations of $\mathbf{K}$? When merging is completed, what location(s) of $\mathbf{K}$ will contain all zeros? What location(s) of $\mathbf{K}$ constitute $\mathbf{K}_{ff}$? $\mathbf{K}$ is shown in Fig. P16.17b.

16.19 Label the nodes for truss a and indicate the sizes of the following matrices: $\mathbf{K}$; the partitions $\mathbf{K}_{ff}$, $\mathbf{K}_{fs}$, $\mathbf{K}_{sf}$, and $\mathbf{K}_{ss}$; the partitions $\mathbf{X}_f$ and $\mathbf{X}_s$; and the partitions $\mathbf{U}_f$ and $\mathbf{U}_s$. What part(s) of the analysis used for truss a changes if trusses b and c are to be analyzed?

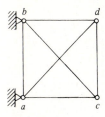

Figure P16.18

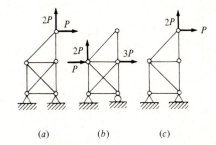

(a)   (b)   (c)

Figure P16.19

16.20 For this truss $A_{ab} = A_{bc} = A$ and $A_{ac} = 2\sqrt{2}\,A$. Calculate the nodal displacements and member forces. If this truss is modified so that $U_b = V_b = U_c = 0$ with the other degrees of freedom unconstrained, what are the nodal displacements and member forces for $X_a = 2P$ and $Y_a = -P$?

16.21 For this truss $A_{ab} = A_{dc} = A_{ad} = A_{bc} = A_{ac} = A$. Calculate the nodal displacements and member forces.

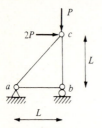

Figure P16.20

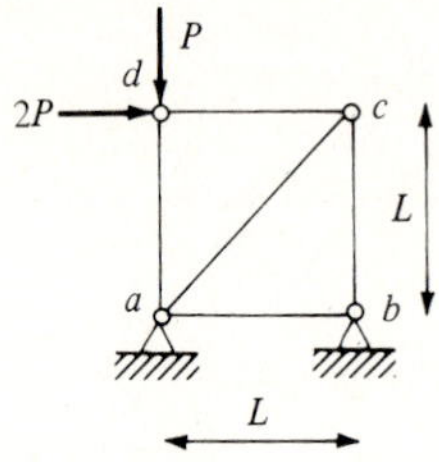

Figure P16.21

16.22 For this truss $A_{ab} = A_{ac} = A_{bc} = A$. Calculate the nodal displacements and member forces. If this truss is modified so that $A_{bc} = \sqrt{2}\,A$ and $U_a = V_b = U_c = V_c = 0$ with the other degrees of freedom unconstrained, what are the nodal displacements and member forces for $Y_a = -P$ and $X_b = 2P$?

16.23 For this truss $A_{ab} = A_{cd} = A_{ad} = A_{bc} = A$ and $A_{bd} = \sqrt{2}\,A$. Calculate the nodal displacements and member forces.

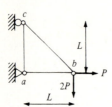

Figure P16.22

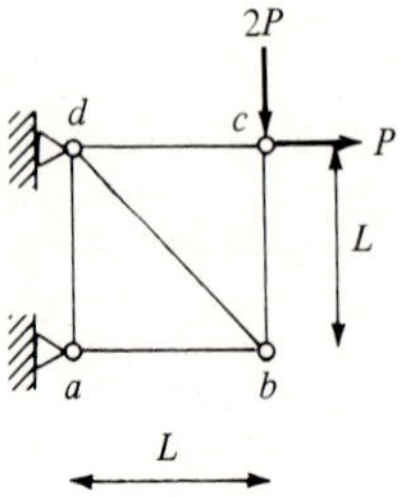

Figure P16.23

16.24 The three-member truss shown in Example 16.2 is constrained so that $U_a = V_a = U_b = U_c = 0$. Calculate the nodal displacements and member forces if the applied loads are $Y_b = 72$ kN and $Y_c = 90$ kN.

16.25 The three-member truss shown in Example 16.2 is constrained so that $V_a = V_b = U_c = 0$. Calculate the nodal displacements and member forces if the only applied load is $X_b = 27$ kN.

16.26 For all members $AE/L = 20,000$ kN/m. Calculate the nodal displacements and member forces.

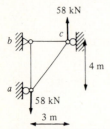

Figure P16.26

16.27 The truss of Prob. 16.26 is modified so that $U_a = V_a = V_c = 0$ with all other degrees of freedom unconstrained. What are the nodal displacements and member forces if the applied loads are $X_b = 45$ kN and $Y_b = -60$ kN?

16.28 The truss of Prob. 16.26 is modified so that $U_a = U_b = V_b = 0$ with all other degrees of freedom unconstrained. What are the nodal displacements and member forces if the applied loads are $X_c = 99$ kN and $Y_c = 132$ kN?

16.29 The truss shown in Example 16.5 is modified by the removal of member cb. Calculate the nodal displacements and member forces.

16.30 Calculate the nodal displacements and member forces for the truss of Example 16.12 if:

(a) It is modified so that $A_{ab} = 10 \times 10^{-4}$ m² and $A_{ad} = 20 \times 10^{-4}$ m².

(b) The loads are removed, and support b moves 5 mm downward.

(c) The loads are removed, and members ab and ac have temperature changes of $+40$ and $-20°C$, respectively ($\alpha = 1.2 \times 10^{-5}$ mm/mm · °C).

(d) The loads are removed and members ac and ae have been fabricated with length errors of -5 and $+5$ mm, respectively.

16.31 AE/L = const for all members. Calculate the displacements and member forces.

16.32 This truss has $A_{ab} = 30 \times 10^{-4}$ m², $A_{ac} = 20 \times 10^{-4}$ m², $A_{ad} = 15 \times 10^{-4}$ m², and $E = 200$ GPa. Calculate the nodal displacements and member forces:

(a) For the illustrated loads.

(b) The loads are removed, and support b moves 7.07 mm downward.

(c) The loads are removed, and member ab increases in temperature by 45°C ($\alpha = 1.2 \times 10^{-5}$ mm/mm · °C).

(d) The loads are removed and members ab and ad have been fabricated with length errors of $+5$ and -5 mm, respectively.

(e) If all member areas in part (a) are doubled, calculate the displacements, reactions, and member forces.

(f) If only member ab has its area doubled in part (a), calculate the displacements, reactions, and member forces.

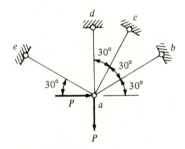

Figure P16.31

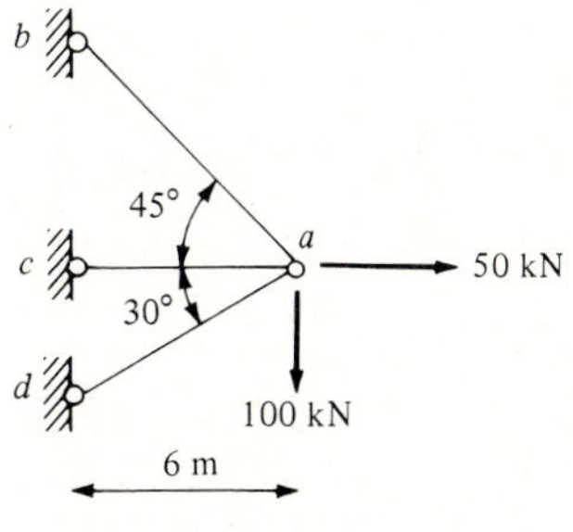

Figure P16.32

16.33 Calculate the nodal displacements and member forces. For all members $A = 4$ in² and $E = 29 \times 10^3$ kips/in².

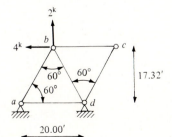

Figure P16.33

16.34 Calculate the nodal displacements and member forces; $A_{ad} = A_{bc} = 8$ in^2, $A_{ab} = A_{cd} = 6$ in^2, $A_{ac} = A_{bd} = 10$ in^2, and $E = 29 \times 10^3$ kips/in^2.

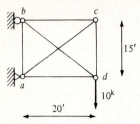

Figure P16.34

16.35 For all chord members $A = 1.00$ in^2, for all vertical members $A = 0.75$ in^2, and for all diagonals $A = 1.25$ in^2, $E = 29 \times 10^3$ kips/in^2. Calculate the nodal displacements and member forces.

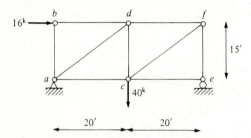

Figure P16.35

16.36 The truss of Prob. 16.35 is modified by adding a member from node b to c with an area of 1.25 in^2. Calculate the displacements and member forces.

16.37 The truss of Prob. 16.35 is modified so that $U_a = V_a = U_e = V_e = U_f = V_f = 0$. Calculate the displacements and member forces for the loads shown.

In Probs. 16.38 to 16.44 calculate the displacements and member forces for the original problem.

16.38 Problem 10.3   **16.39** Problem 11.14
16.40 Problem 10.4   **16.41** Problem 10.6
16.42 Problem 11.11   **16.43** Problem 10.7; all members have
16.44 Problem 11.12 $A = 2000$ mm^2 and $E = 200$ GPa.

ADDITIONAL READING

Gerstle, K. H.: *Basic Structural Analysis,* chap. 13, Prentice-Hall, Englewood Cliffs, N.J., 1974.

Gutkowski, R. M.: *Structures: Fundamental Theory and Behavior,* chap. 7, Van Nostrand Reinhold, New York, 1981.

Harrison, H. B.: *Structural Analysis and Design, pt. 1,* chap. 6, Pergamon, New York, 1980.

Kardestuncer, H.: *Elementary Matrix Analysis of Structures,* chap. 5, McGraw-Hill, New York, 1974.

Laursen, H. I.: *Structural Analysis,* chap. 14, McGraw-Hill, New York, 1978.

McGuire, W., and R. H. Gallagher: *Matrix Structural Analysis,* chaps. 2 and 3, Wiley, New York, 1979.

Martin, H. C.: *Introduction to Matrix Methods of Structural Analysis,* chap. 3, McGraw-Hill, New York, 1966.

Meek, J. L.: *Matrix Structural Analysis,* chap. 6, McGraw-Hill, New York, 1971.

Weaver, W., Jr., and J. M. Gere: *Matrix Analysis of Framed Structures,* 2d ed., chap. 4, Van Nostrand, New York, 1980.

West, H. H.: *Analysis of Structures,* chap. 19, Wiley, New York, 1980.

Willems, N., and W. M. Lucas, Jr.: *Structural Analysis for Engineers,* chap. 9, McGraw-Hill, New York, 1978.

SEVENTEEN

ANALYSIS OF BEAMS USING
THE STIFFNESS METHOD

17.1 INTRODUCTION

The concepts of the stiffness method were described and used to analyze spring assemblages and trusses in Chaps. 15 and 16. These same precepts can also be applied to structures when the state of stress is not simply uniaxial. In this chapter the analysis will focus on structures composed of linear elastic beam segments in which shear and bending moments are the cardinal effects. Figure 17.1 shows a beam structure for which matrix structural analysis is a logical procedure.

The basic governing equations are satisfied for beam structures in a manner similar to that used for spring assemblages and trusses. The material constitutive relations are fulfilled within the individual members, as are the equilibrium and compatibility equations. Merging individual beam element stiffness matrices into the structural stiffness matrix assures that the equilibrium and compatibility conditions will be met at the nodal points.

Beams contain their own unique set of internal forces (shear and bending moment), requiring these generalized forces to be functionally related to rotations and displacements at the nodes through the member stiffness matrix. In addition, the stress matrix that relates the nodal displacements to the stresses within the beam element must be derived. Using these two basic matrices for beams, we can apply the stiffness method to beam structures. Note that for beams it is not necessary to transform quantities from one coordinate system to another since the member and structural coordinates coincide.

A unique problem emerges for the analysis of beam structures: it is not always apparent where the nodal points are located and exactly what constitutes a member. The nodes were well defined for trusses by the constructed panel points, and the members connected these points; but with various factors obscuring the location of the nodes for beams (the presence of loads intervening between support points, etc.), the nodal points must be located by considering the behavioral properties of the element. This process, referred to as *idealization,* must be done carefully to obtain solutions that closely approximate the behavior of the actual beam.

Figure 17.1 Steel framing. (*Setter Leach & Lindstrom; Minneapolis.*)

17.2 THE BEAM MEMBER

It is assumed that the basic beam member has a uniform cross section over its entire length (i.e., it is prismatic) and is constructed from material that is homogeneous, isotropic, and linearly elastic. Furthermore, displacements are assumed to be so small that no geometric nonlinearities are present and conventional beam theory can be used. It will be also assumed that the effects of axial deformation are negligible. This last assumption will be examined in Chap. 18, where the general frame member incorporates axial forces in addition to shears and bending moments. The uniform beam member with moment of inertia I and length L is shown, together with its member coordinate system, in Fig. 17.2. The bending moments $\overline{m}_i$ and $\overline{m}_j$ and the shears $\overline{y}_i$ and $\overline{y}_j$ at the ends of the member are defined as positive if they act in the positive coordinate direction as shown in the figure. In the subsequent discussion, bending moments will be designated as a type of *generalized force*; thus, the shears and bending moments will both be referred to as forces. Analogously, the rotations $\overline{\theta}_i$ and $\overline{\theta}_j$ associated with the bending moments will be designated, along with the displacements $\overline{v}_i$ and $\overline{v}_j$, as *generalized displacements*. The notation used in this chapter is similar to that used previously in that the forces and displacements associated with the member are denoted

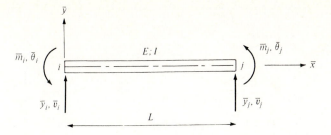

Figure 17.2 The beam member with member coordinates plus generalized forces and displacements.

by lowercase letters and all quantities referred to the member coordinate system will be distinguished with a bar over the symbol.

The stiffness matrix relating the nodal forces and their corresponding displacements is obtained by deforming the beam element into four distinct deflection cases, each corresponding to a nonzero value for one of the four independent displacement components $\bar{v}_i$, $\bar{\theta}_i$, $\bar{v}_j$, and $\bar{\theta}_j$. The nodal forces required to maintain each of these deflection configurations correspond to the various stiffness coefficients of the member stiffness matrix. This procedure is similar to that used for the truss member.

First Displacement Case: $\bar{v}_i \neq 0; \bar{\theta}_i = \bar{v}_j = \bar{\theta}_j = 0$

Figure 17.3a shows the beam member deformed into this configuration. Note that this corresponds to a beam that is fixed against rotation at both ends and given an upward translation at the left end. The methods presented in Part Four can be used to obtain the bending moments and the shear forces at the two ends. This deformed shape corresponds to that used in the moment-distribution method to obtain the fixed-end moments for a member subjected to joint translation. The results of the analysis give the forces

$$\bar{y}_i = -\bar{y}_j = \frac{12EI}{L^3} \bar{v}_i \tag{17.1}$$

and

$$\bar{m}_i = \bar{m}_j = \frac{6EI}{L^2} \bar{v}_i \tag{17.2}$$

Second Displacement Case: $\bar{\theta}_i \neq 0; \bar{v}_i = \bar{v}_j = \bar{\theta}_j = 0$

This deformed shape of the beam member (Fig. 17.3b) corresponds to having the beam supported on a roller at node i and fixed at node j. Again using the methods of analysis of Part Four, we find the forces at nodes i and j to be

$$\bar{y}_i = -\bar{y}_j = \frac{6EI}{L^2} \bar{\theta}_i \tag{17.3}$$

and

$$\bar{m}_i = 2\bar{m}_j = \frac{4EI}{L} \bar{\theta}_i \tag{17.4}$$

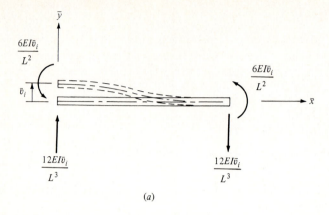

(a)

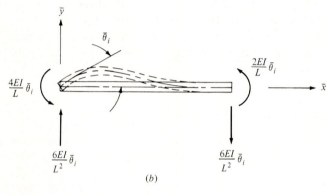

(b)

Figure 17.3 Generalized forces acting on the deformed beam member: (a) $\bar{v}_i \neq 0$ and $\bar{\theta}_i = \bar{v}_j = \bar{\theta}_j = 0$; (b) $\bar{\theta}_i \neq 0$ and $\bar{v}_i = \bar{v}_j = \bar{\theta}_j = 0$.

This deflection state corresponds to the configuration investigated in the moment-distribution method to obtain the rotational stiffness and the carryover factor.

Third Displacement Case: $\bar{v}_j \neq 0; \bar{v}_i = \bar{\theta}_i = \bar{\theta}_j = 0$

Although this deformation of the beam is similar to that for the first displacement case, here the right end of the fixed-fixed beam is given an upward displacement. Therefore the forces can be obtained by extrapolating from the results of the first deflection condition. The nodal forces for the third displacement case are

$$\bar{y}_i = -\bar{y}_j = -\frac{12EI}{L^3}\bar{v}_j \tag{17.5}$$

and
$$\bar{m}_i = \bar{m}_j = -\frac{6EI}{L^2}\bar{v}_j \tag{17.6}$$

Fourth Displacement Case: $\bar{\theta}_j \neq 0; \bar{v}_i = \bar{\theta}_i = \bar{v}_j = 0$

To accomplish the deformations for this condition the beam member is fixed at the left
end and is supported on a roller at the right end. The node j is given a rotation in the
positive coordinate direction, making this case similar to the second displacement case.
The generalized nodal forces can therefore be observed to be

$$\bar{y}_i = -\bar{y}_j = \frac{6EI}{L^2}\bar{\theta}_j \tag{17.7}$$

and
$$2\bar{m}_i = \bar{m}_j = \frac{4EI}{L}\bar{\theta}_j \tag{17.8}$$

General Displacement Case: $\bar{v}_i \neq 0; \bar{\theta}_i \neq 0; \bar{v}_j \neq 0; \bar{\theta}_j \neq 0$

If each of the generalized displacements at the two ends of the beam member is
simultaneously given nonzero values, the forces that maintain equilibrium for a linear
elastic material can be obtained by superposing the four individual displacement cases
previously discussed and analyzed. Thus adding the forces defined in Eqs. (17.1)
through (17.8) gives the general force-deflection relationships for the beam member

$$\bar{y}_i = \frac{EI}{L}\left(\frac{12}{L^2}\bar{v}_i + \frac{6}{L}\bar{\theta}_i - \frac{12}{L^2}\bar{v}_j + \frac{6}{L}\bar{\theta}_j\right)$$

$$\bar{m}_i = \frac{EI}{L}\left(\frac{6}{L}\bar{v}_i + 4\bar{\theta}_i - \frac{6}{L}\bar{v}_j + 2\bar{\theta}_j\right)$$

$$\bar{y}_j = \frac{EI}{L}\left(-\frac{12}{L^2}\bar{v}_i - \frac{6}{L}\bar{\theta}_i + \frac{12}{L^2}\bar{v}_j - \frac{6}{L}\bar{\theta}_j\right) \tag{17.9}$$

$$\bar{m}_j = \frac{EI}{L}\left(\frac{6}{L}\bar{v}_i + 2\bar{\theta}_i - \frac{6}{L}\bar{v}_j + 4\bar{\theta}_j\right)$$

These force-displacement equations can be cast into matrix form

$$\underbrace{\begin{bmatrix}\bar{y}_i \\ \bar{m}_i \\ \bar{y}_j \\ \bar{m}_j\end{bmatrix}}_{\bar{\mathbf{x}}} = \frac{EI}{L}\underbrace{\begin{bmatrix}\dfrac{12}{L^2} & \dfrac{6}{L} & -\dfrac{12}{L^2} & \dfrac{6}{L} \\ \dfrac{6}{L} & 4 & -\dfrac{6}{L} & 2 \\ -\dfrac{12}{L^2} & -\dfrac{6}{L} & \dfrac{12}{L^2} & -\dfrac{6}{L} \\ \dfrac{6}{L} & 2 & -\dfrac{6}{L} & 4\end{bmatrix}}_{\bar{\mathbf{k}}}\underbrace{\begin{bmatrix}\bar{v}_i \\ \bar{\theta}_i \\ \bar{v}_j \\ \bar{\theta}_j\end{bmatrix}}_{\bar{\mathbf{u}}} \tag{17.10}$$

or
$$\bar{\mathbf{x}} = \bar{\mathbf{k}}\,\bar{\mathbf{u}} \tag{17.10a}$$

where $\bar{\mathbf{x}}$ and $\bar{\mathbf{u}}$ are column matrices of nodal forces and nodal displacements, re-spectively, and $\bar{\mathbf{k}}$ is the member stiffness matrix (all three matrices are expressed in member coordinates).

The stiffness matrix in Eq. (17.10) is symmetric, like the stiffness matrices obtained earlier for the spring and truss members. This property is a consequence of the Maxwell-Betti reciprocal theorem (Chap. 10). It can also be observed that the stiffness matrix is singular, i.e., it has no inverse. This stems from the fact that the beam member is not supported against rigid-body motion; therefore, the deflections resulting from an arbitrary set of applied nodal forces will be undetermined. This problem will be avoided when the member is used in conjunction with others to describe a properly supported structure. The columns of the present stiffness matrix do not add to zero, unlike those of member stiffness matrices discussed previously. A given column of the matrix represents the forces necessary to obtain a deformation state corresponding to a unit value for the generalized displacement associated with that column; therefore, any attempt to sum the columns will involve the addition of both forces and moments. This simplistic check on equilibrium will not work in this case because various types of generalized forces are present.

Equation (17.9) can be written in an alternate form, thus for the first two equations,

$$\bar{y}_i = \frac{EI}{L^3}(12\bar{v}_i + 6\bar{\theta}_i L - 12\bar{v}_j + 6\bar{\theta}_j L)$$

$$\frac{\bar{m}_i}{L} = \frac{EI}{L^3}(6\bar{v}_i + 4\bar{\theta}_i L - 6\bar{v}_j + 2\bar{\theta}_j L)$$

The last two equations in Eq. (17.9) will have a similar form, and the resulting four equations can be written in matrix form as

$$\begin{bmatrix} \bar{y}_i \\ \bar{m}_i/L \\ \bar{y}_j \\ \bar{m}_j/L \end{bmatrix} = \frac{EI}{L^3} \begin{bmatrix} 12 & 6 & -12 & 6 \\ 6 & 4 & -6 & 2 \\ -12 & -6 & 12 & -6 \\ 6 & 2 & -6 & 4 \end{bmatrix} \begin{bmatrix} \bar{v}_i \\ \bar{\theta}_i L \\ \bar{v}_j \\ \bar{\theta}_j L \end{bmatrix} \tag{17.11}$$

which defines the alternate form of the stiffness matrix for the beam member. Since the matrix contains only pure numbers and has the scalar multiplier EI/L^3, this form is convenient for use whenever the individual members all have the same cross section and length.

17.3 STRUCTURAL IDEALIZATION AND THE ASSEMBLAGE

The stiffness matrix for the beam member has been derived assuming that: (a) the member is uniform (prismatic) along its length; and (b) that the shear is constant and the bending moment varies linearly between nodal points. A perusal of the equilibrium solutions in Sec. 17.2 shows that the second assumption was implicitly incorporated into the derivation, as can also be shown using basic considerations from Chap. 4.

A free-body diagram of a beam member with the internal shear and bending moment at a generic section is shown in Fig. 17.4. The equilibrium equations for the shear and moment, respectively, are

$$V = \bar{y}_i \qquad \text{and} \qquad M = \bar{y}_i \bar{x} - \bar{m}_i$$

i.e., the member has a constant shear and the moment varies linearly along its length. These observations are also consistent with the differential equations of equilibrium for a beam shown in Eqs. (4.4) and (4.3), stating respectively, that

$$\frac{dM}{dx} = V \qquad \text{and} \qquad \frac{dV}{dx} = q$$

Since $q = 0$ between nodal points, it follows from the integration of the second differential equation that the shear is a constant. Furthermore, by substituting this constant value of shear into the first differential equation and integrating we obtain a linear function for the moment.

These two basic assumptions must be kept in mind when designating nodal points and individual members for a structure. For example, it is possible to analyze the continuous beam shown in Fig. 17.5 using the beam member even though the structure does not have a uniform cross section and is loaded at various points along its length. If the points a to j are each regarded as a node, each individual member conforms to the assumptions, i.e., the shear is constant, the moment varies linearly, and the member is prismatic between nodal points. Thus, the structure is properly *idealized*, which means that the model accurately characterizes the actual behavior of the structure. However, if a member had been defined as extending from a to c, it would not have met the assumptions of constant shear and linearly varying moment. The assumption of a uniform member cross section would have been violated if a single member had been designated between nodes d and f.

Merging member stiffness matrices **k** to obtain the structural stiffness matrix **K** is done for beams in a manner similar to that for spring assemblages and trusses. This interaction of individual members is accomplished by enforcing nodal equilibrium for all the generalized forces, i.e., for both linearly directed (direct) forces and moments.

For beam structures the member and structural coordinates usually coincide since the structural coordinate system is typically selected with the x axis aligned with the beam axis. Since the member stiffness matrices expressed in structural and member coordinates are therefore identical ($\bar{\mathbf{k}} = \mathbf{k}$), the coordinate transformations similar to those used to analyze truss structures are not necessary.

Examples 17.1 and 17.2 illustrate the idealization and merging processes for beams loaded at discrete points.

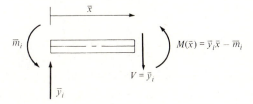

$M(\bar{x}) = \bar{y}_i \bar{x} - \bar{m}_i$

$V = \bar{y}_i$

Figure 17.4 Free-body diagram of a beam member showing the internal shear and bending moment.

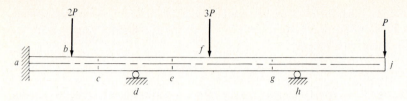

Figure 17.5 Idealization of a continuous beam; $I_{ac} = I_1$, $I_{ce} = I_2$, $I_{eg} = I_3$, $I_{gj} = I_4$.

Example 17.1 Obtain the structural stiffness matrix for this two-member beam structure; $I_{ab} = 200 \times 10^{-6}$ m⁴, $I_{bc} = 300 \times 10^{-6}$ m⁴, and $E = 200$ GPa.

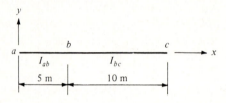

Figure E17.1

SOLUTION For both members, $\bar{\mathbf{k}} = \mathbf{k}$. From Eq. (17.10)

$$\mathbf{k}_{ab} = 8000 \begin{bmatrix} 0.48 & 1.20 & -0.48 & 1.20 \\ 1.20 & 4.00 & -1.20 & 2.00 \\ -0.48 & -1.20 & 0.48 & -1.20 \\ 1.20 & 2.00 & -1.20 & 4.00 \end{bmatrix}$$

$$\mathbf{k}_{bc} = 6000 \begin{bmatrix} 0.12 & 0.60 & -0.12 & 0.60 \\ 0.60 & 4.00 & -0.60 & 2.00 \\ -0.12 & -0.60 & 0.12 & -0.60 \\ 0.60 & 2.00 & -0.60 & 4.00 \end{bmatrix}$$

$$\mathbf{K} = 1000 \begin{bmatrix} V_a & \Theta_a & V_b & \Theta_b & V_c & \Theta_c \\ 3.84 & 9.60 & -3.84 & 9.60 & 0.00 & 0.00 \\ 9.60 & 32.00 & -9.60 & 16.00 & 0.00 & 0.00 \\ -3.84 & -9.60 & 4.56 & -6.00 & -0.72 & 3.60 \\ 9.60 & 16.00 & -6.00 & 56.00 & -3.60 & 12.00 \\ 0.00 & 0.00 & -0.72 & -3.60 & 0.72 & -3.60 \\ 0.00 & 0.00 & 3.60 & 12.00 & -3.60 & 24.00 \end{bmatrix}$$

Thus $\mathbf{X} = \mathbf{KU}$

where

$$\mathbf{U} = \begin{bmatrix} V_a \\ \Theta_a \\ V_b \\ \Theta_b \\ V_c \\ \Theta_c \end{bmatrix} \qquad \mathbf{X} = \begin{bmatrix} Y_a \\ M_a \\ Y_b \\ M_b \\ Y_c \\ M_c \end{bmatrix}$$

DISCUSSION The merge process requires enforcing equilibrium for the forces in the y direction and the moments. Since the member coordinates for ab and bc both coincide with the structural coordinates, no coordinate transformations are required. The structural stiffness matrix is obtained by a simple overlay of $\mathbf{k}_{ab}$ and $\mathbf{k}_{bc}$ similar to that used for two springs in series. Note that merging enforces continuity of deflection and slope at nodes.

Example 17.2 Obtain the structural stiffness matrix for this three-member beam structure; $I = 1034$ in^4, $L = 15$ ft, and $E = 29 \times 10^3$ kips/in^2.

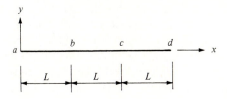

Figure E17.2

SOLUTION For all members $\mathbf{k} = \overline{\mathbf{k}}$. Since all three beams have E, I, and L the same, Eq. (17.11) will be used.

$$\mathbf{k}_{ab} = \mathbf{k}_{bc} = \mathbf{k}_{cd} = (5.144 \text{ kips/in}) \begin{bmatrix} 12 & 6 & -12 & 6 \\ 6 & 4 & -6 & 2 \\ -12 & -6 & 12 & -6 \\ 6 & 2 & -6 & 4 \end{bmatrix}$$

$$\mathbf{K} = 5.144 \begin{bmatrix} 12 & 6 & -12 & 6 & 0 & 0 & 0 & 0 \\ 6 & 4 & -6 & 2 & 0 & 0 & 0 & 0 \\ -12 & -6 & 24 & 0 & -12 & 6 & 0 & 0 \\ 6 & 2 & 0 & 8 & -6 & 2 & 0 & 0 \\ 0 & 0 & -12 & -6 & 24 & 0 & -12 & 6 \\ 0 & 0 & 6 & 2 & 0 & 8 & -6 & 2 \\ 0 & 0 & 0 & 0 & -12 & -6 & 12 & -6 \\ 0 & 0 & 0 & 0 & 6 & 2 & -6 & 4 \end{bmatrix}$$

Thus $\mathbf{X} = \mathbf{KU}$

where

$$\mathbf{U} = \begin{bmatrix} V_a \\ \Theta_a L \\ V_b \\ \Theta_b L \\ V_c \\ \Theta_c L \\ V_d \\ \Theta_d L \end{bmatrix} \qquad \mathbf{X} = \begin{bmatrix} Y_a \\ M_a/L \\ Y_b \\ M_b/L \\ Y_c \\ M_c/L \\ Y_d \\ M_a/L \end{bmatrix}$$

17.4 MEMBER FORCES

The distribution of the internal shear and bending moment along the length of the beam members is required for design purposes and can be calculated using an expression of the form shown in Eq. (16.38a) modified slightly as

$$\mathbf{s}_{ij} = \mathbf{s}\overline{\mathbf{u}}_{ij} \qquad (17.12)$$

where $\mathbf{s}_{ij}$ is a column matrix (in the case of a beam member). Recall from Sec. 17.3 that the shear in the member is constant and the bending moment varies linearly along the length. Hence, it is usually most convenient to calculate: (a) the shear; and (b) the moment at the two endpoints of the beam element, rather than expressing it as a function of the coordinate $\overline{x}$. Thus, by selecting the first, second, and fourth equations from the force-displacement equations (17.9) we obtain the following matrix equation for calculating the internal generalized forces in each individual beam member

$$
\underbrace{\begin{bmatrix} \overline{y}_i \\ \overline{m}_i \\ \overline{m}_j \end{bmatrix}}_{\mathbf{s}_{ij}} = \frac{EI}{L} \underbrace{\begin{bmatrix} \dfrac{12}{L^2} & \dfrac{6}{L} & -\dfrac{12}{L^2} & \dfrac{6}{L} \\ \dfrac{6}{L} & 4 & -\dfrac{6}{L} & 2 \\ \dfrac{6}{L} & 2 & -\dfrac{6}{L} & 4 \end{bmatrix}}_{\mathbf{s}} \underbrace{\begin{bmatrix} \overline{v}_i \\ \overline{\theta}_i \\ \overline{v}_j \\ \overline{\theta}_j \end{bmatrix}}_{\overline{\mathbf{u}}_{ij}} \qquad (17.13)
$$

where the forces for the beam member between nodes i and j are described by the column matrix $\mathbf{s}_{ij}$ and $\mathbf{s}$ is the so-called *stress matrix* for the member. The member forces obtained from Eq. (17.13) are specified with respect to the member coordinates as shown in Fig. 17.2, and the deformation sign convention (Sec. 4.3 and Fig. 4.16) must be used for plotting shear and moment diagrams.

The stiffness matrices obtained in Examples 17.1 and 17.2 have been used for several typical beam structures. These calculations, shown in Examples 17.3 to 17.6, illustrate the usefulness of the stiffness method for calculating deflections and member forces for various beam structures.

Example 17.3 Calculate the displacements and member forces for the beam of Example 17.1 if it is supported as illustrated.

SOLUTION See Example 17.1 for dimensions, etc. From Example 17.1 (using rows and columns 1, 2, and 4 of $\mathbf{K}$)

$$
\mathbf{K}_{ff} = 1000 \begin{bmatrix} 3.84 & 9.60 & 9.60 \\ 9.60 & 32.00 & 16.00 \\ 9.60 & 16.00 & 56.00 \end{bmatrix}
$$

Thus

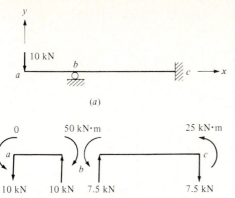

Figure E17.3 (a) The beam with applied loading; (b) free-body diagrams of members.

$$\begin{bmatrix} V_a \\ \Theta_a \\ \Theta_b \end{bmatrix} = 10^{-3}\begin{bmatrix} 3.84 & 9.60 & 9.60 \\ 9.60 & 32.00 & 16.00 \\ 9.60 & 16.00 & 56.00 \end{bmatrix}^{-1}\begin{bmatrix} -10 \\ 0 \\ 0 \end{bmatrix}$$

$$= (125 \times 10^{-6})\begin{bmatrix} 16.67 & -4.17 & -1.67 \\ -4.17 & 1.33 & 0.33 \\ -1.67 & 0.33 & 0.33 \end{bmatrix}\begin{bmatrix} -10 \\ 0 \\ 0 \end{bmatrix} = 10^{-3}\begin{bmatrix} -20.84 \text{ m} \\ 5.21 \text{ rad} \\ 2.09 \text{ rad} \end{bmatrix}$$

Use Eq. (17.13) to calculate the member forces. For member ab

$$\begin{bmatrix} \bar{y}_a \\ \bar{m}_a \\ \bar{m}_b \end{bmatrix}_{ab} = 8000\begin{bmatrix} 0.48 & 1.20 & -0.48 & 1.20 \\ 1.20 & 4.00 & -1.20 & 2.00 \\ 1.20 & 2.00 & -1.20 & 4.00 \end{bmatrix}\left(10^{-3}\begin{bmatrix} -20.84 \\ 5.21 \\ 0.00 \\ 2.09 \end{bmatrix}\right) = \begin{bmatrix} -10.0 \text{ kN} \\ 0.0 \text{ kN} \cdot \text{m} \\ -50.0 \text{ kN} \cdot \text{m} \end{bmatrix}$$

For member bc

$$\begin{bmatrix} \bar{y}_b \\ \bar{m}_b \\ \bar{m}_c \end{bmatrix}_{bc} = 6000\begin{bmatrix} 0.12 & 0.60 & -0.12 & 0.60 \\ 0.60 & 4.00 & -0.60 & 2.00 \\ 0.60 & 2.00 & -0.60 & 4.00 \end{bmatrix}\left(10^{-3}\begin{bmatrix} 0.00 \\ 2.09 \\ 0.00 \\ 0.00 \end{bmatrix}\right) = \begin{bmatrix} 7.5 \text{ kN} \\ 50.0 \text{ kN} \cdot \text{m} \\ 25.0 \text{ kN} \cdot \text{m} \end{bmatrix}$$

See Fig. E17.3b.

DISCUSSION The nodes for this structure were placed so that the individual members have a uniform cross section and the shear is constant (moment varies linearly over the length). $\mathbf{K}_{ff}$ was obtained from $\mathbf{K}$ in Example 17.1, and the equation $\mathbf{U}_f = \mathbf{K}_{ff}^{-1}\mathbf{X}_f$ gives the unknown displacements. The calculated member forces were used to construct the free-body diagrams shown.

Example 17.4 is also a structure that can be solved using $\mathbf{K}$ from Example 17.1 and illustrates how a family of related structures can be solved using variations of the same structural stiffness matrix.

Example 17.4 Calculate the displacements and member forces for the beam of Example 17.1 if it is supported as illustrated.

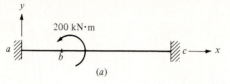

(a)

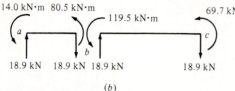

14.0 kN·m 80.5 kN·m

119.5 kN·m

69.7 kN·m

18.9 kN   18.9 kN 18.9 kN   18.9 kN

(b)

Figure E17.4 (a) The beam with applied loading; (b) free-body diagrams of members.

SOLUTION See Example 17.1 for dimensions, etc. From Example 17.1 (using rows and columns 3 and 4 of **K**)

$$\mathbf{K}_{ff} = 1000 \begin{bmatrix} 4.56 & -6.00 \\ -6.00 & 56.00 \end{bmatrix}$$

Thus

$$\begin{bmatrix} V_b \\ \Theta_b \end{bmatrix} = 10^{-3} \begin{bmatrix} 0.2553 & 0.0274 \\ 0.0274 & 0.0208 \end{bmatrix} \begin{bmatrix} 0 \\ 200 \end{bmatrix} = 10^{-3} \begin{bmatrix} 5.48 \text{ m} \\ 4.16 \text{ rad} \end{bmatrix}$$

Use Eq. (17.13) to calculate the member forces. For member ab

$$\begin{bmatrix} \bar{y}_a \\ \bar{m}_a \\ \bar{m}_b \end{bmatrix}_{ab} = 8000 \begin{bmatrix} 0.48 & 1.20 & -0.48 & 1.20 \\ 1.20 & 4.00 & -1.20 & 2.00 \\ 1.20 & 2.00 & -1.20 & 4.00 \end{bmatrix} \left(\begin{bmatrix} 0.00 \\ 0.00 \\ 5.48 \\ 4.16 \end{bmatrix} 10^{-3} \right) = \begin{bmatrix} 18.9 \text{ kN} \\ 14.0 \text{ kN} \cdot \text{m} \\ 80.5 \text{ kN} \cdot \text{m} \end{bmatrix}$$

For member bc

$$\begin{bmatrix} \bar{y}_b \\ \bar{m}_b \\ \bar{m}_c \end{bmatrix}_{bc} = 6000 \begin{bmatrix} 0.12 & 0.60 & -0.12 & 0.60 \\ 0.60 & 4.00 & -0.60 & 2.00 \\ 0.60 & 2.00 & -0.60 & 4.00 \end{bmatrix} \left(\begin{bmatrix} 5.48 \\ 4.16 \\ 0.00 \\ 0.00 \end{bmatrix} 10^{-3} \right) = \begin{bmatrix} 18.9 \text{ kN} \\ 119.5 \text{ kN} \cdot \text{m} \\ 69.7 \text{ kN} \cdot \text{m} \end{bmatrix}$$

See Fig. E17.4b.

Example 17.5 Calculate the displacements and member forces for the beam of Example 17.2 if it is supported as illustrated.

SOLUTION See Example 17.2 for dimensions, etc. From rows and columns 3 through 6 of **K** shown in Example 17.2

$$\mathbf{K}_{ff} = 5.144 \begin{bmatrix} 24 & 0 & -12 & 6 \\ 0 & 8 & -6 & 2 \\ -12 & -6 & 24 & 0 \\ 6 & 2 & 0 & 8 \end{bmatrix}$$

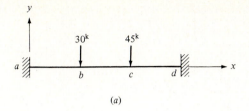

(a)

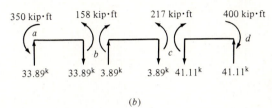

350 kip·ft   158 kip·ft   217 kip·ft   400 kip·ft

33.89^k 33.89^k 3.89^k 3.89^k 41.11^k 41.11^k

(b)

Figure E17.5 (*a*) The beam with applied loading; (*b*) free-body diagrams of members.

Thus,

$$
\begin{bmatrix} V_b \\ \Theta_b L \\ V_c \\ \Theta_c L \end{bmatrix} = 0.1944
\begin{bmatrix} 24 & 0 & -12 & 6 \\ 0 & 8 & -6 & 2 \\ -12 & -6 & 24 & 0 \\ 6 & 2 & 0 & 8 \end{bmatrix}^{-1}
\begin{bmatrix} -30 \\ 0 \\ -45 \\ 0 \end{bmatrix}
=
\begin{bmatrix} -1.170 \\ -1.242 \\ -1.260 \\ 1.188 \end{bmatrix} \text{ in}
$$

The member forces are, for member *ab*

$$
\begin{bmatrix} \bar{y}_a \\ \bar{m}_a/L \\ \bar{m}_b/L \end{bmatrix}_{ab} = 5.144
\begin{bmatrix} 12 & 6 & -12 & 6 \\ 6 & 4 & -6 & 2 \\ 6 & 2 & -6 & 4 \end{bmatrix}
\begin{bmatrix} 0.000 \\ 0.000 \\ -1.170 \\ -1.242 \end{bmatrix}
=
\begin{bmatrix} 33.89 \\ 23.33 \\ 10.56 \end{bmatrix} \text{ kips}
$$

For member *bc*

$$
\begin{bmatrix} \bar{y}_b \\ \bar{m}_b/L \\ \bar{m}_c/L \end{bmatrix}_{bc} = 5.144
\begin{bmatrix} 12 & 6 & -12 & 6 \\ 6 & 4 & -6 & 2 \\ 6 & 2 & -6 & 4 \end{bmatrix}
\begin{bmatrix} -1.170 \\ -1.242 \\ -1.260 \\ 1.188 \end{bmatrix}
=
\begin{bmatrix} 3.89 \\ -10.56 \\ 14.44 \end{bmatrix} \text{ kips}
$$

For member *cd*

$$
\begin{bmatrix} \bar{y}_c \\ \bar{m}_c/L \\ \bar{m}_d/L \end{bmatrix}_{cd} = 5.144
\begin{bmatrix} 12 & 6 & -12 & 6 \\ 6 & 4 & -6 & 2 \\ 6 & 2 & -6 & 4 \end{bmatrix}
\begin{bmatrix} -1.260 \\ 1.188 \\ 0.000 \\ 0.000 \end{bmatrix}
=
\begin{bmatrix} -41.11 \\ -14.44 \\ -26.67 \end{bmatrix} \text{ kips}
$$

See Fig. E17.5*b*.

Example 17.6 Calculate the displacements and member forces for the beam of Example 17.2 if it is supported as illustrated.

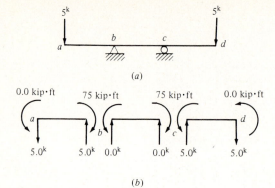

Figure E17.6 (a) The beam with applied loading; (b) free-body diagrams of members.

SOLUTION See Example 17.2 for dimensions, etc. Use **K** from Example 17.2 with $V_b = V_c = 0$. Also, from symmetry, $V_a = V_d$; $\Theta_a = -\Theta_d$; $\Theta_b = -\Theta_c$. These conditions are used to give

$$
\begin{bmatrix} V_a \\ \Theta_a L \\ \Theta_b L \end{bmatrix} = 0.1944 \begin{bmatrix} 12 & 6 & 6 \\ 6 & 4 & 2 \\ 6 & 2 & 6 \end{bmatrix}^{-1} \begin{bmatrix} -5 \\ 0 \\ 0 \end{bmatrix} = \begin{bmatrix} -0.810 \\ 0.972 \\ 0.486 \end{bmatrix} \text{ in}
$$

Note that equations for M_c, M_d, and Y_d have been discarded because of structural symmetry. For member ab

$$
\begin{bmatrix} \bar{y}_a \\ \overline{m}_a/L \\ \overline{m}_b/L \end{bmatrix}_{ab} = 5.144 \begin{bmatrix} 12 & 6 & -12 & 6 \\ 6 & 4 & -6 & 2 \\ 6 & 2 & -6 & 4 \end{bmatrix} \begin{bmatrix} -0.810 \\ 0.972 \\ 0.000 \\ 0.486 \end{bmatrix} = \begin{bmatrix} -5.0 \\ 0.0 \\ -5.0 \end{bmatrix} \text{ kips}
$$

For member bc

$$
\begin{bmatrix} \bar{y}_b \\ \overline{m}_b/L \\ \overline{m}_c/L \end{bmatrix}_{bc} = 5.144 \begin{bmatrix} 12 & 6 & -12 & 6 \\ 6 & 4 & -6 & 2 \\ 6 & 2 & -6 & 4 \end{bmatrix} \begin{bmatrix} 0.000 \\ 0.486 \\ 0.000 \\ -0.486 \end{bmatrix} = \begin{bmatrix} 0.0 \\ 5.0 \\ -5.0 \end{bmatrix} \text{ kips}
$$

See Fig. E17.6b.

DISCUSSION Examples 17.5 and 17.6 treat two structures that can be solved using variations of the structural stiffness matrix obtained in Example 17.2. Using the alternate form of the stiffness matrix (17.11) gives rotations times member length and moments divided by member length.

In Example 17.6, a procedure is used to solve for the deflections that simplifies the calculations if they are being performed by hand, but this approach is questionable if a machine solution is being used. Because of the symmetry of the deflection shape, it is possible to add columns of $\mathbf{K}_{ff}$ if the variables are equal (in magnitude and sign) or subtract columns of $\mathbf{K}_{ff}$ if the variables are equal in magnitude but opposite in sign. Thus, a column of $\mathbf{K}_{ff}$ is replaced by two combined columns. Since these variable dependencies are incorporated, an equal number of dependent equations is created. Thus, the corresponding equations (rows) must be combined in a similar manner and the matrix row replaced by the combined rows. Alternatively, one of the equations (rows) corresponding to the columns combined can be deleted.

17.5 EQUIVALENT NODAL FORCES

The uniform beam member can be used to analyze structures in which the internal shear is constant and the bending moment varies linearly along the member length. The structure shown in Fig. 17.6a can be analyzed using three beam members (one element between each pair of the four labeled nodal points). The beam shown in Fig. 17.6b presents what appears to be an intractable problem to be solved using the beam member. Since the actual shear distribution varies linearly between nodes b and c, the use of a single member between these two nodal points will be expected to give erroneous results. One approach to analyzing this structure would be to use several beam members between nodes b and c; thus the actual linear shear distribution would be approximated by a series of constant shear segments. The more members that are used, the closer the true solution will be approximated. This method would increase the size of the structural stiffness matrix and the time to solve for the displacements. An alternative solution

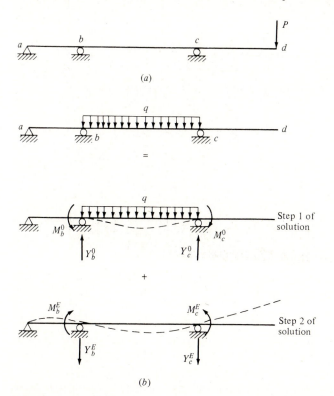

Figure 17.6 Two different load cases for a beam: (a) idealization for discrete loading; (b) solution procedure for distributed loading.

can be obtained by approximating the actual uniform loading with concentrated forces and moments at the nodal points b and c that yield the same nodal displacements as the uniform loading. For this approach, a single element could be used between nodes b and c, and the analysis would be more efficient than the other solution strategy described.

Before obtaining these equivalent forces, we review the matrix displacement solution process briefly for a structure with loads at the nodes. The beam shown in Fig. 17.6a is loaded so that a solution can be obtained using four nodal points and three beam elements. The element stiffness matrices are merged into the structural stiffness matrix, which is partitioned as discussed in Chap. 15 to give

$$\begin{bmatrix} \mathbf{X}_f \\ \hline \mathbf{X}_s \end{bmatrix} = \begin{bmatrix} \mathbf{K}_{ff} & \vdots & \mathbf{K}_{fs} \\ \hline \mathbf{K}_{sf} & \vdots & \mathbf{K}_{ss} \end{bmatrix} \begin{bmatrix} \mathbf{U}_f \\ \hline \mathbf{U}_s \end{bmatrix} \qquad (15.24)$$

where the force matrix, consisting of both the known applied forces $\mathbf{X}_f$ and the unknown reaction forces $\mathbf{X}_s$ for this beam, and the displacement matrix, consisting of both the unknown displacements $\mathbf{U}_f$ and the known supported displacements $\mathbf{U}_s$, are, respectively,

$$\mathbf{X} = \begin{bmatrix} \mathbf{X}_f \\ \hline \mathbf{X}_s \end{bmatrix} = \begin{bmatrix} M_a = 0 \\ M_b = 0 \\ M_c = 0 \\ Y_d = -P \\ M_d = 0 \\ \hline Y_a \\ Y_b \\ Y_c \end{bmatrix} \quad \text{and} \quad \mathbf{U} = \begin{bmatrix} \mathbf{U}_f \\ \hline \mathbf{U}_s \end{bmatrix} = \begin{bmatrix} \Theta_a \\ \Theta_b \\ \Theta_c \\ V_d \\ \Theta_d \\ \hline V_a = 0 \\ V_b = 0 \\ V_c = 0 \end{bmatrix}$$

Since $\mathbf{U}_s$ is null, the matrix equation (15.24) can be solved to give the unknown displacements as

$$\mathbf{U}_f = \mathbf{K}_{ff}^{-1}\mathbf{X}_f \qquad (15.28)$$

and the unknown reaction forces are obtained from

$$\mathbf{X}_s = \mathbf{K}_{sf}\mathbf{U}_f \qquad (15.27)$$

Assume that this same beam is now loaded with a uniformly distributed load between nodes b and c, as shown in Fig. 17.6b. This problem will be solved using the following procedure:

1. Apply initial fixed-end forces to member bc to prevent displacements and rotations at all nodal points when the applied uniform load is on the beam. Note that these initial forces produce shear and bending moment in member bc.

2. Remove the actual distributed loading plus the initial forces used in step 1 and apply forces at nodal points b and c that are equal in magnitude but opposite in direction to the initial fixed-end forces applied in step 1. Calculate the nodal

displacements. Note that these applied forces will introduce displacements throughout the structure.

3. Superpose the results from steps 1 and 2 to obtain the actual solution for the uniformly loaded structure.

This procedure for linear elastic response is similar to that used in Chap. 16 for initial forces on trusses resulting from thermal effects, fabrication errors, etc. It also resembles other stepwise solution methods used in the text, e.g., the moment-distribution method. Since this approach is based on the principle of superposition, its validity need not be discussed in greater detail.

This three-step procedure can be written concisely in the context of the stiffness method using matrix notation. The initial fixed-end forces for each individual beam member relate to the displacements through the stiffness matrix

$$\bar{\mathbf{x}} = \bar{\mathbf{k}}\bar{\mathbf{u}} + \bar{\mathbf{x}}^0 \qquad (17.14)$$

where $\bar{\mathbf{x}}^0$ is the column matrix containing the initial fixed-end forces in the usual order, i.e.,

$$\bar{\mathbf{x}}^0 = \begin{bmatrix} \bar{y}_i^0 \\ \bar{m}_i^0 \\ \bar{y}_j^0 \\ \bar{m}_j^0 \end{bmatrix}$$

For member bc of the beam shown in Fig. 17.6b, these forces, which can be obtained using any of the methods discussed in Part Four, are

$$\bar{m}_b^0 = -\bar{m}_c^0 = \tfrac{1}{12}qL^2 \qquad \text{and} \qquad \bar{y}_b^0 = \bar{y}_c^0 = \tfrac{1}{2}qL$$

Hence for member bc

$$\bar{\mathbf{x}}^0 = \begin{bmatrix} \dfrac{qL}{2} \\[2mm] \dfrac{qL^2}{12} \\[2mm] \dfrac{qL}{2} \\[2mm] -\dfrac{qL^2}{12} \end{bmatrix}$$

These initial fixed-end forces for the individual members are combined at the nodal points when enforcing equilibrium for the assemblage. This process gives the matrix equation for the structure

$$\mathbf{X} = \mathbf{KU} + \mathbf{X}^0 \qquad (17.15)$$

where $\mathbf{X}^0$ is the column matrix containing all the initial fixed-end forces referred to in step 1 of the solution procedure. This matrix equation can be written in partitioned form as

$$\left[\begin{array}{c} \mathbf{X}_f \\ \hline \mathbf{X}_s \end{array}\right] = \left[\begin{array}{c|c} \mathbf{K}_{ff} & \mathbf{K}_{fs} \\ \hline \mathbf{K}_{sf} & \mathbf{K}_{ss} \end{array}\right] \left[\begin{array}{c} \mathbf{U}_f \\ \hline \mathbf{U}_s \end{array}\right] + \left[\begin{array}{c} \mathbf{X}_f^0 \\ \hline \mathbf{X}_s^0 \end{array}\right] \tag{17.16}$$

For the structure shown in Fig. 17.6b with the specific loading given, this equation becomes

$$\left[\begin{array}{c} M_a = 0 \\ M_b = 0 \\ M_c = 0 \\ Y_d = 0 \\ M_d = 0 \\ \hline Y_a \\ Y_b \\ Y_c \end{array}\right] = \left[\begin{array}{c|c} & \\ \mathbf{K}_{ff} & \mathbf{K}_{fs} \\ & \\ \hline & \\ \mathbf{K}_{sf} & \mathbf{K}_{ss} \\ & \end{array}\right] \left[\begin{array}{c} \Theta_a \\ \Theta_b \\ \Theta_c \\ V_d \\ \Theta_d \\ \hline V_a = 0 \\ V_b = 0 \\ V_c = 0 \end{array}\right] + \left[\begin{array}{c} M_a^0 = 0 \\ M_b^0 \\ M_c^0 \\ Y_d^0 = 0 \\ M_d^0 = 0 \\ \hline Y_a^0 = 0 \\ Y_b^0 \\ Y_c^0 \end{array}\right] \tag{17.16a}$$

Equation (17.15) can also be rewritten as

$$\mathbf{X} - \mathbf{X}^0 = \mathbf{X} + \mathbf{X}^E = \mathbf{KU} \tag{17.17}$$

The physical interpretation of subtracting $\mathbf{X}^0$ from both sides of the equation is equivalent to applying the initial fixed-end forces (those required to maintain zero displacement at the nodes) to the structure in the opposite direction (step 2). Therefore, these forces are sometimes referred to as the *equivalent loads* $\mathbf{X}^E$.

For the structure shown in Fig. 17.6b, Eq. (17.17) becomes

$$\left[\begin{array}{c} 0 \\ -\frac{1}{12}qL^2 \\ \frac{1}{12}qL^2 \\ 0 \\ 0 \\ \hline 0 \\ -\frac{1}{2}qL \\ -\frac{1}{2}qL \end{array}\right] = \left[\begin{array}{c|c} & \\ \mathbf{K}_{ff} & \mathbf{K}_{fs} \\ & \\ \hline & \\ \mathbf{K}_{sf} & \mathbf{K}_{ss} \\ & \end{array}\right] \left[\begin{array}{c} \Theta_a \\ \Theta_b \\ \Theta_c \\ V_d \\ \Theta_d \\ \hline 0 \\ 0 \\ 0 \end{array}\right] \tag{17.17a}$$

Solving this equation gives the unknown displacements and reaction forces at the nodal points.

After finding the unknown displacements $\mathbf{U}_f$ and forces at the nodes $\mathbf{X}_s$, the member forces can be obtained. This is step 3 in the solution procedure, and the initial fixed-end forces must be combined with the forces produced by the nodal displacements. This superposition is attained by using the complete force-displacement relationship for the elements [see Eq. (17.14)]; i.e., the forces defined by $\overline{\mathbf{k}}\,\overline{\mathbf{u}}$ are those from step 2 of the solution process and the forces $\overline{\mathbf{x}}^0$ are those associated with step 1 of the solution (see Fig. 17.6*b*). The member forces can also be calculated using the stress matrix and the corresponding initial forces to give

$$\mathbf{s}_{ij} = \mathbf{s}\overline{\mathbf{u}}_{ij} + \overline{\mathbf{s}}^0 \qquad (17.18)$$

where the initial shear $\overline{y}_i^0$ at i and the initial moments $\overline{m}_i^0$ and $\overline{m}_j^0$ are combined into $\overline{\mathbf{s}}^0$

$$\overline{\mathbf{s}}^0 = \begin{bmatrix} \overline{y}_i^0 \\ \overline{m}_i^0 \\ \overline{m}_j^0 \end{bmatrix} \qquad (17.19)$$

A uniformly loaded beam is analyzed in Example 17.7 to illustrate the use of these initial fixed-end forces. The solution obtained for this four-element solution compares favorably with the exact solution.

Example 17.7 Calculate the displacements and member forces for this uniformly loaded beam using equivalent nodal forces. $EI = $ const.

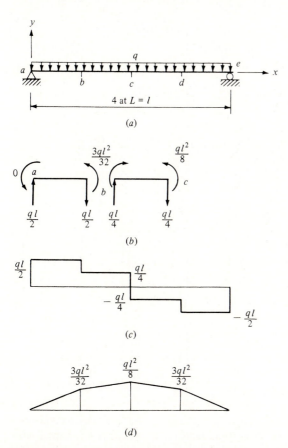

Figure E17.7 (*a*) The beam with applied loading; (*b*) free-body diagrams of *ab* and *bc*; (*c*) shear diagram; (*d*) moment diagram.

SOLUTION For all members $\mathbf{k} = \bar{\mathbf{k}}$

$$
\mathbf{K} = \frac{EI}{L^3}
\begin{bmatrix}
12 & 6 & -12 & 6 & 0 & 0 & 0 & 0 & 0 & 0 \\
6 & 4 & -6 & 2 & 0 & 0 & 0 & 0 & 0 & 0 \\
-12 & -6 & 24 & 0 & -12 & 6 & 0 & 0 & 0 & 0 \\
6 & 2 & 0 & 8 & -6 & 2 & 0 & 0 & 0 & 0 \\
0 & 0 & -12 & -6 & 24 & 0 & -12 & 6 & 0 & 0 \\
0 & 0 & 6 & 2 & 0 & 8 & -6 & 2 & 0 & 0 \\
0 & 0 & 0 & 0 & -12 & -6 & 24 & 0 & -12 & 6 \\
0 & 0 & 0 & 0 & 6 & 2 & 0 & 8 & -6 & 2 \\
0 & 0 & 0 & 0 & 0 & 0 & -12 & -6 & 12 & -6 \\
0 & 0 & 0 & 0 & 0 & 0 & 6 & 2 & -6 & 4
\end{bmatrix}
\qquad
\mathbf{X}^0 = qL
\begin{bmatrix}
1/2 \\
1/12 \\
1 \\
0 \\
1 \\
0 \\
1 \\
0 \\
1/2 \\
-1/12
\end{bmatrix}
$$

Using the support conditions ($V_a = V_e = 0$) and the conditions of symmetry ($V_b = V_d$, $\Theta_a = -\Theta_e$, $\Theta_b = -\Theta_d$, and $\Theta_c = 0$) gives

$$
\begin{bmatrix}
\Theta_a L \\
V_b \\
\Theta_b L \\
V_c
\end{bmatrix}
= \frac{L^3}{EI}
\begin{bmatrix}
4 & -6 & 2 & 0 \\
-6 & 24 & 0 & -12 \\
2 & 0 & 8 & -6 \\
0 & -24 & -12 & 24
\end{bmatrix}^{-1}
\begin{bmatrix}
-qL/12 \\
-qL \\
0 \\
-qL
\end{bmatrix}
= -\frac{qL^4}{EI}
\begin{bmatrix}
8/3 \\
19/8 \\
11/6 \\
10/3
\end{bmatrix}
$$

Since $L = l/4$,

$$
\Theta_a = -\frac{8qL^3}{3EI} = -\frac{ql^3}{24EI} \qquad V_b = -\frac{19qL^4}{8EI} = -\frac{19ql^4}{2048EI}
$$

$$
\Theta_b = -\frac{11qL^3}{6EI} = -\frac{11ql^3}{384EI} \qquad V_c = -\frac{10qL^4}{3EI} = -\frac{5ql^4}{384EI}
$$

Member forces are obtained using Eq. (17.18). For member ab

$$
\begin{bmatrix}
\bar{y}_a \\
\bar{m}_a/L \\
\bar{m}_b/L
\end{bmatrix}_{ab}
= \frac{EI}{L^3}
\begin{bmatrix}
12 & 6 & -12 & 6 \\
6 & 4 & -6 & 2 \\
6 & 2 & -6 & 4
\end{bmatrix}
\left(-\frac{qL^4}{EI}\right)
\begin{bmatrix}
0 \\
8/3 \\
19/8 \\
11/6
\end{bmatrix}
+ qL
\begin{bmatrix}
1/2 \\
1/12 \\
-1/12
\end{bmatrix}
= qL
\begin{bmatrix}
2 \\
0 \\
3/2
\end{bmatrix}
$$

For member bc

$$
\begin{bmatrix}
\bar{y}_b \\
\bar{m}_b/L \\
\bar{m}_c/L
\end{bmatrix}_{bc}
= \frac{EI}{L^3}
\begin{bmatrix}
12 & 6 & -12 & 6 \\
6 & 4 & -6 & 2 \\
6 & 2 & -6 & 4
\end{bmatrix}
\left(-\frac{qL^4}{EI}\right)
\begin{bmatrix}
19/8 \\
11/6 \\
10/3 \\
0
\end{bmatrix}
+ qL
\begin{bmatrix}
1/2 \\
1/12 \\
-1/12
\end{bmatrix}
= qL
\begin{bmatrix}
1 \\
-3/2 \\
2
\end{bmatrix}
$$

The member forces are as shown in Fig. E17.7b noting $L = l/4$. Using the results obtained, we have plotted the shear and moment diagrams in Fig. E17.7c and d, respectively.

DISCUSSION The matrix of initial forces $\mathbf{X}^0$ contains zero entries for the moments at b, c, and d since the left and right beam member at a typical node contribute equal fixed-end moments that are opposite in sign. Because the matrix $\mathbf{X}_f$ is null, the calculations show the first of Eqs. (17.16) as

$$\mathbf{0} = \mathbf{K}_{ff}\mathbf{U}_f + \mathbf{X}_f^0$$

Note that the alternate form of $\mathbf{k}$ was used and the size of $\mathbf{K}_{ff}$ reduced using symmetry arguments similar to those described in Example 17.6. Since the displacements and rotations obtained at the nodes are exact, this idealization using four beam members appears sufficient for these calculations.

The member forces are calculated by the superposition of initial fixed-end forces and the forces resulting from the displacements [see Eq. (17.18)]. The true straight-line shear diagram is approximated by the step function shown, but the value of the shear is exact at both ends. The true moment diagram is a parabola, which is approximated with straight-line segments. The values of the moments at the nodes are exact.

It is important to note that if equivalent loads are used, the nodal displacements obtained will be exact regardless of the number of beam elements. The member end moments will also be exact, but in general there will be a discontinuity in the shear from member to member (across the nodal points). In addition, for each element, the moment is linear and the shear is constant; therefore, it is necessary to use enough beam elements to approximate the exact moment and shear diagrams closely.

17.6 PROBLEMS INVOLVING SETTLEMENT, TEMPERATURE, ETC.

A beam with differential support settlement can be analyzed just like the support settlement of trusses. The force-displacement equations are used in their partitioned form [Eq. (15.24)], and the known support movements are contained in the partition $\mathbf{U}_s$ of the displacement matrix. Hence, the displacements are obtained using

$$\mathbf{U}_f = \mathbf{K}_{ff}^{-1}(\mathbf{X}_f - \mathbf{K}_{fs}\mathbf{U}_s)$$

and the forces corresponding to the known displacements are calculated using the second matrix equation of (15.24)

$$\mathbf{X}_s = \mathbf{K}_{sf}\mathbf{U}_f + \mathbf{K}_{ss}\mathbf{U}_s$$

Finally, Eq. (17.13) is used to calculate the member forces. Example 17.8 illustrates these calculations for a statically indeterminate beam.

Example 17.8 Calculate the nodal displacements and the member forces if the beam of Example 17.4 has the 200-kN · m moment removed and the support at c settles 15 mm.

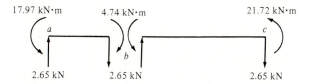

Figure E17.8 Free-body diagrams of members.

SOLUTION

$$\mathbf{U}_f = \mathbf{K}_{ff}^{-1}(\mathbf{X}_f - \mathbf{K}_{fs}\mathbf{U}_s)$$

Since $\mathbf{X}_f$ is null,

$$\begin{bmatrix} V_b \\ \Theta_b \end{bmatrix} = -10^{-3}\begin{bmatrix} 0.2553 & 0.0274 \\ 0.0274 & 0.0208 \end{bmatrix}\left(10^3\begin{bmatrix} -3.84 & -9.60 & -0.72 & 3.60 \\ 9.60 & 16.00 & -3.60 & 12.00 \end{bmatrix}\right)\left(10^{-3}\begin{bmatrix} 0 \\ 0 \\ -15 \\ 0 \end{bmatrix}\right)$$

$$= 10^{-3}\begin{bmatrix} 0.2553 & 0.0274 \\ 0.0274 & 0.0208 \end{bmatrix}\begin{bmatrix} -10.80 \\ -54.00 \end{bmatrix} = 10^{-3}\begin{bmatrix} -4.24 \text{ m} \\ -1.42 \text{ rad} \end{bmatrix}$$

For member ab

$$\begin{bmatrix} \bar{y}_a \\ \bar{m}_a \\ \bar{m}_b \end{bmatrix}_{ab} = 8000\begin{bmatrix} 0.48 & 1.20 & -0.48 & 1.20 \\ 1.20 & 4.00 & -1.20 & 2.00 \\ 1.20 & 2.00 & -1.20 & 4.00 \end{bmatrix}\left(10^{-3}\begin{bmatrix} 0.00 \\ 0.00 \\ -4.24 \\ -1.42 \end{bmatrix}\right) = \begin{bmatrix} 2.65 \text{ kN} \\ 17.97 \text{ kN} \cdot \text{m} \\ -4.74 \text{ kN} \cdot \text{m} \end{bmatrix}$$

For member bc

$$\begin{bmatrix} \bar{y}_b \\ \bar{m}_b \\ \bar{m}_c \end{bmatrix}_{bc} = 6000\begin{bmatrix} 0.12 & 0.60 & -0.12 & 0.60 \\ 0.60 & 4.00 & -0.60 & 2.00 \\ 0.60 & 2.00 & -0.60 & 4.00 \end{bmatrix}\left(10^{-3}\begin{bmatrix} -4.24 \\ -1.42 \\ -15.00 \\ 0.00 \end{bmatrix}\right) = \begin{bmatrix} 2.65 \text{ kN} \\ 4.74 \text{ kN} \cdot \text{m} \\ 21.72 \text{ kN} \cdot \text{m} \end{bmatrix}$$

See Fig. E17.8.

DISCUSSION Since the matrix $\mathbf{X}_f$ is null, the structure is deformed totally by the support movement and the effective forces are attributable to $\mathbf{K}_{fs}\mathbf{U}_s$. The member forces are obtained using Eq. (17.13). Since this is a statically indeterminate beam, it is logical for nonzero member forces to be introduced by differential settlement.

The three-step procedure outlined in Sec. 17.5 can also be used when the structure is subjected to thermal effects, fabrication errors, precambering, etc., which can all be categorized as initial-force problems. This is also the solution procedure used for trusses with initial forces in Chap. 16.

For convenience, the equations from Sec. 17.5 are summarized here. Thus, the initial fixed-end forces for the members $\bar{\mathbf{x}}^0$ are combined, along with the individual stiffness matrices, to give

$$\mathbf{X} = \mathbf{KU} + \mathbf{X}^0 \tag{17.15}$$

or

$$\begin{bmatrix} \mathbf{X}_f \\ \mathbf{X}_s \end{bmatrix} = \begin{bmatrix} \mathbf{K}_{ff} & \mathbf{K}_{fs} \\ \mathbf{K}_{sf} & \mathbf{K}_{ss} \end{bmatrix}\begin{bmatrix} \mathbf{U}_f \\ \mathbf{U}_s \end{bmatrix} + \begin{bmatrix} \mathbf{X}_f^0 \\ \mathbf{X}_s^0 \end{bmatrix} \tag{17.16}$$

After solving for the unknown displacements $\mathbf{U}_f$ and forces $\mathbf{X}_s$, the member forces are obtained using

$$\mathbf{s}_{ij} = \mathbf{s}\bar{\mathbf{u}}_{ij} + \bar{\mathbf{s}}^0 \tag{17.18}$$

The initial fixed-end forces (step 1) for a heated beam member are applied at the nodes to give zero rotations and displacements. They can be calculated using basic considerations. Consider the beam member shown in Fig. 17.7 heated uniformly along its length with a thermal distribution that varies linearly from the lower to the upper surface of the cross section. The temperature at any position on the cross section is

$$T(y) = T_m + \frac{\Delta T\, y}{h} \qquad \text{where} \qquad \begin{aligned} T_m &= \tfrac{1}{2}(T_l + T_u) \\ \Delta T &= T_u - T_l \end{aligned} \qquad (17.20)$$

Thus the strain and the stress are, respectively,

$$\varepsilon^0 = \alpha T(y) \qquad \text{and} \qquad \sigma^0 = E\varepsilon^0 = E\alpha T(y)$$

The axial force that must be applied to each end of the member to maintain the member in an undeformed position is

$$X^0 = \int_{-h/2}^{+h/2} \sigma^0 b\, dy = \int_{-h/2}^{+h/2} E\alpha \left(T_m + \frac{\Delta T}{h} y \right) b\, dy = AE\alpha T_m \qquad (17.21)$$

The moment required to prevent rotation of the cross section is

$$M^0 = \int_{-h/2}^{+h/2} E\alpha T(y) by\, dy = E\alpha \int_{-h/2}^{+h/2} \left(T_m + \frac{\Delta T}{h} y \right) by\, dy = \frac{EI\alpha\, \Delta T}{h} \qquad (17.22)$$

The beam would tend to deform into the shape shown in Fig. 17.7 with a constant curvature; therefore, there will be neither rotations nor displacements if M^0 is applied to the left and right ends in a clockwise and counterclockwise direction, respectively. The force X^0 will be neglected, since it is assumed that the neutral axis remains undeformed for conventional beam theory. Hence, the initial force matrix for the beam member is

$$\overline{\mathbf{x}}^0 = \begin{bmatrix} \overline{y}_i^0 \\ \overline{m}_i^0 \\ \overline{y}_j^0 \\ \overline{m}_j^0 \end{bmatrix} = \frac{EI\alpha\, \Delta T}{h} \begin{bmatrix} 0 \\ -1 \\ 0 \\ 1 \end{bmatrix} \qquad (17.23)$$

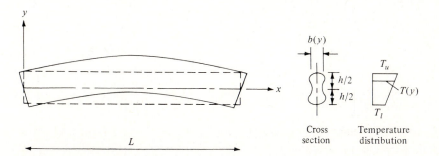

$$y$$

$$L$$

$$b(y)$$

$$h/2$$

$$h/2$$

Cross
section

$$T_u$$

$$T(y)$$

$$T_l$$

Temperature
distribution

Figure 17.7 Beam member with a linear temperature distribution between the lower and upper surfaces.

and the corresponding matrix to be used for calculating member forces is

$$
\bar{s}^0 = \begin{bmatrix} \bar{y}_i^0 \\ \overline{m}_i^0 \\ \overline{m}_j^0 \end{bmatrix} = \frac{EI\alpha\,\Delta T}{h} \begin{bmatrix} 0 \\ -1 \\ 1 \end{bmatrix}
\tag{17.24}
$$

Calculations using the procedure described in conjunction with this initial force matrix are shown in Examples 17.9 and 17.10.

Example 17.9 Calculate the nodal displacements and the member forces if the beam of Example 17.5 has the loads removed and is heated uniformly along its length so that $T_u - T_l = 80°F$; assume that $h = 12$ in; $\alpha = 6.5 \times 10^{-6}$ in/in · °F.

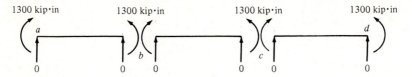

Figure E17.9   Free-body diagrams of members.

SOLUTION

$$
M^0 = \frac{EI\alpha\,\Delta T}{h} = \frac{(29 \times 10^3 \text{ ksi})\,(1034 \text{ in}^4)\,(6.5 \times 10^{-6} \text{ in/in} \cdot {}°\text{F})\,(80°\text{F})}{12 \text{ in}} = 1300 \text{ kip} \cdot \text{in}
$$

$$
\frac{M^0}{L} = 7.22 \text{ kips}
$$

Hence, for each member

$$
\begin{bmatrix} \bar{y}_i^0 \\ \overline{m}_i^0/L \\ \bar{y}_j^0 \\ \overline{m}_j^0/L \end{bmatrix} = 7.22 \begin{bmatrix} 0 \\ -1 \\ 0 \\ 1 \end{bmatrix} \text{ kips}
$$

These initial fixed-end forces are merged to give $\mathbf{X}^0$. It can be observed that $\mathbf{X}_f^0$ is null. Since $\mathbf{X}_f$ is also null, it can be concluded from the first of Eqs. (17.16) that $\mathbf{U}_f = \mathbf{0}$. That is, all nodal points of this beam remain in their undeflected positions. For all members

$$
\begin{bmatrix} \bar{y}_i \\ \overline{m}_i/L \\ \overline{m}_j/L \end{bmatrix} = \mathbf{s0} + \bar{s}^0 = 7.22 \begin{bmatrix} 0 \\ -1 \\ 1 \end{bmatrix} \text{ kips}
$$

See Fig. E17.9.

DISCUSSION   Initial forces and the procedure embodied in Eqs. (17.15), (17.16), and (17.18) are used. This thermal distribution results in zero displacements, but since the structure is statically indeterminate, moments are introduced along its length.

Example 17.10 Calculate the nodal displacements and the member forces if the beam of Example 17.6 has the loads removed and is heated uniformly along its length so that $T_u - T_l = 80°F$; assume that $h = 12$ in; $\alpha = 6.5 \times 10^{-6}$ in/in · °F.

SOLUTION From Example 17.9, $M^0 = 1300$ kip · in, and $\bar{x}^0$ for each of the three members is the same as shown there. For this structure, however, $\mathbf{X}_f^0 = \{0 \ -7.22 \ \ 0 \ \ 0 \ \ 0 \ \ 7.22\}$ kips

$$
\begin{bmatrix} V_a \\ \Theta_a L \\ \Theta_b L \\ \Theta_c L \\ V_d \\ \Theta_d L \end{bmatrix} = 0.1944 \begin{bmatrix} 12 & 6 & 6 & 0 & 0 & 0 \\ 6 & 4 & 2 & 0 & 0 & 0 \\ 6 & 2 & 8 & 2 & 0 & 0 \\ 0 & 0 & 2 & 8 & -6 & 2 \\ 0 & 0 & 0 & -6 & 12 & -6 \\ 0 & 0 & 0 & 2 & -6 & 4 \end{bmatrix}^{-1} \begin{bmatrix} 0.00 \\ 7.22 \\ 0.00 \\ 0.00 \\ 0.00 \\ -7.22 \end{bmatrix}
$$

From symmetry, $V_a = V_d$, $\Theta_a = -\Theta_d$, and $\Theta_b = -\Theta_c$; therefore,

$$
\begin{bmatrix} V_a \\ \Theta_a L \\ \Theta_b L \end{bmatrix} = 0.1944 \begin{bmatrix} 12 & 6 & 6 \\ 6 & 4 & 2 \\ 6 & 2 & 6 \end{bmatrix}^{-1} \begin{bmatrix} 0.00 \\ 7.22 \\ 0.00 \end{bmatrix} = 0.0324 \begin{bmatrix} 5 & -6 & -3 \\ -6 & 9 & 3 \\ -3 & 3 & 3 \end{bmatrix} \begin{bmatrix} 0.00 \\ 7.22 \\ 0.00 \end{bmatrix} = \begin{bmatrix} -1.40 \\ 2.10 \\ 0.70 \end{bmatrix} \text{ in}
$$

For member ab

$$
\begin{bmatrix} \bar{y}_a \\ \bar{m}_a/L \\ \bar{m}_b/L \end{bmatrix}_{ab} = 5.144 \begin{bmatrix} 12 & 6 & -12 & 6 \\ 6 & 4 & -6 & 2 \\ 6 & 2 & -6 & 4 \end{bmatrix} \begin{bmatrix} -1.40 \\ 2.10 \\ 0.00 \\ 0.70 \end{bmatrix} + \begin{bmatrix} 0.00 \\ -7.22 \\ 7.22 \end{bmatrix} = \begin{bmatrix} 0.00 \\ 0.00 \\ 0.00 \end{bmatrix}
$$

For member bc

$$
\begin{bmatrix} \bar{y}_b \\ \bar{m}_b/L \\ \bar{m}_c/L \end{bmatrix}_{bc} = 5.144 \begin{bmatrix} 12 & 6 & -12 & 6 \\ 6 & 4 & -6 & 2 \\ 6 & 2 & -6 & 4 \end{bmatrix} \begin{bmatrix} 0.00 \\ 0.70 \\ 0.00 \\ -0.70 \end{bmatrix} + \begin{bmatrix} 0.00 \\ -7.22 \\ 7.22 \end{bmatrix} = \begin{bmatrix} 0.00 \\ 0.00 \\ 0.00 \end{bmatrix}
$$

DISCUSSION This structure is heated like that in the previous example. The same initial-force solution procedure is used; in addition, the size of $\mathbf{K}_{ff}$ is reduced since there is symmetry of displacements. Since this structure is statically determinate, the nonzero displacements are accompanied by zero member forces.

17.7 DISCUSSION

The general theory of the stiffness method, described in Chap. 15 and extended to trusses in Chap. 16, is applied in this chapter to the analysis of beams. The member stiffness matrix and the corresponding stress matrix are developed using basic considerations and two assumptions:

1. The member is straight, prismatic, and made from linear elastic materials.
2. The shear in the member is constant, implying that the bending moment varies linearly between the nodal points.

Since the beam member behavior includes rotational generalized displacements (and the corresponding moments), it is necessary to enforce equilibrium of both the direct forces and the moments in the merge process. Because the structural coordinate system is generally aligned with the member coordinate system for all the individual members, the various coordinate transformations used in the analysis of trusses are not necessary.

Since nodal points are not a physical feature of beam structures, we must idealize (model) the actual structure; i.e., nodal points and members must be selected so that the resulting individual members display behavior consistent (exactly or approximately) with the assumptions embodied in the member stiffness matrix. Distributed loading requires the use of special initial forces that are incorporated into an analysis using superposition. Initial forces are also used to solve problems involving thermal effects, cambering, fabrication errors, etc.

PROBLEMS

Use the stiffness method for all problems.

17.1 to 17.3 Calculate the unknown displacements and the member forces; for all members $I = 862$ in^4 and $E = 29 \times 10^3$ kips/in^2.

Figure P17.1

Figure P17.2

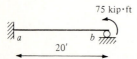

Figure P17.3

17.4 An experiment is being run in the laboratory to establish the force-displacement relationships at the tip of a cantilever beam fixed at the left end. First an upward force of 10 kips is applied at the tip, and a vertical deflection of 0.192 in and a counterclockwise rotation of 0.002 rad are measured at the tip. Then a counterclockwise moment of 200 kip · in is applied at the tip, and an upward deflection of 0.04 in and a counterclockwise rotation of 0.000553 rad are measured at the tip. Calculate the flexibility and stiffness matrices using these data. Comment on the quality of the data.

17.5 to 17.9 Label all the nodes required for an analysis, and indicate the sizes of the matrices **U**, **X**, $\mathbf{K}_{ff}$, $\mathbf{K}_{fs}$, $\mathbf{K}_{sf}$, and $\mathbf{K}_{ss}$.

In Probs. 17.10 to 17.18 calculate the unknown displacements and the member forces.

17.10 $EI = $ const.

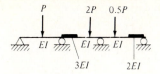

Figure P17.5

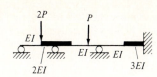

Figure P17.6

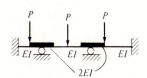

Figure P17.7

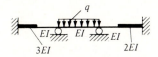

Figure P17.8

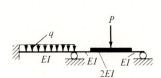

Figure P17.9

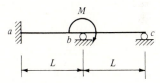

Figure P17.10

17.11 Node b settles a distance Δ; $EI =$ const. What is the reaction at node b?

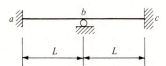

Figure P17.11

17.12 Solve Prob. 17.10 if the structure is modified so that $EI_{bc} = 3EI_{ab}$.

17.13 Solve Prob. 17.11 if the structure is modified so that $EI_{bc} = 3EI_{ab}$.

17.14 Investigate the beam shown in Example 17.1 if it is fixed at nodes a and c, and:

(*a*) A force of 100 kN is applied at node b in the downward direction.

(*b*) Support a settles 10 mm.

(*c*) The upper and lower surfaces of the beam are +30 and 0°C, respectively. The depth of the beam is 360 mm, and $\alpha = 1.2 \times 10^{-5}$ mm/mm · °C.

17.15 Investigate this beam if $I_{ab} = 300$ in^4, $I_{bc} = 200$ in^4, $E = 29 \times 10^3$ kips/in^2, and:

(*a*) The 50-kip load is applied to the structure.

(*b*) Support c settles 0.25 in (the 50-kip load is removed).

(*c*) The upper and lower surfaces of the beam are +120 and +70°F, respectively. The depth of the beam is 8 in, and $\alpha = 6.5 \times 10^{-6}$ in/in · °F (the 50-kip load is removed).

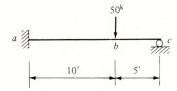

Figure P17.15

17.16 Investigate this beam if $I = 930$ in^4, $E = 29 \times 10^3$ kips/in^2, and:

(a) The two 30-kip loads are applied to the structure.

(b) Support d settles 0.50 in (the loads are removed).

(c) The upper and lower surfaces of the beam are $+150$ and $+50°$F, respectively. The depth of the beam is 14 in, and $\alpha = 6.5 \times 10^{-6}$ in/in · °F (the loads are removed).

17.17 Investigate this beam if $I = 776$ in^4, $E = 29 \times 10^3$ kips/in^2, and:

(a) The illustrated concentrated moments are applied to the structure.

(b) Support d settles 0.33 in (the moments are removed).

(c) The upper and lower surfaces of the beam are $+120$ and $+75°$F, respectively. The depth of the beam is 12 in, and $\alpha = 6.5 \times 10^{-6}$ in/in · °F (the moments are removed).

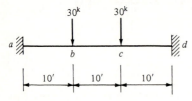

Figure P17.16

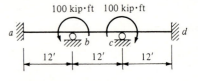

Figure P17.17

17.18 $I = 930$ in^4 and $E = 29 \times 10^3$ kips/in^2.

In Probs. 17.19 to 17.24 calculate the unknown displacements and the reactions.

17.19 $EI = $ const.

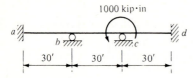

Figure P17.18

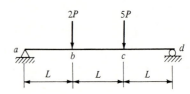

Figure P17.19

17.20 $I = 150$ in^4 and $E = 29 \times 10^3$ kips/in^2.

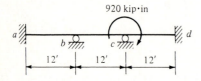

Figure P17.20

17.21 to 17.22 $EI = $ const for the beam and the stiffness of the spring is $12EI/L^3$.

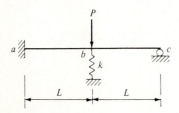

Figure P17.21

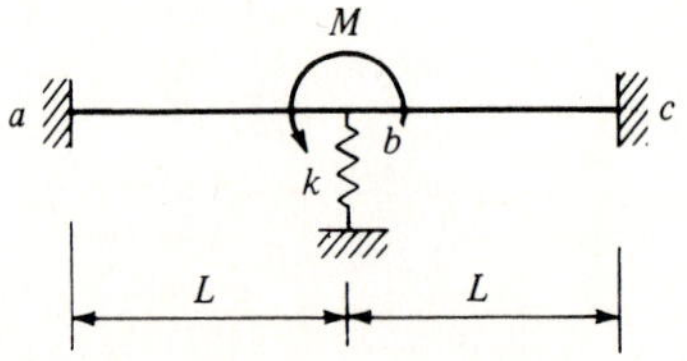

Figure P17.22

17.23 Solve Prob. 17.21 if the structure is modified so that $I_{bc} = 2I$.

17.24 Solve Prob. 17.22 if the structure is modified so that $I_{bc} = 2I$.

17.25 Calculate the unknown displacements and the member forces if $I = 354$ in^4 and $E = 29 \times 10^3$ kips/in^2.

17.26 Calculate the unknown displacements and the member forces if $I_{ab} = 1250 \times 10^{-6}$ m^4, $I_{bc} = 830 \times 10^{-6}$ m^4, and $E = 200$ GPa.

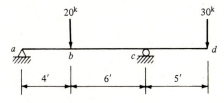

Figure P17.25   **Figure P17.26**

In Probs. 17.27 to 17.37 calculate all the unknown nodal displacements and the member forces for the original problem given.

17.27 Problem 9.14   **17.28** Problem 9.22
17.29 Problem 10.11   **17.30** Problem 10.12
17.31 Problem 11.3   **17.32** Problem 11.5
17.33 Part (*a*) of Prob. 11.24   **17.34** Problem 12.3; $E = 29 \times 10^3$ kips/in^2
17.35 Problem 12.11   **17.36** Problem 12.12
17.37 Problem 12.13

ADDITIONAL READING

Beaufait, F. W.: *Basic Concepts of Structural Analysis,* chap. 9, Prentice-Hall, Englewood Cliffs, N.J., 1977.

Gerstle, K. H.: *Basic Structural Analysis,* chap. 13, Prentice-Hall, Englewood Cliffs, N.J., 1974.

Gutkowski, R. M.: *Structures: Fundamental Theory and Behavior,* chap. 7, Van Nostrand Reinhold, New York, 1981.

Harrison, H. B.: *Structural Analysis and Design, pt. 1,* chap. 8, Pergamon, New York, 1980.

Kardestuncer, H.: *Elementary Matrix Analysis of Structures,* chaps. 5 and 6, McGraw-Hill, New York, 1974.

Laursen, H. I.: *Structural Analysis,* chap. 14, McGraw-Hill, New York, 1978.

McGuire, W., and R. H. Gallagher: *Matrix Structural Analysis,* chaps. 4 and 5, Wiley, New York, 1979.

Martin, H. C.: *Introduction to Matrix Methods of Structural Analysis,* chap. 4, McGraw-Hill, New York, 1966.

Meek, J. L.: *Matrix Structural Analysis,* chap. 7, McGraw-Hill, New York, 1971.

Weaver, W., Jr., and J. M. Gere: *Matrix Analysis of Framed Structures,* 2d ed., chaps. 4 and 6, Van Nostrand Reinhold, New York, 1980.

West, H. H.: *Analysis of Structures,* chap. 19, Wiley, New York, 1980.

Willems, N., and W. M. Lucas, Jr.: *Structural Analysis for Engineers,* chap. 9, McGraw-Hill, New York, 1978.

EIGHTEEN

THE STIFFNESS METHOD

18.1 INTRODUCTION

The concepts described in the previous chapters of Part Five can be extended and applied to rigid frames. In this case we must transform the properties of a set of generally oriented beam members into the common structural coordinate system using a procedure similar to that described in Chap. 16 for truss members. With this accomplished, the general solution algorithm for the stiffness method, outlined in Fig. 15.8, can be used to yield the nodal displacements and member forces.

The primary objective of this chapter is to enhance the stiffness method so that solutions embody the same assumptions as classical analyses of linear elastic plane rigid frames. That is, the members are inextensible along their longitudinal axes, the deformed and undeformed member lengths are nominally equal, and the governing equations apply to both the initial and displaced geometry. An ancillary objective is to formulate the general approach to include the effect of axial deformations on the behavior of rigid-frame structures. This is accomplished using $\bar{\mathbf{k}}$ obtained by combining the frame stiffness matrix with that of the truss member. This effect can be significant in high-rise structures, an example of which is illustrated in Fig. 18.1.

A brief description of the analysis of three-dimensional frames near the end of the chapter demonstrates how easily the stiffness method can be extended to this more general class of structures. The reader with additional interests in this application is urged to consult references devoted exclusively to matrix structural analysis.

18.2 THE BEAM MEMBER IN RIGID-FRAME ANALYSIS

A plane rigid frame can be envisioned as being composed of individual beam members, oriented in various directions and connected together at their ends. Thus, the properties of the linear elastic prismatic beam member described in Chap. 17, when properly modified, can be used to analyze plane rigid frames. The rigid frame shown in

Figure 18.1 The World Trade Center, New York. *(The Port Authority of New York and New Jersey.)*

Fig. 18.2 consists of four beam members joined at node a by a moment connection. For this structure there is only one generalized unknown displacement Θ_a that must be considered if the members are all axially inextensible. When $I = I_{ab} = I_{ac} = I_{ad} = I_{ae}$ and $L = L_{ab} = L_{ac} = L_{ad} = L_{ae}$, the stiffness terms contributed by each of these four members can be computed using the stiffness matrix in Eq. (17.10), giving

$$k_{ab} = k_{ac} = k_{ad} = k_{ae} = \frac{4EI}{L}$$

In this case the merged stiffness of the structure with boundary conditions enforced is the scalar

$$K = \frac{16EI}{L}$$

The single force-displacement equation is

$$M = \frac{16EI}{L} \Theta_a$$

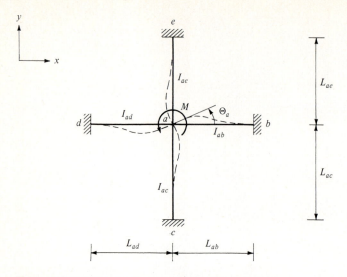

Figure 18.2 Rigid frame with one generalized displacement.

which gives

$$\Theta_a = \frac{ML}{16EI}$$

The member forces for a typical member such as ab are obtained using Eq. (17.13), yielding

$$
\begin{bmatrix} \bar{y}_a \\ \bar{m}_a \\ \bar{m}_b \end{bmatrix} = \frac{EI}{L}
\begin{bmatrix}
\dfrac{12}{L^2} & \dfrac{6}{L} & -\dfrac{12}{L^2} & \dfrac{6}{L} \\[2mm]
\dfrac{6}{L} & 4 & -\dfrac{6}{L} & 2 \\[2mm]
\dfrac{6}{L} & 2 & -\dfrac{6}{L} & 4
\end{bmatrix}
\begin{bmatrix} 0 \\ \dfrac{ML}{16EI} \\ 0 \\ 0 \end{bmatrix}
=
\begin{bmatrix} \dfrac{3M}{8L} \\[2mm] \dfrac{M}{4} \\[2mm] \dfrac{M}{8} \end{bmatrix}
$$

These results satisfy equilibrium and could have been calculated using one of the classical methods, e.g., moment distribution.

In this elementary example of rigid-frame analysis all the member rotations are about the implied z axis; therefore, no transformation of the member forces or the displacements is required even though the individual member axes do not coincide with the structural axis for the entire frame. Generally, each nodal point in a plane rigid frame has three independent generalized displacements that must be considered (two translations and one rotation), and usually it is necessary to transform the generalized member forces and displacements into the common structural coordinate system. For the beam member shown in Fig. 18.3 the generalized forces at node i are transformed as follows

$$\bar{x}_i = x_i \cos \phi + y_i \sin \phi \qquad \bar{y}_i = -x_i \sin \phi + y_i \cos \phi \qquad \bar{m}_i = m_i \qquad (18.1)$$

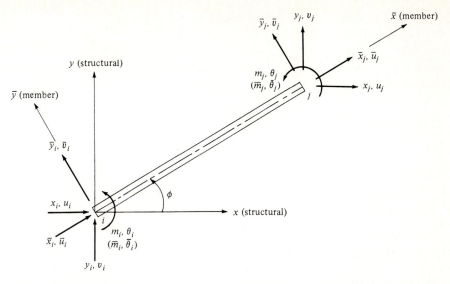

Figure 18.3 The generalized nodal forces and displacements for an arbitrarily oriented beam member.

Similar transformations can be expressed for the generalized forces at node j; thus, for both nodes of the member

$$
\begin{bmatrix} \bar{x}_i \\ \bar{y}_i \\ \bar{m}_i \\ \bar{x}_j \\ \bar{y}_j \\ \bar{m}_j \end{bmatrix} = \begin{bmatrix} \cos\phi & \sin\phi & 0 & 0 & 0 & 0 \\ -\sin\phi & \cos\phi & 0 & 0 & 0 & 0 \\ 0 & 0 & 1 & 0 & 0 & 0 \\ 0 & 0 & 0 & \cos\phi & \sin\phi & 0 \\ 0 & 0 & 0 & -\sin\phi & \cos\phi & 0 \\ 0 & 0 & 0 & 0 & 0 & 1 \end{bmatrix} \begin{bmatrix} x_i \\ y_i \\ m_i \\ x_j \\ y_j \\ m_j \end{bmatrix}
$$

$$\bar{\mathbf{x}} \qquad\qquad\qquad \mathbf{T} \qquad\qquad\qquad \mathbf{x}$$

(18.2)

or
$$\bar{\mathbf{x}} = \mathbf{T}\mathbf{x} \tag{18.2a}$$

where $\bar{\mathbf{x}}$ = column matrix containing nodal force components in member coordinate system

$\mathbf{x}$ = column matrix containing nodal force components in structural coordinate system

$\mathbf{T}$ = coordinate transformation matrix for beam member

$\mathbf{T}$ is an orthogonal matrix, similar to the transformation matrix obtained for truss members [Eq. (16.25)]. That is, $\mathbf{T}^{-1} = \mathbf{T}^T$, which implies that

$$\mathbf{x} = \mathbf{T}^T\bar{\mathbf{x}} \tag{18.3}$$

This same transformation can be applied to the corresponding column matrices of the generalized displacements $\bar{\mathbf{u}}$ and $\mathbf{u}$. That is,

$$\bar{\mathbf{u}} = \mathbf{T}\mathbf{u} \tag{18.4}$$

and the transformation from member to structural coordinates is

$$\mathbf{u} = \mathbf{T}^T \overline{\mathbf{u}} \tag{18.5}$$

where the generalized nodal displacements in the member and structural coordinate systems are represented, respectively, by the column matrices

$$\overline{\mathbf{u}} = \begin{bmatrix} \overline{u}_i \\ \overline{v}_i \\ \overline{\theta}_i \\ \overline{u}_j \\ \overline{v}_j \\ \overline{\theta}_j \end{bmatrix} \quad \text{and} \quad \mathbf{u} = \begin{bmatrix} u_i \\ v_i \\ \theta_i \\ u_j \\ v_j \\ \theta_j \end{bmatrix} \tag{18.6}$$

In the analysis of beams in Chap. 17 it was assumed that the member is inextensible, i.e., the axial deformation is identically zero. For a girder in a rigid frame the floor system can be imagined as inhibiting axial deformations of the member so that the girder will displace in a direction parallel to its longitudinal axis with no accompanying deformation. Several different solution procedures can accommodate this type of behavior. The assumption of the longitudinal inextensibility of the members can be enforced by operating on $\mathbf{K}_{ff}$. A heuristic argument outlining this approach is presented in this section, and the method is demonstrated, but the equations for formal execution of this procedure are not developed until Chap. 19. An alternative to assuming zero axial deformation is to include the actual axial stiffness of members in each individual stiffness matrix, and this approach is described in Sec. 18.3.

Before the zero member length changes can be incorporated into the merged stiffness matrix, it is necessary to acknowledge the presence of the displacements $\overline{u}_i$ and $\overline{u}_j$ in each stiffness matrix (expressed in member coordinates) without changing the informational content of Eq. (17.10)

$$\begin{bmatrix} \overline{x}_i \\ \overline{y}_i \\ \overline{m}_i \\ \overline{x}_j \\ \overline{y}_j \\ \overline{m}_j \end{bmatrix} = \frac{EI}{L} \begin{bmatrix} 0 & 0 & 0 & 0 & 0 & 0 \\ 0 & \dfrac{12}{L^2} & \dfrac{6}{L} & 0 & -\dfrac{12}{L^2} & \dfrac{6}{L} \\ 0 & \dfrac{6}{L} & 4 & 0 & -\dfrac{6}{L} & 2 \\ 0 & 0 & 0 & 0 & 0 & 0 \\ 0 & -\dfrac{12}{L^2} & -\dfrac{6}{L} & 0 & \dfrac{12}{L^2} & -\dfrac{6}{L} \\ 0 & \dfrac{6}{L} & 2 & 0 & -\dfrac{6}{L} & 4 \end{bmatrix} \begin{bmatrix} \overline{u}_i \\ \overline{v}_i \\ \overline{\theta}_i \\ \overline{u}_j \\ \overline{v}_j \\ \overline{\theta}_j \end{bmatrix} \tag{18.7}$$

$$\overline{\mathbf{x}} \qquad\qquad \overline{\mathbf{k}} \qquad\qquad \overline{\mathbf{u}}$$

In this form $\overline{\mathbf{k}}$ is conformable (with respect to multiplication) with $\mathbf{T}$ and $\mathbf{T}^T$, and the member stiffness matrix can be calculated in terms of the structural coordinates in the usual manner. That is,

$$\mathbf{k} = \mathbf{T}^T \bar{\mathbf{k}} \mathbf{T} \tag{18.8}$$

The form of $\bar{\mathbf{k}}$ contained in Eq. (18.7) expresses the fact that the member has zero axial stiffness and not the rigid axial property desired; therefore, when using this form of $\bar{\mathbf{k}}$ it is imperative that the member inextensibilities be enforced before the solution is attempted in order to avoid wrong results. As demonstrated in Example 18.1, this operation is typically performed after enforcing the displacement boundary conditions on $\mathbf{K}$ by a combination of the rows and columns of $\mathbf{K}_{ff}$.

Example 18.1 Calculate the nodal displacements and member forces for this rigid frame. Assume that the members do not deform along their longitudinal axes; $EI = \text{const.}$

SOLUTION

$$\mathbf{T}_{ab} = \mathbf{T}_{dc} = \begin{bmatrix} 0 & 1 & 0 & 0 & 0 & 0 \\ -1 & 0 & 0 & 0 & 0 & 0 \\ 0 & 0 & 1 & 0 & 0 & 0 \\ 0 & 0 & 0 & 0 & 1 & 0 \\ 0 & 0 & 0 & -1 & 0 & 0 \\ 0 & 0 & 0 & 0 & 0 & 1 \end{bmatrix}$$

Hence,

$$\mathbf{k}_{ab} = \mathbf{k}_{dc} = \frac{EI}{L} \begin{bmatrix} \dfrac{12}{L^2} & 0 & -\dfrac{6}{L} & -\dfrac{12}{L^2} & 0 & -\dfrac{6}{L} \\ 0 & 0 & 0 & 0 & 0 & 0 \\ -\dfrac{6}{L} & 0 & 4 & \dfrac{6}{L} & 0 & 2 \\ -\dfrac{12}{L^2} & 0 & \dfrac{6}{L} & \dfrac{12}{L^2} & 0 & \dfrac{6}{L} \\ 0 & 0 & 0 & 0 & 0 & 0 \\ -\dfrac{6}{L} & 0 & 2 & \dfrac{6}{L} & 0 & 4 \end{bmatrix}$$

$$\mathbf{k}_{bc} = \frac{EI}{L} \begin{bmatrix} 0 & 0 & 0 & 0 & 0 & 0 \\ 0 & \dfrac{12}{L^2} & \dfrac{6}{L} & 0 & -\dfrac{12}{L^2} & \dfrac{6}{L} \\ 0 & \dfrac{6}{L} & 4 & 0 & -\dfrac{6}{L} & 2 \\ 0 & 0 & 0 & 0 & 0 & 0 \\ 0 & -\dfrac{12}{L^2} & -\dfrac{6}{L} & 0 & \dfrac{12}{L^2} & -\dfrac{6}{L} \\ 0 & \dfrac{6}{L} & 2 & 0 & -\dfrac{6}{L} & 4 \end{bmatrix}$$

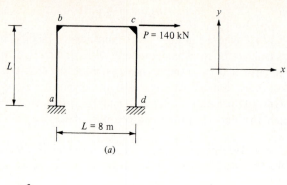

(a)

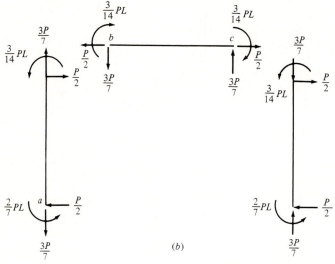

(b)

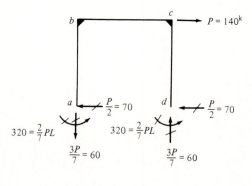

(c)

Figure E18.1 (a) The rigid frame with applied loading; (b) end forces for each member; (c) free-body diagram of the rigid frame.

Thus,
$$\mathbf{K}_{ff} = \frac{EI}{L}
\begin{array}{cccccc}
U_b & V_b & \Theta_b & U_c & V_c & \Theta_c
\end{array}
\begin{bmatrix}
\dfrac{12}{L^2} & 0 & \dfrac{6}{L} & 0 & 0 & 0 \\[2ex]
0 & \dfrac{12}{L^2} & \dfrac{6}{L} & 0 & -\dfrac{12}{L^2} & \dfrac{6}{L} \\[2ex]
\dfrac{6}{L} & \dfrac{6}{L} & 8 & 0 & -\dfrac{6}{L} & 2 \\[2ex]
0 & 0 & 0 & \dfrac{12}{L^2} & 0 & \dfrac{6}{L} \\[2ex]
0 & -\dfrac{12}{L^2} & -\dfrac{6}{L} & 0 & \dfrac{12}{L^2} & -\dfrac{6}{L} \\[2ex]
0 & \dfrac{6}{L} & 2 & \dfrac{6}{L} & -\dfrac{6}{L} & 8
\end{bmatrix}$$

Since $V_b = V_c = 0$ (axial inextensibility of ab and dc)

$$\mathbf{K}_{ff} = \frac{EI}{L}
\begin{array}{cccc}
U_b & \Theta_b & U_c & \Theta_c
\end{array}
\begin{bmatrix}
\dfrac{12}{L^2} & \dfrac{6}{L} & 0 & 0 \\[2ex]
\dfrac{6}{L} & 8 & 0 & 2 \\[2ex]
0 & 0 & \dfrac{12}{L^2} & \dfrac{6}{L} \\[2ex]
0 & 2 & \dfrac{6}{L} & 8
\end{bmatrix}$$

Since member bc does not elongate, $U_b = U_c$ and

$$\mathbf{K}_{ff} = \frac{EI}{L}
\begin{array}{ccc}
U_b & \Theta_b & \Theta_c
\end{array}
\begin{bmatrix}
\dfrac{24}{L^2} & \dfrac{6}{L} & \dfrac{6}{L} \\[2ex]
\dfrac{6}{L} & 8 & 2 \\[2ex]
\dfrac{6}{L} & 2 & 8
\end{bmatrix}
\qquad
\begin{bmatrix} U_b \\ \Theta_b \\ \Theta_c \end{bmatrix}
= \frac{L}{EI}
\begin{bmatrix}
\dfrac{24}{L^2} & \dfrac{6}{L} & \dfrac{6}{L} \\[2ex]
\dfrac{6}{L} & 8 & 2 \\[2ex]
\dfrac{6}{L} & 2 & 8
\end{bmatrix}^{-1}
\begin{bmatrix} P \\ 0 \\ 0 \end{bmatrix}$$

$$\begin{bmatrix} U_b \\ \Theta_b \\ \Theta_c \end{bmatrix}
= \frac{L^3}{84EI}
\begin{bmatrix}
5 & -\dfrac{3}{L} & -\dfrac{3}{L} \\[2ex]
-\dfrac{3}{L} & \dfrac{13}{L^2} & -\dfrac{1}{L^2} \\[2ex]
-\dfrac{3}{L} & -\dfrac{1}{L^2} & \dfrac{13}{L^2}
\end{bmatrix}
\begin{bmatrix} P \\ 0 \\ 0 \end{bmatrix}
= \frac{PL^3}{84EI}
\begin{bmatrix}
5 \\[2ex]
-\dfrac{3}{L} \\[2ex]
-\dfrac{3}{L}
\end{bmatrix}$$

DISCUSSION If the length changes of the columns are to be ignored, this implies that $V_b = V_c = 0$. These conditions can be introduced in the same manner as zero-support generalized displacements, and this yields the 4×4 matrix $\mathbf{K}_{ff}$, with $\mathbf{U}_f$ containing U_b, Θ_b, U_c, and Θ_c. The fact that member bc does not change length implies that $U_b = U_c$, and this can be carried out by replacing column 1 by the sum of columns 1 and 3 and discarding column 3. If a corresponding set of operations is carried out on the rows of $\mathbf{K}_{ff}$, a symmetric matrix results. Note that the operations on the rows are equivalent to combining the equations and that it is imperative for the right side of the equations (the forces) to be operated on in a corresponding manner. Alternatively, the first equation could have been retained (after combining the columns) and the third discarded, but this would yield a stiffness matrix that is not symmetric.

The member forces for this expanded version of $\overline{\mathbf{k}}$ (Eq. 18.7) are obtained in the usual manner according to Eq. (15.32) after the nodal generalized displacements have been calculated. The presence of the additional nodal displacements $\overline{u}_i$ and $\overline{u}_j$ poses no difficulties provided the matrix $\mathbf{s}$ for the beam member shown in Eq. (17.13) is augmented with the proper columns of zeros as follows

$$
\mathbf{s} = \frac{EI}{L}
\begin{bmatrix}
0 & \dfrac{12}{L^2} & \dfrac{6}{L} & 0 & -\dfrac{12}{L^2} & \dfrac{6}{L} \\[2ex]
0 & \dfrac{6}{L} & 4 & 0 & -\dfrac{6}{L} & 2 \\[2ex]
0 & \dfrac{6}{L} & 2 & 0 & -\dfrac{6}{L} & 4
\end{bmatrix}
\tag{18.9}
$$

Thus, the multiplication

$$
\mathbf{s}_{ij} = \mathbf{s}\overline{\mathbf{u}}_{ij}
\tag{18.10}
$$

will yield the assumed member response; i.e., if the member is inextensible, there is no longitudinal deformation; hence, the force of deformation is zero. Note that

$$
\mathbf{s}_{ij} =
\begin{bmatrix}
\overline{y}_i \\
\overline{m}_i \\
\overline{m}_j
\end{bmatrix}
\qquad
\overline{\mathbf{u}}_{ij} =
\begin{bmatrix}
\overline{u}_i \\
\overline{v}_i \\
\overline{\theta}_i \\
\overline{u}_j \\
\overline{v}_j \\
\overline{\theta}_j
\end{bmatrix}
\tag{18.11}
$$

These member forces are calculated for Example 18.1. Example 18.2 demonstrates the complete analysis procedure for the displacements and member forces for a plane rigid linear elastic frame in which not all the members are orthogonal. Example 18.3 illustrates the method for a rigid frame with a load applied between the member joints and requires the use of an additional node under the applied load.

Example 18.1 (continued) Calculate the member forces.

SOLUTION For member ab note that the displacements must be transformed from structural to member coordinates

$$
\begin{bmatrix} \bar{y}_a \\ \bar{m}_a \\ \bar{m}_b \end{bmatrix}_{ab} = \frac{EI}{L} \begin{bmatrix} -\dfrac{12}{L^2} & \dfrac{6}{L} \\[2mm] -\dfrac{6}{L} & 2 \\[2mm] -\dfrac{6}{L} & 4 \end{bmatrix} \frac{PL^3}{84EI} \begin{bmatrix} -5 \\[1mm] -\dfrac{3}{L} \end{bmatrix} = \begin{bmatrix} \dfrac{P}{2} \\[2mm] \dfrac{2PL}{7} \\[2mm] \dfrac{3PL}{14} \end{bmatrix}
$$

For member bc

$$
\begin{bmatrix} \bar{y}_b \\ \bar{m}_b \\ \bar{m}_c \end{bmatrix}_{bc} = \frac{EI}{L} \begin{bmatrix} \dfrac{12}{L^2} & \dfrac{6}{L} & -\dfrac{12}{L^2} & \dfrac{6}{L} \\[2mm] \dfrac{6}{L} & 4 & -\dfrac{6}{L} & 2 \\[2mm] \dfrac{6}{L} & 2 & -\dfrac{6}{L} & 4 \end{bmatrix} \frac{PL^3}{84EI} \begin{bmatrix} 0 \\[1mm] -\dfrac{3}{L} \\[1mm] 0 \\[1mm] -\dfrac{3}{L} \end{bmatrix} = \begin{bmatrix} -\dfrac{3P}{7} \\[2mm] -\dfrac{3PL}{14} \\[2mm] -\dfrac{3PL}{14} \end{bmatrix}
$$

For member cd the same calculations as those for member ab will give these member forces.

DISCUSSION The zero columns in **s** corresponding to zero $\bar{u}_i$ and $\bar{u}_j$ have been disregarded. The displacements obtained are used in the calculation of the member forces, and these results are equivalent to those in Example 12.9, where the frame was analyzed by the moment-distribution method. The shear and bending moments at the ends of the members (Fig. E18.1b) were obtained from the stiffness method; the member axial forces were calculated from Newton's third law and member equilibrium.

Example 18.2 Calculate (a) the nodal displacements and (b) member forces for this rigid frame. Assume that the members do not deform along their longitudinal axes; $I_{ab} = 2.0I$, $I_{bc} = I$, $I_{cd} = 2.5I$, $I = 600$ in^4, and $E = 29 \times 10^3$ kips/in^2.

SOLUTION (a)

$$
\mathbf{k}_{ab} = \frac{EI_{ab}}{L_{ab}}
\begin{array}{c}
\begin{matrix} U_a & V_a & \Theta_a & U_b & V_b & \Theta_b \end{matrix} \\
\begin{bmatrix}
\dfrac{12}{L_{ab}^2} & 0 & -\dfrac{6}{L_{ab}} & -\dfrac{12}{L_{ab}^2} & 0 & -\dfrac{6}{L_{ab}} \\[3mm]
0 & 0 & 0 & 0 & 0 & 0 \\[3mm]
-\dfrac{6}{L_{ab}} & 0 & 4 & \dfrac{6}{L_{ab}} & 0 & 2 \\[3mm]
-\dfrac{12}{L_{ab}^2} & 0 & \dfrac{6}{L_{ab}} & \dfrac{12}{L_{ab}^2} & 0 & \dfrac{6}{L_{ab}} \\[3mm]
0 & 0 & 0 & 0 & 0 & 0 \\[3mm]
-\dfrac{6}{L_{ab}} & 0 & 2 & \dfrac{6}{L_{ab}} & 0 & 4
\end{bmatrix}
\end{array}
$$

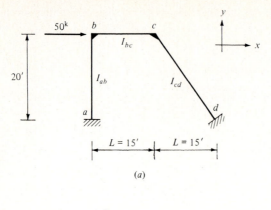

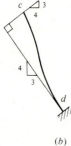

Figure E18.2 (*a*) The rigid frame with applied loading; (*b*) deformed geometry of member *dc*.

$$
\mathbf{k}_{bc} = \frac{EI_{bc}}{L_{bc}}
\begin{array}{c}
\begin{array}{cccccc}
U_b & V_b & \Theta_b & U_c & V_c & \Theta_c
\end{array} \\
\begin{bmatrix}
0 & 0 & 0 & 0 & 0 & 0 \\
0 & \dfrac{12}{L_{bc}^2} & \dfrac{6}{L_{bc}} & 0 & -\dfrac{12}{L_{bc}^2} & \dfrac{6}{L_{bc}} \\
0 & \dfrac{6}{L_{bc}} & 4 & 0 & -\dfrac{6}{L_{bc}} & 2 \\
0 & 0 & 0 & 0 & 0 & 0 \\
0 & -\dfrac{12}{L_{bc}^2} & -\dfrac{6}{L_{bc}} & 0 & \dfrac{12}{L_{bc}^2} & -\dfrac{6}{L_{bc}} \\
0 & \dfrac{6}{L_{bc}} & 2 & 0 & -\dfrac{6}{L_{bc}} & 4
\end{bmatrix}
\end{array}
$$

$$
\mathbf{T}_{dc} = \frac{1}{5}
\begin{bmatrix}
-3 & 4 & 0 & 0 & 0 & 0 \\
-4 & -3 & 0 & 0 & 0 & 0 \\
0 & 0 & 5 & 0 & 0 & 0 \\
0 & 0 & 0 & -3 & 4 & 0 \\
0 & 0 & 0 & -4 & -3 & 0 \\
0 & 0 & 0 & 0 & 0 & 5
\end{bmatrix}
$$

$$
\mathbf{k}_{dc} = \frac{EI_{dc}}{25L_{dc}}
\begin{array}{c}
\begin{array}{cccccc} U_d & V_d & \Theta_d & U_c & V_c & \Theta_c \end{array} \\
\begin{bmatrix}
\dfrac{192}{L_{dc}^2} & \dfrac{144}{L_{dc}^2} & -\dfrac{120}{L_{dc}} & -\dfrac{192}{L_{dc}^2} & -\dfrac{144}{L_{dc}^2} & -\dfrac{120}{L_{dc}} \\[8pt]
\dfrac{144}{L_{dc}^2} & \dfrac{108}{L_{dc}^2} & -\dfrac{90}{L_{dc}} & -\dfrac{144}{L_{dc}^2} & -\dfrac{108}{L_{dc}^2} & -\dfrac{90}{L_{dc}} \\[8pt]
-\dfrac{120}{L_{dc}} & -\dfrac{90}{L_{dc}} & 100 & \dfrac{120}{L_{dc}} & \dfrac{90}{L_{dc}} & 50 \\[8pt]
-\dfrac{192}{L_{dc}^2} & -\dfrac{144}{L_{dc}^2} & \dfrac{120}{L_{dc}} & \dfrac{192}{L_{dc}^2} & \dfrac{144}{L_{dc}^2} & \dfrac{120}{L_{dc}} \\[8pt]
-\dfrac{144}{L_{dc}^2} & -\dfrac{108}{L_{dc}^2} & \dfrac{90}{L_{dc}} & \dfrac{144}{L_{dc}^2} & \dfrac{108}{L_{dc}^2} & \dfrac{90}{L_{dc}} \\[8pt]
-\dfrac{120}{L_{dc}} & -\dfrac{90}{L_{dc}} & 50 & \dfrac{120}{L_{dc}} & \dfrac{90}{L_{dc}} & 100
\end{bmatrix}
\end{array}
$$

$$
\mathbf{K}_{ff} = E
\begin{array}{c}
\begin{array}{cccccc} U_b & V_b & \Theta_b & U_c & V_c & \Theta_c \end{array} \\
\begin{bmatrix}
\dfrac{12I_{ab}}{L_{ab}^3} & 0 & \dfrac{6I_{ab}}{L_{ab}^2} & 0 & 0 & 0 \\[8pt]
0 & \dfrac{12I_{bc}}{L_{bc}^3} & \dfrac{6I_{bc}}{L_{bc}^2} & 0 & -\dfrac{12I_{bc}}{L_{bc}^3} & \dfrac{6I_{bc}}{L_{bc}^2} \\[8pt]
\dfrac{6I_{ab}}{L_{ab}^2} & \dfrac{6I_{bc}}{L_{bc}^2} & \left(\dfrac{4I_{ab}}{L_{ab}}+\dfrac{4I_{bc}}{L_{bc}}\right) & 0 & -\dfrac{6I_{bc}}{L_{bc}^2} & \dfrac{2I_{bc}}{L_{bc}} \\[8pt]
0 & 0 & 0 & \dfrac{192I_{dc}}{25L_{dc}^3} & \dfrac{144I_{dc}}{25L_{dc}^3} & \dfrac{120I_{dc}}{25L_{dc}^2} \\[8pt]
0 & -\dfrac{12I_{bc}}{L_{bc}^3} & -\dfrac{6I_{bc}}{L_{bc}^2} & \dfrac{144I_{dc}}{25L_{dc}^3} & \left(\dfrac{12I_{bc}}{L_{bc}^3}+\dfrac{108I_{dc}}{25L_{dc}^3}\right) & \left(-\dfrac{6I_{bc}}{L_{bc}^2}+\dfrac{90I_{dc}}{25L_{dc}^2}\right) \\[8pt]
0 & \dfrac{6I_{bc}}{L_{bc}^2} & \dfrac{2I_{bc}}{L_{bc}} & \dfrac{120I_{dc}}{25L_{dc}^2} & \left(-\dfrac{6I_{bc}}{L_{bc}^2}+\dfrac{90I_{dc}}{25L_{dc}^2}\right) & \left(\dfrac{4I_{bc}}{L_{bc}}+\dfrac{4I_{dc}}{L_{dc}}\right)
\end{bmatrix}
\end{array}
$$

Let $L_{bc} = L$, $L_{ab} = \frac{4}{3}L$, $L_{dc} = \frac{5}{3}L$, $I_{bc} = I$, $I_{ab} = 2.0I$, and $I_{dc} = 2.5I$ and note that $V_b = 0$; then

$$
\mathbf{K}_{ff} = \frac{EI}{L}
\begin{array}{c}
\begin{array}{ccccc} U_b & \Theta_b & U_c & V_c & \Theta_c \end{array} \\
\begin{bmatrix}
\dfrac{81}{8L^2} & \dfrac{27}{4L} & 0 & 0 & 0 \\[8pt]
\dfrac{27}{4L} & 10 & 0 & -\dfrac{6}{L} & 2 \\[8pt]
0 & 0 & \dfrac{96(27)}{625L^2} & \dfrac{72(27)}{625L^2} & \dfrac{12(9)}{25L} \\[8pt]
0 & -\dfrac{6}{L} & \dfrac{72(27)}{625L^2} & \dfrac{8958}{625L^2} & -\dfrac{69}{25L} \\[8pt]
0 & 2 & \dfrac{12(9)}{25L} & -\dfrac{69}{25L} & 10
\end{bmatrix}
\end{array}
$$

Since member bc is inextensible, $U_b = U_c$ and

$$
\mathbf{K}_{ff} = \frac{EI}{L}
\begin{array}{cccc}
U_b & \Theta_b & V_c & \Theta_c
\end{array}
\begin{bmatrix}
\dfrac{14.2722}{L^2} & \dfrac{6.75}{L} & \dfrac{3.1104}{L^2} & \dfrac{4.32}{L} \\[2mm]
\dfrac{6.75}{L} & 10 & -\dfrac{6}{L} & 2 \\[2mm]
\dfrac{3.1104}{L^2} & -\dfrac{6}{L} & \dfrac{14.3328}{L^2} & -\dfrac{2.76}{L} \\[2mm]
\dfrac{4.32}{L} & 2 & -\dfrac{2.76}{L} & 10
\end{bmatrix}
$$

Noting that $V_c = \frac{3}{4}U_c$ since member cd does not change length (see Fig. E18.2b), we have

$$
\mathbf{K}_{ff} = \frac{EI}{L}
\begin{array}{ccc}
U_b & \Theta_b & \Theta_c
\end{array}
\begin{bmatrix}
\dfrac{27.00}{L^2} & \dfrac{2.25}{L} & \dfrac{2.25}{L} \\[2mm]
\dfrac{2.25}{L} & 10 & 2 \\[2mm]
\dfrac{2.25}{L} & 2 & 10
\end{bmatrix}
= \frac{EI}{L}
\begin{bmatrix}
0.12 & 0.15 & 0.15 \\
0.15 & 10 & 2 \\
0.15 & 2 & 10
\end{bmatrix}
$$

$$
\begin{bmatrix}
U_b \\ \Theta_b \\ \Theta_c
\end{bmatrix}
= \frac{L}{EI}
\begin{bmatrix}
8.602151 & -0.107527 & -0.107527 \\
-0.107527 & 0.105511 & -0.019489 \\
-0.107527 & -0.019489 & 0.105511
\end{bmatrix}
\begin{bmatrix}
50 \\ 0 \\ 0
\end{bmatrix}
$$

$$
= [124.1379 \times 10^{-6}\ (\text{kip}\cdot\text{ft})^{-1}]
\begin{bmatrix}
430.1076 \\ -5.3764 \\ -5.3764
\end{bmatrix}
= 10^{-3}
\begin{bmatrix}
53.39\ \text{ft} \\ -0.6674\ \text{rad} \\ -0.6674\ \text{rad}
\end{bmatrix}
$$

(b) For ab, $I_{ab} = 2I$, and $L_{ab} = \frac{4}{3}L = 20$

$$
\begin{bmatrix}
\bar{y}_a \\ \bar{m}_a \\ \bar{m}_b
\end{bmatrix}_{ab}
= \frac{EI_{ab}}{L_{ab}}
\begin{bmatrix}
\dfrac{12}{L_{ab}^2} & \dfrac{6}{L_{ab}} & -\dfrac{12}{L_{ab}^2} & \dfrac{6}{L_{ab}} \\[2mm]
\dfrac{6}{L_{ab}} & 4 & -\dfrac{6}{L_{ab}} & 2 \\[2mm]
\dfrac{6}{L_{ab}} & 2 & -\dfrac{6}{L_{ab}} & 4
\end{bmatrix}
\frac{L}{EI}
\begin{bmatrix}
0 \\ 0 \\ -430.1076 \\ -5.3764
\end{bmatrix}
$$

$$
= \frac{3}{2}
\begin{bmatrix}
11.2903 \\ 118.2795 \\ 107.5267
\end{bmatrix}
=
\begin{bmatrix}
16.94\ \text{kips} \\ 177.42\ \text{kip}\cdot\text{ft} \\ 161.29\ \text{kip}\cdot\text{ft}
\end{bmatrix}
$$

For bc

$$
\begin{bmatrix} \bar{y}_b \\ \bar{m}_b \\ \bar{m}_c \end{bmatrix}_{bc} = \frac{EI}{L} \begin{bmatrix} \dfrac{12}{L^2} & \dfrac{6}{L} & -\dfrac{12}{L^2} & \dfrac{6}{L} \\[2mm] \dfrac{6}{L} & 4 & -\dfrac{6}{L} & 2 \\[2mm] \dfrac{6}{L} & 2 & -\dfrac{6}{L} & 4 \end{bmatrix} \frac{L}{EI}\begin{bmatrix} 0 \\ -5.3764 \\ 322.5807 \\ -5.3764 \end{bmatrix} = \begin{bmatrix} -21.50 \text{ kips} \\ -161.29 \text{ kip} \cdot \text{ft} \\ -161.29 \text{ kip} \cdot \text{ft} \end{bmatrix}
$$

For dc, $I_{dc} = 2.5I$ and $L_{dc} = \tfrac{5}{3}L$

$$
\begin{bmatrix} \bar{u}_d \\ \bar{v}_d \\ \bar{\theta}_d \\ \bar{u}_c \\ \bar{v}_c \\ \bar{\theta}_c \end{bmatrix}_{dc} = \frac{1}{5} \begin{bmatrix} -3 & 4 & 0 & & & \\ -4 & -3 & 0 & & \mathbf{0} & \\ 0 & 0 & 5 & & & \\ \hline & & & -3 & 4 & 0 \\ & \mathbf{0} & & -4 & -3 & 0 \\ & & & 0 & 0 & 5 \end{bmatrix} \frac{L}{EI}\begin{bmatrix} 0 \\ 0 \\ 0 \\ 430.1076 \\ 322.5807 \\ -5.3764 \end{bmatrix} = \frac{L}{EI}\begin{bmatrix} 0 \\ 0 \\ 0 \\ 0 \\ -537.6345 \\ -5.3764 \end{bmatrix}
$$

$$
\begin{bmatrix} \bar{y}_d \\ \bar{m}_d \\ \bar{m}_c \end{bmatrix}_{dc} = \frac{EI_{dc}}{L_{dc}} \begin{bmatrix} \dfrac{12}{L_{dc}^2} & \dfrac{6}{L_{dc}} & -\dfrac{12}{L_{dc}^2} & \dfrac{6}{L_{dc}} \\[2mm] \dfrac{6}{L_{dc}} & 4 & -\dfrac{6}{L_{dc}} & 2 \\[2mm] \dfrac{6}{L_{dc}} & 2 & -\dfrac{6}{L_{dc}} & 4 \end{bmatrix} \frac{L}{EI}\begin{bmatrix} 0 \\ 0 \\ -537.6345 \\ -5.3764 \end{bmatrix}
$$

$$
= \frac{3}{2}\begin{bmatrix} 9.03225 \\ 118.27948 \\ 107.52668 \end{bmatrix} = \begin{bmatrix} 13.55 \text{ kips} \\ 177.42 \text{ kip} \cdot \text{ft} \\ 161.29 \text{ kip} \cdot \text{ft} \end{bmatrix}
$$

DISCUSSION $\mathbf{K}_{ff}$ is expressed in terms of I and L of member bc for computational convenience. The condition that $V_b = 0$ ensures that member ab will not change length. The inextensibility of member bc implies that $U_b = U_c$, and this is imposed as described in Example 18.1. From a sketch of the deformed member dc (Fig. E18.2b) it can be observed that if member dc does not change length, $V_c = 3U_c/4$. This latter condition requires a combination of the appropriate rows and columns of $\mathbf{K}_{ff}$ in this ratio. The final stiffness matrix to be inverted for this problem is 3×3, and the solution gives U_b, Θ_b, and Θ_c.

In calculating the member forces note that in each case the generalized displacements must be transformed into their respective individual member coordinate systems. A comparison with the results in Example 12.10 reveals that approximately the same member forces are obtained by moment distribution.

Example 18.3 Calculate (a) the displacements at nodes b, c, and d and (b) the member forces for this rigid frame. Assume that the members do not deform along their longitudinal axes. For all members $E = 200$ GPa and $I = 50 \times 10^{-6}$ m^4.

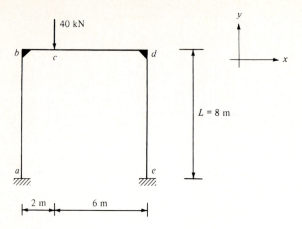

Figure E18.3

SOLUTION (a) For bc, $L_{bc} = \frac{1}{4}L$

$$
\mathbf{k}_{bc} = \frac{4EI}{L}
\begin{array}{c}
\begin{array}{cccccc} U_b & V_b & \Theta_b & U_c & V_c & \Theta_c \end{array} \\
\begin{bmatrix}
0 & 0 & 0 & 0 & 0 & 0 \\
0 & \dfrac{192}{L^2} & \dfrac{24}{L} & 0 & -\dfrac{192}{L^2} & \dfrac{24}{L} \\
0 & \dfrac{24}{L} & 4 & 0 & -\dfrac{24}{L} & 2 \\
0 & 0 & 0 & 0 & 0 & 0 \\
0 & -\dfrac{192}{L^2} & -\dfrac{24}{L} & 0 & \dfrac{192}{L^2} & -\dfrac{24}{L} \\
0 & \dfrac{24}{L} & 2 & 0 & -\dfrac{24}{L} & 4
\end{bmatrix}
\end{array}
= \frac{EI}{L}
\begin{bmatrix}
0 & 0 & 0 & 0 & 0 & 0 \\
0 & 12 & 12 & 0 & -12 & 12 \\
0 & 12 & 16 & 0 & -12 & 8 \\
0 & 0 & 0 & 0 & 0 & 0 \\
0 & -12 & -12 & 0 & 12 & -12 \\
0 & 12 & 8 & 0 & -12 & 16
\end{bmatrix}
$$

For cd, $L_{cd} = \frac{3}{4}L$

$$
\mathbf{k}_{cd} = \frac{4EI}{3L}
\begin{array}{c}
\begin{array}{cccccc} U_c & V_c & \Theta_c & U_d & V_d & \Theta_d \end{array} \\
\begin{bmatrix}
0 & 0 & 0 & 0 & 0 & 0 \\
0 & \dfrac{64}{3L^2} & \dfrac{8}{L} & 0 & -\dfrac{64}{3L^2} & \dfrac{8}{L} \\
0 & \dfrac{8}{L} & 4 & 0 & -\dfrac{8}{L} & 2 \\
0 & 0 & 0 & 0 & 0 & 0 \\
0 & -\dfrac{64}{3L^2} & -\dfrac{8}{L} & 0 & \dfrac{64}{3L^2} & -\dfrac{8}{L} \\
0 & \dfrac{8}{L} & 2 & 0 & -\dfrac{8}{L} & 4
\end{bmatrix}
\end{array}
= \frac{EI}{L}
\begin{bmatrix}
0 & 0 & 0 & 0 & 0 & 0 \\
0 & \dfrac{4}{9} & \dfrac{4}{3} & 0 & -\dfrac{4}{9} & \dfrac{4}{3} \\
0 & \dfrac{4}{3} & \dfrac{16}{3} & 0 & -\dfrac{4}{3} & \dfrac{8}{3} \\
0 & 0 & 0 & 0 & 0 & 0 \\
0 & -\dfrac{4}{9} & -\dfrac{4}{3} & 0 & \dfrac{4}{9} & -\dfrac{4}{3} \\
0 & \dfrac{4}{3} & \dfrac{8}{3} & 0 & -\dfrac{4}{3} & \dfrac{16}{3}
\end{bmatrix}
$$

For the unconstrained generalized displacements at the top of members ab and ed

$$\mathbf{k}_{ab} = \mathbf{k}_{ed} = \frac{EI}{L}\begin{bmatrix} 0.1875 & 0 & 0.75 \\ 0 & 0 & 0 \\ 0.75 & 0 & 4 \end{bmatrix}$$

The merged stiffness matrix is

$$\mathbf{K}_{ff} = \frac{EI}{L}$$

	U_b	V_b	Θ_b	U_c	V_c	Θ_c	U_d	V_d	Θ_d
	0.1875	0	0.75	0	0	0	0	0	0
	0	12	12	0	−12	12	0	0	0
	0.75	12	20	0	−12	8	0	0	0
	0	0	0	0	0	0	0	0	0
	0	−12	−12	0	12.4444	−10.6667	0	−0.4444	1.3333
	0	12	8	0	−10.6667	21.3333	0	−1.3333	2.6667
	0	0	0	0	0	0	0.1875	0	0.75
	0	0	0	0	−0.4444	−1.3333	0	0.4444	−1.3333
	0	0	0	0	1.3333	2.6667	0.75	−1.3333	9.3333

Since $V_b = V_d = 0$ and $U_b = U_c = U_d$, we have

$$\mathbf{K}_{ff} = \frac{EI}{L}$$

	U_b	Θ_b	V_c	Θ_c	Θ_d
	0.3750	0.75	0	0	0.75
	0.75	20	−12	8	0
	0	−12	12.4444	−10.6667	1.3333
	0	8	−10.6667	21.3333	2.6667
	0.75	0	1.3333	2.6667	9.3333

$$\begin{bmatrix} U_b \\ \Theta_b \\ V_c \\ \Theta_c \\ \Theta_d \end{bmatrix} = \frac{L}{EI}\begin{bmatrix} 3.809533 & -0.285717 & -0.214291 & 0.035713 & -0.285716 \\ -0.285717 & 0.154764 & 0.178575 & 0.032740 & -0.011905 \\ -0.214291 & 0.178575 & 0.368312 & 0.126120 & -0.071430 \\ 0.035713 & 0.032740 & 0.126120 & 0.103983 & -0.050596 \\ -0.285716 & -0.011905 & -0.071430 & -0.050596 & 0.154763 \end{bmatrix}\begin{bmatrix} 0 \\ 0 \\ -40 \\ 0 \\ 0 \end{bmatrix}$$

$$= 8 \times 10^{-4}\frac{1}{kN \cdot m}\begin{bmatrix} 8.5716 \\ -7.1430 \\ -14.7325 \\ -5.0448 \\ 2.8572 \end{bmatrix} = 10^{-3}\begin{bmatrix} 6.86 \text{ m} \\ -5.71 \text{ rad} \\ -11.79 \text{ m} \\ -4.04 \text{ rad} \\ 2.29 \text{ rad} \end{bmatrix}$$

(b) For ab

$$
\begin{bmatrix} \bar{y}_a \\ \bar{m}_a \\ \bar{m}_b \end{bmatrix}_{ab} = \frac{EI}{L} \begin{bmatrix} \dfrac{12}{L^2} & \dfrac{6}{L} & -\dfrac{12}{L^2} & \dfrac{6}{L} \\ \dfrac{6}{L} & 4 & -\dfrac{6}{L} & 2 \\ \dfrac{6}{L} & 2 & -\dfrac{6}{L} & 4 \end{bmatrix} \frac{L}{EI} \begin{bmatrix} 0 \\ 0 \\ -8.5716 \\ -7.1430 \end{bmatrix}
$$

$$
= \begin{bmatrix} 0.1875 & 0.75 & -0.1875 & 0.75 \\ 0.75 & 4 & -0.75 & 2 \\ 0.75 & 2 & -0.75 & 4 \end{bmatrix} \begin{bmatrix} 0 \\ 0 \\ -8.5716 \\ -7.1430 \end{bmatrix} = \begin{bmatrix} -3.75 \text{ kN} \\ -7.86 \text{ kN} \cdot \text{m} \\ -22.14 \text{ kN} \cdot \text{m} \end{bmatrix}
$$

For *ed*

$$
\begin{bmatrix} \bar{y}_e \\ \bar{m}_e \\ \bar{m}_d \end{bmatrix}_{ed} = \frac{EI}{L} \begin{bmatrix} 0.1875 & 0.75 & -0.1875 & 0.75 \\ 0.75 & 4 & -0.75 & 2 \\ 0.75 & 2 & -0.75 & 4 \end{bmatrix} \frac{L}{EI} \begin{bmatrix} 0 \\ 0 \\ -8.5716 \\ 2.8572 \end{bmatrix} = \begin{bmatrix} 3.75 \text{ kN} \\ 12.14 \text{ kN} \cdot \text{m} \\ 17.86 \text{ kN} \cdot \text{m} \end{bmatrix}
$$

For *bc*

$$
\begin{bmatrix} \bar{y}_b \\ \bar{m}_b \\ \bar{m}_c \end{bmatrix}_{bc} = \frac{4EI}{L} \begin{bmatrix} 3 & 3 & -3 & 3 \\ 3 & 4 & -3 & 2 \\ 3 & 2 & -3 & 4 \end{bmatrix} \frac{L}{EI} \begin{bmatrix} 0 \\ -7.1430 \\ -14.7325 \\ -5.0448 \end{bmatrix} = \begin{bmatrix} 30.54 \text{ kN} \\ 22.14 \text{ kN} \cdot \text{m} \\ 38.93 \text{ kN} \cdot \text{m} \end{bmatrix}
$$

For *cd*

$$
\begin{bmatrix} \bar{y}_c \\ \bar{m}_c \\ \bar{m}_d \end{bmatrix}_{cd} = \frac{4EI}{3L} \begin{bmatrix} 0.3333 & 1 & -0.3333 & 1 \\ 1 & 4 & -1 & 2 \\ 1 & 2 & -1 & 4 \end{bmatrix} \frac{L}{EI} \begin{bmatrix} -14.7325 \\ -5.0448 \\ 0 \\ 2.8572 \end{bmatrix} = \begin{bmatrix} -9.46 \text{ kN} \\ -38.93 \text{ kN} \cdot \text{m} \\ -17.86 \text{ kN} \cdot \text{m} \end{bmatrix}
$$

DISCUSSION This problem could have been simplified by replacing the load at node *c* with equivalent fixed-end forces at nodes *b* and *d* (see Sec. 17.5 for a description of equivalent forces); however, this approach would not yield the generalized displacements at node *c*. $\mathbf{K}_{ff}$ is expressed in terms of I and L of member *ab* for convenience, and the condition of axial rigidity of the members is enforced in the usual manner. Note that in this case it is necessary to combine three columns (and rows) since the U displacements of nodes *b*, *c*, and *d* are equal. The inverse of the final 5×5 stiffness matrix was carried out on a computer. The calculated member forces compare favorably with those obtained in Example 12.11 using moment distribution.

18.3 THE BEAM MEMBER WITH AXIAL DEFORMATION

The beam member discussed in Sec. 18.2 can be modified and used for the analysis of rigid frames in which the elastic deformations of the members must be calculated. As noted in Chap. 10, it is typically assumed that the axial and bending effects are uncoupled; i.e., the force in the longitudinal direction will induce neither shear nor bending into the member, and the bending does not result in axial forces. In so-called beam-column behavior these two phenomena are coupled, but a discussion of this behavior is beyond the scope of this book.

In Chap. 16 we observed that for a linear elastic truss member the axial nodal deflections and forces are related by Hooke's law [see Eq. (16.35)] as follows:

$$\bar{x}_i = \frac{AE}{L}(\bar{u}_i - \bar{u}_j) \qquad \bar{x}_j = \frac{AE}{L}(-\bar{u}_i + \bar{u}_j)$$

Incorporating these into the force-deflection relations of Eq. (18.7) for the prismatic beam member, we obtain (if the elastic axial deformations are to be considered)

$$
\begin{bmatrix} \bar{x}_i \\ \bar{y}_i \\ \bar{m}_i \\ \bar{x}_j \\ \bar{y}_j \\ \bar{m}_j \end{bmatrix}
= \frac{E}{L}
\begin{bmatrix}
A & 0 & 0 & -A & 0 & 0 \\
0 & \dfrac{12I}{L^2} & \dfrac{6I}{L} & 0 & -\dfrac{12I}{L^2} & \dfrac{6I}{L} \\
0 & \dfrac{6I}{L} & 4I & 0 & -\dfrac{6I}{L} & 2I \\
-A & 0 & 0 & A & 0 & 0 \\
0 & -\dfrac{12I}{L^2} & -\dfrac{6I}{L} & 0 & \dfrac{12I}{L^2} & -\dfrac{6I}{L} \\
0 & \dfrac{6I}{L} & 2I & 0 & -\dfrac{6I}{L} & 4I
\end{bmatrix}
\begin{bmatrix} \bar{u}_i \\ \bar{v}_i \\ \bar{\theta}_i \\ \bar{u}_j \\ \bar{v}_j \\ \bar{\theta}_j \end{bmatrix}
\tag{18.12}
$$

Thus, this stiffness matrix can be used in conjunction with the coordinate transformation matrix $\mathbf{T}$ in Eq. (18.2) since it has the generalized forces and displacements properly ordered and the two matrices are conformable with respect to multiplication. Therefore, $\bar{\mathbf{k}}$ can be transformed from member to structural coordinates using the usual approach shown in Eq. (18.8).

The member forces are computed using the nodal displacements, i.e., $\mathbf{s}_{ij} = \mathbf{su}_{ij}$, or

$$
\begin{bmatrix} \bar{x}_i \\ \bar{y}_i \\ \bar{m}_i \\ \bar{m}_j \end{bmatrix}
= \frac{E}{L}
\begin{bmatrix}
-A & 0 & 0 & A & 0 & 0 \\
0 & \dfrac{12I}{L^2} & \dfrac{6I}{L} & 0 & -\dfrac{12I}{L^2} & \dfrac{6I}{L} \\
0 & \dfrac{6I}{L} & 4I & 0 & -\dfrac{6I}{L} & 2I \\
0 & \dfrac{6I}{L} & 2I & 0 & -\dfrac{6I}{L} & 4I
\end{bmatrix}
\begin{bmatrix} \bar{u}_i \\ \bar{v}_i \\ \bar{\theta}_i \\ \bar{u}_j \\ \bar{v}_j \\ \bar{\theta}_j \end{bmatrix}
\tag{18.13}
$$

Example 18.4 illustrates the use of a mixture of beam and truss members that is necessary for the solution of some problems. Example 18.5 explores the use of the combined axial and bending behavior for the rigid frame analyzed previously in which the members were assumed to be axially rigid.

Example 18.4 Compute U_b, V_b, and Θ_b and the forces in the tie rod and beam.

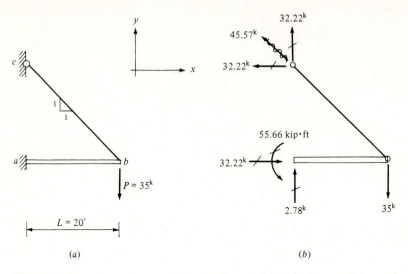

Figure E18.4 (*a*) The structure with applied loading; (*b*) free-body diagrams of the members.

For beam *ab*

$$I = 455 \text{ in}^4 = 21.94252 \times 10^{-3} \text{ ft}^4$$

$$A = 22.6 \text{ in}^2 = 156.94444 \times 10^{-3} \text{ ft}^2$$

For tie rod *bc*

$$A = 0.7854 \text{ in}^2 = 5.45417 \times 10^{-3} \text{ ft}^2$$

For both members

$$E = 29 \times 10^3 \text{ kips/in}^2 = 4176 \times 10^3 \text{ kips/ft}^2$$

SOLUTION For member *bc*

$$\phi = 135° \qquad \cos \phi = -\frac{1}{\sqrt{2}} \qquad \sin \phi = +\frac{1}{\sqrt{2}}$$

$$\mathbf{k}_{bc} = \frac{1}{\sqrt{2}}\begin{bmatrix} -1 & -1 & 0 & 0 \\ 1 & -1 & 0 & 0 \\ 0 & 0 & -1 & -1 \\ 0 & 0 & 1 & -1 \end{bmatrix} \frac{AE}{\sqrt{2}L}\begin{bmatrix} 1 & 0 & -1 & 0 \\ 0 & 0 & 0 & 0 \\ -1 & 0 & 1 & 0 \\ 0 & 0 & 0 & 0 \end{bmatrix} \frac{1}{\sqrt{2}}\begin{bmatrix} -1 & 1 & 0 & 0 \\ -1 & -1 & 0 & 0 \\ 0 & 0 & -1 & 1 \\ 0 & 0 & -1 & -1 \end{bmatrix}$$

$$\qquad\qquad U_b \qquad\qquad V_b \qquad\qquad U_c \qquad\qquad V_c$$

$$= \frac{10^{-3}E}{L}\begin{bmatrix} 1.92835 & -1.92835 & -1.92835 & 1.92835 \\ -1.92835 & 1.92835 & 1.92835 & -1.92835 \\ -1.92835 & 1.92835 & 1.92835 & -1.92835 \\ 1.92835 & -1.92835 & -1.92835 & 1.92835 \end{bmatrix}$$

For member ab

$$
\mathbf{k}_{ab} = \frac{10^{-3}E}{L}
\begin{array}{cccccc}
U_a & V_a & \Theta_a & U_b & V_b & \Theta_b
\end{array}
\begin{bmatrix}
156.94444 & 0 & 0 & -156.94444 & 0 & 0 \\
0 & 0.65828 & 6.58276 & 0 & -0.65828 & 6.58276 \\
0 & 6.58276 & 87.77008 & 0 & -6.58276 & 43.88504 \\
-156.94444 & 0 & 0 & 156.94444 & 0 & 0 \\
0 & -0.65828 & -6.58276 & 0 & 0.65828 & -6.58276 \\
0 & 6.58276 & 43.88504 & 0 & -6.58276 & 87.77008
\end{bmatrix}
$$

Thus,

$$
\mathbf{K}_{ff} = \frac{10^{-3}E}{L}
\begin{array}{ccc}
U_b & V_b & \Theta_b
\end{array}
\begin{bmatrix}
158.87279 & -1.92835 & 0 \\
-1.92835 & 2.58663 & -6.58276 \\
0 & -6.58276 & 87.77008
\end{bmatrix}
$$

$$
\begin{bmatrix} U_b \\ V_b \\ \Theta_b \end{bmatrix}
= \frac{10^3 L}{E}
\begin{bmatrix}
0.006366 & 0.005865 & 0.000440 \\
0.005865 & 0.483205 & 0.036240 \\
0.000440 & 0.036240 & 0.014111
\end{bmatrix}
\begin{bmatrix} 0 \\ -35 \\ 0 \end{bmatrix}
$$

$$
= 4.789 \times 10^{-3}
\begin{bmatrix} -0.205275 \\ -16.912159 \\ -1.268413 \end{bmatrix}
=
\begin{bmatrix} -0.00098 \text{ ft} \\ -0.08099 \text{ ft} \\ -0.00607 \text{ rad} \end{bmatrix}
=
\begin{bmatrix} -0.0118 \text{ in} \\ -0.9719 \text{ in} \\ -0.00607 \text{ rad} \end{bmatrix}
$$

The member forces for ab are

$$
\begin{bmatrix} \overline{x} \\ \overline{y}_a \\ \overline{m}_a \\ \overline{m}_b \end{bmatrix}_{ab}
= \frac{10^{-3}E}{L}
\begin{bmatrix}
-156.9444 & 0 & 0 & 156.9444 & 0 & 0 \\
0 & 0.65828 & 6.58276 & 0 & -0.65828 & 6.58276 \\
0 & 6.58276 & 87.77008 & 0 & -6.58276 & 43.88504 \\
0 & 6.58276 & 43.88504 & 0 & -6.58276 & 87.77008
\end{bmatrix}
$$

$$
\times \frac{10^3 L}{E}
\begin{bmatrix} 0 \\ 0 \\ 0 \\ -0.205275 \\ -16.912159 \\ -1.268413 \end{bmatrix}
=
\begin{bmatrix} -32.22 \text{ kips} \\ 2.78 \text{ kips} \\ 55.66 \text{ kip} \cdot \text{ft} \\ 0.00 \text{ kip} \cdot \text{ft} \end{bmatrix}
$$

For bc, $s_{bc} = \mathbf{sTu}$

$$
s_{bc} = \frac{10^{-3}E}{L}
\begin{bmatrix} -3.857284 & 0 & 3.857284 & 0 \end{bmatrix}
\frac{1}{\sqrt{2}}
\left[
\begin{array}{cc:cc}
-1 & 1 & & \\
-1 & -1 & \multicolumn{2}{c}{\mathbf{0}} \\
\hdashline
 & & -1 & 1 \\
\multicolumn{2}{c:}{\mathbf{0}} & -1 & -1
\end{array}
\right]
\frac{10^3 L}{E}
$$

$$\times \begin{bmatrix} -0.205275 \\ -16.912159 \\ 0 \\ 0 \end{bmatrix} = \begin{bmatrix} -2.727924 & 0 & 2.727924 & 0 \end{bmatrix} \begin{bmatrix} -16.706884 \\ 17.117434 \\ 0 \\ 0 \end{bmatrix} = 45.57 \text{ kips}$$

DISCUSSION The stiffness matrix for the truss member from Eq. (16.16) is combined with that for the beam with axial deformation, as shown in Eq. (18.12), and this gives $\mathbf{K}_{ff}$. The values in $\mathbf{K}_{ff}$ have all been calculated using units of feet. The beam is a standard steel rolled shape (W 10 × 77); therefore, the relative magnitudes of the numbers in this matrix are indicative of those encountered in steel construction. Note that the stiffness contributed to the (1,1) term in $\mathbf{K}_{ff}$ by the tie rod is relatively small. Since the axial deformation of the beam (U_b) turns out to be much smaller than V_b, the structure could have been solved assuming that the beam is axially rigid. The reactions calculated from the member forces are displayed in Fig. E18.4b.

Example 18.5 Calculate the nodal displacements for the rigid frame of Example 18.1 considering the axial elastic deformations of the members. For all members $I = 308 \times 10^{-6}$ m^4, $A = 16,517 \times 10^{-6}$ m^2, and $E = 200$ GPa.

SOLUTION From Example 18.1 the stiffness matrix without axial effects is

$$\mathbf{K}_{ff}^{(B)} = \frac{E \times 10^{-6}}{L} \begin{array}{cccccc} U_b & V_b & \Theta_b & U_c & V_c & \Theta_c \end{array}$$

$$\mathbf{K}_{ff}^{(B)} = \frac{E \times 10^{-6}}{L} \begin{bmatrix} 57.75 & 0 & 231.00 & 0 & 0 & 0 \\ 0 & 57.75 & 231.00 & 0 & -57.75 & 231.00 \\ 231.00 & 231.00 & 2464.00 & 0 & -231.00 & 616.00 \\ 0 & 0 & 0 & 57.75 & 0 & 231.00 \\ 0 & -57.75 & -231.00 & 0 & 57.75 & -231.00 \\ 0 & 231.00 & 616.00 & 231.00 & -231.00 & 2464.00 \end{bmatrix}$$

The merged stiffness matrix for only the axial effects is

$$\mathbf{K}_{ff}^{(A)} = \frac{E \times 10^{-6}}{L} \begin{array}{cccccc} U_b & V_b & \Theta_b & U_c & V_c & \Theta_c \end{array}$$

$$\mathbf{K}_{ff}^{(A)} = \frac{E \times 10^{-6}}{L} \begin{bmatrix} 16,517 & 0 & 0 & -16,517 & 0 & 0 \\ 0 & 16,517 & 0 & 0 & 0 & 0 \\ 0 & 0 & 0 & 0 & 0 & 0 \\ -16,517 & 0 & 0 & 16,517 & 0 & 0 \\ 0 & 0 & 0 & 0 & 16,517 & 0 \\ 0 & 0 & 0 & 0 & 0 & 0 \end{bmatrix}$$

Thus the combined stiffness matrix is

$$\mathbf{K}_{ff} = \mathbf{K}_{ff}^{(B)} + \mathbf{K}_{ff}^{(A)}$$

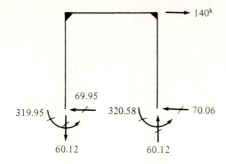

140^k

69.95

319.95   320.58   70.06

60.12   60.12   **Figure E18.5**

$$
= \frac{E \times 10^{-6}}{L} \times
\begin{array}{cccccc}
U_b & V_b & \Theta_b & U_c & V_c & \Theta_c \\
\end{array}
$$

$$
\begin{bmatrix}
16{,}574.75 & 0 & 231.00 & -16{,}517.00 & 0 & 0 \\
0 & 16{,}574.75 & 231.00 & 0 & -57.75 & 231.00 \\
231.00 & 231.00 & 2464.00 & 0 & -231.00 & 616.00 \\
-16{,}517.00 & 0 & 0 & 16{,}574.75 & 0 & 231.00 \\
0 & -57.75 & -231.00 & 0 & 16{,}574.75 & -231.00 \\
0 & 231.00 & 616.00 & 231.00 & -231.00 & 2464.00
\end{bmatrix}
$$

$$
\mathbf{K}_{ff}^{-1} = \frac{10^6 L}{E}
\begin{array}{cccccc}
U_b & V_b & \Theta_b & U_c & V_c & \Theta_c \\
\end{array}
$$

$$
\begin{bmatrix}
0.012406 & 0.000026 & -0.000935 & 0.012376 & -0.000026 & -0.000931 \\
0.000026 & 0.000061 & -0.000006 & 0.000026 & 0.000000 & -0.000006 \\
-0.000935 & -0.000006 & 0.000504 & -0.000931 & 0.000006 & -0.000038 \\
0.012376 & 0.000026 & -0.000931 & 0.012406 & -0.000026 & -0.000935 \\
-0.000026 & 0.000000 & 0.000006 & -0.000026 & 0.000061 & 0.000006 \\
-0.000931 & -0.000006 & -0.000038 & -0.000935 & 0.000006 & 0.000504
\end{bmatrix}
$$

$$
\begin{bmatrix}
U_b \\
V_b \\
\Theta_b \\
U_c \\
V_c \\
\Theta_c
\end{bmatrix}
= \frac{10^6 L}{E}
\begin{bmatrix}
1.732640 \\
0.003640 \\
-0.130340 \\
1.736840 \\
-0.003640 \\
-0.130900
\end{bmatrix}
=
\begin{bmatrix}
0.069306 \text{ m} \\
0.000146 \text{ m} \\
-0.005214 \text{ rad} \\
0.069474 \text{ m} \\
-0.000146 \text{ m} \\
-0.005236 \text{ rad}
\end{bmatrix}
$$

For the member forces use Eq. (18.13), $\mathbf{s}_{ij} = \mathbf{su}_{ij}$. For ab

$$
\begin{bmatrix}
\bar{x}_b \\
\bar{y}_a \\
\overline{m}_a \\
\overline{m}_b
\end{bmatrix}_{ab}
= \frac{10^{-6} E}{L}
\begin{bmatrix}
-16{,}517 & 0 & 0 & 16{,}517 & 0 & 0 \\
0 & 57.75 & 231.00 & 0 & -57.75 & 231.00 \\
0 & 231.00 & 1232.00 & 0 & -231.00 & 616.00 \\
0 & 231.00 & 616.00 & 0 & -231.00 & 1232.00
\end{bmatrix}
$$

$$
\times \frac{10^6 L}{E}
\begin{bmatrix}
0 \\
0 \\
0 \\
0.003640 \\
-1.732640 \\
-0.130340
\end{bmatrix}
=
\begin{bmatrix}
60.122 \text{ kips} \\
69.951 \text{ kips} \\
319.950 \text{ kip} \cdot \text{ft} \\
239.661 \text{ kip} \cdot \text{ft}
\end{bmatrix}
$$

For bc

$$
\begin{bmatrix}
\bar{x}_c \\
\bar{y}_b \\
\overline{m}_b \\
\overline{m}_c
\end{bmatrix}_{bc}
= \mathbf{s}\frac{10^6 L}{E}
\begin{bmatrix}
1.732640 \\
0.003640 \\
-0.130340 \\
1.736840 \\
-0.003640 \\
-0.130900
\end{bmatrix}
=
\begin{bmatrix}
69.371 \text{ kips} \\
-59.926 \text{ kips} \\
-239.532 \text{ kip} \cdot \text{ft} \\
-239.877 \text{ kip} \cdot \text{ft}
\end{bmatrix}
$$

For dc

$$
\begin{bmatrix}
\bar{x}_c \\
\bar{y}_d \\
\overline{m}_d \\
\overline{m}_c
\end{bmatrix}_{dc}
= \mathbf{s}\frac{10^6 L}{E}
\begin{bmatrix}
0 \\
0 \\
0 \\
-0.003640 \\
-1.736840 \\
-0.130900
\end{bmatrix}
=
\begin{bmatrix}
-60.122 \text{ kips} \\
70.065 \text{ kips} \\
320.576 \text{ kip} \cdot \text{ft} \\
239.941 \text{ kip} \cdot \text{ft}
\end{bmatrix}
$$

DISCUSSION For convenience, $\mathbf{K}_{ff}$ is formed by adding the terms contributed by the axial stiffness of the members to the 6×6 stiffness matrix obtained in Example 18.1. The generalized displacements (obtained by computer) reveal that the axial deformations of the three members are small. The elongation and shortening of the left and right columns, respectively, appear in this solution; however, a comparison of these displacements with those calculated in Example 18.1 discloses a difference of less than 0.5 percent. This small change results in a correspondingly small change in the member forces. The reactions are shown in Fig. E18.5. Although the axial deformations can generally be ignored for most low structures, this effect can be significant in high-rise structures.

The floor or roof system can inhibit axial deformations of the girders, and one is tempted to increase greatly the longitudinal stiffness of the girders. From Eq. (18.12) it would appear that a member could be described to be axially rigid by inserting very large entries in locations (1,1), (1,4), (4,1), and (4,4); however, this approach can pose significant problems. When such a member matrix (with very large values for A) is incorporated into the merged stiffness matrix, these large numbers can make $\mathbf{K}_{ff}$ ill conditioned with respect to solution. That is, numerical difficulties can be encountered when the generalized displacements are being computed; however, the results depend upon the precision of the computer being used and the magnitudes of the areas relative to the other stiffness terms.

18.4 THE THREE-DIMENSIONAL BISYMMETRIC BEAM MEMBER

The stiffness method can be generalized for structures composed of discrete members oriented in three-dimensional space by extending the previous discussion. Consider the prismatic beam member in Fig. 18.4; its cross section is symmetric with respect to each of its two principal axes $\bar{y}$ and $\bar{z}$. For three-dimensional behavior of this member, at each node we must consider six generalized forces $\bar{x}$, $\bar{y}$, $\bar{z}$, $\overline{m}_{\bar{x}}$, $\overline{m}_{\bar{y}}$, and $\overline{m}_{\bar{z}}$ and their corresponding generalized displacements $\bar{u}$, $\bar{v}$, $\bar{w}$, $\bar{\theta}_{\bar{x}}$, $\bar{\theta}_{\bar{y}}$, and $\bar{\theta}_{\bar{z}}$. All these quantities are shown in Fig. 18.4 directed in their respective position coordinate directions, with the moments and their corresponding rotations indicated as vectors with double arrowheads. This linear elastic member can be envisioned to have four distinct response modes: (1) tension-compression in the x direction, (2) flexure in the $\bar{x}\bar{y}$ plane, (3) flexure in the $\bar{x}\bar{z}$ plane, and (4) torsion about the $\bar{x}$ axis. For small displacements these four individual phenomena are uncoupled; i.e., the generalized forces from each individual response mode produce neither forces nor displacements in the other modes. The generalized force-displacement relationships for the first two phenomena were obtained in Eq. (18.12), but the moment of inertia used in these equations must now be noted more explicitly as $I_{\bar{z}}$, where

$$ I_{\bar{z}} = \int_{\text{area}} \bar{y}^2 \, dA $$

and the moments and rotations must also be subscripted with a $\bar{z}$.

The generalized force-displacement equations for flexure in the $\bar{x}\bar{z}$ plane (see Fig. 18.5) are similar to those expressed for the $\bar{x}\bar{y}$ plane; thus

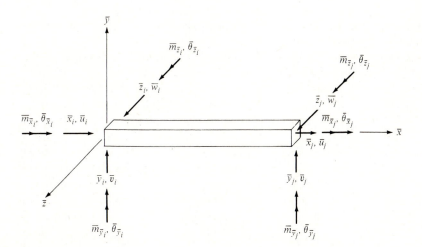

Figure 18.4 The beam member with six generalized forces per node.

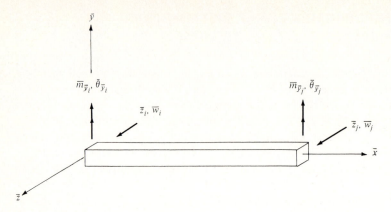

Figure 18.5 Flexure in the $\bar{x}\bar{z}$ plane.

$$\bar{z}_i = \frac{EI_{\bar{y}}}{L}\left(\frac{12}{L^2}\bar{w}_i - \frac{6}{L}\bar{\theta}_{\bar{y}_i} - \frac{12}{L^2}\bar{w}_j - \frac{6}{L}\bar{\theta}_{\bar{y}_j}\right)$$

$$\bar{m}_{\bar{y}_i} = \frac{EI_{\bar{y}}}{L}\left(-\frac{6}{L}\bar{w}_i + 4\bar{\theta}_{\bar{y}_i} + \frac{6}{L}\bar{w}_j + 2\bar{\theta}_{\bar{y}_j}\right)$$

$$\bar{z}_j = \frac{EI_{\bar{y}}}{L}\left(-\frac{12}{L^2}\bar{w}_i + \frac{6}{L}\bar{\theta}_{\bar{y}_i} + \frac{12}{L^2}\bar{w}_j + \frac{6}{L}\bar{\theta}_{\bar{y}_j}\right)$$

$$\bar{m}_{\bar{y}_j} = \frac{EI_{\bar{y}}}{L}\left(-\frac{6}{L}\bar{w}_i + 2\bar{\theta}_{\bar{y}_i} + \frac{6}{L}\bar{w}_j + 4\bar{\theta}_{\bar{y}_j}\right)$$

(18.14)

where

$$I_{\bar{y}} = \int_{area} \bar{z}^2\, dA$$

It is left as an exercise for the reader to verify the sign changes that occur between the sets of equations for flexural behavior in the two planes.

All the equations for bending have been obtained using the assumption that plane sections remain plane after distortion; however, this can be altered by the shear deformations in certain cases, e.g., those involving beams with perforated or laced webs or when the depth-to-span ratio is large. This effect has been ignored because for most civil engineering structures the additional displacements due to web shear strain are negligible.

The generalized force-displacement relationships for the torsional effects of the member can be investigated using basic considerations from strength of materials. The member shown in Fig. 18.6 is subjected to torsional end moments $\bar{m}_{\bar{x}_i}$ and $\bar{m}_{\bar{x}_j}$. A uniform member that is not restrained against longitudinal deformation (warping) and is subjected to pure torsion along its length, experiences a twist per unit length of

$$\frac{d\bar{\theta}_{\bar{x}}}{dx} = \frac{m_{\bar{x}}}{G\kappa_T}$$

(a)

Figure 18.6 Torsion about the $\bar{x}$ axis.

where $\overline{m}_{\bar{x}}$ = twisting moment at a generic section
$\quad\quad G$ = modulus of rigidity = $E/2(1 + \nu)$
$\quad\quad \nu$ = Poisson's ratio
$\quad\quad \kappa_T$ = torsion constant

The torsion constant, also referred to as the *St. Venant torsion constant,* depends upon the geometry of the cross section. For a circular cross section κ_T is the polar moment of inertia, but for other shapes κ_T must be computed using methods from the theory of elasticity. Although the derivation of κ_T is beyond the scope of this book, approximate expressions for several common shapes are indicated in Fig. 18.7. Integrating Eq. (*a*) over the length of the member gives

$$\bar{\theta}_{\bar{x}_j} - \bar{\theta}_{\bar{x}_i} = \frac{L}{G\kappa_T}\,\overline{m}_{\bar{x}_j} \tag{b}$$

Equilibrium requires that

$$\overline{m}_{\bar{x}_i} = -\overline{m}_{\bar{x}_j} \tag{c}$$

Thus, the relations between the generalized forces and displacements for torsion are

$$\overline{m}_{\bar{x}_i} = \frac{G\kappa_T}{L}\left(\bar{\theta}_{\bar{x}_i} - \bar{\theta}_{\bar{x}_j}\right) \qquad \overline{m}_{\bar{x}_j} = \frac{G\kappa_T}{L}\left(-\bar{\theta}_{\bar{x}_i} + \bar{\theta}_{\bar{x}_j}\right) \tag{18.15}$$

Torsion in circular sections results in only the uniform St. Venant stresses, and the distortions occur in the planes normal to the member axis. However, for other cross sections deformations will occur in the longitudinal direction due to the shearing strains, and the cross sections will warp. We can envision that if a member with a propensity for warping is rigidly confined at each end, considerable axial stresses will be produced. For closed cylindrical or rectangular tubes and solid rectangular sections, e.g., concrete members, normal stresses due to restraint of warping can generally be neglected; however, this may not be the case for member cross sections such as rolled-steel W sections and light-gage steel beams. In these cases, restraint of warping can produce significant stresses depending upon the support conditions and the torsional moments introduced. Typically for most civil engineering structures, the applied torsion is small, and we can ignore the effects of restrained warping.

The structural engineer must decide when warping is important, but this raises the more general question of how to approach the analysis of complex structures to obtain an economical and accurate solution. Problem solution methods can be broadly divided into two approaches: (1) all assumptions are made at the beginning of the investigation to give a simplified model or (2) no simplifying assumptions are made, and all factors, no matter how small their effect, are included in the analysis. The structural engineer with sufficient experience is capable of making assumptions to yield a solution that characterizes the behavior of the structure with the required degree of accuracy. Making rational initial assumptions should not be confused with the totally unacceptable approach of neglecting effects that the engineer neither understands nor is able to calculate.

The general 12×12 stiffness matrix for the beam member in Fig. 18.4 can be assembled by combining the force-displacement relations for the four individual response modes. Thus, by combining Eqs. (18.12), (18.14), and (18.15) we get the $\bar{\mathbf{k}}$ matrix in member coordinates for the three-dimensional beam member as shown in Eq. (18.16).

$$
\begin{bmatrix} \bar{x}_i \\ \bar{y}_i \\ \bar{z}_i \\ \bar{m}_{x_i} \\ \bar{m}_{y_i} \\ \bar{m}_{z_i} \\ \bar{x}_j \\ \bar{y}_j \\ \bar{z}_j \\ \bar{m}_{x_j} \\ \bar{m}_{y_j} \\ \bar{m}_{z_j} \end{bmatrix}
= \frac{E}{L}
\begin{bmatrix}
A & 0 & 0 & 0 & 0 & 0 & -A & 0 & 0 & 0 & 0 & 0 \\
0 & \dfrac{12I_z}{L^2} & 0 & 0 & 0 & \dfrac{6I_z}{L} & 0 & -\dfrac{12I_z}{L^2} & 0 & 0 & 0 & \dfrac{6I_z}{L} \\
0 & 0 & \dfrac{12I_{\bar y}}{L^2} & 0 & -\dfrac{6I_{\bar y}}{L} & 0 & 0 & 0 & -\dfrac{12I_{\bar y}}{L^2} & 0 & -\dfrac{6I_{\bar y}}{L} & 0 \\
0 & 0 & 0 & \dfrac{\kappa_T}{2(1+\nu)} & 0 & 0 & 0 & 0 & 0 & -\dfrac{\kappa_T}{2(1+\nu)} & 0 & 0 \\
0 & 0 & -\dfrac{6I_{\bar y}}{L} & 0 & 4I_{\bar y} & 0 & 0 & 0 & \dfrac{6I_{\bar y}}{L} & 0 & 2I_{\bar y} & 0 \\
0 & \dfrac{6I_z}{L} & 0 & 0 & 0 & 4I_z & 0 & -\dfrac{6I_z}{L} & 0 & 0 & 0 & 2I_z \\
-A & 0 & 0 & 0 & 0 & 0 & A & 0 & 0 & 0 & 0 & 0 \\
0 & -\dfrac{12I_z}{L^2} & 0 & 0 & 0 & -\dfrac{6I_z}{L} & 0 & \dfrac{12I_z}{L^2} & 0 & 0 & 0 & -\dfrac{6I_z}{L} \\
0 & 0 & -\dfrac{12I_{\bar y}}{L^2} & 0 & \dfrac{6I_{\bar y}}{L} & 0 & 0 & 0 & \dfrac{12I_{\bar y}}{L^2} & 0 & \dfrac{6I_{\bar y}}{L} & 0 \\
0 & 0 & 0 & -\dfrac{\kappa_T}{2(1+\nu)} & 0 & 0 & 0 & 0 & 0 & \dfrac{\kappa_T}{2(1+\nu)} & 0 & 0 \\
0 & 0 & -\dfrac{6I_{\bar y}}{L} & 0 & 2I_{\bar y} & 0 & 0 & 0 & \dfrac{6I_{\bar y}}{L} & 0 & 4I_{\bar y} & 0 \\
0 & \dfrac{6I_z}{L} & 0 & 0 & 0 & 2I_z & 0 & -\dfrac{6I_z}{L} & 0 & 0 & 0 & 4I_z
\end{bmatrix}
\begin{bmatrix} \bar{u}_i \\ \bar{v}_i \\ \bar{w}_i \\ \bar{\theta}_{x_i} \\ \bar{\theta}_{y_i} \\ \bar{\theta}_{z_i} \\ \bar{u}_j \\ \bar{v}_j \\ \bar{w}_j \\ \bar{\theta}_{x_j} \\ \bar{\theta}_{y_j} \\ \bar{\theta}_{z_j} \end{bmatrix}
$$

(18.16)

18.5 COORDINATE TRANSFORMATIONS IN THREE-DIMENSIONAL SPACE

The three-dimensional beam, along with the member coordinate system, is shown oriented in the structural coordinate system in Fig. 18.8. There are three independent

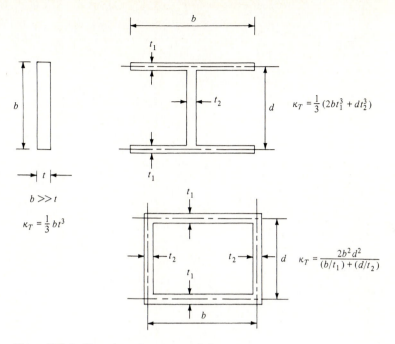

Figure 18.7 St. Venant's torsion constants (κ_T).

orthogonal components of force and a similar number of moment components at each node, each of which must be transformed from the member to the structural coordinate system. For clarity, only one component of force is displayed in Fig. 18.9a to c. When we denote the angles between the force $\bar{x}_i$ and the x, y, and z axes as $\alpha_{\bar{x}}$, $\beta_{\bar{x}}$, and $\gamma_{\bar{x}}$, respectively, the three components of this force in the respective structural x, y, and z directions are $\bar{x}_i \cos \alpha_{\bar{x}}$, $\bar{x}_i \cos \beta_{\bar{x}}$, and $\bar{x}_i \cos \gamma_{\bar{x}}$. The components of the forces $\bar{y}_i$ and $\bar{z}_i$ in the structural coordinates can be expressed in a similar fashion using the notation of Fig. 18.9b and c. Combining the components of $\bar{x}_i$, $\bar{y}_i$, and $\bar{z}_i$ gives

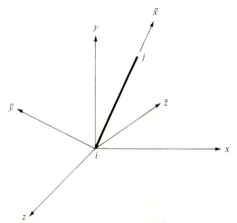

Figure 18.8 The beam in three-dimensional space.

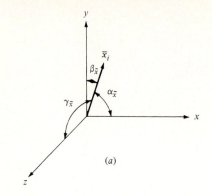

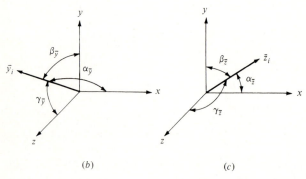

Figure 18.9 Transformation of forces from member to structural coordinates: $(a)\ \bar{x}_i$; $(b)\ \bar{y}_i$; $(c)\ \bar{z}_i$.

$$x_i = \bar{x}_i \cos \alpha_{\bar{x}} + \bar{y}_i \cos \alpha_{\bar{y}} + \bar{z}_i \cos \alpha_{\bar{z}}$$

$$y_i = \bar{x}_i \cos \beta_{\bar{x}} + \bar{y}_i \cos \beta_{\bar{y}} + \bar{z}_i \cos \beta_{\bar{z}} \qquad (18.17)$$

$$z_i = \bar{x}_i \cos \gamma_{\bar{x}} + \bar{y}_i \cos \gamma_{\bar{y}} + \bar{z}_i \cos \gamma_{\bar{z}}$$

Let λ, μ, and ν with the appropriate subscripts denote the direction cosines; then these transformation equations can be written

$$x_i = \bar{x}_i \lambda_{\bar{x}} + \bar{y}_i \lambda_{\bar{y}} + \bar{z}_i \lambda_{\bar{z}}$$

$$y_i = \bar{x}_i \mu_{\bar{x}} + \bar{y}_i \mu_{\bar{y}} + \bar{z}_i \mu_{\bar{z}} \qquad (18.18)$$

$$z_i = \bar{x}_i \nu_{\bar{x}} + \bar{y}_i \nu_{\bar{y}} + \bar{z}_i \nu_{\bar{z}}$$

or

$$\mathbf{x}_i = \mathbf{\Gamma}^T \bar{\mathbf{x}}_i \qquad (18.19)$$

where

$$\mathbf{x}_i = \begin{bmatrix} x_i \\ y_i \\ z_i \end{bmatrix} \qquad \bar{\mathbf{x}}_i = \begin{bmatrix} \bar{x}_i \\ \bar{y}_i \\ \bar{z}_i \end{bmatrix} \qquad \mathbf{\Gamma}^T = \begin{bmatrix} \lambda_{\bar{x}} & \lambda_{\bar{y}} & \lambda_{\bar{z}} \\ \mu_{\bar{x}} & \mu_{\bar{y}} & \mu_{\bar{z}} \\ \nu_{\bar{x}} & \nu_{\bar{y}} & \nu_{\bar{z}} \end{bmatrix} \qquad (18.20)$$

Since $\mathbf{\Gamma}^T$ is an orthogonal transformation,

$$(\boldsymbol{\Gamma}^T)^T = \boldsymbol{\Gamma} = (\boldsymbol{\Gamma}^T)^{-1}$$

or
$$\boldsymbol{\Gamma}^T = \boldsymbol{\Gamma}^{-1} \tag{18.21}$$

Therefore,
$$\bar{\mathbf{x}}_i = \boldsymbol{\Gamma}\mathbf{x}_i \tag{18.22}$$

This same transformation can be carried out on the moments at node i and on all the generalized forces at node j, yielding

$$
\begin{bmatrix} \bar{x}_i \\ \bar{y}_i \\ \bar{z}_i \\ \hline \bar{m}_{\bar{x}_i} \\ \bar{m}_{\bar{y}_i} \\ \bar{m}_{\bar{z}_i} \\ \hline \bar{x}_j \\ \bar{y}_j \\ \bar{z}_j \\ \hline \bar{m}_{\bar{x}_j} \\ \bar{m}_{\bar{y}_j} \\ \bar{m}_{\bar{z}_j} \end{bmatrix}
=
\begin{bmatrix} \boldsymbol{\Gamma} & \mathbf{0} & \mathbf{0} & \mathbf{0} \\ \mathbf{0} & \boldsymbol{\Gamma} & \mathbf{0} & \mathbf{0} \\ \mathbf{0} & \mathbf{0} & \boldsymbol{\Gamma} & \mathbf{0} \\ \mathbf{0} & \mathbf{0} & \mathbf{0} & \boldsymbol{\Gamma} \end{bmatrix}
\begin{bmatrix} x_i \\ y_i \\ z_i \\ \hline m_{x_i} \\ m_{y_i} \\ m_{z_i} \\ \hline x_j \\ y_j \\ z_j \\ \hline m_{x_j} \\ m_{y_j} \\ m_{z_j} \end{bmatrix}
\tag{18.23}
$$

$$\bar{\mathbf{x}} \qquad\qquad\qquad \mathbf{T} \qquad\qquad\qquad \mathbf{x}$$

where $\mathbf{0}$ indicates a 3×3 null matrix. In succinct form

$$\bar{\mathbf{x}} = \mathbf{Tx} \tag{18.24}$$

Just as $\boldsymbol{\Gamma}$ is orthogonal, so also is $\mathbf{T}$. Thus,

$$\mathbf{T}^T = \mathbf{T}^{-1} \qquad \text{and} \qquad \mathbf{x} = \mathbf{T}^T\bar{\mathbf{x}}$$

In addition, the generalized displacements transform in the usual fashion with this 12×12 version of $\mathbf{T}$ [see Eqs. (18.4) and (18.5)]; furthermore, the stiffness matrix in member coordinates as given in Eq. (18.16) can be transformed into structural coordinates using Eq. (18.8).

To demonstrate the use of these equations, a simple three-dimensional structure, the balcony frame, is presented in Example 18.6. Although in this structure the bending of one member induces torsion in the other, and vice versa, if it is assumed that the members do not deform longitudinally, we do not need all 6 generalized displacements to describe the behavior at each of the nodes.

Example 18.6 Calculate the displacements at b plus the member forces for this structure. Member ab is ST $14 \times 10 \times \frac{3}{8}$ (structural tubing), member bc is ST $14 \times 10 \times \frac{1}{2}$ and for both members $E = 29 \times 10^3$ kips/in^2 and $\nu = 0.30$.

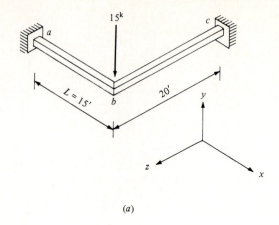

(a)

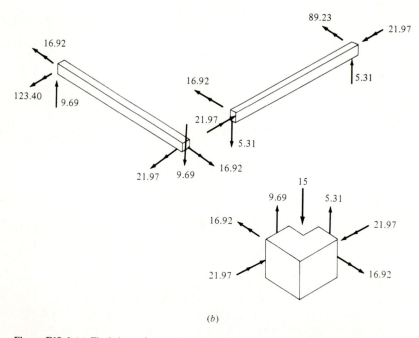

(b)

Figure E18.6 (a) The balcony frame with applied loading; (b) free-body diagrams of the members and joint b.

SOLUTION For member ab (ST $14 \times 10 \times \frac{3}{8}$)

$$I_{ab} = 476 \text{ in}^4 = 22.955 \times 10^{-3} \text{ ft}^4$$

$$\kappa_{T_{ab}} = \frac{2b^2d^2t}{b + d} = \frac{2(9.625)^2(13.625)^2(0.375)}{23.25} = 555 \text{ in}^4 = 26.765 \times 10^{-3} \text{ ft}^4$$

$$
\mathbf{k}_{ab} = \frac{E}{L_{ab}}
\begin{array}{ccc}
V_b & \Theta_{xb} & \Theta_{zb}
\end{array}
\begin{bmatrix}
\dfrac{12I_{ab}}{L_{ab}^2} & 0 & -\dfrac{6I_{ab}}{L_{ab}} \\[2ex]
0 & \dfrac{\kappa_{Tab}}{2(1+\nu)} & 0 \\[2ex]
-\dfrac{6I_{ab}}{L_{ab}} & 0 & 4I_{ab}
\end{bmatrix}
= \frac{E \times 10^{-3}}{L}
\begin{bmatrix}
1.2243 & 0 & -9.1820 \\
0 & 10.2942 & 0 \\
-9.1820 & 0 & 91.8200
\end{bmatrix}
$$

For member bc (ST $14 \times 10 \times \frac{1}{2}$)

$$
I_{bc} = 608 \text{ in}^4 = 29.321 \times 10^{-3} \text{ ft}^4 \qquad L_{bc} = \tfrac{4}{3}L
$$

$$
\kappa_{Tbc} = \frac{2b^2d^2t}{b+d} = \frac{2(9.5)^2(13.5)^2(0.5)}{23.0} = 715 \text{ in}^4 = 34.481 \times 10^{-3} \text{ ft}^4
$$

$$
\mathbf{k}_{bc} = \frac{E}{L_{bc}}
\begin{array}{ccc}
V_b & \Theta_{xb} & \Theta_{zb}
\end{array}
\begin{bmatrix}
\dfrac{12I_{bc}}{L_{bc}^2} & \dfrac{6I_{bc}}{L_{bc}} & 0 \\[2ex]
\dfrac{6I_{bc}}{L_{bc}} & 4I_{bc} & 0 \\[2ex]
0 & 0 & \dfrac{\kappa_{Tbc}}{2(1+\nu)}
\end{bmatrix}
= \frac{E \times 10^{-3}}{L}
\begin{array}{ccc}
V_b & \Theta_{xb} & \Theta_{zb}
\end{array}
\begin{bmatrix}
0.6597 & 6.5972 & 0 \\
6.5972 & 87.9630 & 0 \\
0 & 0 & 9.9464
\end{bmatrix}
$$

$$
\mathbf{K}_{ff} = \frac{10^{-3}E}{L}
\begin{array}{ccc}
V_b & \Theta_{xb} & \Theta_{zb}
\end{array}
\begin{bmatrix}
1.8840 & 6.5972 & -9.1820 \\
6.5972 & 98.2572 & 0 \\
-9.1820 & 0 & 101.7664
\end{bmatrix}
$$

$$
\begin{bmatrix} V_b \\ \Theta_{xb} \\ \Theta_{zb} \end{bmatrix}
= \frac{10^3L}{E}
\begin{bmatrix}
1.632407 & -0.109602 & 0.147286 \\
-0.109603 & 0.017536 & -0.009889 \\
0.147286 & -0.009889 & 0.023115
\end{bmatrix}
\begin{bmatrix} -15 \\ 0 \\ 0 \end{bmatrix}
$$

$$
= 3.591954 \times 10^{-3}
\begin{bmatrix} -24.486105 \\ 1.644045 \\ -2.209290 \end{bmatrix}
=
\begin{bmatrix} -0.08795 \text{ ft} \\ 0.005905 \text{ rad} \\ -0.007936 \text{ rad} \end{bmatrix}
=
\begin{bmatrix} -1.0554 \text{ in} \\ 0.005905 \text{ rad} \\ -0.007936 \text{ rad} \end{bmatrix}
$$

The member forces are, for ab

$$
\begin{bmatrix} \bar{y}_a \\ \bar{m}_{xa} \\ \bar{m}_{za} \\ \bar{m}_{zb} \end{bmatrix}_{ab}
= \frac{10^{-3}E}{L}
\begin{bmatrix}
-1.2243 & 0 & 9.1820 \\
0 & -10.2942 & 0 \\
-9.1820 & 0 & 45.9100 \\
-9.1820 & 0 & 91.8200
\end{bmatrix}
\frac{10^3L}{E}
\begin{bmatrix} -24.486105 \\ 1.644045 \\ -2.209290 \end{bmatrix}
$$

$$
=
\begin{bmatrix} 9.69 \text{ kips} \\ -16.92 \text{ kip} \cdot \text{ft} \\ 123.40 \text{ kip} \cdot \text{ft} \\ 21.97 \text{ kip} \cdot \text{ft} \end{bmatrix}
$$

For *bc*

$$
\begin{bmatrix} \bar{y}_b \\ \bar{m}_{\bar{x}b} \\ \bar{m}_{\bar{z}b} \\ \bar{m}_{\bar{z}c} \end{bmatrix}_{bc} = \frac{10^{-3}E}{L} \begin{bmatrix} 0.6597 & 0 & 6.5972 \\ 0 & 9.9464 & 0 \\ 6.5972 & 0 & 87.9630 \\ 6.5972 & 0 & 43.9815 \end{bmatrix} \frac{10^3 L}{E} \begin{bmatrix} -24.486105 \\ 2.209290 \\ 1.644045 \end{bmatrix}
$$

$$
= \begin{bmatrix} -5.31 \text{ kips} \\ 21.97 \text{ kip} \cdot \text{ft} \\ -16.92 \text{ kip} \cdot \text{ft} \\ -89.23 \text{ kip} \cdot \text{ft} \end{bmatrix}
$$

DISCUSSION These two members, each made from rectangular structural tubing, are connected at node b in such a way that moments and shear can be transmitted at this joint. If the axial deformations of both members are ignored, only Θ_{xb}, Θ_{zb}, and V_b must be investigated. The St. Venant torsion constants are computed using the expression in Fig. 18.7. In general, these closed sections are relatively efficient for torsion; nevertheless we note from the member stiffness matrices that the torsional stiffness term is only about 10 percent of the bending stiffness. All dimensions in $\mathbf{K}_{ff}$ are in feet for convenience.

The structural displacements must be transformed back into their respective member coordinates before the member forces are calculated. The matrix **s** was constructed for each member by selecting the appropriate entries from the general stiffness matrix in Eq. (18.16). It is suggested as an exercise that the reader calculate the complete matrix **s** corresponding to the general beam member.

A free-body diagram of the members and joint b (Fig. E18.6b) reveals that the components are in equilibrium; e.g., the bending moment in member ab at b equals the negative of the torsional moment in member bc.

18.6 DISCUSSION

The material in this chapter reveals the potential of the stiffness method for analyzing rigid-frame structures. Although the purpose is simply to introduce the topic, an understanding of the formulations enables one to investigate various types of typical rigid frames. The reader with additional interests should consult texts devoted exclusively to matrix structural analysis before attempting to analyze complex structures subjected to general loading conditions.

To avoid duplication, a number of phenomena have not been explicitly described in this chapter; material from the previous chapters in Part Five should be incorporated with the analysis of rigid-frame structures as the need arises. In situations involving *distributed loads*, the concepts presented in Sec. 17.5 on *equivalent nodal loads* must be used. Rigid frames subjected to *initial forces* induced by temperature, lack of fit, etc., must be analyzed using the appropriate theory from Sec. 17.6. The analysis for *support settlements* and/or *specified nodal displacements* involves the use of partitioning the merged stiffness matrix (Sec. 15.6).

Considerations of *idealization* for rigid frames have not been discussed in this chapter, i.e., nodal-point locations, member lengths, and section properties, etc. For relatively simple structures it is sufficient to refer to the guidelines in Chap. 17, but since this process is extremely important, a more thorough study of idealization should be made for complex structures.

PROBLEMS

Use the stiffness method for all problems. Unless noted otherwise, ignore member axial deformations and regard the material as steel with $E = 200$ GPa or $E = 29 \times 10^3$ kips/in².

In Probs. 18.1 to 18.7 calculate nodal generalized displacements and member forces for the original problem given.

18.1 Problem 12.19

18.2 Problem 11.18

18.3 Problem 12.16

18.4 Problem 12.18

18.5 Problem 12.41

18.6 Part (*a*) of Problem 12.43

18.7 Problem 12.44

In Probs. 18.8 to 18.16 replace the loading applied between the ends of members in the original problem with equivalent nodal loads. Calculate the generalized displacements at nodal points and the member forces.

18.8 Problem 11.29

18.9 Problem 12.7

18.10 Problem 12.8

18.11 Problem 12.9

18.12 Problem 12.10

18.13 Problem 12.17

18.14 Problem 12.47

18.15 Problem 12.48; let $L_a = 17.6$ ft.

18.16 Problem 12.49

In Probs. 18.17 to 18.19 calculate generalized displacements and member forces for the original problem.

18.17 Problem 12.50

18.18 Problem 12.51

18.19 Problem 12.52

18.20 Calculate generalized displacements and member forces for Prob. 11.19 assuming axial deformation of the beam to be (*a*) zero and (*b*) elastic.

18.21 Calculate generalized displacements and member forces for Prob. 10.21 assuming axial deformation of the beam to be (*a*) zero and (*b*) elastic.

18.22 Solve Prob. 18.4 including the effects of member axial deformations; $A_{ab} = A_{cd} = 20$ in² and $A_{bc} = 25$ in².

18.23 Solve Prob. 18.7 including the effects of member axial deformations, $A_{ab} = 25 \times 10^{-3}$ m², $A_{bc} = 10 \times 10^{-3}$ m², and $A_{cd} = 5 \times 10^{-3}$ m².

18.24 Remove the load from the rigid frame of Prob. 18.4 and compute the generalized displacements and member forces if: (*a*) support *d* settles 1.0 in; (*b*) the upper and lower surfaces of member *bc* are +120 and +75°F, respectively. The depth of the beam is 14 in and $\alpha = 6.5 \times 10^{-6}$ in/in · °F.

18.25 Remove the load from the rigid frame of Prob. 18.6 and compute the generalized displacements and member forces if: (*a*) support *d* settles 10 mm; (*b*) the upper and lower surfaces of member *bc* are +50 and 25°C, respectively. The depth of the beam is 200 mm and $\alpha = 1.2 \times 10^{-5}$ m/m · °C.

18.26 Remove the load from the rigid frame of Prob. 18.7 and compute the generalized displacements and member forces if: (*a*) support *a* moves 15 mm toward the right; (*b*) members *ab* and *cd* both experience a uniform temperature increase of 65°C ($\alpha = 1.2 \times 10^{-5}$ m/m · °C).

18.27 Remove the load from the rigid frame of Prob. 18.16 and compute the generalized displacements and member forces if: (*a*) support *d* moves 1 in to the left; (*b*) member *ab* experiences a uniform temperature increase of 100°F ($\alpha = 6.5 \times 10^{-6}$ in/in · °F).

18.28 Remove the concentrated force from the balcony frame of Example 18.6 and apply a uniformly distributed load of 1 kip/ft in the $-y$ direction along both members. Calculate the generalized displacements at joint b and the member forces.

18.29 This rigid steel frame is constructed of rectangular structural tubing $14 \times 10 \times \frac{1}{2}$ in with $L = 16$ ft, $E = 29 \times 10^6$ lb/in^2, and $\nu = 0.3$. Calculate the generalized displacements and member forces if: (a) a concentrated force of 12 kips ($-y$ direction) is placed at point a; (b) a uniformly distributed load of 1 kip/ft ($-y$ direction) is placed on member ab.

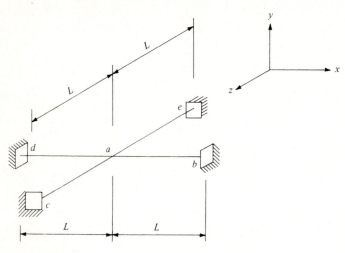

Figure P18.29

18.30 This rigid steel frame is constructed of rectangular structural tubing $30 \times 15 \times 1$ cm, with $L = 3$ m, $E = 200$ GPa, and $\nu = 0.3$. Calculate the generalized displacements at the nodal points indicated if: (a) a concentrated force of 10 kN ($-y$ direction) is applied at point c; (b) a uniformly distributed load of 2 kN/m ($-y$ direction) is applied to member bd.

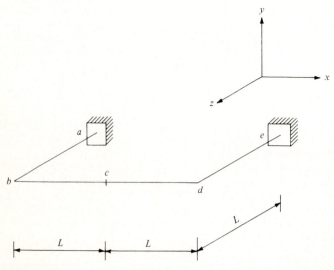

Figure P18.30

ADDITIONAL READING

Harrison, H. B.: *Structural Analysis and Design,* pt. 1, chaps. 8 and 9, Pergamon, New York, 1980.

Kardestuncer, H.: *Elementary Matrix Analysis of Structures,* chap. 5, McGraw-Hill, New York, 1974.

McGuire, W., and R. H. Gallagher: *Matrix Structural Analysis,* chaps. 4 and 5, Wiley, New York, 1979.

Martin, H. C.: *Introduction to Matrix Methods of Structural Analysis,* chap. 8, McGraw-Hill, New York, 1966.

Weaver, W., Jr., and J. M. Gere: *Matrix Analysis of Framed Structures,* chaps. 4 to 6, Van Nostrand, New York, 1980.

NINETEEN

SPECIAL CONSIDERATIONS FOR THE STIFFNESS METHOD

19.1 INTRODUCTION

Although the basic operations for analyzing trusses, beams, and rigid frames by the stiffness method have been presented in the preceding chapters of Part Five, some of these structures can present special problems. *Matrix condensation* is a useful technique for reducing the number of equations to be solved and investigating members with specific generalized forces given as zero, e.g., a beam with an internal hinge. In Chap. 18 the constraint condition that displacements are dependent because of the inextensibility of members was imposed by operating on the merged stiffness matrix in an ad hoc fashion. A more general method is described in this chapter for treating *constraints*. Another common structural feature is that of a displacement support condition skewed with respect to the structural coordinates, e.g., a roller with its support surface inclined to the horizontal, which can be analyzed using *nodal coordinates*. Finally, some comments are advanced for reducing the problem size for systems that are *symmetric* and/or *antisymmetric*.

Since the materials in Part Five are merely introductory, further study should be devoted to matrix structural analysis before attempting to solve large production-type problems.

19.2 MATRIX CONDENSATION

Sometimes it is either necessary or desirable to reduce the order of the force-displacement equations while retaining the effect of all the generalized forces and displacements. However, the generalized displacements (and forces) that do not explicitly appear in the final equations must implicitly affect the response of the structure. This can be done by partitioning the equations

$$\mathbf{X}_f = \mathbf{K}_{ff}\mathbf{U}_f \tag{19.1}$$

as

$$\begin{bmatrix} \mathbf{X}_\alpha \\ --- \\ \mathbf{X}_\beta \end{bmatrix} = \begin{bmatrix} \mathbf{K}_{\alpha\alpha} & \vdots & \mathbf{K}_{\alpha\beta} \\ ---- & \vdots & ----- \\ \mathbf{K}_{\beta\alpha} & \vdots & \mathbf{K}_{\beta\beta} \end{bmatrix} \begin{bmatrix} \mathbf{U}_\alpha \\ --- \\ \mathbf{U}_\beta \end{bmatrix} \qquad (19.2)$$

$$\mathbf{X}_f \qquad\qquad \mathbf{K}_{ff} \qquad\qquad \mathbf{U}_f$$

In general, $\mathbf{X}_\alpha$ and $\mathbf{X}_\beta$ are both matrices of known forces, and $\mathbf{U}_\alpha$ and $\mathbf{U}_\beta$ are matrices of unknown displacements. Expanding Eq. (19.2) gives the matrix equations

$$\mathbf{X}_\alpha = \mathbf{K}_{\alpha\alpha}\mathbf{U}_\alpha + \mathbf{K}_{\alpha\beta}\mathbf{U}_\beta \qquad \mathbf{X}_\beta = \mathbf{K}_{\beta\alpha}\mathbf{U}_\alpha + \mathbf{K}_{\beta\beta}\mathbf{U}_\beta \qquad (19.3)$$

Solving the second of these yields

$$\mathbf{U}_\beta = \mathbf{K}_{\beta\beta}^{-1}(\mathbf{X}_\beta - \mathbf{K}_{\beta\alpha}\mathbf{U}_\alpha) \qquad (19.4)$$

When this result is substituted into the first of Eqs. (19.3) and terms are rearranged, the combined equations become

$$\mathbf{X}_\alpha - \mathbf{K}_{\alpha\beta}\mathbf{K}_{\beta\beta}^{-1}\mathbf{X}_\beta = (\mathbf{K}_{\alpha\alpha} - \mathbf{K}_{\alpha\beta}\mathbf{K}_{\beta\beta}^{-1}\mathbf{K}_{\beta\alpha})\mathbf{U}_\alpha \qquad (19.5)$$

or the so-called *condensed* force-displacement equations are

$$\mathbf{X}_c = \mathbf{K}_{cc}\mathbf{U}_c \qquad (19.6)$$

where $\mathbf{U}_c = \mathbf{U}_\alpha$ and the *condensed force* and *stiffness matrices* are, respectively,

$$\mathbf{X}_c = \mathbf{X}_\alpha - \mathbf{K}_{\alpha\beta}\mathbf{K}_{\beta\beta}^{-1}\mathbf{X}_\beta \qquad (19.7)$$

and

$$\mathbf{K}_{cc} = \mathbf{K}_{\alpha\alpha} - \mathbf{K}_{\alpha\beta}\mathbf{K}_{\beta\beta}^{-1}\mathbf{K}_{\beta\alpha} \qquad (19.8)$$

This *matrix condensation* or *static condensation* can be very useful when it is desirable to reduce the order of the stiffness matrix. That is, since the order of $\mathbf{K}_{cc}$ is less than that of $\mathbf{K}_{ff}$, the solution is obtained by inverting the two smaller matrices $\mathbf{K}_{\alpha\alpha}$ and $\mathbf{K}_{\beta\beta}$. This approach can be used whether these matrices are explicitly inverted or the solution is found by a method such as Gauss elimination. Handling these reduced-order matrices can be extremely useful when investigating large problems or in working with a computer with limited core capacity. With this approach we solve smaller sets of equations, but since we must also perform a number of matrix multiplications which can consume a great amount of computer effort, the trade-off must be carefully considered. Note that matrix condensation is analogous to solving a pair of simultaneous equations by the elimination of one unknown.

The matrix $\mathbf{K}_{cc}$ contains the implicit effect of the generalized displacements in $\mathbf{U}_\beta$ even though their stiffness coefficients do not appear explicitly. Thus, by condensing $\mathbf{K}_{ff}$, solving Eq. (19.6), and substituting back into Eq. (19.4) all the generalized displacements are calculated. This is quite different from the solution obtained by solving the equation $\mathbf{X}_\alpha = \mathbf{K}_{\alpha\alpha}\mathbf{U}_\alpha$ [see Eq. (19.3)] since this formulation implies that $\mathbf{U}_\beta$ is null, i.e., that all the generalized displacements contained therein are constrained.

If the force matrix $\mathbf{X}_f$ is partitioned so that $\mathbf{X}_\beta$ is null, we note from Eq. (19.7) that $\mathbf{X}_c = \mathbf{X}_\alpha$, and Eq. (19.5) gives the condensed force-displacement equations as

$$\mathbf{X}_\alpha = \mathbf{K}_{cc}\mathbf{U}_\alpha \qquad (19.9)$$

Thus, in this case, the force matrix does not have to be condensed, which reduces the number of manipulations required to obtain the governing equations. Example 19.1 illustrates the application of matrix condensation in the form of Eq. (19.9) to reduce the computational effort required to calculate the generalized displacements.

Example 19.1 Calculate the generalized displacements for Example 17.5 by matrix condensation.

SOLUTION

$$
\begin{bmatrix} -30 \\ -45 \\ ---- \\ 0 \\ 0 \end{bmatrix} = 5.144
\begin{bmatrix} 24 & -12 & \vdots & 0 & 6 \\ -12 & 24 & \vdots & -6 & 0 \\ ----- & ----- & & ----- & ----- \\ 0 & -6 & \vdots & 8 & 2 \\ 6 & 0 & \vdots & 2 & 8 \end{bmatrix}
\begin{bmatrix} V_b \\ V_c \\ ---- \\ \Theta_b L \\ \Theta_c L \end{bmatrix}
$$

$$
\mathbf{K}_{cc} = 5.144 \begin{bmatrix} 24 & -12 \\ -12 & 24 \end{bmatrix} - 5.144 \begin{bmatrix} 0 & 6 \\ -6 & 0 \end{bmatrix} \frac{1}{60} \begin{bmatrix} 8 & -2 \\ -2 & 8 \end{bmatrix} \begin{bmatrix} 0 & -6 \\ 6 & 0 \end{bmatrix}
$$

$$
= 5.144 \begin{bmatrix} 24 & -12 \\ -12 & 24 \end{bmatrix} - \frac{5.144(6)}{5} \begin{bmatrix} 4 & 1 \\ 1 & 4 \end{bmatrix} = 5.144 \begin{bmatrix} 19.2 & -13.2 \\ -13.2 & 19.2 \end{bmatrix}
$$

$$
\mathbf{K}_{cc}^{-1} = \frac{1}{5.144(194.4)} \begin{bmatrix} 19.2 & 13.2 \\ 13.2 & 19.2 \end{bmatrix}
$$

$$
\begin{bmatrix} V_b \\ V_c \end{bmatrix} = \frac{1}{1000} \begin{bmatrix} 19.2 & 13.2 \\ 13.2 & 19.2 \end{bmatrix} \begin{bmatrix} -30 \\ -45 \end{bmatrix} = \begin{bmatrix} -1.170 \\ -1.260 \end{bmatrix} \text{ in}
$$

$$
\begin{bmatrix} \Theta_b L \\ \Theta_c L \end{bmatrix} = -\frac{1}{60} \begin{bmatrix} 8 & -2 \\ -2 & 8 \end{bmatrix} \begin{bmatrix} 0 & -6 \\ 6 & 0 \end{bmatrix} \begin{bmatrix} -1.170 \\ -1.260 \end{bmatrix} = -\frac{12}{60} \begin{bmatrix} -1 & -4 \\ 4 & 1 \end{bmatrix} \begin{bmatrix} -1.170 \\ -1.260 \end{bmatrix}
$$

$$
= \begin{bmatrix} -1.242 \\ 1.188 \end{bmatrix} \text{ in}
$$

DISCUSSION The generalized displacements and forces from Example 17.5 have been rearranged in the equations so that the moments and the associated rotations can be eliminated. Since $M_b = M_c = 0$, Eq. (19.9) can be used and the solution requires the inverse of the 2×2 matrices $\mathbf{K}_{\beta\beta}$ and $\mathbf{K}_{cc}$, instead of that of the larger $\mathbf{K}_{ff}$.

19.3 RELEASE OF GENERALIZED MEMBER NODAL FORCES

An ideal internal hinge in a beam introduces a discontinuity in the slope of the elastic curve and supports zero bending moment. This behavior is an example of a broad class of conditions in which one or more of the generalized member forces at a node is specified as zero — sometimes referred to as the *end release* of the generalized force. These special conditions can be treated by starting with the usual member stiffness matrix and eliminating the known zero generalized force using Eq. (19.9). This yields a modified (condensed) member stiffness matrix with the appropriate generalized force identical to zero and the corresponding displacement eliminated.

For example, consider the beam member in Fig. 19.1a with an internal hinge at node j. To obtain the stiffness matrix incorporating this condition the member stiffness matrix from Eq. (17.10) is first partitioned as

$$
\bar{\mathbf{k}} = \frac{EI}{L}
\left[
\begin{array}{ccc:c}
\dfrac{12}{L^2} & \dfrac{6}{L} & -\dfrac{12}{L^2} & \dfrac{6}{L} \\[2ex]
\dfrac{6}{L} & 4 & -\dfrac{6}{L} & 2 \\[2ex]
-\dfrac{12}{L^2} & -\dfrac{6}{L} & \dfrac{12}{L^2} & -\dfrac{6}{L} \\[1ex]
\hdashline \\[-2ex]
\dfrac{6}{L} & 2 & -\dfrac{6}{L} & 4
\end{array}
\right]
$$

The condensed stiffness matrix is calculated using Eq. (19.8).

$$
\bar{\mathbf{k}}_{cc} = \frac{EI}{L}
\begin{bmatrix}
\dfrac{12}{L^2} & \dfrac{6}{L} & -\dfrac{12}{L^2} \\[2ex]
\dfrac{6}{L} & 4 & -\dfrac{6}{L} \\[2ex]
-\dfrac{12}{L^2} & -\dfrac{6}{L} & \dfrac{12}{L^2}
\end{bmatrix}
- \frac{EI}{L}
\begin{bmatrix}
\dfrac{6}{L} \\[2ex]
2 \\[2ex]
-\dfrac{6}{L}
\end{bmatrix}
\frac{1}{4}
\begin{bmatrix}
\dfrac{6}{L} & 2 & -\dfrac{6}{L}
\end{bmatrix}
$$

Thus

$$
\bar{\mathbf{k}}_{cc} = \frac{3EI}{L}
\begin{bmatrix}
\dfrac{1}{L^2} & \dfrac{1}{L} & -\dfrac{1}{L^2} \\[2ex]
\dfrac{1}{L} & 1 & -\dfrac{1}{L} \\[2ex]
-\dfrac{1}{L^2} & -\dfrac{1}{L} & \dfrac{1}{L^2}
\end{bmatrix}
\tag{19.10}
$$

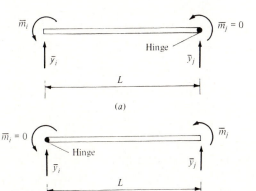

Figure 19.1 Beam member with hinge at (a) right node; and (b) left node.

and the force-displacement matrix equations are

$$
\begin{bmatrix} \bar{y}_i \\ \bar{m}_i \\ \bar{y}_j \end{bmatrix} = \frac{3EI}{L} \begin{bmatrix} \dfrac{1}{L^2} & \dfrac{1}{L} & -\dfrac{1}{L^2} \\[2mm] \dfrac{1}{L} & 1 & -\dfrac{1}{L} \\[2mm] -\dfrac{1}{L^2} & -\dfrac{1}{L} & \dfrac{1}{L^2} \end{bmatrix} \begin{bmatrix} \bar{v}_i \\ \bar{\theta}_i \\ \bar{v}_j \end{bmatrix} \tag{19.11}
$$

Note that in this case with the *hinge at node j* the generalized displacement $\bar{\theta}_j$ has been eliminated and will not be calculated when nodal displacements are computed, but this fact does not imply that this generalized displacement is zero.

Similarly, the condensed force-displacement relations for a beam member with a *hinge at node i* (Fig. 19.1b) is obtained by eliminating the zero generalized moment $\bar{m}_i$ and its corresponding rotation, to give

$$
\begin{bmatrix} \bar{y}_i \\ \bar{y}_j \\ \bar{m}_j \end{bmatrix} = \frac{3EI}{L} \begin{bmatrix} \dfrac{1}{L^2} & -\dfrac{1}{L^2} & \dfrac{1}{L} \\[2mm] -\dfrac{1}{L^2} & \dfrac{1}{L^2} & -\dfrac{1}{L} \\[2mm] \dfrac{1}{L} & -\dfrac{1}{L} & 1 \end{bmatrix} \begin{bmatrix} \bar{v}_i \\ \bar{v}_j \\ \bar{\theta}_j \end{bmatrix} \tag{19.12}
$$

Both forms of the hinged beam member in Eqs. (19.11) and (19.12) have a corresponding matrix **s** that must be used in computing member forces. Example 19.2 illustrates the use of these special member stiffness matrices for a beam.

Example 19.2 Calculate the generalized displacement at b and the member forces for this uniform beam with an internal hinge; EI = const.

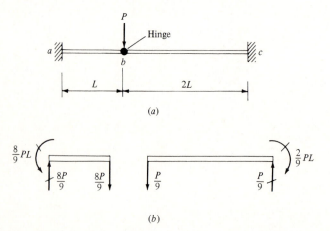

(a)

(b)

Figure E19.2 (a) Beam with an internal hinge; (b) free-body diagrams of beam members.

SOLUTION For the hinge on the left member use Eq. (19.10) for ab

$$\mathbf{K}_{ff} = \frac{EI}{L} \begin{bmatrix} \left(\dfrac{3}{L^2} + \dfrac{12}{8L^2}\right) & \dfrac{6}{4L} \\ \dfrac{6}{4L} & 2 \end{bmatrix} = \frac{EI}{L} \begin{bmatrix} \dfrac{9}{2L^2} & \dfrac{3}{2L} \\ \dfrac{3}{2L} & 2 \end{bmatrix}$$

$$\begin{bmatrix} V_b \\ \Theta_{b+} \end{bmatrix} = \frac{4L^3}{27EI} \begin{bmatrix} 2 & -\dfrac{3}{2L} \\ -\dfrac{3}{2L} & \dfrac{9}{2L^2} \end{bmatrix} \begin{bmatrix} -P \\ 0 \end{bmatrix} = \frac{4PL^3}{27EI} \begin{bmatrix} -2 \\ \dfrac{3}{2L} \end{bmatrix}$$

where Θ_{b+} = rotation to the right of b.

For member ab

$$\begin{bmatrix} \bar{y}_a \\ \bar{m}_a \end{bmatrix} = \frac{3EI}{L} \begin{bmatrix} \dfrac{1}{L^2} & \dfrac{1}{L} & -\dfrac{1}{L^2} \\ \dfrac{1}{L} & 1 & -\dfrac{1}{L} \end{bmatrix} \frac{4PL^3}{27EI} \begin{bmatrix} 0 \\ 0 \\ -2 \end{bmatrix} = \frac{8PL^2}{9} \begin{bmatrix} \dfrac{1}{L^2} \\ \dfrac{1}{L} \end{bmatrix}$$

For member bc

$$\begin{bmatrix} \bar{y}_b \\ \bar{m}_b \\ \bar{m}_c \end{bmatrix} = \frac{EI}{2L} \begin{bmatrix} \dfrac{12}{4L^2} & \dfrac{6}{2L} & -\dfrac{12}{4L^2} & \dfrac{6}{2L} \\ \dfrac{6}{2L} & 4 & -\dfrac{6}{2L} & 2 \\ \dfrac{6}{2L} & 2 & -\dfrac{6}{2L} & 4 \end{bmatrix} \frac{4PL^3}{27EI} \begin{bmatrix} -2 \\ \dfrac{3}{2L} \\ 0 \\ 0 \end{bmatrix} = \frac{2PL}{9} \begin{bmatrix} -\dfrac{1}{2L} \\ 0 \\ -1 \end{bmatrix}$$

With a hinge on both members

$$\mathbf{k} = \frac{3EI}{L^3} + \frac{3EI}{8L^3} = \frac{27EI}{8L^3} \qquad V_b = -\frac{8L^3}{27EI} P$$

With a hinge on the right member use Eq. (19.12) for bc.

$$\mathbf{K}_{ff} = \frac{EI}{L} \begin{bmatrix} \left(\dfrac{12}{L^2} + \dfrac{3}{8L^2}\right) & -\dfrac{6}{L} \\ -\dfrac{6}{L} & 4 \end{bmatrix} = \frac{EI}{L} \begin{bmatrix} \dfrac{99}{8L^2} & -\dfrac{6}{L} \\ -\dfrac{6}{L} & 4 \end{bmatrix}$$

$$\begin{bmatrix} V_b \\ \Theta_{b-} \end{bmatrix} = \frac{L}{EI} \frac{2L^2}{27} \begin{bmatrix} 4 & \dfrac{6}{L} \\ \dfrac{6}{L} & \dfrac{99}{8L^2} \end{bmatrix} \begin{bmatrix} -P \\ 0 \end{bmatrix}$$

$$\begin{bmatrix} V_b \\ \Theta_{b-} \end{bmatrix} = \frac{4PL^3}{27EI} \begin{bmatrix} -2 \\ -\dfrac{3}{L} \end{bmatrix}$$

where Θ_{b-} = rotation to the left of b.

DISCUSSION This structure can be analyzed using three different combinations of conventional and hinged beam members, and all have been demonstrated for computing the generalized displacements. In using two hinged members, only V_b is obtained, but in each of the other approaches the rotation at point b that is associated with the conventional beam member is obtained from the calculations.

Only the computation of member forces has been demonstrated for a hinged member between nodes a and b and a conventional beam member between nodes b and c (Fig. E19.2b). This required the **s** matrix corresponding to the stiffness matrix in each case. It is suggested as an exercise that the member forces for the other two choices of stiffness matrices be calculated.

19.4 DISCUSSION OF TRANSFORMATIONS

The analyses of both constraints (Sec. 19.5) and nodal coordinates (Sec. 19.6) require a more general understanding of transformations. In the ensuing discussion the properties of transformations will be developed using energy considerations from Chap. 10.

The transformation from structural to member coordinates for the generalized nodal displacements of a member can be expressed as

$$\bar{\mathbf{u}} = \mathbf{Tu} \tag{19.13}$$

Considering these both to be orthogonal coordinate axes, the work by the various generalized nodal forces can be determined by examining a typical component. For example, the generalized force $\bar{x}_i$ from $\bar{\mathbf{x}}$ produces the work $\frac{1}{2}\bar{x}_i\bar{u}_i$, with no work yielded in conjunction with any other displacement in $\bar{\mathbf{u}}$. That is, the nodal displacement and force components are coupled in the energy sense, and the force and displacement vectors are called *conjugate vectors*. The term "vector" is used for convenience as a mathematical generalization to denote a one-dimensional array and should not be confused with a physical vector, e.g., force or displacement. Hence, a mathematical vector may contain components from several physical vectors. The concept of forces and displacements coupled in an energy sense was described in a heuristic manner in Sec. 10.2. If force and displacement are conjugate vector sets expressed in both member and structural coordinates, the work done by the nodal forces (during linear elastic response) in each case is

$$W_e = \tfrac{1}{2}\bar{\mathbf{u}}^T\bar{\mathbf{x}} \tag{19.14}$$

and

$$W_e = \tfrac{1}{2}\mathbf{u}^T\mathbf{x} \tag{19.15}$$

Since work is a scalar which is invariant under a rotation of coordinates,

$$\tfrac{1}{2}\bar{\mathbf{u}}^T\bar{\mathbf{x}} = \tfrac{1}{2}\mathbf{u}^T t\,\mathbf{x} \tag{19.16}$$

Substituting Eq. (19.13) into (19.16) gives

$$\mathbf{u}^T \mathbf{T}^T \overline{\mathbf{x}} = \mathbf{u}^T \mathbf{x}$$

Thus
$$\mathbf{x} = \mathbf{T}^T \overline{\mathbf{x}} \qquad (19.17)$$

That is, we have shown that the transformation of displacements in Eq. (19.13) implies the corresponding transformation of forces in Eq. (19.17). Transformations of this type are called *contragredient*.

The transformation of the stiffness matrix from member to structural coordinates can also be accomplished using the argument that the work is invariant with respect to the coordinate system. Thus, for a linear elastic member

$$W_e = \tfrac{1}{2}\overline{\mathbf{u}}^T \overline{\mathbf{x}} = \tfrac{1}{2}\overline{\mathbf{u}}^T \overline{\mathbf{k}}\,\overline{\mathbf{u}} = \tfrac{1}{2}\mathbf{u}^T \mathbf{T}^T \overline{\mathbf{k}}\mathbf{T}\mathbf{u} \qquad (19.18)$$

or alternatively,
$$W_e = \tfrac{1}{2}\mathbf{u}^T \mathbf{x} = \tfrac{1}{2}\mathbf{u}^T \mathbf{k}\mathbf{u} \qquad (19.19)$$

Comparing these two expressions and noting that work must be invariant with respect to a coordinate transformation, we conclude that

$$\mathbf{k} = \mathbf{T}^T \overline{\mathbf{k}}\mathbf{T} \qquad (19.20)$$

which is the same form obtained previously in Eq. (16.37) for the special case in which $\mathbf{T}$ is an orthogonal transformation. Note that if the displacements and forces are described by the contragredient transformations,

$$\mathbf{u} = \mathbf{T}^T \overline{\mathbf{u}} \qquad (19.21)$$

and
$$\overline{\mathbf{x}} = \mathbf{T}\mathbf{x} \qquad (19.22)$$

Therefore,
$$\overline{\mathbf{k}} = \mathbf{T}\mathbf{k}\mathbf{T}^T \qquad (19.23)$$

Verification of Eqs. (19.22) and (19.23) is left as an exercise for the reader.

The transformations discussed in Chaps. 16 and 18 are all orthogonal, i.e.,

$$\mathbf{T}^T = \mathbf{T}^{-1} \qquad \text{or} \qquad \mathbf{T}^T \mathbf{T} = \mathbf{I}$$

and Eqs. (16.33) and (16.37) were derived using this property. Thus, we note that an orthogonal transformation is contragredient to itself.

Using the developments in this section, the displacements, forces, and stiffness matrix can be transformed from member to structural coordinates and vice versa even though the transformation is not orthogonal. For example, if it were of interest to retain only the axial displacements and corresponding forces for the truss member, shown in Fig. 19.2, we would have

$$\begin{bmatrix} \overline{u}_i \\ \overline{u}_j \end{bmatrix} = \begin{bmatrix} \cos\phi & \sin\phi & 0 & 0 \\ 0 & 0 & \cos\phi & \sin\phi \end{bmatrix} \begin{bmatrix} u_i \\ v_i \\ u_j \\ v_j \end{bmatrix} \qquad (19.24)$$

and the contragredient transformation for the forces is

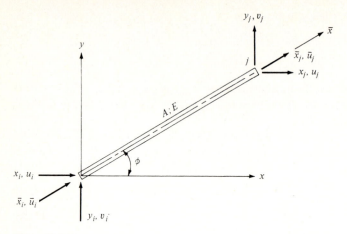

Figure 19.2 The truss member.

$$
\begin{bmatrix} x_i \\ y_i \\ x_j \\ y_j \end{bmatrix} = \begin{bmatrix} \cos \phi & 0 \\ \sin \phi & 0 \\ 0 & \cos \phi \\ 0 & \sin \phi \end{bmatrix} \begin{bmatrix} \bar{x}_i \\ \bar{x}_j \end{bmatrix} \tag{19.25}
$$

Using Eq. (19.20) with

$$
\bar{\mathbf{k}} = \frac{AE}{L} \begin{bmatrix} 1 & -1 \\ -1 & 1 \end{bmatrix}
$$

gives the stiffness matrix in structural coordinates identical to that derived in Eq. (16.16).

19.5 CONSTRAINT EQUATIONS

In this section a *constraint* will indicate any type of relationship between generalized displacements. For example, an unyielding support is a constraint that is analytically described by putting that generalized displacement into the partition $\mathbf{U}_s$, and its effect does not appear in the partition $\mathbf{K}_{ff}$; or there may be a dependency between several generalized displacements, which must be imposed before solving for the displacements. For example, in Fig. 19.3a the structure has experienced a support settlement such that the relationships for the vertical displacements of the foundation are prescribed. In Fig. 19.3b the girders are assumed to be rigid, and it is required to indicate this fact by noting that all the horizontal nodal displacements at a given floor level are equal. Constraints can be introduced into the force-displacement relations if we partition the displacement column matrix as

$$
\mathbf{U}_f = \begin{bmatrix} \mathbf{U}_r \\ --- \\ \mathbf{U}_d \end{bmatrix} \tag{19.26}
$$

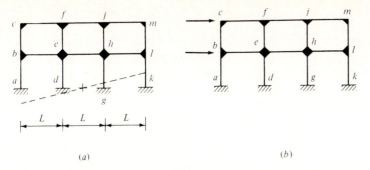

Figure 19.3 (a) Rigid frame with support settlements, $V_a = -V_k$, $V_d = -V_g$, and $V_d = \frac{1}{3}V_a$; (b) rigid frame with lateral loads, $U_c = U_f = U_j = U_m$ and $U_b = U_e = U_h = U_l$.

where $\mathbf{U}_r$ contains the generalized displacements to be retained and $\mathbf{U}_d$ contains those which are constrained, i.e., have a prescribed dependency. The constraint conditions can be expressed in the form of a simple transformation

$$\mathbf{U}_d = \mathbf{TU}_r \qquad (19.27)$$

For example, for the structure in Fig. 19.3a this transformation is

$$
\begin{bmatrix} V_d \\ V_g \\ V_k \end{bmatrix} =
\begin{bmatrix} \frac{1}{3} & 0 & 0 & \cdots & 0 \\ -\frac{1}{3} & 0 & 0 & \cdots & 0 \\ -1 & 0 & 0 & \cdots & 0 \end{bmatrix}
\begin{bmatrix} V_a \\ U_b \\ V_b \\ . \\ U_m \\ V_m \end{bmatrix}
$$

The force-displacement equations are partitioned as implied by Eq. (19.26) to yield

$$
\begin{bmatrix} \mathbf{X}_r \\ --- \\ \mathbf{X}_d \end{bmatrix} =
\begin{bmatrix} \mathbf{K}_{rr} & \vdots & \mathbf{K}_{rd} \\ --- & --- & --- \\ \mathbf{K}_{dr} & \vdots & \mathbf{K}_{dd} \end{bmatrix}
\begin{bmatrix} \mathbf{U}_r \\ --- \\ \mathbf{U}_d \end{bmatrix} \qquad (19.28)
$$

$$\mathbf{X}_f \qquad\qquad \mathbf{K}_{ff} \qquad\qquad \mathbf{U}_f$$

Expanding Eq. (19.28) gives the two matrix equations

$$\mathbf{X}_r = \mathbf{K}_{rr}\mathbf{U}_r + \mathbf{K}_{rd}\mathbf{U}_d \qquad \mathbf{X}_d = \mathbf{K}_{dr}\mathbf{U}_r + \mathbf{K}_{dd}\mathbf{U}_d \qquad (19.29)$$

Substituting Eq. (19.27) for the constrained generalized displacements into Eqs. (19.29) yields

$$\mathbf{X}_r = (\mathbf{K}_{rr} + \mathbf{K}_{rd}\mathbf{T})\mathbf{U}_r \qquad \mathbf{X}_d = (\mathbf{K}_{dr} + \mathbf{K}_{dd}\mathbf{T})\mathbf{U}_r \qquad (19.30)$$

The first of Eqs. (19.30) can be solved to yield $\mathbf{U}_r$, but note that the stiffness matrix in this form is not generally symmetric.

In Sec. 19.4 it was pointed out that for a given displacement transformation there is a corresponding contragredient force transformation. That is, the force transformation for the constrained degrees of freedom is

$$\overline{\mathbf{X}}_r = \mathbf{T}^T \mathbf{X}_d$$

where $\overline{\mathbf{X}}_r$ denotes that the forces are transferred to the location of the r generalized displacements, and we shall observe that these augment $\mathbf{X}_r$. For the structure in Fig. 19.3a this contragredient force transformation is

$$
\begin{bmatrix} \overline{Y}_a \\ \overline{X}_b \\ \overline{Y}_b \\ \vdots \\ \overline{X}_m \\ \overline{Y}_m \end{bmatrix}
=
\begin{bmatrix} \tfrac{1}{3} & -\tfrac{1}{3} & -1 \\ 0 & 0 & 0 \\ 0 & 0 & 0 \\ \cdot & \cdot & \cdot \\ 0 & 0 & 0 \\ 0 & 0 & 0 \end{bmatrix}
\begin{bmatrix} Y_d \\ Y_g \\ Y_k \end{bmatrix}
$$

Multiplying the second matrix equation in Eq. (19.30) by $\mathbf{T}^T$ and adding the two equations gives

$$\mathbf{X}_r + \mathbf{T}^T \mathbf{X}_d = (\mathbf{K}_{rr} + \mathbf{K}_{rd}\mathbf{T} + \mathbf{T}^T\mathbf{K}_{dr} + \mathbf{T}^T\mathbf{K}_{dd}\mathbf{T})\mathbf{U}_r \tag{19.31}$$

or

$$\hat{\mathbf{X}}_r = \hat{\mathbf{K}}_{rr}\mathbf{U}_r \tag{19.32}$$

where

$$\hat{\mathbf{X}}_r = \mathbf{X}_r + \mathbf{T}^T\mathbf{X}_d \tag{19.33}$$

and

$$\hat{\mathbf{K}}_{rr} = \mathbf{K}_{rr} + \mathbf{K}_{rd}\mathbf{T} + \mathbf{T}^T\mathbf{K}_{dr} + \mathbf{T}^T\mathbf{K}_{dd}\mathbf{T} \tag{19.34}$$

This gives the symmetric matrix $\hat{\mathbf{K}}_{rr}$, which can be used to obtain a solution realizing the usual efficiencies. Symmetry can be demonstrated by transposing $\hat{\mathbf{K}}_{rr}$ and noting that the matrix is equal to its transpose. Example 19.3 demonstrates this approach for the structure of Example 18.2.

Example 19.3 Solve for the generalized displacements of Example 18.2 using constraint equations.

SOLUTION The constraint conditions (Eq. 19.27) are

$$
\begin{bmatrix} U_c \\ V_c \end{bmatrix}
=
\begin{bmatrix} 1 & 0 & 0 \\ 0.75 & 0 & 0 \end{bmatrix}
\begin{bmatrix} U_b \\ \Theta_b \\ \Theta_c \end{bmatrix}
$$

From Example 18.2

$$
\begin{bmatrix} X_b \\ M_b \\ M_c \\ --- \\ X_c \\ Y_c \end{bmatrix}
= \frac{EI}{L}
\begin{bmatrix}
\dfrac{10.125}{L^2} & \dfrac{6.75}{L} & 0 & \vdots & 0 & 0 \\[2mm]
\dfrac{6.75}{L} & 10 & 2 & \vdots & 0 & -\dfrac{6}{L} \\[2mm]
0 & 2 & 10 & \vdots & \dfrac{4.32}{L} & -\dfrac{2.76}{L} \\
\multicolumn{6}{c}{\text{--------------------------}} \\
0 & 0 & \dfrac{4.32}{L} & \vdots & \dfrac{4.1472}{L^2} & \dfrac{3.1104}{L^2} \\[2mm]
0 & -\dfrac{6}{L} & -\dfrac{2.76}{L} & \vdots & \dfrac{3.1104}{L^2} & \dfrac{14.3328}{L^2}
\end{bmatrix}
\begin{bmatrix} U_b \\ \Theta_b \\ \Theta_c \\ --- \\ U_c \\ V_c \end{bmatrix}
$$

$$\mathbf{K}_{rd}\mathbf{T} = \frac{EI}{L}\begin{bmatrix} 0 & 0 & 0 \\ -\frac{4.5}{L} & 0 & 0 \\ \frac{2.25}{L} & 0 & 0 \end{bmatrix} \qquad \mathbf{T}^T\mathbf{K}_{dr} = \frac{EI}{L}\begin{bmatrix} 0 & -\frac{4.5}{L} & \frac{2.25}{L} \\ 0 & 0 & 0 \\ 0 & 0 & 0 \end{bmatrix}$$

$$\mathbf{T}^T\mathbf{K}_{dd}\mathbf{T} = \frac{EI}{L}\begin{bmatrix} \frac{16.875}{L^2} & 0 & 0 \\ 0 & 0 & 0 \\ 0 & 0 & 0 \end{bmatrix}$$

Using Eq. (19.34) gives

$$\hat{\mathbf{K}}_{rr} = \mathbf{K}_{rr} + \mathbf{K}_{rd}\mathbf{T} + \mathbf{T}^T\mathbf{K}_{dr} + \mathbf{T}^T\mathbf{K}_{dd}\mathbf{T} = \frac{EI}{L}\begin{bmatrix} \frac{27}{L^2} & \frac{2.25}{L} & \frac{2.25}{L} \\ \frac{2.25}{L} & 10 & 2 \\ \frac{2.25}{L} & 2 & 10 \end{bmatrix}$$

DISCUSSION The constraint condition $U_c = U_b$ affirms the inextensibility of member bc; the fact that $V_c = 0.75U_c$ can be observed by noting the kinematics of the displaced joint c. $\mathbf{K}_{ff}$ from Example 18.2 is partitioned according to Eq. (19.28) and components of $\hat{\mathbf{K}}_{rr}$ as prescribed by Eq. (19.34) are computed. In this case $\hat{\mathbf{X}}_r = \mathbf{X}_r$ since $\mathbf{X}_d = 0$. Note that the solution of $\mathbf{X}_r = \hat{\mathbf{K}}_{rr}\mathbf{U}_r$ yields the same generalized displacements as Example 18.2.

19.6 NODAL COORDINATES

It is sometimes convenient to express the forces, displacements, and stiffness matrix in terms of one (or more) unique coordinate system(s) associated with an individual nodal point. For example, since the support surface at point c for the truss portrayed in Fig. 19.4 is not aligned with the xy structural coordinate system, the conventional method of enforcing zero support displacements cannot be used. However, if the generalized forces and displacements at c are transformed into the $x'y'$ coordinate system, we can enforce the condition that $V_c' = 0$ in the usual fashion. Typically, it is preferable to express the nodal displacements and forces in terms of the xy structural coordinate directions; furthermore, it would be fruitless to transform all nodal degrees of freedom to the $x'y'$ coordinates since in this case the zero support displacement at point b could not be enforced in this system. Thus, there is a need for expressing the structural response of this truss in terms of both the structural coordinates and a single nodal coordinate system. The structure can be analyzed by (1) transforming only the displacements at node c into a unique nodal coordinate system while maintaining all others in the general xy structural system, (2) enforcing the zero displacement support conditions in the force-displacement equations, and (3) proceeding with the solution in the usual fashion for the stiffness method.

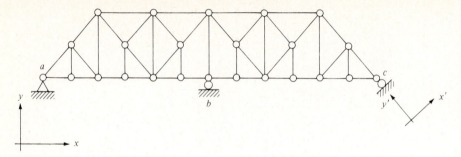

Figure 19.4 Truss with a skew support.

Nodal coordinates may also be necessary in some other situations. For example, Fig. 19.5 illustrates the end of a truss bridge showing the usual portal bracing spanning perpendicular to the two primary trusses. This portal bracing is composed of a truss that has zero stiffness perpendicular to its plane, and the displacements in the structural x direction must be retained in the total problem since the primary trusses can display such deflections. We must use a space-truss idealization to study the interaction of the main trusses and the portal bracing. Since the individual stiffnesses of the main trusses and the portal bracing are planar, it is necessary to constrain the displacements for nodes c, d, e, and f normal to the plane of the portal truss. This can be accomplished by using the coordinate system $x'y'z'$ for these four nodes and enforcing the condition that W' (displacement in the z' direction) be zero for each of them. These constraints perpendicular to the portal bracing are required to prevent a singular stiffness matrix, a major numerical difficulty.

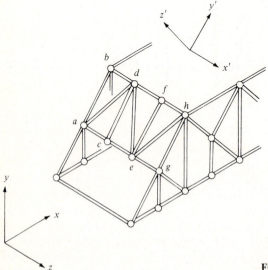

Figure 19.5 Portal bracing for a truss bridge.

The displacements in the nodal coordinates $\mathbf{U}'$ can be expressed in terms of those described in the structural coordinates $\mathbf{U}$ using a transformation which is expressed in the usual manner

$$\mathbf{U} = \mathbf{T}\mathbf{U}' \tag{19.35}$$

and there is a corresponding transformation for the forces such that

$$\mathbf{X} = \mathbf{T}\mathbf{X}' \tag{19.36}$$

Since the column matrices of displacements and forces expressed in both coordinate systems contain all nodal quantities of the structure, $\mathbf{T}$ is a square matrix that is also orthogonal ($\mathbf{T}^T = \mathbf{T}^{-1}$). Usually, only one or two nodal coordinate systems need be used for a given structure; therefore, $\mathbf{T}$ will generally resemble the identity matrix except for those rows and columns which are associated with a nodal coordinate system. For these degrees of freedom the transformation from nodal to structural coordinates is expressed in terms of the appropriate direction cosines. For example, for the truss of Fig. 19.4

$$\mathbf{T} = \begin{bmatrix} 1 & 0 & 0 & \cdots & 0 & 0 \\ 0 & 1 & 0 & \cdots & 0 & 0 \\ 0 & 0 & 1 & \cdots & 0 & 0 \\ \cdots & \cdots & \cdots & \cdots & \cdots & \cdots \\ 0 & 0 & 0 & \cdots & \lambda'_x & \mu'_x \\ 0 & 0 & 0 & \cdots & \lambda'_y & \mu'_y \end{bmatrix} \tag{a}$$

where λ'_x and μ'_x are the direction cosines of the structural x axis with respect to the $x'y'$ nodal axis at c and λ'_y and μ'_y are the direction cosines of the structural y axis.

The force-displacement relations in structural coordinates

$$\mathbf{X} = \mathbf{K}\mathbf{U} \tag{19.37}$$

are transformed into the nodal coordinate system to give

$$\mathbf{X}' = \mathbf{K}'\mathbf{U}' \tag{19.38}$$

where

$$\mathbf{K}' = \mathbf{T}^T\mathbf{K}\mathbf{T} \tag{19.39}$$

which can be written in partitioned form as

$$\begin{bmatrix} \mathbf{X}'_s \\ \hline \mathbf{X}'_n \end{bmatrix} = \begin{bmatrix} \mathbf{K}'_{ss} & \mathbf{K}'_{sn} \\ \hline \mathbf{K}'_{ns} & \mathbf{K}'_{nn} \end{bmatrix} \begin{bmatrix} \mathbf{U}'_s \\ \hline \mathbf{U}'_n \end{bmatrix} \tag{19.40}$$

Quantities with a subscript s are expressed in structural coordinates and those with an n are in nodal coordinates. Note that $\mathbf{X}'_s = \mathbf{X}_s$ and $\mathbf{U}'_s = \mathbf{U}_s$, where $\mathbf{X}_s$ and $\mathbf{U}_s$ are the corresponding partitions of $\mathbf{X}$ and $\mathbf{U}$, respectively. With this partitioning scheme Eq. (19.39) becomes

$$\mathbf{K}' = \begin{bmatrix} \mathbf{K}'_{ss} & \vdots & \mathbf{K}'_{sn} \\ \cdots & \vdots & \cdots \\ \mathbf{K}'_{ns} & \vdots & \mathbf{K}'_{nn} \end{bmatrix} = \begin{bmatrix} \mathbf{I} & \vdots & \mathbf{0} \\ \cdots & \vdots & \cdots \\ \mathbf{0}^T & \vdots & \boldsymbol{\Gamma}^T \end{bmatrix} \begin{bmatrix} \mathbf{K}_{ss} & \vdots & \mathbf{K}_{sn} \\ \cdots & \vdots & \cdots \\ \mathbf{K}_{ns} & \vdots & \mathbf{K}_{nn} \end{bmatrix}$$

$$\times \begin{bmatrix} \mathbf{I} & \vdots & \mathbf{0} \\ \cdots & \vdots & \cdots \\ \mathbf{0}^T & \vdots & \boldsymbol{\Gamma} \end{bmatrix} = \begin{bmatrix} \mathbf{K}_{ss} & \vdots & \mathbf{K}_{sn}\boldsymbol{\Gamma} \\ \cdots & \vdots & \cdots \\ \boldsymbol{\Gamma}^T\mathbf{K}_{ns} & \vdots & \boldsymbol{\Gamma}^T\mathbf{K}_{nn}\boldsymbol{\Gamma} \end{bmatrix}$$

$$(19.41)$$

where $\mathbf{I}$ = identity matrix

$\quad\quad \mathbf{0}$ = null matrix

$\quad\quad \mathbf{0}^T$ = transpose of null matrix

$\quad\quad \boldsymbol{\Gamma}$ = transformation matrix containing appropriate direction cosines

For example, for the truss of Fig. 19.4

$$\boldsymbol{\Gamma} = \begin{bmatrix} \lambda'_x & \mu'_x \\ \lambda'_y & \mu'_y \end{bmatrix} \qquad (b)$$

From Eq. (19.41) we can observe that $\mathbf{K}'$ is obtained with only a few multiplications, and it is not necessary to carry out the complete multiplication of $\mathbf{T}^T\mathbf{K}\mathbf{T}$ to obtain $\mathbf{K}'$ since a considerable portion of $\mathbf{K}$ is completely unaltered if only a limited number of nodal coordinates are used. That is, for the truss of Fig. 19.4, $\mathbf{K}'_{sn}$ is a matrix with only two columns, $\mathbf{K}'_{ns}$ is the transpose of $\mathbf{K}'_{sn}$, and $\mathbf{K}'_{nn}$ is merely a 2×2 matrix.

Upon calculating $\mathbf{K}'$ the condition of the zero nodal support displacements can be invoked to give $\mathbf{K}'_{ff}$ (the partition of the merged stiffness matrix containing the effect of the unconstrained degrees of freedom), and the solution for the generalized displacements can be obtained by the stiffness method. This procedure is illustrated in Example 19.4 for a simple truss with one skew support condition.

Example 19.4 Solve this problem using nodal coordinates. For all members AE/L = 20,000 kN/m.

SOLUTION With the boundary conditions at b and c enforced (see Example 16.2)

$$\begin{bmatrix} X_a \\ Y_a \\ Y_b \\ Y_c \end{bmatrix} = (2 \times 10^4) \begin{bmatrix} 1.00 & 0.00 & 0.00 & 0.00 \\ 0.00 & 1.00 & 0.00 & -1.00 \\ 0.00 & 0.00 & 0.64 & -0.64 \\ 0.00 & -1.00 & -0.64 & 1.64 \end{bmatrix} \begin{bmatrix} U_a \\ V_a \\ V_b \\ V_c \end{bmatrix}$$

Equation (19.35) gives

$$\begin{bmatrix} U_a \\ V_a \\ V_b \\ V_c \end{bmatrix} = \begin{bmatrix} 0.707 & 0.707 & 0 & 0 \\ -0.707 & 0.707 & 0 & 0 \\ 0 & 0 & 1 & 0 \\ 0 & 0 & 0 & 1 \end{bmatrix} \begin{bmatrix} U'_a \\ V'_a \\ V_b \\ V_c \end{bmatrix}$$

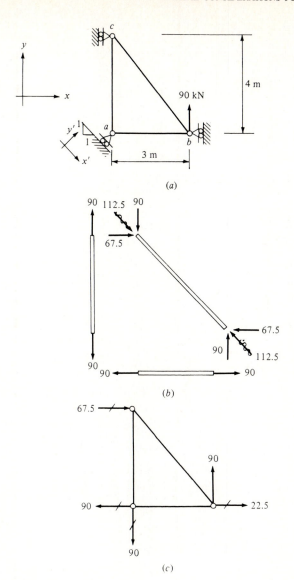

Figure E19.4 (a) Truss with a skew support; (b) free-body diagrams of the members; (c) free-body diagram of the truss.

From Eq. (19.39)

$$
\mathbf{K}' = (2 \times 10^4)
\begin{array}{c}
\begin{array}{cccc} U_a' & V_a' & V_b & V_c \end{array} \\
\begin{bmatrix}
1.000 & 0.000 & 0.000 & 0.707 \\
0.000 & 1.000 & 0.000 & -0.707 \\
0.000 & 0.000 & 0.640 & -0.640 \\
0.707 & -0.707 & -0.640 & 1.640
\end{bmatrix}
\end{array}
$$

Imposing the condition that $V'_a = 0$ gives

$$\mathbf{K}_{ff} = (2 \times 10^4) \begin{bmatrix} 1.000 & 0.000 & 0.707 \\ 0.000 & 0.640 & -0.640 \\ 0.707 & -0.640 & 1.640 \end{bmatrix}$$

Hence,

$$\begin{bmatrix} U'_a \\ V_b \\ V_c \end{bmatrix} = 10^{-4} \begin{bmatrix} 1.000 & -0.707 & -0.707 \\ -0.707 & 1.781 & 1.000 \\ -0.707 & 1.000 & 1.000 \end{bmatrix} \begin{bmatrix} 0 \\ 90 \\ 0 \end{bmatrix} = 10^{-3} \begin{bmatrix} -6.36 \\ 16.03 \\ 9.00 \end{bmatrix} \text{m}$$

From Eq. (19.35)

$$\begin{bmatrix} U_a \\ V_a \\ V_b \\ V_c \end{bmatrix} = \begin{bmatrix} 0.707 & 0.707 & 0 & 0 \\ -0.707 & 0.707 & 0 & 0 \\ 0 & 0 & 1 & 0 \\ 0 & 0 & 0 & 1 \end{bmatrix} \left(10^{-3} \begin{bmatrix} -6.36 \\ 0.00 \\ 16.03 \\ 9.00 \end{bmatrix} \right) = 10^{-3} \begin{bmatrix} -4.50 \\ 4.50 \\ 16.03 \\ 9.00 \end{bmatrix}$$

For member ab

$$s_{ab} = (2 \times 10^4) \begin{bmatrix} -1 & 0 & 1 & 0 \end{bmatrix} \left(10^{-3} \begin{bmatrix} -4.50 \\ 4.50 \\ 0.00 \\ 16.03 \end{bmatrix} \right) = 90 \text{ kN (t)}$$

For member bc:

$$\bar{\mathbf{u}}_{bc} = \begin{bmatrix} -0.6 & 0.8 & 0 & 0 \\ -0.8 & -0.6 & 0 & 0 \\ 0 & 0 & -0.6 & 0.8 \\ 0 & 0 & -0.8 & -0.6 \end{bmatrix} \left(10^{-3} \begin{bmatrix} 0.00 \\ 16.03 \\ 0.00 \\ 9.00 \end{bmatrix} \right) = 10^{-3} \begin{bmatrix} 12.82 \\ -9.62 \\ 7.20 \\ -5.40 \end{bmatrix}$$

$$s_{bc} = (2 \times 10^4) \begin{bmatrix} -1 & 0 & 1 & 0 \end{bmatrix} \left(10^{-3} \begin{bmatrix} 12.82 \\ -9.62 \\ 7.20 \\ -5.40 \end{bmatrix} \right) = -112.5 \text{ kN (c)}$$

For member ac

$$\bar{\mathbf{u}}_{ac} = \begin{bmatrix} 0 & 1 & 0 & 0 \\ -1 & 0 & 0 & 0 \\ 0 & 0 & 0 & 1 \\ 0 & 0 & -1 & 0 \end{bmatrix} \left(10^{-3} \begin{bmatrix} -4.50 \\ 4.50 \\ 0.00 \\ 9.00 \end{bmatrix} \right) = 10^{-3} \begin{bmatrix} 4.50 \\ 4.50 \\ 9.00 \\ 0.00 \end{bmatrix}$$

$$s_{ac} = (2 \times 10^4) \begin{bmatrix} -1 & 0 & 1 & 0 \end{bmatrix} \left(10^{-3} \begin{bmatrix} 4.50 \\ 4.50 \\ 9.00 \\ 0.00 \end{bmatrix} \right) = 90 \text{ kN (t)}$$

DISCUSSION The merged stiffness matrix from Example 16.2 is used, and the support conditions $U_b = U_c = 0$ have been enforced to give the **K** matrix. In this case the entire **T** matrix is used in calculating **K′** since the multiplication is straightforward. For the equation $\mathbf{X'} = \mathbf{K'U'}$ with $\mathbf{U'} = \{U'_a \quad V'_a \quad V_b \quad V_c\}$, the condition $V'_a = 0$ is enforced to give the 3×3 stiffness matrix. The solution is obtained yielding $\mathbf{U'}$, and for completeness all the displacements in the structural coordinates, plus the member forces, are calculated. Note that the member-oriented displacements $(\bar{\mathbf{u}})$ must be evaluated in each case before the member forces can be calculated.

Alternatively, skew support conditions can be imposed by using the constraint conditions described in Sec. 19.5. For example, for the truss of Fig. 19.4 the dependency of the u and v displacements at point c could be described using Eq. (19.27) and this condition could be introduced into the stiffness matrix by using Eq. (19.34) to yield $\hat{\mathbf{K}}_{rr}$.

Another method for treating this situation is to use a special boundary element consisting of a fictitious spring, as illustrated in Fig. 19.6. This approach should be used with discretion since in order to accomplish the desired zero displacement (along the axis of the replacement member) it is necessary to use an extremely stiff element. If this stiffness is many orders of magnitude greater than the other nodal stiffness values for the structure, a stiffness matrix ill conditioned with respect to solution may result.

19.7 SYMMETRY AND ANTISYMMETRY

Symmetric or antisymmetric behavior occurs if the structure is symmetric and the loads are symmetric or antisymmetric (see Sec. 12.8). The principles of symmetry and antisymmetry can be extended to the analysis of linear elastic structures by the stiffness method to reduce the problem size and minimize the computational effort. That is, the problem size can often be reduced by invoking symmetry or antisymmetry or replacing an arbitrary load by its equivalent symmetric and antisymmetric parts. For some of the example problems in Part Five we have contracted the size of the stiffness matrix using principles of symmetry and antisymmetry, but this was done on an ad hoc basis. Since it is useful to understand how it can be done rigorously, these arguments will be formalized in this section using some of the principles described earlier in the chapter.

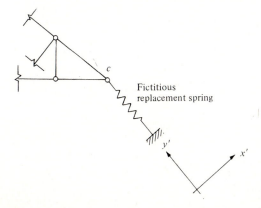

Fictitious
replacement spring

Figure 19.6 Alternate method for analyzing skew supports.

The symmetrically loaded structure in Fig. 19.7a deflects in such a manner that the slope at node c is zero. As a result, the solution can be obtained by analyzing only half the structure with the appropriate support conditions; this alternate structure is portrayed in Fig. 19.7a. Note that $\mathbf{K}_{ff}$ for the actual structure is 5×5 whereas, that for the alternate structure is merely 2×2. If this approach is extended to larger structures, a considerable computational and corresponding economic savings will be realized.

Another method for reducing the size of the problem for the beam of Fig. 19.7a is to use the constraint equations presented in Sec. 19.5. After enforcing all zero generalized displacements, including that of Θ_c, the generalized displacements and forces to be retained (matrices subscripted with r) and those which are constrained (matrices subscripted with d) are

$$\mathbf{U}_r = \begin{bmatrix} \Theta_a \\ \Theta_b \end{bmatrix} \qquad \mathbf{X}_r = \begin{bmatrix} M_a \\ M_b \end{bmatrix} \qquad \mathbf{U}_d = \begin{bmatrix} \Theta_d \\ \Theta_e \end{bmatrix} \qquad \mathbf{X}_d = \begin{bmatrix} M_d \\ M_e \end{bmatrix}$$

The corresponding constraint transformation is

$$\mathbf{U}_d = \begin{bmatrix} 0 & -1 \\ -1 & 0 \end{bmatrix} \mathbf{U}_r$$

Thus, $\hat{\mathbf{K}}_{rr}$ can be obtained using Eq. (19.34) and the solution executed using the usual approach for the stiffness method.

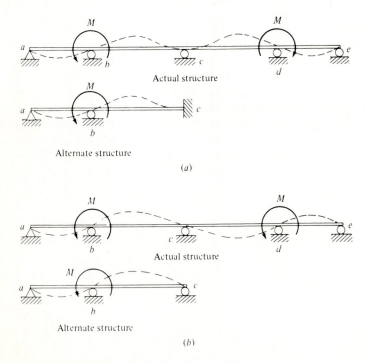

Figure 19.7 (a) Symmetric and (b) antisymmetric behavior.

The antisymmetric deflection pattern displayed in Fig. 19.7b implies that the moment at c is zero; therefore, analyzing the alternate structure with a simple support at c will yield the solution for the actual structure. In this case, if the structure is analyzed using the constraint equations, we have

$$\begin{bmatrix} \Theta_d \\ \Theta_e \end{bmatrix} = \begin{bmatrix} 0 & 1 & 0 \\ 1 & 0 & 0 \end{bmatrix} \begin{bmatrix} \Theta_a \\ \Theta_b \\ \Theta_c \end{bmatrix}$$

Although these two approaches for invoking symmetry (or antisymmetry) naturally yield the same solution, the stiffness matrices obtained will not necessarily be identical. That is, the constraint equations effectively combine both columns and rows of **K**, whereas analyzing an alternate structure implies that the columns and rows of the matrix have not been manipulated. The analysis of a symmetric beam is presented in Example 19.5 using both an alternate structure and the constraint equations. A perusal of these results will reveal the difference in the two stiffness matrices generated.

Example 19.5 Calculate the generalized displacements for the beam of Example 17.7 using (a) an alternate structure and (b) constraint equations; EI = const.

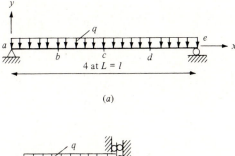

(a)

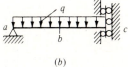

(b)

Figure E19.5 (a) Symmetric beam with symmetric loading; (b) alternate structure.

SOLUTION (a) For the alternate structure in Fig. E19.5b

$$\begin{bmatrix} M_a/L \\ Y_b \\ M_b/L \\ Y_c \end{bmatrix} = \frac{EI}{L^3} \begin{bmatrix} 4 & -6 & 2 & 0 \\ -6 & 24 & 0 & -12 \\ 2 & 0 & 8 & -6 \\ 0 & -12 & -6 & 12 \end{bmatrix} \begin{bmatrix} \Theta_a L \\ V_b \\ \Theta_b L \\ V_c \end{bmatrix}$$

$$\begin{bmatrix} \Theta_a L \\ V_b \\ \Theta_b L \\ V_c \end{bmatrix} = \frac{L^3}{EI} \begin{bmatrix} 2.000 & 1.500 & 1.000 & 2.000 \\ 1.500 & 1.333 & 1.000 & 1.833 \\ 1.000 & 1.000 & 1.000 & 1.500 \\ 2.000 & 1.833 & 1.500 & 2.667 \end{bmatrix} qL \begin{bmatrix} -1/12 \\ -1 \\ 0 \\ -1/2 \end{bmatrix} = \frac{qL^4}{EI} \begin{bmatrix} -2.667 \\ -2.375 \\ -1.833 \\ -3.333 \end{bmatrix}$$

(b) The constraint equations for Fig. E19.5a are

$$\begin{bmatrix} V_d \\ \Theta_d L \\ \Theta_e L \end{bmatrix} = \begin{bmatrix} 0 & 1 & 0 & 0 \\ 0 & 0 & -1 & 0 \\ -1 & 0 & 0 & 0 \end{bmatrix} \begin{bmatrix} \Theta_a L \\ V_b \\ \Theta_b L \\ V_c \end{bmatrix}$$

From Example 17.7

$$\begin{bmatrix} M_a/L \\ Y_b \\ M_b/L \\ Y_c \\ ---- \\ Y_d \\ M_d/L \\ M_e/L \end{bmatrix} = \frac{EI}{L^3} \begin{bmatrix} 4 & -6 & 2 & 0 & \vdots & 0 & 0 & 0 \\ -6 & 24 & 0 & -12 & \vdots & 0 & 0 & 0 \\ 2 & 0 & 8 & -6 & \vdots & 0 & 0 & 0 \\ 0 & -12 & -6 & 24 & \vdots & -12 & 6 & 0 \\ ------ & & & & & & & \\ 0 & 0 & 0 & -12 & \vdots & 24 & 0 & 6 \\ 0 & 0 & 0 & 6 & \vdots & 0 & 8 & 2 \\ 0 & 0 & 0 & 0 & \vdots & 6 & 2 & 4 \end{bmatrix} \begin{bmatrix} \Theta_a L \\ V_b \\ \Theta_b L \\ V_c \\ ---- \\ V_d \\ \Theta_d L \\ \Theta_e L \end{bmatrix}$$

$$\mathbf{K}_{rd}\mathbf{T} = \frac{EI}{L^3} \begin{bmatrix} 0 & 0 & 0 & 0 \\ 0 & 0 & 0 & 0 \\ 0 & 0 & 0 & 0 \\ 0 & -12 & -6 & 0 \end{bmatrix} \qquad \mathbf{T}^T\mathbf{K}_{dr} = \frac{EI}{L^3} \begin{bmatrix} 0 & 0 & 0 & 0 \\ 0 & 0 & 0 & -12 \\ 0 & 0 & 0 & -6 \\ 0 & 0 & 0 & 0 \end{bmatrix}$$

$$\mathbf{T}^T\mathbf{K}_{dd}\mathbf{T} = \frac{EI}{L^3} \begin{bmatrix} 4 & -6 & 2 & 0 \\ -6 & 24 & 0 & 0 \\ 2 & 0 & 8 & 0 \\ 0 & 0 & 0 & 0 \end{bmatrix}$$

$$\hat{\mathbf{K}}_{rr} = \mathbf{K}_{rr} + \mathbf{K}_{rd}\mathbf{T} + \mathbf{T}^T\mathbf{K}_{dr} + \mathbf{T}^T\mathbf{K}_{dd}\mathbf{T} = \frac{EI}{L^3} \begin{bmatrix} 8 & -12 & 4 & 0 \\ -12 & 48 & 0 & -24 \\ 4 & 0 & 16 & -12 \\ 0 & -24 & -12 & 24 \end{bmatrix}$$

$$\hat{\mathbf{X}}_r = \mathbf{X}_r + \mathbf{T}^T\mathbf{X}_d = -qL \begin{bmatrix} 1/12 \\ 1 \\ 0 \\ 1 \end{bmatrix} + \begin{bmatrix} 0 & 0 & -1 \\ 1 & 0 & 0 \\ 0 & -1 & 0 \\ 0 & 0 & 0 \end{bmatrix} qL \begin{bmatrix} -1 \\ 0 \\ +1/12 \end{bmatrix} = \begin{bmatrix} -\dfrac{qL}{6} \\ -2qL \\ 0 \\ -qL \end{bmatrix}$$

From $\mathbf{U}_r = \hat{\mathbf{K}}_{rr}^{-1}\hat{\mathbf{X}}_r$,

$$\begin{bmatrix} \Theta_a L \\ V_b \\ \Theta_b L \\ V_c \end{bmatrix} = \frac{L^3}{EI} \begin{bmatrix} 1.000 & 0.750 & 0.500 & 1.000 \\ 0.750 & 0.667 & 0.500 & 0.917 \\ 0.500 & 0.500 & 0.500 & 0.750 \\ 1.000 & 0.917 & 0.750 & 1.333 \end{bmatrix} \left(qL \begin{bmatrix} -1/6 \\ -2 \\ 0 \\ -1 \end{bmatrix} \right) = \frac{qL^4}{EI} \begin{bmatrix} -2.667 \\ -2.375 \\ -1.833 \\ -3.333 \end{bmatrix}$$

DISCUSSION The stiffness matrix for the alternate structure is obtained by merging the individual member stiffness matrices in the usual fashion. Note that the alternate form of the beam stiffness matrices is being used; hence, moments are divided by L and the rotations multiplied by L. This yields the same displacements and member forces as Example 17.7.

The constraint equations give a $\hat{\mathbf{K}}_{rr}$ that differs from that obtained using the alternate structure; however, the constrained forces in $\hat{\mathbf{X}}_r$ are also different. This problem involves initial fixed-end forces, which have been transferred to the opposite side of the force-displacement equation for the purpose of calculating $\hat{\mathbf{X}}_r$. The solution of $\hat{\mathbf{X}}_r = \hat{\mathbf{K}}_{rr}\mathbf{U}_r$ yields the same generalized displacements obtained previously.

PROBLEMS

Use the stiffness method for all problems.

In Probs. 19.1 to 19.6 condense (eliminate) all unloaded generalized displacements in the original problem and calculate all nodal displacements.

19.1 Problem 16.23   **19.2** Problem 16.33   **19.3** Problem 16.34
19.4 Problem 17.10   **19.5** Part (*a*) of Prob. 17.15   **19.6** Problem 17.18

In Probs. 19.7 to 19.10 calculate all generalized nodal displacements. Use the member stiffness matrix for a beam with the moment at one end equal to zero; i.e., use Eqs. (19.11) and (19.12).

19.7 $I = 600$ in^4 and $E = 29 \times 10^3$ kips/in^2.

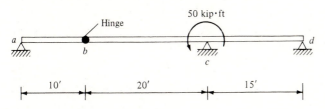

Figure P19.7

19.8 $I = 300 \times 10^{-6}$ m^4 and $E = 200$ GPa.

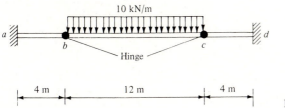

Figure P19.8

19.9 $A = 60$ in^2, $I = 900$ in^4, and $E = 1.6 \times 10^6$ lb/in^2. *Hint:* The member stiffness must include both flexural and axial contributions.

19.10 Solve Prob. 10.16. Use equivalent nodal forces.

In Probs. 19.11 to 19.14 calculate all nodal displacements. Use the concept of nodal coordinates.

19.11 $AE/L = 20,000$ kN/m for all members.

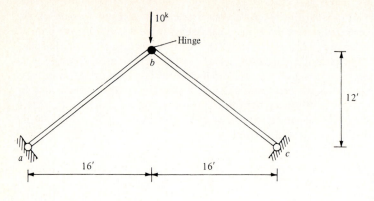

Figure P19.9

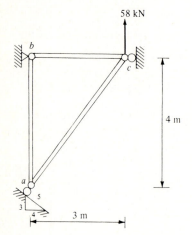

Figure P19.11

19.12 $A_{ad} = A_{bc} = 8$ in^2, $A_{ab} = A_{cd} = 6$ in^2, $A_{ac} = A_{bd} = 10$ in^2, and $E = 29 \times 10^3$ kips/in^2.

19.13 $I_{ab} = 600 \times 10^{-6}$ m^4, $I_{bc} = 250 \times 10^{-6}$ m^4, $I_{cd} = 150 \times 10^{-6}$ m^4, and $E = 200$ GPa. Ignore member axial deformations.

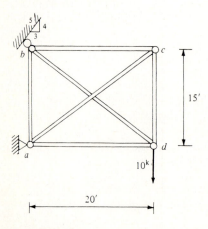

Figure P19.12

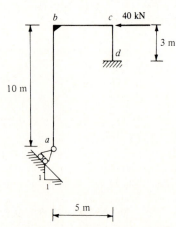

Figure P19.13

19.14 $I_{ab} = I_{cd} = I$ and $I_{bc} = 2I$. Replace the applied loading with equivalent nodal loads at b and c. Ignore member axial deformations.

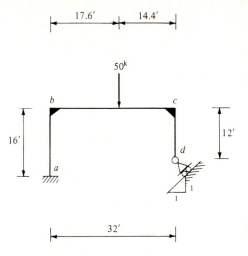

Figure P19.14

In Probs. 19.15 to 19.19 calculate the nodal displacements and member forces for the original problem number using the principles of symmetry and/or antisymmetry.

19.15 Problem 12.6   **19.16** Part (a) of Prob. 17.16
19.17 Part (a) of Prob. 17.17   **19.18** Problem 17.19
19.19 Problem 12.50; $E = 1.6 \times 10^6$ lb/in²;
          ignore member axial deformations.

ADDITIONAL READING

Kardestuncer, H.: *Elementary Matrix Analysis of Structures,* chap. 10, McGraw-Hill, New York, 1974.

McGuire, W., and R. H. Gallagher: *Matrix Structural Analysis,* chap. 10, Wiley, New York, 1979.

Martin, H. C.: *Introduction to Matrix Methods of Structural Analysis,* chap. 7, McGraw-Hill, New York, 1966.

Weaver, W., Jr., and J. M. Gere: *Matrix Analysis of Framed Structures,* chap. 6, Van Nostrand, New York, 1980.

SYSTEMS OF MEASUREMENT

Although the structural engineer in the United States must work with products, documents, institutes, and industries that use U. S. Customary System (USCS) units almost exclusively, the civil engineering profession is working toward the official adoption and use of the System International (SI) units. It therefore behooves the student to develop a facility for working with both systems. Tables A.1 through A.3 provide a ready reference for the SI units used in this book. More complete information on the entire SI system is available in many references.

Table A.1 Selected SI units

Quantity	Unit	SI symbol
Base units:		
Length	meter	m
Mass	kilogram	kg
Time	second	s
Derived units:		
Area	meter^2	m^2
Density, mass	kilogram/meter^3	kg/m^3
Force	newton	N $(= \text{kg} \cdot \text{m/s}^2)$
Moment of force	newton-meter	N $\cdot$ m
Moment of inertia, area	meter^4	m^4
Pressure, stress	pascal	Pa $(= \text{N/m}^2)$
Spring constant	newton/meter	N/m
Temperature[†]	Celsius	°C
Work, energy	joule	J $(= \text{N} \cdot \text{m})$
Supplementary units:		
Plane angle	radian	rad

[†]The SI base unit for thermodynamic temperature, the kelvin, is not used in this book.

Table A.2 SI unit prefixes

Multiple factor	SI prefix	Symbol	Multiple factor	SI prefix	Symbol
10^{12}	tera	T	10^{-1}	deci†	d
10^{9}	giga	G	10^{-2}	centi†	c
10^{6}	mega	M	10^{-3}	milli	m
10^{3}	kilo	k	10^{-6}	micro	μ
10^{2}	hecto†	h	10^{-9}	nano	n
10	deka†	da	10^{-12}	pico	p

†Use should generally be avoided.

Table A.3 Some conversion factors

Length

1 in = 0.0254 m	1 m = 39.37 in
1 ft = 0.3048 m	1 m = 3.281 ft

Area

$1\ in^2 = 6.452 \times 10^{-4}\ m^2$	$1\ m^2 = 1.550 \times 10^3\ in^2$
$1\ ft^2 = 9.290 \times 10^{-2}\ m^2$	$1\ m^2 = 1.076 \times 10\ ft^2$

Moment of inertia, area

$1\ in^4 = 4.162 \times 10^{-7}\ m^4$	$1\ m^4 = 2.402 \times 10^6\ in^4$

Force and force per unit length

1 lb = 4.448 N	1 N = 0.2248 lb
1 lb/ft = 14.59 N/m	$1\ N/m = 6.852 \times 10^{-2}\ lb/ft$

Bending moment

1 lb · in = 0.1130 N · m	1 N · m = 8.851 lb · in
1 lb · ft = 1.3558 N · m	1 N · m = 0.7376 lb · ft

Stress

$1\ lb/in^2 = 6.895 \times 10^3\ Pa$	$1\ Pa = 0.1450 \times 10^{-3}\ lb/in^2$
$1\ lb/ft^2 = 47.88\ Pa$	$1\ Pa = 2.089 \times 10^{-2}\ lb/ft^2$

Temperature

$T_F = 1.8T_C + 32$	$T_C = \dfrac{T_F - 32}{1.8}$

SURVEY OF STRUCTURAL ENGINEERING

B.1 HISTORICAL SUMMARY

Since many of the basic principles of structural engineering are derived from solid mechanics and strength of materials, a history of the field must acknowledge the contributions of Archimedes (287–212 B.C.), Leonardo da Vinci (1452–1519), Galileo Galilei (1564–1642), Robert Hooke (1635–1703), Isaac Newton (1642–1727), and others. Many of the theories and methods described in this book, together with their discoverers, are noted in chronological order in Table B.1.†

Improved structural theory and significant advances in structural design frequently occur simultaneously. For example, in the late 1940s the stiffness method was developed for the delta-wing aircraft, and today, advanced aircraft, as well as many complex buildings and bridges, owe their existence to this method. Thus, we can get a better understanding of the development of structural theory through the history of structures. The following brief historical sketch of some notable bridges and buildings is offered to stimulate the readers' interest.§

The Pont du Gard in southwest France is an arch structure 270 m long completed by the Romans in 18 B.C. Wooden trusses were also built by the Romans as evidenced by the multiple-span bridge over the Danube, built in 104 A.D. (depicted on Trajan's Column in Rome) with individual spans of 37 m. Wooden-truss bridge technology became highly developed, e.g., the 119-m bridge over the Limmat near Wettingen, Switzerland, built in 1778.

In 1776–1779 the first cast-iron arch bridge was constructed by A. Darby III in England. The cantilever design of the Forth railway bridge in Scotland with two 521-m spans was the longest bridge in the world when completed in 1890. The Brooklyn

---

†For more complete information on the history of mechanics see S. P. Timoshenko, *History of Strength of Materials*, McGraw-Hill, New York, 1953.

§One of several references on the history of structures is H. Straub, *A History of Civil Engineering*, MIT Press, Cambridge, Mass., 1964.

Table B.1 Some structural theories and engineers

Development or theory	Engineer[§]
Introduction of modern truss	Andrea Palladio (1518–1580)
Principle of virtual displacements	Johann Bernoulli (1667–1748)
Theory of elastic curves; strain energy for a beam	Daniel Bernoulli (1700–1782)
Application of minimal principles; energy method vs. direct method	Leonhard Euler (1707–1783)
Generalized forces and displacements	J. L. Lagrange (1736–1813)
Theorems on rigid trusses (1837)	August F. Möbius (1790–1868)
Analysis of determinate trusses (1839–1851)	Squire Whipple (1804–1888), Karl Culmann (1821–1881), J. W. Schwedler (1823–1894)
Three-moment equation (1849); conservation of energy (1852)	B. P. E. Clapeyron (1799–1864)
Consistent displacements (1864);[†] reciprocal theorem of deflections (1872)	James Clerk Maxwell (1831–1879)
Influence lines (1868); method of elastic weights (1870); general three-moment equation (1892)	Otto Mohr (1835–1918)
Castigliano's theorems (1873)	Alberto Castigliano (1847–1884)
Moment-area method (1873)	Charles E. Greene (1842–1903)
Müller-Breslau principle (1886)	H. Müller-Breslau (1851–1925)
Complementary strain energy (1889)	Friedrich Engesser (1848–1931)
Analysis of space trusses (1892)	August Föppl (1854–1924)
Rayleigh-Ritz method (1909)	Lord Rayleigh (1842–1919), Walter Ritz (1878–1909)
Slope-deflection equations (1915)	George A. Maney (1888–1947)
Moment-distribution method (1932)	Hardy Cross (1885–1959)
Relaxation methods (1940)	Richard Vynne Southwell (1888–1970)

[†]Also discovered independently by Mohr (1874). Although the Maxwell-Mohr method is the method of virtual forces, both discoverers applied it to the solution of indeterminate structures.

[§]With the exception of A. F. Möbius who was a mathematician.

Bridge, completed in 1883 with a center span of 468 m, was the first of many wire cable-suspension bridges. The first modern cable-stayed bridge was the Strömsund Bridge in Sweden, completed in 1955.

The 47 bridges designed by Robert Maillert in Switzerland between 1896 and 1940 are elegant examples of the use of reinforced concrete. Prestressed concrete is also used extensively for bridges, and the Swiss Felsenau Bridge (1975) by C. Menn demonstrates the use of segmental cantilever construction.

The Great Pyramid (ca. 2700 B.C.) is representative of many contributions to the history of buildings by the Egyptians, who also used post and lintel construction. The Parthenon, completed in 438 B.C., illustrates further development of this type of construction made by the Greeks. Vaults and domes were erected by the Romans, and the 43-m-diameter ribbed masonry dome of the Pantheon was completed about 123 A.D. Dome and arch construction was developed and refined during the Middle Ages.

A prominent example of early wrought-iron building construction is the 300-m Eiffel Tower, completed in 1889. The Tacoma Building in Chicago, completed in 1889, was the first iron and steel skyscraper. Further refinements in these designs yielded the Empire State Building (1931), which rises 375 m and has 85 stories above ground. The 442-m Sears Tower in Chicago, completed in 1975, has 109 stories and is now the tallest building in the world.

B.2 THE ELECTRONIC DIGITAL COMPUTER

The computer has significantly changed the work and methods of the structural engineer. The first electronic digital computer, ENIAC, was completed in 1946 at the University of Pennsylvania. The computer of today bears little resemblance to that first 30-ton behemoth because of such discoveries as the transistor (1947) and the silicon chip (1959). Over the past 25 years, computational speed has increased by a factor of 200, while the cost, energy consumption, and size of computers of comparable power have decreased by a factor of 10,000. The powerful desktop computer, combined with proper software and structural engineering programs, has significantly reduced the time between problem description and final results.

B.3 ANALYSIS METHODS

Methods of structural analysis can be broadly categorized as classical, approximate, or computer-oriented. The underlying concepts for each category may be similar, and most methods can be programmed for the digital computer. However, the distinction is clarified in at least one case if we contrast moment distribution with the stiffness method. Although both are classified as equilibrium methods, the latter is oriented toward computer implementation while the former is primarily a hand-computational method.

Since the stiffness method has enabled the engineer to analyze a broad spectrum of structures conveniently and accurately, the student should understand it thoroughly. Classical and approximate methods must also be mastered because (1) many are basic to the study of structural analysis, (2) they can frequently be used to investigate preliminary structural designs, (3) they represent alternative methods for solving structural problems, and (4) they provide the necessary means of checking computer-generated solutions for plausibility.

B.4 THE RELATIONSHIP BETWEEN ANALYSIS AND DESIGN

The objective of most structural engineering work is to produce a system that will function in a prescribed manner and provide for the safety and welfare of the occupants and the general public. Structural analysis and design are the disciplines that must be used to attain this goal. The analysis and design function are closely interrelated and both must be mastered. For example, in a statically indeterminate structure, the individual member stiffnesses influence the distribution of the loads throughout the total assemblage. Thus, a preliminary member sizing and first analysis is typically followed by additional design-analysis cycles to obtain an acceptable structure.

B.5 CODES AND SPECIFICATIONS

Whether it is a building, bridge, airplane, ship, or offshore drilling platform, a structure is probably under the jurisdiction of a municipality, bureau, or agency, each of which typically has prescribed criteria for judging structural acceptability. For example, a high-rise building in a city is likely to be scrutinized in accordance with the *Uniform Building Code*. In addition, the acceptable behavior of structures and their components are usually prescribed in a specification issued by the appropriate material institute (e.g., American Institute of Steel Construction, American Concrete Institute). Although the structural engineer must observe these codes and specifications, they do not constitute a substitute for engineering judgment. For example, a building code may prescribe a minimum design snow load, but it is the responsibility of the structural engineer to determine whether this minimum will be exceeded by the local conditions.

B.6 LEGAL AND ECONOMIC CONSIDERATIONS

The safety and structural integrity of constructed facilities are obvious design requirements that the users assume have been met. As a guarantee to the client and the general public, the engineer responsible for a project must ordinarily be legally certified to practice structural engineering. Certification is usually conferred by a legally constituted panel of peers after the candidate has demonstrated competence through testing and shown evidence of sufficient experience. The engineer must also have an understanding of the principles of economics in order to complete the structure on time and at the prescribed price. Thus, any design must be both structurally and economically efficient. Since the cost is frequently the single most important criterion for acceptability, costs must be considered during all phases of the design.

B.7 THE ROLE OF THE STRUCTURAL ENGINEER

Typically, the structural engineer is a member of a design team charged with responsibility for the total project. For example, the disciplines of architecture and structural, mechanical, and electrical engineering are represented on a major building design team. The team members should work as equal partners, and the technical demands of one member should not be allowed to jeopardize the integrity of the overall project. Thus, the structural engineer must understand all the systems and materials in order to be an effective team member.

ANSWERS TO SELECTED PROBLEMS

Chapter 2

2.1 (*b*) R_{ay} = 2.75 kips (↑), R_{bx} = 1.00 kip (↔), R_{by} = 0.25 kip (↑). (*e*) R_{ax} = 0 kN, R_{ay} = 15 kN (↓), M_a = 90 kN · m (↶), R_{cy} = 33 kN (↑).

2.3 $x = L/3$.

2.4 (*a*) Determinate; (*b*) indeterminate; (*e*) determinate; (*f*) determinate.

Chapter 3

3.1 (*a*) Simple; (*c*) compound; (*e*) simple; (*g*) simple; (*i*) simple.

3.2 (*a*) Stable, indeterminate (1 redundant); (*c*) stable, determinate; (*e*) stable, determinate; (*g*) stable, indeterminate (4 redundants); (*i*) stable, indeterminate (2 redundants); (*k*) geometrically unstable; (*m*) stable, determinate; (*o*) stable, determinate; (*q*) geometrically unstable.

3.3 (*a*) F_{hl} = 500 kN (t), F_{gk} = 1000 kN (c), $F_{hj} = F_{jk}$ = 1414 kN (t), $F_{gj} = F_{jl}$ = 707 kN (c).

3.5 $F_{ac} = F_{ce}$ = 28 kN (c), F_{bd} = 70 kN (t), F_{de} = 35 kN (t), F_{ad} = 35 kN (c), F_{ab} = 21 kN (t), F_{cd} = 42 kN (t).

3.6 F_{bd} = 13.5 kips (t), $F_{df} = F_{fh}$ = 1.5 kips (t), $F_{ac} = F_{ce}$ = 6 kips (c), $F_{eg} = F_{ef}$ = 0, F_{ab} = 10 kips (t), F_{cd} = 4 kips (t), F_{gh} = 2 kips (t), F_{ad} = 12.5 kips (c), F_{de} = 7.5 kips (t), F_{eh} = 2.5 kips (c).

3.9 $F_{ac} = F_{ce}$ = 13.94 kips (t), $F_{eg} = F_{gh}$ = 5.06 kips (t), $F_{cb} = F_{gf}$ = 0, F_{ed} = 4.76 kips (t), F_{ab} = 8.44 kips (t), F_{bd} = 4.02 kips (c), F_{df} = 8.35 kips (c), F_{fh} = 8.44 kips (c).

3.11 $F_{ad} = F_{df} = F_{be}$ = 4 kN (c), $F_{bc} = F_{ab}$ = 2 kN (c), F_{ce} = 5.66 kN (t), $F_{ae} = F_{ef}$ = 2.83 kN (t), $F_{fh} = F_{hl}$ = 2.83 kN (c), $F_{de} = F_{fg} = F_{gj} = F_{hk} = F_{gh} = F_{jk} = F_{kl} = F_{hj}$ = 0.

3.13 $F_{bd} = F_{df}$ = 30 kN (t), $F_{fh} = F_{hk}$ = 4.5 kN (t), $F_{ac} = F_{ce} = F_{eg}$ = 18 kN (c), F_{ab} = 24 kN (c), F_{gh} = 12 kN (c), F_{fg} = 22.5 kN (t), F_{gk} = 7.5 kN (c), $F_{ad} = F_{cf} = F_{cd} = F_{ef} = F_{jk} = F_{gj}$ = 0.

3.15 F_{dh} = 60 kips (c), $F_{ac} = F_{ce} = F_{eg} = F_{gj}$ = 30 kips (t), $F_{ab} = F_{bd}$ = 50 kips (c), F_{df} = 50 kips (t), $F_{fh} = F_{fj}$ = 25 kips (t), F_{fg} = 40 kips (t), F_{hj} = 40 kips (c), $F_{bc} = F_{be} = F_{de}$ = 0.

3.17 $F_{bc} = F_{ad} = F_{de}$ = 24 kN (t), F_{ab} = 8 kN (t), F_{be} = 25.3 kN (c), F_{ac} = 33.94 kN (c), $F_{cf} = F_{df} = F_{ef}$ = 0.

3.20 F_{df} = 21.03 kN (c), F_{ef} = 7.70 kN (c), F_{de} = 17.18 kN (t), F_{cb} = 21.04 kN (c), F_{ac} = 28.73 kN (t), F_{ab} = 15.40 kN (t), F_{ad} = 6.90 kN (c), F_{cf} = 25.76 kN (t), F_{be} = 18.86 kN (c).

3.23 $R_{ax} = R_{bx}$ = 0, R_{rx} = 96 kN (↔), R_{sx} = 96 kN (↔), R_{sy} = 64 kN (↑).

3.25 (*a*) F_{ce} = 120 kN (c), $F_{ad} = F_{ab}$ = 60 kN (c), $F_{ae} = F_{bd}$ = 67.08 kN (t), F_{bc} = 67.08 kN (c).

Chapter 4

4.1 (b) $F_{a-} = -1$ kip, $V_{a-} = -0.25$ kip, $M_a = 1$ kip $\cdot$ ft; (e) $F_{a-} = 0$, $V_{a-} = -15$ kN, $M_a = +90$ kN $\cdot$ m.

4.3 (a) Shear at: $b = L_2P/L$; left of $a = L_2P/L$; right of $a = -L_1P/L$; $c = -L_1P/L$. Moment at: $b = 0$; $a = L_1L_2P/L$; $c = 0$. (e) Shear (kN) at: $a = -15$; $b = -15$; left of $c = -15$; right of $c = +18$; $d = 12$. Moment (kN $\cdot$ m) at: $a = +90$; $b = 0$; $c = -90$; $d = 0$.

4.5 Thrust (kN) at: $a = +10$; b (on the column below the beam) $= +10$; b (on the column above the beam) $= 0$; $c = 0$; b (on the beam) $= -10$; $d = -10$. Shear (kN) at: $a = 0$; b (on the column below the beam) $= 0$; b (on the column above the beam) $= +10$; $c = +10$; b (on the beam) $= -10$; $d = -10$. Moment (kN $\cdot$ m) at: $a = 0$; b (on the column below the beam) $= 0$; b (on the column above the beam) $= +10$; $c = 0$; b (on the beam) $= +50$; $d = 0$.

4.7 (a) Shear (lb) at: $a = +160$; left of $b = +160$; right of $b = -240$; left of $c = -240$; right of $c = +400$; $d = 0$. Beam: simply supported at a and c; concentrated load $= -400$ lb at b; uniform load $= -100$ lb/ft from c to d.

Chapter 5

5.3 R_{ax}: Influence line varies linearly from $+1.00$ at a to $-\frac{2}{3}$ at d. V_b: Influence line varies linearly from 0 at a to $-\frac{2}{3}$ to the left of b and from $+\frac{1}{3}$ to the right of b to $-\frac{2}{3}$ at d. M_b: Influence line varies linearly from 0 at a to $\frac{10}{3}$ at b to $-\frac{20}{3}$ at d.

5.4 (a) 354.17 kN; (b) $+40.83$ kN; -126.67 kN; (c) $+558.33$ kN $\cdot$ m; -966.67 kN $\cdot$ m; (d) $+1016.67$ kN $\cdot$ m.

5.7 (a) R_{ax}: Influence line varies linearly from 0 at b to $+1.0$ at c. R_{ay}: Influence line has a constant value of $+1.0$. (b) R_c: Influence line varies linearly from 0 at b to $+1.0$ at c. (c) M_d: Influence line varies linearly from 0 at b to $+5.0$ at c. V_d: Influence line varies linearly from 0 at b to -1.0 at c. F_d: Influence line has a constant value of -1.0.

5.9 (a) R_a: Influence line varies linearly from $+0.943$ at a to -0.471 at c. (b) R_{ex}: Influence line varies linearly from $+0.667$ at a to -0.333 at c. R_{ey}: Influence line varies linearly from $+0.333$ at a to $+1.333$ at c. (c) V_d: Influence line varies linearly from -0.667 at a to $+0.333$ at c. M_d: Influence line varies linearly from $+3.333$ at a to -1.667 at c.

5.12 R_a: Influence line varies linearly from $+1.0$ at a to -0.50 at e to 0 at c. R_b: Influence line varies linearly from 0 at a to $+1.5$ at e to 0 at c. V_d: Influence line varies linearly from 0 at a to -0.5 to the left of d, linearly from $+0.5$ to the right of d to -0.5 at e to 0 at c. M_d: Influence line varies linearly from 0 at a to $+2$ at d to -2 at e to 0 at c.

5.14 (a) R_a: Influence line has a constant value of $+1.0$ from a to c, varies linearly to 0 at e, and is 0 from e to f. M_a: Influence line varies linearly from 0 at a to -10 at c to 0 at e, and is 0 from e to f. (b) V_b: Influence line is 0 from a to the left of b, has a constant value of $+1.0$ from b to c, varies linearly from c to 0 at e and is 0 from e to f. M_b: Influence line is 0 from a to b, varies linearly to -5.0 at c and to 0 at e, is 0 from e to f. (c) V_d: Influence line is 0 from a to c and e to f, varies linearly from 0 at c to -0.5 to the left of d, and varies linearly from $+0.5$ to the right of d to 0 at e. M_d: Influence line is 0 from a to c and e to f, varies linearly from 0 at c to $+5.0$ at d to 0 at e.

5.20 (a) V_{cd}: Influence line varies linearly from -2.000 at b to $+0.625$ at a to 0 at f. (b) M_d: Influence line varies linearly from -12 at b to $+15$ at a to 0 at f.

5.22 V_g: Influence line varies linearly from $+1.00$ at a to -0.33 at d. M_g: Influence line varies linearly from $+6$ at a to -2 at d.

5.25 (a) V_{bc}: Influence line is 0 from a to b, and varies linearly from -0.25 at b to $+0.25$ at c to 0 at e. (b) M_b: Influence line is 0 from a to b, and varies linearly from $+3.75$ at b to $+1.25$ at c to 0 at e. (c) M_c: Influence line is 0 from a to b, and varies linearly from $+1.25$ at b to $+3.75$ at c to 0 at e.

5.29 $(M_e)_{max} = -50$ kip $\cdot$ ft.

5.32 F_{dg}: Influence line varies linearly from 0 at a to -0.57 at e to $+0.57$ at g to 0 at k. F_{fg}: Influence line is zero. F_{eg}: Influence line varies linearly from 0 at a to $+1.2$ at e to 0 at k.

5.34 F_{ce}: Influence line varies linearly from 0 at a to -0.707 at f to 0 at j. F_{de}: Influence line varies linearly from 0 at a to -1.0 at d to 0 at f and is 0 from f to j. F_{df}: Influence line varies linearly from 0 at a to -0.50 at d to 0 at f and is 0 from f to j.

5.37 F_{bc}: Influence line varies linearly from 0 at a to $+1.0$ at c to 0 at e and is 0 from e to h. F_{ce}: Influence line varies linearly from 0 at a to $+0.56$ at c to 0 at h. F_{be}: Influence line varies linearly from 0 at a to -0.46 at c to $+0.33$ at e to 0 at h.

5.40 F_{kl}: Influence line is 0 from a to j and n to r; varies linearly from 0 at j to $+1.0$ at l to 0 at n. F_{hk}: Influence line is 0 from a to j and n to r; varies linearly from 0 at j to $+0.83$ at l to 0 at n. F_{hm}: Influence line varies linearly from 0 at a to -0.75 at j to 0 at r.

5.43 F_{gk}: Influence line is 0. F_{fe}: Influence line is 0 from a to c and g to j; varies linearly from 0 at c to $+1.0$ at e to 0 at g. F_{fg}: Influence line is 0 from a to e, varies linearly to $+1.25$ at g and all ordinates are $+1.25$ from g to j. F_{bd}: Influence line varies linearly from 0 at a to -1.67 at j.

5.47 $V_{\max} = \pm 116$ kips; $M_{\max+} = +1073$ kip $\cdot$ ft; $M_{\max-} = -600$ kip $\cdot$ ft.

Chapter 6

6.2 Critical: $D + L + S$. Member: $F_{ab} = -10,317$ lb; $F_{ac} = +8253$ lb; $F_{bc} = -1810$ lb; $F_{bd} = -8375$ lb; $F_{cd} = +3084$ lb.

6.4 Critical: $D + L + I$. Member: $F_{bd} = -129.3$ kips; $F_{cd} = -23.6$ kips ($+16.1$ kips); $F_{de} = -23.6$ kips ($+16.1$ kips); $F_{ce} = +122.9$ kips.

Chapter 7

7.2 (a) $h_c = 13.26$ m, $h_d = 20.21$ m, $h_e = 24.00$ m, $h_f = 21.47$ m; (b) $y_c = 10.86$ m, $y_d = 15.41$ m, $y_e = 16.80$ m, $y_f = 11.87$ m; (c) $T_{fb} = T_{\max} = 3873.5$ kN; (d) $L = 302.5$ m.

7.4 (a) $y = (hx^2/2L^3)(3 - x/L)$; (b) $y = 12hx^2/11L^2$, $0 \le x \le L/2$; $y = (h/11L^2)(-8x^3/L + 24x^2 - 6Lx + L^2)$, $L/2 \le x \le L$; (c) $y = 8hx^2/7L^2$, $0 \le x \le L/2$, $y = (h/7L^2)(4x^2 + 4Lx - L^2)$, $L/2 \le x \le L$.

7.8 (a) M (kN $\cdot$ m): $a = 0$, $b = +45$, $c = +60$, $d = +45$, $e = 0$, $f = +520$, $g = 0$; (b) $T_f = -1131$ kN (c), $V_f = 0$, $M_f = 0$; (c) $T_f = -24.8$ kN (c), $V_f = +3.6$ kN, $M_f = -200$ kN $\cdot$ m.

Chapter 8

8.1 $F_{ad} = F_{bd} = 166.67$ kips (t), $F_{de} = 160.00$ kips (t), $F_{cd} = F_{ce} = 200.00$ kips (c).

8.5 $F_{ab} = F_{ad} = 8.54$ kN (c), $F_{bc} = 20.61$ kN (c), $F_{cd} = 3.54$ kN (c), $F_{ac} = 41.23$ kN (t), $F_{d0} = F_{a'd} = -F_{a0} = 7.07$ kN (t), $F_{b0} = F_{bc'} = -F_{c0} = 17.07$ kN (t), $F_{ac'} = 29.57$ kN (c), $F_{cd'} = 24.14$ kN (c), $F_{aa'} = 12.07$ kN (t), $F_{cc'} = 5.0$ kN (t).

8.7 (a) Simple: $F_{ad} = -F_{bd} = -F_{cf} = 28.28$ kN (t), $F_{df} = 20.0$ kN (t), all others zero. (b) Complex: $F_{ad} = F_{ae} = -F_{bf} = -F_{cf} = 14.14$ kN (t), $F_{df} = F_{ef} = -F_{be} = -F_{de} = -F_{cd} = 10.0$ kN (t).

8.9 Complex: (a) $F_{a_1b_2} = 1.94$ kN (c), $F_{a_1h_2} = 0.65$ kN (c), $F_{a_1a_2} = 2.83$ kN (c), $F_{a_2a_3} = 3.64$ kN (c), $F_{a_2b_2} = F_{a_2h_2} = 0.94$ kN (t), $F_{a_3b_2} = 3.64$ kN (t), $F_{a_3h_2} = F_{a_3g_3} = 0$, $F_{a_3c_3} = 3.75$ kN (c).

Chapter 9

9.1 $u_b = 4$ mm ($\rightarrow$), $v_b = 13.33$ mm ($\downarrow$).

9.5 $v = (P/48EI)(4x^3 - 3L^2x)$.

9.13 $x = 22.36$ ft, $v_{\max} = 0.493$ in ($\downarrow$), $\theta_{\max} = \theta_c = 0.00386$ rad ($\circlearrowright$).

9.15 $\theta_c = 620 \times 10^{-6}$ rad ($\circlearrowleft$), $v_c = 0.028$ in ($\uparrow$), $\theta_f = 0.059$ rad ($\circlearrowright$), $v_f = 0.838$ in ($\downarrow$).

9.17 $v_b = 0.490$ in ($\downarrow$), $v_c = 0.919$ in ($\downarrow$), $v_e = 0$.

9.21 $x = \sqrt{3}L/3$, $v_{\max} = (\sqrt{3}PL^3)/54EI$.

9.23 $P_2 = 3P_1$.

9.25 $L_1 = 0.3039L$.

9.27 (a) $P = 18.75$ kips; (b) $P = 20.0$ kips.

9.29 $d = 18.77$ in.

9.31 $x = 4.62$ m (from point a), $v_{\max} = 5.13$ mm ($\downarrow$).

9.33 $v_a = 7.78$ mm ($\downarrow$).

9.45 $x = 5.10$ ft (right of hinge), $v_{\max} = 0.147$ in ($\downarrow$).

Chapter 10

10.1 $v_c = 2.5$ mm ($\downarrow$), $u_c = 0.6$ mm ($\rightarrow$), $v_d = 2.8$ mm ($\downarrow$), $u_d = 0.2$ mm ($\rightarrow$).

10.3 (a) $u_d = 0.335$ in ($\rightarrow$); (b) $u_d = 0.100$ in ($\rightarrow$).

10.5 (a) $v_e = 0.234$ in ($\downarrow$) $v_{d-g} = 0.066$ in (together); (b) $v_e = 0.088$ in ($\uparrow$); (c) $v_e = 0.375$ in ($\uparrow$).

10.7 (a) $u_f = 7.5$ mm ($\rightarrow$); (b) $u_f = 0$.

10.9 (a) $v_b = 0.781$ in ($\downarrow$); (b) $v_b = 0.536$ in ($\uparrow$), $u_b = 0.358$ in ($\leftarrow$); (c) $v_b = 1.25$ in ($\uparrow$).

10.11 $\Theta_a = 0.00690$ rad ($\circlearrowright$), $\Theta_b = 0.00603$ rad ($\circlearrowleft$).

10.13 $\Theta_b = 4.64 \times 10^{-3}$ rad ($\circlearrowright$), $v_b = 0.883$ in ($\downarrow$).

10.16 $q = 3.0$ kips/ft.

10.21 $v_c = 0.353$ in ($\downarrow$).

10.23 (a) $v_b = 0$; (b) $v_b = 0.128$ in ($\downarrow$); (c) $v_d = 8.681 \times 10^{-4}$ in ($\downarrow$).

10.43 $v_{bb} = 0.179$ in, $v_{bc} = 0.139$ in, $P_b = 4.30$ kips ($\downarrow$), $P_c = 1.94$ kips ($\uparrow$).

Chapter 11

11.1 $R_a = 3P/32$ ($\updownarrow$), $R_b = 11P/16$ ($\uparrow$), $R_d = 13P/32$ ($\uparrow$).

11.3 (a) $R_b = R_d = 22.5$ kips ($\uparrow$), $R_c = 25$ kips ($\downarrow$); (b) $R_b = R_d = 2.91$ kips ($\uparrow$), $R_c = 5.82$ kips ($\updownarrow$).

11.5 (a) $R_b = 32.5$ kN ($\uparrow$), $R_d = 37.5$ kN ($\uparrow$), $M_d = 80$ kN $\cdot$ m ($\curvearrowright$); (b) $R_b = 3.52$ kN ($\downarrow$), $R_d = 3.52$ kN ($\uparrow$), $M_d = 28.16$ kN $\cdot$ m ($\curvearrowright$).

11.7 $R_b = 7.375$ kN ($\uparrow$), $R_c = 1.75$ kN ($\uparrow$), $R_e = 3.875$ kN ($\uparrow$), $M_b = -20$ kN $\cdot$ m, $M_c = -1$ kN $\cdot$ m, $M_d = +15.5$ kN $\cdot$ m.

11.9 (a) $x = 0.2899L$; (b) $x = 0.3055L$.

11.11 (a) $R_{ay} = 0$, $R_{ax} = 22.5$ kips ($\rightarrow$), $R_{ey} = 60.0$ kips ($\uparrow$), $R_{ex} = 22.5$ kips ($\leftarrow$), $F_{ce} = -F_{ac} = 22.5$ kips (t), $F_{bd} = 45$ kips (t), $F_{bc} = 60$ kips (t), $F_{be} = F_{cd} = 75$ kips (c), $F_{ab} = 0$; (b) $R_{ax} = 145$ kips ($\leftarrow$), $R_{ex} = 145$ kips ($\rightarrow$), $F_{ac} = F_{ce} = 145$ kips (t), all others zero.

11.13 (a) $F_{ab} = 111.7$ kN (t), $F_{ac} = 7.2$ kN (t), $F_{ad} = 42.1$ kN (c); (b) $F_{ab} = 39.5$ kN (c), $F_{ac} = 76.3$ kN (t), $F_{ad} = 55.8$ kN (c).

11.15 $F_{df} = 1.604P$ (t), $F_{ce} = 1.397P$ (c), $F_{de} = 0.560P$ (t), $F_{cf} = 0.854P$ (c), $F_{cd} = F_{ef} = 0.397P$ (c).

11.17 $R_{ax} = 17.1$ kips ($\leftrightarrow$), $R_{ay} = 47.2$ kips ($\uparrow$), $R_{dx} = 17.1$ kips ($\leftarrow$), $R_{dy} = 72.8$ kips ($\uparrow$), $M_b = 566$ kip $\cdot$ ft, $M_c = 308$ kip $\cdot$ ft.

11.19 $R_{dx} = 4R_{dy}/3 = 8.99$ kips ($\leftarrow$), $R_{ax} = 8.99$ kips ($\rightarrow$), $R_{ay} = 17.26$ kips ($\uparrow$), $M_a = 105.12$ kip $\cdot$ ft ($\curvearrowright$).

11.21 $R_{ay} = 14.62$ kips ($\updownarrow$), $M_a = 46.2$ kip $\cdot$ ft ($\curvearrowright$), $R_{by} = 24.62$ kips ($\uparrow$).

11.23 $R_{ay} = 8.30$ kips ($\uparrow$), $M_a = 3.5$ kip $\cdot$ ft ($\curvearrowright$), $R_b = 45.7$ kips ($\uparrow$), $M_{\max} = +13.72$ kip $\cdot$ ft (4.15 ft from a), $M_b = -104$ kip $\cdot$ ft.

11.25 (a) $R_a = 13.12$ kips ($\uparrow$), $R_b = 18.77$ kips ($\uparrow$), $R_c = 1.25$ kips ($\uparrow$), $R_e = 13.73$ kips ($\uparrow$), $R_g = 3.14$ kips ($\uparrow$); (b) $R_a = R_g = 7.15$ kips ($\downarrow$), $R_b = R_e = 26.22$ kips ($\uparrow$), $R_c = 38.15$ kips ($\updownarrow$).

11.27 $R_{ax} = 16$ kN $(\leftrightarrow)$, $R_{ay} = 5.22$ kN $(\updownarrow)$, $R_{ey} = 4.45$ kN $(\uparrow)$, $R_{hy} = 0.78$ kN $(\uparrow)$, $F_{dg} = 1.3$ kN (c), $F_{ef} = 0$, $F_{df} = 1.0$ kN (c), $F_{eg} = 2.1$ kN (t).

11.29 $R_{ax} = 5.88$ kips $(\leftrightarrow)$, $R_{ay} = 24.15$ kips $(\uparrow)$, $R_{dx} = 18.12$ kips $(\leftrightarrow)$, $R_{dy} = 26.25$ kips $(\uparrow)$, $M_d = 95.5$ kip · ft $(\curvearrowright)$, $M_b = 35.3$ kip · ft, $M_c = -73.4$ kip · ft.

11.59 $M_b = -114.3$ kN · m.

11.61 $M_b = -120$ kN · m, $M_c = -84$ kN · m.

11.63 $M_c = -185.8$ kip · ft.

11.65 $q = 1.751$ kips/ft.

11.67 $x = 13L/56$.

11.69 $M_a = M_d = 4.68$ kip · ft, $M_b = M_c = -56.25$ kip · ft.

Chapter 12

12.1 $M_{ab} = -17.2$ kN · m, $M_{ba} = -M_{bc} = 55.6$ kN · m, $M_{cb} = 73.0$ kN · m.

12.3 $M_{ab} = -85.2$ kip · ft, $M_{ba} = -M_{bc} = 60.0$ kip · ft.

12.5 $M_{ba} = -M_{bc} = 268.1$ kN · m, $M_{cb} = -M_{cd} = 119.2$ kN · m, $M_{dc} = 372.4$ kN · m.

12.7 $M_{ba} = 194.2$ kN · m, $M_{bc} = -200.6$ kN · m, $M_{bd} = 6.4$ kN · m, $M_{cb} = 143.9$ kN · m.

12.9 $M_{ab} = -139.12$ kN · m, $M_{ba} = 171.77$ kN · m, $M_{bd} = -186.3$ kN · m, $M_{db} = -M_{de} = 106.75$ kN · m, $M_{ed} = -53.38$ kN · m, $M_{bc} = 14.51$ kN · m, $M_{cb} = 7.26$ kN · m.

12.11 $M_{ab} = -132$ kip · ft, $M_{ba} = -M_{bc} = -169$ kip · ft, $M_{cb} = -M_{cd} = 287$ kip · ft, $M_{dc} = -M_{de} = -378$ kip · ft, $M_{ed} = 390$ kip · ft.

12.13 $M_{ba} = M_{cd} = -M_{bc} = -M_{cb} = -100.7$ kip · ft.

12.15 $M_{ab} = -48$ kip · ft, $M_{ba} = -M_{bc} = -42.9$ kip · ft, $M_{cb} = 40$ kip · ft.

12.17 $M_{ab} = -49.74$ kip · ft, $M_{ba} = -M_{bc} = -33.38$ kip · ft, $M_{cb} = -M_{cd} = 32.65$ kip · ft, $M_{dc} = -41.12$ kip · ft.

12.19 (a) $M_{ab} = M_{ac} = M_{ad} = M_{ae} = M/4$; (b) $M_{ab} = M_{ac'} = M_{ae} = 2M/7$, $M_{ad} = M/7$; (c) $M_{ab} = M_{ac} = M_{ae} = M/5$, $M_{ad} = 2M/5$; (d) $M_{ab} = M_{ac} = M_{ae} = M/5$, $M_{ad} = 2M/5$; (e) $M_{ab} = M_{ac'} = M_{ae} = 2M/7$, $M_{ad} = M/7$.

12.21 (a) $M_{ba} = M_{dc} = 0.08PL$, $M_{cb} = 0.055PL$; (b) $M_{ba} = M_{dc} = 0.053qL^2$, $M_{cb} = 0.036qL^2$.

12.29 $I_{bc} = 1500$ in⁴.

12.31 (a) $x = 0.145L$; (b) $x = 0.153L$.

12.41 $M_{ab} = -337.2$ kip · ft, $M_{ba} = -339.6$ kip · ft, $M_{bc} = 293.3$ kip · ft, $M_{bd} = 46.3$ kip · ft, $M_{db} = -M_{de} = 98.9$ kip · ft.

12.43 (a) $M_{ab} = -258.7$ kN · m, $M_{bc} = 164.7$ kN · m, $M_{cd} = 41.3$ kN · m, $M_{dc} = 152.8$ kN · m; (b) $M_{ab} = -313.5$ kN · m, $M_{bc} = 175.8$ kN · m, $M_{cd} = -12.0$ kN · m, $M_{dc} = 163.0$ kN · m.

12.47 $M_{ab} = 9.4$ kip · ft, $M_{bc} = -28.1$ kip · ft, $M_{cb} = 37.5$ kip · ft, $M_{dc} = 0$.

12.49 $M_{ab} = +83.3$ kip · ft, $M_{bc} = -106.9$ kip · ft, $M_{cd} = +27.1$ kip · ft, $M_{dc} = +43.1$ kip · ft.

12.51 $M_{ab} = -98.6$ kN · m, $M_{ba} = -81.4$ kN · m, $M_{be} = 105.2$ kN · m, $M_{bc} = -23.8$ kN · m, $M_{cb} = -M_{cd} = -36.1$ kN · m.

Chapter 13

13.1 (a) $F_{df} = 7.5$ kN (t), $F_{ce} = 15.0$ kN (c), $F_{de} = 25.0$ kN (t), $F_{cf} = 0$, $F_{ef} = -F_{cd} = 10.0$ kN (t); (b) $F_{df} = -F_{ce} = 15.0$ kN (t), $F_{de} = -F_{cf} = 12.5$ kN (t), $F_{ef} = -F_{cd} = 5.0$ kN (t); (c) $F_{df} = 7.5$ kN (t), $F_{ce} = 22.5$ kN (c), $F_{de} = 25.0$ kN (t), $F_{cf} = 0$.

13.5 (a) $F_{eg} = 0.72$ kN (t), $F_{fh} = 2.60$ kN (c), $F_{eh} = 3.25$ kN (t), $F_{fg} = 0$, $F_{gh} = 2.89$ kN (c), $F_{ef} = 4.60$ kN (c); (b) $F_{eg} = -F_{fh} = 1.65$ kN (t), $F_{eh} = -F_{fg} = 1.62$ kN (t), $F_{gh} = F_{ef} = 1.00$ kN (c); (c) $F_{eg} = 2.02$ kN (t), $F_{fh} = 1.30$ kN (c), $F_{eh} = 1.00$ kN (t), $F_{fg} = 2.52$ kN (c).

13.7 $F_{bd} = -F_{eg} = 98$ kN (t), $M_e = -M_d = 87.5$ kN $\cdot$ m.

13.9 $F_{bd} = 8.74$ kips (t), $F_{cd} = 13.34$ kips (c), $F_{ce} = 6.22$ kips (t), $F_{pn} = 0.40$ kip (t), $F_{pm} = 5.03$ kips (t), $F_{mq} = 12.07$ kips (c), $M_{ab} = M_{rq} = -M_{ba} = -M_{qr} = -27$ kip $\cdot$ ft.

13.11 $F_{ce} = 60.00$ kN (c), $F_{be} = 22.53$ kN (t), $F_{bd} = 20.00$ kN (c), $F_{ps} = 30.00$ kN (t), $F_{pr} = 22.53$ kN (c), $F_{nr} = 40.00$ kN (c), $F_{gj} = -F_{be} = 22.53$ kN (c), $F_{jm} = -F_{pr} = 22.53$ kN (t), $F_{fj} = 70$ kN (c), $F_{ji} = 10$ kN (t), $M_{ab} = -M_{ba} = 90$ kN $\cdot$ m, $M_{hj} = M_{jh} = 180$ kN $\cdot$ m.

13.13 (a) $M = 414$ kip $\cdot$ ft; (b) $M = -232.88$ kip $\cdot$ ft; (c) $V = 86.25$ kips; (d) $F = 517.5$ kips; (e) $F = 258.75$ kips; (f) and (g) $M = 698.64$ kip $\cdot$ ft (base).

13.15 (a) $M_{ba} = -M_{ab} = 90$ kN $\cdot$ m, $M_{cb} = -M_{bc} = 30$ kN $\cdot$ m, $M_{be} = +120$ kN $\cdot$ m; (b) same as (a).

13.17 (a) 10-11, $M_{ext} = 120$ kN $\cdot$ m, $M_{int} = 240$ kN $\cdot$ m, 3-4, $M_{ext} = 228$ kN $\cdot$ m, $M_{int} = 456$ kN $\cdot$ m, $M_{\mathbb{C}} = 456$ kN $\cdot$ m; (b) 10th, $M_{ext} = 288$ kN $\cdot$ m, 3rd, $M_{int} = 3 M_{ext} = 576$ kN $\cdot$ m.

Chapter 14

14.3 ILs vary linearly from a to b and cubically from b to c to d. R_b: $a = +1.625$, $b = +1.000$, -0.0962 at 4.62 m from d. R_c: $a = -0.75$, $b = 0.00$, $c = +1.00$. R_d: $a = +0.125$, -0.0962 at 4.62 m right of b, $d = +1.000$.

14.5 (a) R_{dx}: IL varies linearly from 0 at a and d to $+0.375$ at c. F_{ab}: IL varies linearly from 0 at a and d to -0.625 at c. F_{bc}: IL varies linearly from 0 at a and d to $+1.00$ at c. (b) $F_{ac} = -F_{cd}$: IL varies linearly from 0 at a and d to $+0.50$ at c, all other members have null IL.

14.7 F_{df}: $IL = 0$ from a to c, varies linearly to $+0.460$ at e to $+1.604$ at g. F_{de}: $IL = 0$ from a to c, varies linearly to $+0.763$ at e to $+0.560$ at g. F_{cf}: $IL = 0$ from a to c, varies linearly to -0.651 at e to -0.854 at g.

14.9 IL ordinates at lower panel points. (a) F_{be}: 0, 0, +0.75, +0.50, +0.25, 0; F_{bd}: 0, -0.6, -0.9, -0.6, -0.3, 0; F_{dg}: 0, 0, 0, +0.50, +0.25, 0; F_{df}: 0, -0.45, -0.90, -0.90, -0.45, 0. (b) F_{be}: 0, -0.125, +0.375 (linear variation to 0 at k); F_{bd}: 0, -0.525, -0.675 (linear variation to 0 at k); F_{dg}: 0, -0.125, -0.250, +0.250, +0.125, 0; F_{df}: 0, -0.375, -0.75, -0.75, -0.375, 0.

14.11 ILs vary cubically. F_{cd}: 0 at a, +1.50 at c; R_{ax}: 0 at a, +1.20 at c; R_{ay}: +1.00 at a, -0.50 at c; M_a: 0 at a, -4.38 at 9.83 ft from a, -2.01 at c.

Chapter 15

15.3 $U_b = 12$ mm, $U_c = 28$ mm, $U_d = 31$ mm, $X_a = -576$ kN.

15.5 $U_c = 8.8$ mm, $X_b = 1517$ kN, $X_a = -1152$ kN, $X_d = -554$ kN.

15.9 $Y_e = 1575$ kN, $U_e = 16.67$ mm, $s_{ae} = 500$ kN (t), $s_{be} = 675$ kN (t), $s_{ce} = 250$ kN (c), $s_{de} = 900$ kN (c).

15.11 $Y_c = 86.4$ kN, $U_c = 12.5$ mm, $s_{ac} = 22$ kN (c), $s_{bc} = 122$ kN (t).

15.13 $U_a = 1.679$ in, $U_b = 0.679$ in, $V_b = -1.569$ in, $s_{ab} = 100$ kips (c), $s_{bc} = 88.74$ kips (t), $s_{bd} = 274.51$ kips (c).

15.15 $U_b = -13.3$ mm, $U_d = -30$ mm, $X_a = -20$ kN, $X_c = -60$ kN, $s_{ab} = 20$ kN (t), $s_{cd} = 60$ kN (t).

15.17 $U_b = 0.067$ in, $V_b = -1.21$ in, $X_a = -3.33$ kips, $X_c = -46.67$ kips, $Y_c = 35.00$ kips, $s_{ab} = 3.33$ kips (t), $s_{bc} = 58.33$ kips (t), $s_{ac} = 0$.

15.19 $U_b = 0.80$ in., $U_c = 2.01$ in, $Y_c = -17.50$ kips, $X_a = -40$ kips, $Y_a = -12.5$ kips, $Y_b = 30.06$ kips, $s_{ab} = 40$ kips (t), $s_{ac} = 12.5$ kips (t), $s_{bc} = 50$ kips (c).

15.21 $U_c = 1.42$ in, $V_c = 0.50$ in, $U_d = 1.93$ in, $V_d = -0.40$ in, $X_a = 0$, $Y_a = -12.5$ kips, $X_b = -50$ kips, $Y_b = 47.5$ kips, $s_{ab} = 0$, $s_{ac} = 12.5$ kips (t), $s_{bc} = 62.5$ kips (c), $s_{bd} = 10$ kips (c), $s_{cd} = 25$ kips (t).

15.23 $U_c = 0.562$ in, $V_c = 0.389$ in, $U_d = 0.821$ in, $V_d = -0.761$ in, $X_a = -12.03$ kips, $Y_a = -18.75$ kips, $X_b = -12.97$ kips, $Y_b = 28.75$ kips, $s_{ab} = 0$, $s_{ac} = 9.73$ kips (t), $s_{ad} = 15.04$ kips (t), $s_{bc} = 16.21$ kips (c), $s_{bd} = 19.02$ kips (c), $s_{cd} = 12.97$ kips (t).

Chapter 16

16.5 $Y_a = 115.21$ kN.

16.7 $\{X_b \quad Y_b\} = \{546.20 \quad 409.65\}$ kN.

16.9 $X_b = 76.91$ kips.

16.11 $\{X_a \quad Y_a \quad X_b \quad Y_b\} = \{128.18 \quad -128.18 \quad -128.18 \quad 128.18\}$ kips.

16.13 (a) **K**: 16×16; (b) $\mathbf{K}_{ff}$: 13×13, $\mathbf{K}_{fs}$: 13×3, $\mathbf{K}_{sf}$: 3×13, $\mathbf{K}_{ss}$: 3×3; (c) $\mathbf{X}_f$: 13×1, $\mathbf{X}_s$: 3×1; (d) $\mathbf{U}_f$: 13×1, $\mathbf{U}_s$: 3×1.

16.15 (a) **K**: 16×16; (b) $\mathbf{K}_{ff}$: 12×12, $\mathbf{K}_{fs}$: 12×4, $\mathbf{K}_{sf}$: 4×12, $\mathbf{K}_{ss}$: 4×4; (c) $\mathbf{X}_f$: 12×1, $\mathbf{X}_s$: 4×1; (d) $\mathbf{U}_f$: 12×1, $\mathbf{U}_s$: 4×1; $\{X_b \quad Y_b \quad X_c \quad Y_c \quad X_d \quad Y_d \quad X_e \quad Y_e \quad X_f \quad Y_f \quad X_g \quad Y_g\} = \{0 \quad 0 \quad 0 \quad -P \quad 0 \quad 0 \quad 0 \quad 0 \quad 0 \quad 0 \quad 0 \quad -2P\}$.

16.17 $\mathbf{k}_{bd}$: F, H, N, Q; Null partitions: B and E; $\mathbf{K}_{ff}$: K, L, P, Q.

16.21 $\{U_c \quad V_c \quad U_d \quad V_d\} = (PL/AE)\{7.656 \quad -2.000 \quad 9.650 \quad -1.000\}$; $s_{ab} = 0$, $s_{bc} = 2P$ (c), $s_{ad} = P$ (c), $s_{cd} = 2P$ (c), $s_{ac} = 2.83$ (t).

16.23 $\{U_b \quad V_b \quad U_c \quad V_c\} = (PL/AE)\{-2 \quad -6 \quad 1 \quad -8\}$; $s_{ab} = s_{bc} = 2P$ (c), $s_{ad} = 0$, $s_{cd} = P$ (t), $s_{bd} = 2.83P$ (t).

16.25 $\{U_a \quad U_b \quad V_c\} = \{6.15 \quad 6.15 \quad -1.80\}$ mm; $s_{ab} = 0$, $s_{bc} = 45$ kN (t), $s_{ac} = 36$ kN (c).

16.29 $\{U_c \quad V_c \quad U_d \quad V_d\} = \{0.956 \quad 0 \quad 0.956 \quad -0.250\}$ in; $s_{bd} = 50$ kips (c), $s_{ad} = 70.62$ kips (t), all others zero.

16.31 $\{U_a \quad V_a\} = (PL/AE)\{0.715 \quad -0.582\}$; $s_{ab} = 0.328P$ (c), $s_{ac} = 0.146P$ (t), $s_{ad} = 0.582P$ (t), $s_{ae} = 0.910P$ (t).

16.32 (a) $\{U_a \quad V_a\} = \{0.109 \quad -2.128\}$ mm; (b) $\{U_a \quad V_a\} = \{1.245 \quad -4.970\}$ mm; (c) $\{U_a \quad V_a\} = \{1.141 \quad -4.555\}$ mm, $s_{ab} = 39.2$ kN (c), $s_{ac} = 76.1$ kN (t), $s_{ad} = 55.8$ kN (c); (d) $\{U_a \quad V_a\} = \{-0.517 \quad -7.950\}$ mm, $s_{ab} = 18.1$ kN (t), $s_{ac} = 34.5$ kN (c), $s_{ad} = 25.0$ kN (t); (e) $\{U_a \quad V_a\} = \{0.054 \quad -1.064\}$ mm, member forces same as in part (a); (f) $\{U_a \quad V_a\} = \{-0.10 \quad -1.29\}$ mm, $s_{ab} = 119.1$ kN (t), $s_{ac} = 6.7$ kN (c), $s_{ad} = 31.7$ kN (c).

16.35 $\{U_b \quad V_b \quad U_c \quad V_c \quad U_d \quad V_d \quad U_e \quad U_f \quad V_f\} = \{0.559 \quad 0 \quad 0.286 \quad -1.006 \quad 0.427 \quad -0.891 \quad 0.286 \quad 0.140 \quad -0.215\}$ in.

16.37 $\{U_b \quad V_b \quad U_c \quad V_c \quad U_d \quad V_d\} = \{0.414 \quad 0 \quad 0.146 \quad -0.800 \quad 0.282 \quad -0.688\}$ in.

16.38 (a) $\{U_b \quad V_b \quad U_c \quad U_d \quad V_d \quad U_e \quad V_e\} = \{0.268 \quad 0.056 \quad 0.033 \quad 0.334 \quad -0.037 \quad -0.033 \quad -0.698\}$ in.

16.42 $\{U_b \quad V_b \quad U_c \quad V_c \quad U_d \quad V_d\} = \{0.129 \quad -0.097 \quad -0.047 \quad -0.221 \quad 0.222 \quad -0.617\}$ in.

16.43 (a) $\{U_b \quad V_b \quad U_c \quad V_c \quad U_d \quad V_d \quad U_e \quad U_f \quad V_f\} = \{-7.53 \quad 0 \quad 0 \quad 10.04 \quad 0 \quad -10.04 \quad 0 \quad 7.53 \quad 0\}$ mm; (b) $\{U_b \quad V_b \quad U_c \quad V_c \quad U_d \quad V_d \quad U_e \quad U_f \quad V_f\} = (1.80)\{0 \quad 1 \quad 0 \quad 1 \quad 0 \quad 1 \quad 0 \quad 0 \quad 1\}$ mm.

Chapter 17

17.1 $\{\Theta_a \quad \Theta_b\} = (10^{-3})\{-3.686 \quad 3.917\}$ rad, $\{\bar{y}_a \quad \bar{m}_a \quad \bar{m}_b\}_{ab} = \{0.60 \text{ kip} \quad -720 \text{ kip} \cdot \text{in} \quad 864 \text{ kip} \cdot \text{in}\}$.

17.3 $\Theta_b = 2.16 \times 10^{-3}$ rad, $\{\bar{y}_a \quad \bar{m}_a \quad \bar{m}_b\}_{ab} = \{5.625 \text{ kips} \quad 450 \text{ kip} \cdot \text{in} \quad 900 \text{ kip} \cdot \text{in}\}$.

17.11 $Y_b = -24EI\Delta/L^3$.

17.13 $Y_b = -39EI\Delta/L^3$.

17.15 (a) $\{V_b \quad \Theta_b \quad \Theta_c\} = \{-0.361 \text{ in} \quad 0.752 \times 10^{-3} \text{ rad} \quad 8.652 \times 10^{-3} \text{ rad}\}$; (b) $\{V_b \quad \Theta_b \quad \Theta_c\} = \{-0.127 \text{ in} \quad -1.818 \times 10^{-3} \text{ rad} \quad -2.159 \times 10^{-3} \text{ rad}\}$; (c) $\{V_b \quad \Theta_b \quad \Theta_c\} = \{0.043 \text{ in} \quad -0.087 \times 10^{-3} \text{ rad} \quad -1.632 \times 10^{-3} \text{ rad}\}$.

17.17 (a) $\{\Theta_b \quad \Theta_c\} = (1.279 \times 10^{-3})\{1 \quad -1\}$ rad; (b) $\{\Theta_b \quad \Theta_c\} = (10^{-4})\{4.582 \quad -18.327\}$ rad; (c) $\{\Theta_b \quad \Theta_c\} = \{0 \quad 0\}$.

17.19 $\{\Theta_a L \quad V_b \quad \Theta_b L \quad V_c \quad \Theta_c L \quad \Theta_d L\} = (PL^3/EI)\{-1.963 \quad -1.463 \quad -0.463 \quad -1.370 \quad 0.704 \quad 1.704\}$

17.21 $\{V_b \quad \Theta_b L \quad \Theta_c L\} = (PL^3/EI)\{-0.0389 \quad -0.0167 \quad 0.0667\}$.

17.23 $\{V_b \quad \Theta_b L \quad \Theta_c L\} = (PL^3/EI)\{-0.0333 \quad 0.0000 \quad 0.0500\}$.

17.25 $\{\Theta_a \quad V_b \quad \Theta_b \quad \Theta_c \quad V_d \quad \Theta_d\} = (10^{-3})\{1.730 \text{ rad} \quad 77.53 \text{ in} \quad 1.385 \text{ rad} \quad -5.482 \text{ rad} \quad -540.60 \text{ in} \quad -10.774 \text{ rad}\}$.

17.29 $\{\Theta_a \quad \Theta_b\} = (10^{-4})\{-68.97 \quad 60.34\} \text{ rad}; \{\bar{y}_a \quad \bar{m}_a \quad \bar{m}_b\}_{ab} = \{31.66 \text{ kips} \quad 0 \quad 0\}$.

17.31 (*a*) $\{V_a \quad \Theta_a \quad \Theta_b\} = (10^{-3})\{-1886.9 \text{ in} \quad 19.862 \text{ rad} \quad 7.448 \text{ rad}\}$; (*b*) $\{V_a \quad \Theta_a \quad \Theta_b\} = (10^{-3})\{625.15 \text{ in} \quad -5.2088 \text{ rad} \quad -5.2088 \text{ rad}\}$.

17.32 (*a*) $\{V_a \quad \Theta_a \quad \Theta_b \quad V_c \quad \Theta_c\} = (10^{-3})\{4.4444 \text{ m} \quad -2.0000 \text{ rad} \quad -2.6667 \text{ rad} \quad -8.0000 \text{ m} \quad 0.6667 \text{ rad}\}$; (*b*) $\{V_a \quad \Theta_a \quad \Theta_b \quad V_c \quad \Theta_c\} = (10^{-3})\{-27.5 \text{ m} \quad 3.75 \text{ rad} \quad 3.75 \text{ rad} \quad -6.25 \text{ m} \quad 2.813 \text{ rad}\}$.

17.34 $\{\Theta_b \quad \Theta_c\} = (10^{-4})\{-0.2028 \quad 2.6878\} \text{ rad}$.

17.35 $\{\Theta_b \quad \Theta_c \quad \Theta_d\} = (10^{-4})\{14.778 \quad -21.943 \quad 3.102\} \text{ rad}$.

17.37 $\{\Theta_a \quad \Theta_b \quad \Theta_c \quad \Theta_d\} = (10^{-3})\{12.5 \quad 0 \quad 0 \quad 12.5\} \text{ rad}$.

Chapter 18

18.1 (*a*) $\Theta_a = -ML/16EI$; (*b*) $\Theta_a = -ML/14EI$; (*c*) $\Theta_a = -ML/20EI$; (*d*) $\Theta_a = -ML/20EI$; (*e*) $\Theta_a = -ML/14EI$.

18.5 $\{V_b \quad \Theta_b \quad \Theta_c \quad V_d \quad \Theta_d \quad \Theta_e\} = (10^3/EI)\{-6952 \quad 0.9471 \quad 96.07 \quad -6952 \quad -21.77 \quad -85.66\}$.

18.6 $\{U_b \quad V_b \quad \Theta_b \quad V_c \quad \Theta_c\} = (10^{-3})\{-35.32 \text{ m} \quad -47.11 \text{ m} \quad -9.420 \text{ rad} \quad -35.32 \text{ m} \quad 11.179 \text{ rad}\}$.

18.8 $\{\Theta_c \quad \Theta_b\} = (1/EI)\{-69.1782 \quad -77.6400\} \text{ rad}$.

18.9 $\{\Theta_a \quad \Theta_b \quad \Theta_d\} = (1/EI)\{-426.31 \quad 102.62 \quad -129.44\} \text{ rad}$.

18.11 $\{\Theta_b \quad \Theta_d\} = (1/EI)\{-21.77 \quad 160.13\} \text{ rad}$.

18.12 $\{\Theta_b \quad \Theta_c \quad \Theta_d\} = (1/EI)\{26.224 \quad -51.556 \quad 70.780\} \text{ rad}$.

18.13 $\{\Theta_b \quad \Theta_c \quad U_a\} = (1/EI)\{-81.8304 \quad -28.2144 \quad 1101.7859\}$.

18.16 $\{U_b \quad \Theta_b \quad U_c \quad V_c \quad \Theta_c\} = \{-1.179 \text{ in} \quad -0.005887 \text{ rad} \quad -1.179 \text{ in} \quad -0.08846 \text{ in} \quad 0.004044 \text{ rad}\}$.

18.18 $\{U_b \quad \Theta_b \quad U_c \quad \Theta_c\} = (10^3/EI)\{0.698 \quad -0.05270 \quad 1.090 \quad 0.01790\}$.

18.21 (*a*) $\{\Theta_a \quad V_b \quad \Theta_b \quad V_c \quad \Theta_c\} = (10^{-3})\{-0.2500 \text{ rad} \quad -129.31 \text{ in} \quad -2.7328 \text{ rad} \quad -342.93 \text{ in} \quad -3.9741 \text{ rad}\}$; (*b*) $\{\Theta_a \quad U_b \quad V_b \quad \Theta_b \quad U_c \quad V_c \quad \Theta_c\} = (10^{-3})\{-0.30517 \text{ rad} \quad -4.9655 \text{ in} \quad -135.93 \text{ in} \quad -2.7879 \text{ rad} \quad -4.9655 \text{ in} \quad -352.86 \text{ in} \quad -4.0293 \text{ rad}\}$.

18.25 $\{U_b \quad V_b \quad \Theta_b \quad U_c \quad V_c \quad \Theta_c\} = (a)$: $(10^{-3})\{-1.544 \text{ m} \quad -2.059 \text{ m} \quad -1.912 \text{ rad} \quad -1.544 \text{ m} \quad -11.544 \text{ m} \quad -1.011 \text{ rad}\}$; (b): $(10^{-3})\{-3.089 \text{ m} \quad -4.119 \text{ m} \quad -1.824 \text{ rad} \quad -3.089 \text{ m} \quad -3.089 \text{ m} \quad 1.980 \text{ rad}\}$.

18.27 (*a*) $\{U_b \quad \Theta_b \quad U_c \quad V_c \quad \Theta_c\} = \{-0.639 \text{ in} \quad 2.6256 \times 10^{-3} \text{ rad} \quad -0.639 \text{ in} \quad 0.271 \text{ in} \quad -0.49936 \times 10^{-3} \text{ rad}\}$.

18.29 (*a*) $\{V_a \quad \Theta_{xa} \quad \Theta_{za}\} = \{-0.1004 \text{ in} \quad 0.0 \quad 0.0\}$; (*b*) $\{V_a \quad \Theta_{xa} \quad \Theta_{za}\} = \{-0.0669 \text{ in} \quad 0.0 \quad -0.3298 \times 10^{-3} \text{ rad}\}$.

Chapter 19

19.8 $\{V_b \quad \Theta_{b-} \quad \Theta_{b+}\} = (10^{-3})\{-21.33 \text{ in} \quad -8.00 \text{ rad} \quad -12.00 \text{ rad}\}$.

19.9 $\{\Theta_a \quad V_b \quad \Theta_{b+} \quad \Theta_c\} = (10^{-3})\{-0.1157 \text{ rad} \quad -34.72 \text{ in} \quad 0.1157 \text{ rad} \quad 0.1157 \text{ rad}\}$.

19.11 $V_c = 4.53 \text{ mm}$ (all others zero), $s_{ac} = 72.5 \text{ kips (t)}$ (all others zero).

19.13 $\{U_a \quad V_a \quad \Theta_a \quad U_b \quad V_b \quad \Theta_b \quad U_c \quad \Theta_c\} = (10^{-3})\{23.80 \text{ m} \quad -23.80 \text{ m} \quad 3.245 \text{ rad} \quad -11.18 \text{ m} \quad -23.80 \text{ m} \quad 4.002 \text{ rad} \quad -11.18 \text{ m} \quad 5.364 \text{ rad}\}$.

INDEX